Val Raghavan • **Double Fertilization**

Val Raghavan

Double Fertilization

Embryo and Endosperm Development
in Flowering Plants

With 75 Illustrations, Including 16 Color Plates

Springer

Professor Dr. Val Raghavan
Department of Plant Cellular & Molecular Biology
The Ohio State University
318 West 12th Avenue
Columbus
Ohio 43210
USA

e-mail: raghavan.1@osu.edu

Library of Congress Control Number: 2005930635

ISBN-10 3-540-27791-9   Springer-Verlag Berlin Heidelberg New York
ISBN-13 978-3-540-27791-0 Springer-Verlag Berlin Heidelberg New York

This work is subject to copyright. All rights are reserved, whether the whole or part of the material is concerned, specifically the rights of translation, reprinting, reuse of illustrations, recitation, broadcasting, reproduction on microfilm or in any other way, and storage in data banks. Duplication of this publication or parts thereof is permitted only under the provisions of the German Copyright Law of September 9, 1965, in its current version, and permissions for use must always be obtained from Springer-Verlag. Violations are liable for prosecution under the German Copyright Law.

Springer-Verlag is a part of Springer Science+Business Media
springeronline.com

© Springer-Verlag Berlin Heidelberg 2006

Printed in Germany

The use of general descriptive names, registered names, trademarks, etc. in this publication does not imply, even in the absence of a specific statement, that such names are exempt from the relevant protective laws and regulations and therefore free for general use.

Cover design: design&production, Heidelberg, Germany
Typesetting and production: LE-TeX Jelonek, Schmidt & Vöckler GbR, Leipzig, Germany
31/3150 YL 5 4 3 2 1 0 – Printed on acid-free paper

For
Lakshmi Raghavan and Anita Raghavan

About the Author

Since 1978, Val Raghavan has been a Professor at The Ohio State University, Columbus, Ohio (USA), where he is currently affiliated with the Department of Plant Cellular and Molecular Biology. After obtaining a Ph.D. degree from Princeton University, Dr. Raghavan held post-doctoral positions at Harvard University, Rockefeller University, and Dartmouth College, and faculty appointments at the University of Malaya, Kuala Lumpur and the National University of Singapore. He has published extensively on various aspects of the development of vascular plants, especially on zygotic and asexual embryogenesis in flowering plants.

Preface

Double fertilization is hailed as a unique event in the life cycle of flowering plants. Defined as the union of one sperm with the egg on the one hand, and of a second sperm with the diploid fusion nucleus on the other, double fertilization sets in motion the chain of events that result in the formation of the embryo and endosperm. Whereas recognition of the importance of these two fusion episodes in seed formation in flowering plants is as old as the discovery of double fertilization itself, their central role in the development of the embryo and endosperm in seeds and grains of crop plants used widely in human and animal nutrition came to be recognized only in later years.

The study of the development of the embryo and endosperm from their single-celled beginnings under the rubric of embryology has occupied an important position in the multifaceted investigations on the reproductive biology of flowering plants undertaken during most of the past century. In recent years, descriptive studies of embryo and endosperm development have been overshadowed by the increasing use of genetic and molecular approaches to study flowering plant embryology, led by the work in the model plant *Arabidopsis thaliana*. Although some of these studies have been reviewed periodically in multiauthored volumes, my objective in writing this book is to provide an overview of past accomplishments in the field, and a sense of the outstanding future problems as they relate to the products of double fertilization. Admittedly, molecular and genetic studies in conjunction with screening of mutants, isolation of genes, and identification of their protein products, are emphasized to some extent at the expense of structural and developmental studies. The main reason for this is that I have tried to write a book on investigations that reflect a rethinking of the way that we have viewed embryogenesis and endosperm development as the end product of a series of stereotypical divisions. In my opinion, these recent studies with molecular overtones have brought us close to an understanding of the critical details that control the transformation of these cellular domains of the ovule into mature tissues of the seed or grain.

The book begins with an account of the history of the discovery of double fertilization, which must surely find a place in a volume dealing with that topic. The details of how body plans of eudicot and monocot embryos are established occupy the next chapter, which sets the stage, in the following three chapters, for a discussion of notable advances made in the identification of the genetic and molecular factors that control the development of the embryo (Chaps. 3, 5) and suspensor (Chap. 4) during progressive embryogenesis. The last chapter to deal wholly with embryos (Chap. 6) describes their general strategies during quiescence or dormancy. The main body of the book concludes with accounts of the developmental, genetic, and molecular studies on the endosperm covered in Chaps. 7 and 8, and, in the final chapter, descriptions of apomixis, somatic embryogenesis, and pollen embryogenesis illustrating embryogenesis and partial endosperm development in the absence of double fertilization. The level of exposition of the topics in the different chapters is considered suitable for graduate students who want get a coherent view of the current perspectives on embryogenesis and endosperm development in flowering plants, and for researchers in the field who plan fresh attacks on unsolved problems on the topics covered.

In conclusion, I would like to thank the many publishers/authors who gave me permission to use illustrations from published articles in my book. Besides myself, no one contributed more to the preparation of the final manuscript than Mr. Eduardo Acosta, Webmaster of my Department. He transformed my rough pencil sketches into professional black and white drawings or into images in gorgeous colors, and was also responsible for transferring all of the illustrations into their electronic versions suitable for printing. It is my pleasure to acknowledge my indebtedness to Eduardo for this help. On the producing side at Springer, Heidelberg, I appreciate the editorial advice and suggestions given from time to time by Dr. Jutta Lindenborn, desk editor, and the professional expertise, critical judgments, and interest in the subject matter of the book that Dr. Helen Rothnie, copy editor, brought to the job. Last, but not least, I thank my wife Lakshmi for her appreciation of my interests, which allowed me to spend long hours in my office and laboratory where I felt comfortable to pursue scholarly activities. My daughter, Anita, was generous with her sense of good humor, often transmitted by remote control from London, England, during the preparation of this book.

Columbus, Ohio   V. Raghavan
August 2005

Contents

Abbreviations XIII
Illustration Credits XVII

1 Double Fertilization – A Defining Feature of Flowering Plants
1.1 Discovery of Double Fertilization 2
1.1.1 Who Discovered Double Fertilization? 3
1.1.2 Universality of Double Fertilization in Flowering Plants 5
1.2 Seed Development without Double Fertilization 7
1.3 A Case for Double Fertilization in Gymnosperms 9
1.4 Structural and Cytological Perspectives on Double Fertilization 11
1.4.1 Cellular Nature of the Sperm and the Male Germ Unit 11
1.4.2 Pollen Tube Guidance and Sperm Entry into Embryo Sac 13
1.4.3 Nuclear Fusions 15
1.5 In vitro Double Fertilization 17
1.6 Double Fertilization and the Coming of Age of Plant Embryology 19
1.6.1 The Changing Scene 20
1.6.2 Genetic and Molecular Studies of Embryogenesis and Endosperm Development 21
1.6.3 Problems and Prospects 22
1.7 Concluding Comments 22
References 23

2 Establishment of the Embryo Body Plan – A Reassessment of Cell Lineage and Cell Fate
2.1 Organization of the Egg and Zygote 30
2.2 From the Zygote to the Embryo 34
2.2.1 A Model of Embryogenesis in Eudicots 35
2.2.2 A Model of Embryogenesis in Monocots 41
2.2.3 Are Embryonic Organs and Tissues Lineage-restricted Compartments? 43
2.2.4 Abnormal Embryo Types 44
2.3 Physiological Considerations of Embryogenesis 45
2.3.1 A Role for Auxin Polar Transport in Embryogenesis 46
2.3.2 Embryo Nutrition 47
2.3.3 Embryo Culture Investigations 48
2.4 Concluding Comments 51
References 52

3 Pattern Formation in Embryos – Interpretation of Positional Information
3.1 Initiation and Maintenance of Embryo Meristems 58
3.1.1 Shoot Apical Meristem 58
3.1.2 Root Apical Meristem 66
3.2 Genetic and Molecular Control of Embryo Pattern Formation 69
3.2.1 Apicobasal Patterning of the Embryo 69
3.2.2 Radial Patterning of the Embryo 72
3.3 Concluding Comments 75
References 76

4 Life and Times of the Suspensor – Cell Signaling between the Embryo and Suspensor
4.1 Morphological and Physiological Considerations 82
4.1.1 Subcellular Morphology of the Suspensor 85
4.1.2 Nuclear Cytology of the Suspensor 88
4.1.3 Functional Physiology of the Suspensor 90
4.1.4 Developmental Physiology and Programmed Death of Suspensor Cells 92
4.2 Genetic Control of Suspensor Form 94
4.3 Concluding Comments 97
References 97

5 Genetic and Molecular Control of Embryogenesis – Role of Nonzygotic and Zygotic Genes
5.1 Asymmetry in Parental Genome Contributions 102
5.1.1 Evidence for Maternal-effect Genes 103
5.1.2 Silencing of Paternal Genes 105
5.2 Gene Activity during Progressive Embryogenesis 106
5.2.1 Gene Expression during Early Embryogenesis 108
5.2.2 Gene Expression during Late Embryogenesis and Transition to Germination 112
5.3 Embryo Gene Expression Program Studied by Mutant Screening 115
5.3.1 Embryo-lethal Mutants of *Arabidopsis* 115
5.3.2 The World of Cytokinesis-Defective Mutants 117
5.3.3 Embryo-defective Mutants of *Arabidopsis* 120
5.3.4 Embryo-Defective Mutants of Maize and Rice 121
5.4 Concluding Comments 123
References 123

6	**Maturation and Dormancy – Survival Strategies of the Embryo**	**8**	**Genetics and Molecular Biology of the Endosperm – A Tale of Two Model Systems**
6.1	Embryo Maturation 132	8.1	Specification of Form in the Endosperm of *Arabidopsis* 174
6.1.1	Synthesis of Maturation Proteins 132		
6.1.2	Is Embryo Maturation in the Seed Developmentally Regulated by ABA? 134	8.1.1	Endosperm Development without Double Fertilization 175
6.1.3	Genetic Regulation of Embryo Maturation by ABA 136	8.1.2	Parental Gene Dosage in Endosperm Development 176
6.2	Embryo Dormancy 138	8.2	Genetics and Molecular Biology of the Cereal Endosperm 178
6.2.1	Carbohydrates in Desiccation Tolerance 141		
6.2.2	Proteins in Desiccation Tolerance 143	8.2.1	Embryo-surrounding Region and Transfer Layer 178
6.3	Concluding Comments 145 References 146	8.2.2	Aleurone Cells and Starchy Endosperm 180
		8.3	Concluding Comments 182 References 182
7	**Developmental and Functional Biology of the Endosperm – A Medley of Cellular Interactions**	**9**	**Non-zygotic Embryo Development – Embryogenesis without Sex**
7.1	Cellular Organization of the Endosperm 152	9.1	Apomixis 188
7.1.1	The Odyssey of Free Nuclei to a Cellular Tissue 154	9.1.1	Case Studies of Diplosporous and Aposporous Apomicts 189
7.1.2	Development of the Endosperm in *Arabidopsis* 156	9.1.2	Adventive Embryogenesis 190
		9.1.3	Molecular Genetics of Apomixis 192
7.2	Biochemical Organization of the Endosperm 158	9.2	Somatic Embryogenesis 192
7.2.1	DNA Amplification 158	9.2.1	A History of the Recent Past 192
7.2.2	Accumulation of Storage Products 160	9.2.2	Somatic Embryogenesis in Carrot and other Model Systems 194
7.2.3	Programmed Cell Death of the Endosperm 161		
7.3	Role of the Endosperm in Embryo Nutrition 162	9.2.3	Embryonic Proteins and Regulation of Gene Expression 199
7.3.1	Structural Modifications of the Endosperm 162	9.3	Pollen Embryogenesis 200
7.3.2	Physiological Considerations 164	9.3.1	Responsive Stage of Pollen Development and Pollen Embryogenic Potential 201
7.3.3	Genetic Considerations 165		
7.4	Concluding Comments 167 References 168	9.3.2	Cytology of Pollen Embryogenesis 203
		9.3.3	Molecular Biology of Pollen Embryogenesis 204
		9.4	Concluding Comments 205 References 206

Color Plates 213
Index 229

Abbreviations

GENERAL ABBREVIATIONS

ABA	abscisic acid
APC	anaphase-promoting-complex
CaMV	cauliflower mosaic virus
cDNA	complementary DNA
2,4-D	2,4-dichlorophenoxyacetic acid
EMS	ethylmethane sulfonate
ER	endoplasmic reticulum
GA	gibberellic acid
GABA	γ-aminobutyric acid
GFP	green fluorescent protein
GlyRS	glycyl tRNA synthetase
GUS	β-glucuronidase
IAA	indoleacetic acid
ICL	isocitrate lyase
JIM8	a monoclonal antibody
MS	malate synthase
MYB	recognition site in the genome identified with myeloblastosis-associated viruses
NAA	naphthaleneacetic acid
NPA	naphthylphthalamic acid
pcd	programmed cell death
rRNA	ribosomal RNA
RT-PCR	reverse transcription polymerase chain reaction
T-DNA	transferred DNA
TIBA	triiodobenzoic acid
TUNEL	terminal deoxyribonucleotidyl transferase-mediated dUTP-fluorescein nick end labeling

LIST OF cDNA CLONES, GENES, MUTANTS, AND PROTEIN PRODUCTS

Listed below are the cDNA clones, genes, mutants, and protein products and their abbreviations in the form in which they are first mentioned in the text. With a few exceptions, abbreviations and names of wild-type genes are given here and in the text in italicized capital letters; mutants are indicated in italicized lowercase letters. Abbreviations of protein products are given in capital letters.

AAP	*AMINO ACID PERMEASE*
aba	*ABA-deficient*
ABC	*ATP-binding cassette*
ABI	*ABA-INSENSITIVE*
abp	*auxin-binding protein*
Ac	*Activator*
adl	*Arabidopsis dynamin-like proteins*
AGL	*AGAMOUS-Like*
AGO	*ARGONAUTE*
AHAP3	*Arabidopsis HAP3*
ALDP	adrenoleukodystrophy protein
ALE	*ABNORMAL LEAF SHAPE*
aml	*Arabidopsis Minute-like*
ANT	*AINTEGUMENTA*
AP2	*APETALA2*
ARF	ADP-ribosylation factor; auxin response factor
ARL2	a relative of the ARF-family of proteins
AS	*ASYMMETRIC LEAVES*
ask	*Arabidopsis thaliana Skip-like1*
ASKη	*Arabidopsis shaggy*-related protein kinase *etha*
ASKζ	*Arabidopsis shaggy*-related protein kinase *dzeta*
Atcul	*Arabidopsis thaliana cullin*
AtEm	*Arabidopsis thaliana Em*
Athb	*Arabidopsis thaliana HOMEOBOX*
AtLTP	*Arabidopsis thaliana LIPID TRANSFER PROTEIN*
ATML	*Arabidopsis thaliana MERISTEM L1 LAYER*

AtpA	*atp1, ATPase1*; a mitochondrial gene	*cts*	*comatose*
AtPIN	*Arabidopsis thaliana* PIN-FORMED	*CUC*	*CUP-SHAPED COTYLEDON*
		CUL	*CULLIN*
AtRPS5	mutated gene of *aml1*	*CYCD*	*CYCLIN D*
ATS	*ARABIDOPSIS THALIANA SEED*	*CycZme1*	*Zea mays* mitotic cyclin belonging to the subgroup *Zeama;CycB1*
AtSERK	*Arabidopsis thaliana* SERK	*cyd*	*cytokinesis-defective* mutant of pea
At2S3	*Arabidopsis thaliana* 2S ALBUMIN	*cyt*	*cytokinesis-defective* mutant of *Arabidopsis*
AX92	a gene of *Brassica napus* embryos and seedlings	*DcSERK*	*Daucus carota* SERK
		DDM	*DECREASE IN DNA METHYLATION*
axr	*auxin-resistant*		
B22E	a barley endosperm gene	*dek*	*defective kernel*
BAP	*BASAL LAYER ANTIFUNGAL PROTEINS*	*DEM*	*DEFECTIVE EMBRYO AND MERISTEMS*
BBM	*BABY BOOM*	*des*	*defective seedling*
bdl	*bodenlos*	*dex*	*defective endosperm expressing xenia*
BETL	*BASAL ENDOSPERM TRANSFER LAYER*		
		dgr	*distorted growth*
bga	*borgia*	*dme*	*demeter*
bio	*biotin mutant*	*DOM*	*DOMINO*
BIO2	biotin synthase gene	*dsc*	*discolored*
BOP	*BLADE-ON-PETIOLE*	*dzr*	a post-transcriptional regulator of zein
BP	*BREVIPEDICELLUS*		
bt	*brittle*	E1, E2	embryonic proteins
bZIP	basic leucine zipper class of transcriptional regulators	*edd*	*embryo-defective development*
		EED	*EMBRYONIC ECTODERM DEVELOPMENT*
C1	a gene in the anthocyanin pathway of maize		
		EEL	*ENHANCED Em LEVEL*
cab	gene encoding chlorophyll *a/b* binding protein	*Em*	*EARLY METHIONINE-LABELED*
cap	*capulet*	*EMB*	*EMBRYO-DEFECTIVE*
CBF	CCAAT-box-binding transcription factor	*emb*	*embryo-specific*
		eml	*embryoless*
CDC	*CELL DIVISION CYCLIN*	*emp*	*empty pericarp*
CDK	cyclin-dependent kinase	*END*	*ENDOSPERM*
C-ESE	*CARROT EARLY SOMATIC EMBRYOGENESIS*	*EP*	*EXTRACELLLULAR PROTEIN*
		ERG	*ERA-RELATED GTPases*
CHAPERONIN-60α	an *Arabidopsis* gene	*ESC*	*EXTRA SEX COMBS*
		Esr	*EMBRYO SURROUNDING REGION*
CHI	*CHITINASE*		
CHO	*CHAMPIGNON*	*F644*	an *Arabidopsis* gene
CLE	*CLAVATA-Like*	*FBP*	*FLORAL BINDING PROTEIN*
clv	*clavata*	*fer*	*feronia*
CNA	*CORONA*	*FIE*	*FERTILIZATION-INDEPENDENT ENDOSPERM*
cox	gene of cytochrome-*c* subunit		
CPC	*CAPRICE*	*FIL*	*FILAMENTOUS FLOWER*
cph	*cephalopod*	*FIS*	*FERTILIZATION-INDEPENDENT SEEDS*
cr	*crinkly*		
CRC	*CRUCIFERIN*	*fist*	an *Arabidopsis* embryo mutant

FK	FACKEL	MET1 a/s	METHYL TRANSFERASE anti-sense
fs	fass		
FUS	FUSCA	mic	mickey
FWA	a late-flowering *Arabidopsis* gene	mgo	mgoun
GAI	GIBBERELLIN-INSENSITIVE	mp	monopteros
gcs	glucosidase	msi	multicopy suppressor of IRA (inhibitory regulator of Harvey sarcoma virus oncogene RAS-cAMP pathway)
gk	gurke		
GL	GLABRA		
GLA	GLOBULAR ARREST		
GLM	GOLLUM	MtSERK	*Medicago truncatula* SERK
glo	globby	Mu	Mutator
GlyRS	glycyl-tRNA synthetase	nam	no apical meristem
gn	gnom	NRP	NO APICAL MERISTEM (NAM)-RELATED PROTEIN
GRAS	transcription factors encoded by *SHR*, *SCR*, *GAI* and *RGA* genes		
		OLEO	OLEOSIN
GRP94	a chaperone protein	ORG	ORIGIN RECOGNITION COMPLEX
HAL	HALLIMASCH		
HAP3	heme-activated protein 3	OSH	ORYZA SATIVA HOMEOBOX
hbt	hobbit	OsKn1	*Oryza sativa* KNOTTED1-like
hik	hinkel	PAP85	an *Arabidopsis* gene encoding a vicilin-like protein
HMG	high mobility group protein		
HOS	HOMEOBOX GENE OF ORYZA SATIVA	PAS	PASTICCINO
		PEI	an *Arabidopsis* gene
HSP	heat shock protein	PER	PEROXIREDOXIN
HYD	HYDRA	PFI	PFIFFERLING
ig	indeterminate gametophyte	PGA	PLANT GROWTH ACTIVATOR
iku	haiku		
JAG	JAGGED	PHB	PHABULOSA
KAN	KANADI	PHE	PHERES
KAPP	kinase associated protein phosphatase	PIC	PINOCCHIO
		PID	PINOID
keu	keu	PILZ	a group of *Arabidopsis* genes
KIS	KIESEL	PIN	PIN-FORMED
kn	knolle	pkl	pickle
KN	KNOTTED	PLS	POLARIS
knf	knopf	PLT	PLETHORA
KTi	Kunitz trypsin inhibitor	PNH	PINHEAD
lachrima	a maize gene	pol	poltergeist
LEA	LATE EMBRYOGENESIS ABUNDANT	POR	PORCINO
		PP2C	PROTEIN PHOSPHATASE 2C
LEC	LEAFY COTYLEDON	PRL	PROLIFERA
L1L	LEC1-LIKE	pt	primordial timing
LLP	ligand-like protein	pZE40	a barley endosperm gene
LTP	LIPID TRANSFER PROTEIN	R	RED (a gene controlling pigmentation of maize aleurone cells)
MADS-box	floral organ identity genes		
		RAB	RESPONSIVE TO ABA
MAT	MATURATION	rbcL	gene of the large subunit of Rubisco
MEA	MEDEA		
MEG	MATERNALLY EXPRESSED GENE	REV	REVOLUTA
		RGA	REPRESSOR OF GA
MET	METHYL TRANSFERASE	rgf	reduced grain filling

RINO	myo-inositol-1-phosphate synthase gene	su	sugary
Roc	rice outermost cell-specific	sus	suspensor
Rop	Rho-like GTPase	TCP	Teosinte branched1, Cycloidea, and PCF1 genes which encode transcription factors
RPS16	ribosomal protein S16		
RSH	ROOT-SHOOT-HYPOCOTYL-DEFECTIVE	TFC	tubulin folding cofactor
		ton	tonneau
rsw	radially swollen	TOR	TARGET OF RAPAMYCIN
rsy	raspberry	tpl	topless
sal	supernumerary aleurone	tps	trehalose phosphate synthase
SCF	SKP1 [SUPPRESSOR OF KINETOCHORE PROTEINS1]/CDC53 [or CULLIN], F-box protein	TTG	TRANSPARENT TESTA GLABRA
		TTN	TITAN
		twn	twin
SCR	SCARECROW	vcl	vacuoleless
SCZ	SCHIZORIZA	VP1	VIVIPAROUS1
SCE7	a member of the ARF nucleotide exchange factors	Vp1-R	wild type viviparous gene
		vp1-R	mutant allele of Vp1-R
seg	shrunken endosperm caused by the maternal genotype	Vpp	a gene that encodes a type of vacuolar H+-translocating inorganic pyrophosphatase
SERK	SOMATIC EMBRYOGENESIS RECEPTOR KINASE		
		Wee1	a protein kinase
SET domain	proteins encoded by SUPPRESSION OF VARIEGATION, ENHANCER OF ZEST, and TRITHORAX genes	WER	WEREWOLF
		WOL	WOODEN LEG
		WOX	WUSCHEL-related homeobox
		wus	wuschel
		wx	waxy
sex	shrunken endosperm expressing xenia	XTC	EXTRA COTYLEDON
		YAB	YABBY
sh	shrunken	YEC2	yeast protein of unknown function
SHAGGY	a gene that encodes a protein kinase in Drosophila		
		Zeama; CycA1, B1, B2	groups of the Zea mays mitotic cyclin gene
SHD	SHEPHERD		
shl	shootless		
SHR	SHORT ROOT	ZLL	ZWILL
sin	short integument	ZmAE	Zea mays ANDROGENIC EMBRYOS
slp	schlepperless		
sml	shootmeristemless	ZmEBE	Zea mays embryo sac/basal endosperm transfer layer/embryo surrounding region
smt	sterol methyl transferase		
SNAP	a vesicle trafficking gene		
SNARE	soluble N-ethylmaleimide-sensitive factor attachment protein receptors	ZmHox	Zea mays homeobox
		ZmMRP	Zea mays MYB-RELATED PROTEIN
SOL	SUPPRESSOR OF LLP		
SPÄTZLE	a maize gene involved in endosperm cellularization	ZmOCL	Zea mays OUTER CELL LAYER
		ZmPRPL 35	Zea mays PLASTID RIBOSOMAL PROTEIN L35
SPL	SPOROCYTELESS		
srn	siréne		
SSR16	SMALL SUBUNIT RIBOSOMAL PROTEIN S16	ZmSERK	Zea mays SERK
		ZmWee1	a maize homolog of Wee1
stm	shoot meristemless		

Illustration Credits

Illustrations not credited herein have been prepared by the author.

FIGURES

Fig. 1.2a,b Nawaschin S (1898) Resultate einer Revision der Befruchtungsvorgänge bei *Lilium martagon* und *Fritillaria tenella*. Bulletin de l'Académie Impériale des Sciences de St.-Pétersbourg Ser 5, 9:377–382

Fig. 1.3a–c Guignard L (1899) Sur les anthérozoïdes et la double copulation sexuelle chez les végétaux angiospermes. Comptes Rendus des Séances de l'Académie des Sciences 128:864–871

Fig. 1.4 Higashiyama T, Kuroiwa H, Kawano S, Kuroiwa T (1997) Kinetics of double fertilization in *Torenia fournieri* based on direct observations of the naked embryo sac. Planta 203:101–110. © Springer, Berlin Heidelberg New York

Fig. 1.5a–f Friedman WE (1991) Double fertilization in *Ephedra trifurca*, a non-flowering seed plant: the relationship between fertilization events and the cell cycle. Protoplasma 165:106–120

Fig. 1.6a,b Huang B-Q, Russell SD (1994) Fertilization in *Nicotiana tabacum*: cytoskeletal modifications in the embryo sac during synergid degeneration. Planta 194:200–214 © Springer, Berlin Heidelberg New York

Fig. 1.7 Carmichael JS, Friedman WE (1995) Double fertilization in *Gnetum gnemon*: the relationship between the cell cycle and sexual reproduction. Plant Cell 7:1975–1988 © American Society of Plant Biologists

Fig. 1.8a–u Kranz E (2001) In vitro fertilization. In: Bhojwani SS, Soh WY (editors) Current trends in the embryology of angiosperms. Kluwer Academic Publishers, Dordrecht, pp 143–166. Reprinted with kind permission of Springer Science and Business Media

Fig. 2.1 Webb MC, Gunning BES (1991) The microtubular cytoskeleton during development of the zygote, prembryo and free-nuclear endosperm in *Arabidopsis thaliana* (L.) Heynh. Planta 184:187–195 © Springer, Berlin Heidelberg New York

Fig. 2.2a–c Kuroiwa H, Nishimura Y, Higashiyama T, Kuroiwa T (2002) *Pelargonium* embryogenesis: cytological investigations of organelles in early embryogenesis from the egg to the two-celled embryo. Sexual Plant Reproduction 15:1–12 © Springer, Berlin Heidelberg New York

Fig. 2.3a–f Sheridan WF, Clark JK (1994) Fertilization and embryogeny in maize. In: Freeling M, Walbot V (editors) The maize handbook. Springer-Verlag, New York, pp 3–10 © Springer, Berlin Heidelberg New York

Fig. 2.4a–c Yakovlev MS, Yoffe MD (1957) On some peculiar features in the embryogeny of *Paeonia* L. Phytomorphology 7:74–82

Fig. 3.1 Sentoku N, Sato Y, Kurata N, Ito Y, Kitano H, Matsuoka M (1999) Regional expression of the rice *KN1*-type homeobox gene family during embryo, shoot, and flower development. Plant Cell 11:1651–1663 © American Society of Plant Biologists

Fig. 3.2 Mayer KFX, Schoof H, Haecker A, Lenhard M, Jürgens G, Laux T (1998) Role of *WUSCHEL* in regulating stem cell fate in the *Arabidopsis* shoot meristem. Cell 95:805–815 © 1998, reprinted with permission from Elsevier

Fig. 3.3a–d Moussian B, Schoof H, Haecker A, Jürgens G, Laux T (1998) Role of the *ZWILLE* gene in the regulation of central shoot meristem cell fate during *Arabidopsis* embryogenesis. EMBO J 17:1799–1809 © 1998, reprinted by permission, Macmillan Publishers Ltd

Fig. 3.4 Schoof H, Lenhard M, Haecker A, Mayer KFX, Jürgens G, Laux T (2000) The stem cell population of *Arabidopsis* shoot meristems is maintained by a regulatory loop between the *CLAVATA* and *WUSCHEL* genes. Cell 100:635–644 © 2000, reprinted with permission from Elsevier

Fig. 3.5a–f Assaad FF, Mayer U, Wanner G, Jürgens G (1996) The *KEULE* gene is involved in cytokinesis in *Arabidopsis*. Molecular and General Genetics 253:267–277 © Springer, Berlin Heidelberg New York

Fig. 4.1a–d Swamy BGL (1949) Embryological studies in the Orchidaceae. II. Embryogeny. American Midland Naturalist 41:202–232

Fig. 4.1e Maheshwari P, Singh B (1952) Embryology of *Macrosolen cochinchinensis*. Botanical Gazette 114:20–32

Fig. 4.1f Prakash S (1960) Morphological and embryological studies in the family Loranthaceae – VI. *Peraxilla tetrapetala* (Linn. F.) van Tiegh. Phytomorphology 10:224–234

Fig. 4.1g Schaffner M (1906) The embryology of the shepherd's purse. Ohio Naturalist 7:1–8

Fig. 4.1h Simoncioli C (1974) Ultrastructural characteristics of "*Diplotaxis erucoides* (L.) DC" suspensor. Giornale Botanico Italiano 108:175–189

Fig. 4.2a–f Guignard L (1881) Recherches d'embryogénie végétale comparée. 1st Mémoire: Légumineuses. Annales des Sciences Naturelles Botanique Série 6, 12:5–166

Fig. 4.2g Rau MA (1950) The suspensor haustoria of some species of *Crotalaria* Linn. Annals of Botany 14:557–562

Fig. 4.2h Nagl W (1962) Über Endopolyploidie, Restitutionskernbildung und Kernstrukturen im Suspensor von Angiospermen und einer Gymnosperme. Österreichische Botanische Zeitschrift 109:431–494

Fig. 4.2i Mercy ST, Kakar SN, Varghese TM (1974) Embryology of *Cicer arietinum* and *C. soongaricum*. Bulletin of the Torrey Botanical Club 101:26–30

Fig. 4.3a Nagl W, Kühner S (1976) Early embryogenesis in *Tropaeolum majus* L.: diversification of plastids. Planta 133:15–19 © Springer, Berlin Heidelberg New York

Fig. 4.3b Subramanyam K (1963) Embryology of *Sedum ternatum* Michx. Journal of the Indian Botanical Society (Maheshwari Commemoration Volume) 52A:259–275

Fig. 4.3c Swamy BGL (1942) Female gametophyte and embryogeny in *Cymbidium bicolor* Lindl. Proceedings of the Indian Academy of Sciences 15B:194–201

Fig. 4.4 Schulz P, Jensen WA (1969) *Capsella* embryogenesis: the suspensor and the basal cell. Protoplasma 67:139–163

Fig. 4.5a–d Yeung EC, Clutter ME (1978) Embryogeny of *Phaseolus coccineus*: growth and microanatomy. Protoplasma 94:19–40

Fig. 4.6 Lima-de-Faria A, Pero R, Avanzi S, Durante M, Ståhle U, D'Amato F, Granström H (1975) Relation between ribosomal RNA genes and the DNA satellites of *Phaseolus coccineus*. Hereditas 79:5–20

Fig. 4.7a–d Gerlach-Cruse D (1969) Embryo- und Endospermentwicklung nach einer Röntgenbestrahlung der Fruchtknoten von *Arabidopsis thaliana* (L.) Heynh. Radiation Botany 9:433–442 © 1969, reprinted with permission from Elsevier

Fig. 4.8a–c Schwartz BW, Yeung EC, Meinke DW (1994) Disruption of morphogenesis and transformation of the suspensor in abnormal suspensor mutants of *Arabidopsis*. Development 120:3235–3245 © Company of Biologists

Fig. 4.9a,b Schwartz BW, Yeung EC, Meinke DW (1994) Disruption of morphogenesis and transformation of the suspensor in abnormal *sus*pensor mutants of *Arabidopsis*. Development 120:3235–3245 © Company of Biologists

Fig. 5.1 Goldberg RB, Barker SJ, Perez-Grau L (1989) Regulation of gene expression during plant embryogenesis. Cell 56:149–160 © 1989, reprinted with permission from Elsevier

Fig. 5.2 Yoshida KT, Wada T, Koyama H, Mizobuchi-Fukuoka R, Naito S (1999) Temporal and spatial patterns of accumulation of the transcript of *myo*-inositol-1-phosphate synthase and phytin-containing particles during seed development in rice. Plant Physiology 119:65–72 © American Society of Plant Biologists

Fig. 5.3 Zhang JZ, Santes CM, Engel ML, Gasser CS, Harada JJ (1996) DNA sequences that activate isocitrate lyase gene expression during late embryogenesis and during postgerminative growth. Plant Physiology 110:1069–1079 © American Society of Plant Biologists

Fig. 5.4 Tzafrir I, McElver JA, Liu C, Yang LJ, Wu JQ, Martinez A, Patton DA, Meinke DW (2002) Diversity of TITAN functions in *Arabidopsis* seed development. Plant Physiology 128:38–51 © American Society of Plant Biologists

Fig. 5.5a–d Mayer U, Herzog U, Berger F, Inzé D, Jürgens G (1999) Mutations in the *PILZ* group genes disrupt the microtubule cytoskeleton and uncouple cell cycle progression from cell division in *Arabidopsis* embryo and endosperm. European Journal of Cell Biology 78:100–108 © 1999, reprinted with permission from Elsevier

Fig. 6.1 Parcy F, Valon C, Raynal M, Gaubier-Comella P, Delseny M, Giraudat J (1994) Regulation of gene expression programs during *Arabidopsis* seed development: roles of the *ABI3* locus and of endogenous abscisic acid. Plant Cell 6:1567–1582 © American Society of Plant Biologists

Fig. 6.2 Choinski JS Jr, Trelease RN, Doman DC (1981) Control of enzyme activities in cotton cotyledons during maturation and germination. III. In-vitro embryo development in the presence of abscisic acid. Planta 152:428–435 © Springer, Berlin Heidelberg New York

Fig. 6.3 Sánchez-Martínez D, Puigdomènech P, Pagès M (1986) Regulation of gene expression in developing *Zea mays* embryos. Protein synthesis during embryogenesis and early germination of maize. Plant Physiology 82:543–549 © American Society of Plant Biologists

Fig. 6.4a–c Raghavan V (2002) Induction of vivipary in *Arabidopsis* by silique culture: implications for seed dormancy and germination. American Journal of Botany 89:766–776

Fig. 6.5a–j Brenac P, Smith MF, Obendorf RL (1997) Raffinose accumulation in maize embryos in the absence of a fully functional *Vp1* gene product. Planta 203:222–228 © Springer, Berlin Heidelberg New York

Fig. 6.6a,b Parcy F, Valon C, Raynal M, Gaubier-Comella P, Delseny M, Giraudat J (1994) Regulation of gene expression programs during *Arabidopsis* seed development: roles of the *ABI3* locus and of endogenous abscisic acid. Plant Cell 6:1567–1582 © American Society of Plant Biologists

Fig. 7.2a–d Olsen O-A, Brown RC, Lemmon BE (1995) Pattern and process of wall formation in developing endosperm. Bioessays 17:803–812. Reprinted with permission of Wiley-Liss, Inc. a subsidiary of John Wiley & Sons, Inc

Fig. 7.3a,b Kowles RV, Phillips RL (1988) Endosperm development in maize. International Review of Cytology 112:97–136 © 1988, reprinted with permission from Elsevier

Fig. 7.4 Young TE, Gallie DR (2000) Programmed cell death during endosperm development. Plant Molecular Biology 44:283–301. Reprinted with kind permission of Springer Science and Business Media

Fig. 7.5 Davis RW, Smith JD, Cobb BG (1990) A light and electron microscope investigation of the transfer cell region of maize caryopses. Canadian Journal of Botany 68:471–479

Fig. 8.1a–e Sørensen MB, Mayer U, Lukowitz W, Robert H, Chambrier P, Jürgens G, Somerville C, Lepiniec L, Berger F (2002) Cellularisation in the endosperm of *Arabidopsis thaliana* is coupled to mitosis and shares multiple components with cytokinesis. Development 129:5567–5576 © Company of Biologists

Fig. 8.2a,b Becraft PW, Li K, Dey N, Asuncion-Crabb Y (2002) The maize *dek1* gene functions in embryonic pattern formation and cell fate specification. Development 129:5217–5225 © Company of Biologists

Fig. 9.1 Spielman M, Vinkenoog R, Scott RJ (2003) Genetic mechanisms of apomixis. Philosophical Transactions of the Royal Society Series B 358:1095–1103

Fig. 9.2 Koltunow AM, Soltys K, Nito N, McClure S (1995) Anther, ovule, seed, and nucellar embryo development in *Citrus sinensis* cv. Valencia. Canadian Journal of Botany 73:1567–1582

Fig. 9.3a–l McCabe PF, Valentine TA, Forsberg LS, Pennell RI (1997) Soluble signals from cells identified at the cell wall establish a developmental pathway in carrot. Plant Cell 9:2225–2241© American Society of Plant Biologists

Fig. 9.4 Raghavan V (1986) Embryogenesis in angiosperms. A developmental and experimental study. Cambridge University Press, New York

PLATES

Plate 1, Fig. a–d Williams JH, Friedman WE (2002) Identification of diploid endosperm in an early angiosperm lineage. Nature 415:522–526 © 2002, reprinted by permission, Macmillan Publishers Ltd

Plate 1, Fig. e–i Fu Y, Yuan M, Huang B-Q, Yang, H-Y, Zee S-Y, O'Brien TP (2000) Changes in actin organization in the living egg apparatus of *Torenia fournieri* during fertilization. Sexual Plant Reproduction 12:315–322 © Springer, Berlin Heidelberg New York

Plate 1, Fig. j–l Weterings K, Apuya NR, Bi Y, Fischer RL, Harada JJ, Goldberg RB (2001) Regional localization of suspensor mRNAs during early embryo development. Plant Cell 13:2409–2425 © American Society of Plant Biologists

Plate 5, Fig. a–i Smith LG, Jackson D, Hake S (1995) Expression of *knotted1* marks shoot meristem formation during maize embryogenesis. Developmental Genetics 16:344–348. Reprinted with permission of Wiley-Liss, Inc. a subsidiary of John Wiley & Sons, Inc

Plate 6, Fig. a,b van den Berg C, Willemsen V, Hage W, Weisbeek P, Scheres B (1995) Cell fate in the *Arabidopsis* root meristem determined by directional signalling. Nature 378:62–65 © 1995, reprinted by permission, Macmillan Publishers Ltd

Plate 6, Fig. c–e Dolan L, Janmaat K, Willemsen V, Linstead P, Poethig S, Roberts K, Scheres K (1993) Cellular organisation of the *Arabidopsis thaliana* root. Development 119:71–84 © Company of Biologists

Plate 7, Fig. a–r Wysocka-Diller JW, Helariutta Y, Fukaki H, Malamy JE, Benfey PN (2000) Molecular analysis of SCARECROW function reveals a radial patterning mechanism common to root and shoot. Development 127:595–603 © Company of Biologists

Plate 8, Fig. a–c Grossniklaus U, Spillane C, Page DR, Köhler C (2001) Genomic imprinting and seed development: endosperm formation with and without sex. Current Opinion in Plant Biology 4:21–27 © 2001, reprinted with permission from Elsevier

Plate 8, Fig. d–g Perry SE, Nichols KW, Fernandez DE (1996) The MADS domain protein AGL15 localizes to the nucleus during early stages of seed development. Plant Cell 8:1977–1989 © American Society of Plant Biologists

Plate 9, Fig. a,b Li Z, Thomas TL (1998) *PEI1*, an embryo-specific zinc finger protein gene required for heart-stage embryo formation in *Arabidopsis*. Plant Cell 10:383–398 © American Society of Plant Biologists

Plate 9, Fig. c–j Elster R, Bommert P, Sheridan WF, Werr W (2000) Analysis of four *embryo-specific* mutants in *Zea mays* reveals that incomplete radial organization of the proembryo interferes with subsequent development. Development Genes and Evolution 210: 300–310 © 2000, Springer Berlin Heidelberg New York

Plate 10, Fig. a–h Nambara E, Keith K, McCourt P, Naito S (1995) A regulatory role for the *ABI3* gene in the establishment of embryo maturation in *Arabidopsis thaliana*. Development 121:629–636 © Company of Biologists

Plate 11, Fig. a Raz V, Bergervoet JHW, Koornneef M (2001) Sequential steps for developmental arrest in *Arabidopsis thaliana* seeds. Development 128:243–252 © Company of Biologists

Plate 11, Fig. b–d Brown RC, Lemmon BE, Olsen O-A (1994) Endosperm development in barley: microtubule involvement in the morphogenetic pathway. Plant Cell 6:1241–1251 © American Society of Plant Biologists

Plate 12, Fig. a–f Luo M, Bilodeau P, Dennis ES, Peacock WJ, Chaudhury A (2000) Expression and parent-of-origin effects for *FIS2*, *MEA*, and *FIE* in the endosperm and embryo of developing *Arabidopsis* seeds. Proceedings of the National Academy of Sciences, USA 97:10637–10642 © 2000, National Academy of Sciences USA

Plate 12, Fig. g,h Ohad N, Margossian L, Hsu Y, Williams C, Repetti P, Fischer RL (1996) A mutation that allows endosperm development without fertilization. Proceedings of the National Academy of Sciences, USA 93:5319–5324 © 1996, National Academy of Sciences USA

Plate 13, Fig. a,b Adams S, Vinkenoog R, Spielman M, Dickinson HG, Scott RJ (2000) Parent-of-origin effects on seed development in *Arabidopsis thaliana* require DNA methylation. Development 127:2493–2502 © Company of Biologists

Plate 13, Fig. c–f Opsahl-Ferstad H-G, le Deunff E, Dumas C, Rogowsky PM (1997) *ZmEsr*, a novel endosperm-specific gene expressed in a restricted region around the maize embryo. Plant Journal 12:235–246 © Blackwell Publishing Co

Plate 14, Fig. a,b Olsen O-A, Linnestad C, Nichols SE (1999) Developmental biology of the cereal endosperm. Trends in Plant Science 4:253–257 © 1999, reprinted with permission from Elsevier

Plate 14, Fig. c–e Hueros G, Varotto S, Salamini F, Thompson RD (1995) Molecular characterization of *BET1*, a gene expressed in the endosperm transfer cells of maize. Plant Cell 7:747–757 © American Society of Plant Biologists

Plate 14, Fig. f–h Shen B, Li C, Min Z, Meeley RB, Tarczynski MC, Olsen O-A (2003) *sal1* determines the number of aleurone cell layers in maize endosperm and encodes a class E vacuolar sorting protein. Proceedings of the National Academy of Sciences, USA 100:6552–6557 © 2003, National Academy of Sciences USA

Plate 16, Fig. a–g Schmidt EDL, Guzzo F, Toonen MAJ, de Vries SC (1997) A leucine-rich repeat containing receptor-like kinase marks somatic plant cells competent to form embryos. Development 124:2049–2062 © Company of Biologists

Plate 16, Fig. h–k Touraev A, Ilham A, Vicente O, Heberle-Bors E (1990) Stress-induced microspore embryogenesis in tobacco: an optimized system for molecular studies. Plant Cell Reports 15:561–565 © Springer, Berlin Heidelberg New York

CHAPTER 1

1 Double Fertilization – A Defining Feature of Flowering Plants

The expression fertilization may be used in an abstract or a concrete sense. In the abstract it denotes the process by which characters from two individuals are transmitted to a single organism in the succeeding generation. This phenomenon is almost universal throughout the animal and vegetable kingdoms, and its effects have been observed by many successive generations of breeders both of animals and of plants. In this way a considerable body of evidence has accumulated, and it has been found that certain laws are universally true of organisms which thus spring from a double stock. Such an organism passes through its complete life history, which may include more than one cycle of development. It exhibits a combination of characters drawn from both parents. The offspring of the same pair differ from each other: some resemble one parent, some the other, and those of mixed appearance may lean to either side. But a balance is maintained in each generation between the two stocks, so that neither parent has on the whole greater weight than the other.

E. Sargant 1900

1.1 Discovery of Double Fertilization 2
1.1.1 Who Discovered Double Fertilization? 3
1.1.2 Universality of Double Fertilization in Flowering Plants 5
1.2 Seed Development without Double Fertilization 7
1.3 A Case for Double Fertilization in Gymnosperms 9
1.4 Structural and Cytological Perspectives on Double Fertilization 11
1.4.1 Cellular Nature of the Sperm and the Male Germ Unit 11
1.4.2 Pollen Tube Guidance and Sperm Entry into Embryo Sac 13
1.4.3 Nuclear Fusions 15
1.5 In vitro Double Fertilization 17
1.6 Double Fertilization and the Coming of Age of Plant Embryology 19
1.6.1 The Changing Scene 20
1.6.2 Genetic and Molecular Studies of Embryogenesis and Endosperm Development 21
1.6.3 Problems and Prospects 22
1.7 Concluding Comments 22
References 23

This book is about post-fertilization reproductive development in the most evolutionarily successful and wonderfully diverse group of plants on the face of the earth: angiosperms or flowering plants. Angiosperms, along with four different groups of living representatives of gymnosperms, namely, cycads, Ginkgoales (which includes the monotypic *Ginkgo biloba*), conifers, and Gnetales, are also known as seed plants. Seeds of angiosperms are enclosed within a fruit instead of being produced as exposed units on the surface of sporophylls or similar structures as they are in gymnosperms. Although study of the reproductive biology of angiosperms has a long history, sustained cellular and molecular investigations of this topic constitute a modern development.

Fertilization, besides its obvious role in genetic recombination, essentially denotes the fusion of the egg and sperm to form a zygote and, as will soon become clear, the word does not capture the full scope of events that occur in flowering plants. The traditional setting for fertilization in flowering plants is the sanctum sanctorum of the female gametophyte – more popularly known as the embryo sac – which itself is wrapped in several layers of cells of the nucellus and integuments constituting the ovule. A typical embryo sac initially has two groups of four haploid nuclei embedded within it, one at the micropylar end and the other at the opposite, chalazal end. The demarcation of groups of three nuclei at each end, each nucleus surrounded by its own cytoplasmic domain as a distinct, compartmentalized, membrane-bound cell, is the primary determinant of form of the mature embryo sac. The three cells at the micropylar pole are organized as the egg apparatus, consisting of a large egg cell flanked on either side by a cellular synergid. The three cells at the opposite pole become the antipodals. The main body of the embryo sac remaining after the egg apparatus and antipodals are cut off is the central cell consisting of the two orphaned nuclei from either pole, which may remain separate, side-by-side, as unfused haploid nuclei, or fuse to form a diploid polar fusion nucleus. The mature embryo sac is thus a seven-celled, eight-nucleate supercell in which fertilization occurs (Fig. 1.1). This type of embryo sac development, which is prevalent in about 70% of angiosperms, is known as the 'normal' type, and, because it was first described in *Polygonum divaricatum* (Polygonaceae), it is conventionally designated as the 'Polygonum' type (Maheshwari 1950). In the context of fertilization, the term female germ unit has been proposed for the egg apparatus and the central cell (Dumas et al. 1984), but it is not widely used.

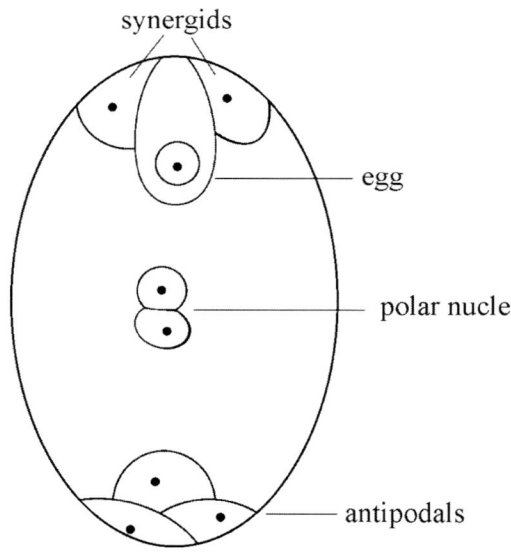

Fig. 1.1 Diagram of a 'Polygonum' type of embryo sac showing the disposition of cells

The process of fertilization in flowering plants, including the encounter of the male and female gametes and the actual fusion of gametic nuclei, presents a degree of complexity not found in other groups of plants. Pollination, resulting in the transfer of pollen grains from the anther to the stigmatic surface of the appropriate flower type, is the beginning of a cascade of events that delivers the male gamete to the vicinity of the egg. Following germination of pollen grains on the stigma, the resulting pollen tubes carrying the two male gametes (produced by a mitotic division of the generative cell of the pollen grain) navigate through the carbohydrate-rich matrix of the stigma, style, and the ovular tissues, and reach the vicinity of the embryo sac. Fusion of the male and female gametes takes place when the pollen tube enters the embryo sac and releases the sperm. Hitherto partially or totally uncharacterized extracellular matrix components of the stigma and style spring into action to sustain pollen tube growth, and the ever-present signaling molecules generated by the diploid cells of the ovule or the haploid cells of the embryo sac for pollen tube attraction contribute to successful fertilization (Johnson and Preuss 2002). Following fertilization, the ovule develops into the seed enclosed in the ovary, which becomes the fruit. Although these facts – the bare bones of the reproductive biology of flowering plants – have long been known, perspectives on the molecular genetics of the individual phases involved have come from recent cell biological studies and analyses of female gametophytic mutants of *Arabidopsis thaliana* (Brassicaceae; hereafter referred to by genus name only). The purpose of this chapter is to present an overview of the peripheral and central events of fertilization in flowering plants with a focus on both old and new literature.

1.1
Discovery of Double Fertilization

Unambiguous proof of the actual fusion of the male and female gametes embodied in fertilization in flowering plants can be traced to a monographic publication of Strasburger (1884). This work was devoted mostly to the nuclear cytology of pollen grains and pollen tubes of plants belonging to a wide range of families, and to the fate of male gametes delivered by pollen tubes in the embryo sacs

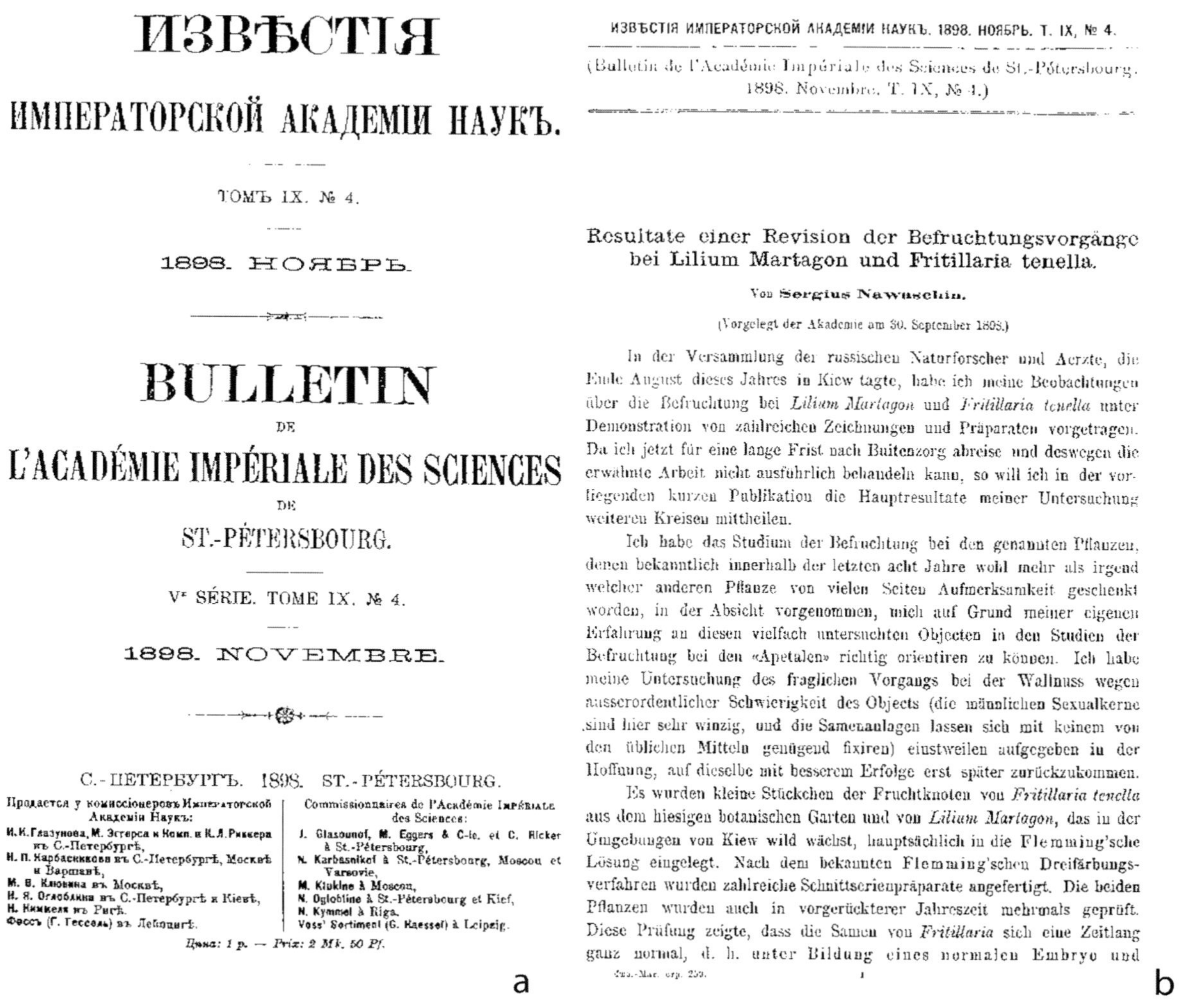

ИЗВѢСТІЯ

ИМПЕРАТОРСКОЙ АКАДЕМІИ НАУКЪ.

ТОМЪ IX. № 4.

1898. НОЯБРЬ.

BULLETIN

DE

L'ACADÉMIE IMPÉRIALE DES SCIENCES

DE

ST.-PÉTERSBOURG.

V^E SÉRIE. TOME IX. № 4.

1898. NOVEMBRE.

С.-ПЕТЕРБУРГЪ. 1898. ST.-PÉTERSBOURG.

Продается у коммиссіонеровъ Императорской Академіи Наукъ:
И. К. Глазунова, М. Эггерса и Комп. и К. Л. Риккера въ С.-Петербургѣ,
Н. П. Карбасникова въ С.-Петербургѣ, Москвѣ и Варшавѣ,
М. В. Клюкина въ Москвѣ,
Н. Я. Оглоблина въ С.-Петербургѣ и Кіевѣ,
Н. Киммеля въ Ригѣ,
Фоссъ (Г. Гессель) въ Лейпцигѣ.

Commissionnaires de l'Académie Impériale des Sciences:
J. Glasounof, M. Eggers & C-ie. et C. Ricker à St.-Pétersbourg,
N. Karbasnikof à St.-Pétersbourg, Moscou et Varsovie,
M. Klukine à Moscou,
N. Oglobline à St.-Pétersbourg et Kief,
N. Kymmel à Riga,
Voss' Sortiment (G. Haessel) à Leipzig.

Цѣна: 1 р. — Prix: 2 Mk. 50 Pf.

a

ИЗВѢСТІЯ ИМПЕРАТОРСКОЙ АКАДЕМІИ НАУКЪ. 1898. НОЯБРЬ. Т. IX, № 4.

(Bulletin de l'Académie Impériale des Sciences de St.-Pétersbourg. 1898. Novembre. T. IX, № 4.)

Resultate einer Revision der Befruchtungsvorgänge bei Lilium Martagon und Fritillaria tenella.

Von Sergius Nawaschin.

(Vorgelegt der Akademie am 30. September 1898.)

In der Versammlung der russischen Naturforscher und Aerzte, die Ende August dieses Jahres in Kiew tagte, habe ich meine Beobachtungen über die Befruchtung bei *Lilium Martagon* und *Fritillaria tenella* unter Demonstration von zahlreichen Zeichnungen und Präparaten vorgetragen. Da ich jetzt für eine lange Frist nach Buitenzorg abreise und deswegen die erwähnte Arbeit nicht ausführlich behandeln kann, so will ich in der vorliegenden kurzen Publikation die Hauptresultate meiner Untersuchung weiteren Kreisen mittheilen.

Ich habe das Studium der Befruchtung bei den genannten Pflanzen, denen bekanntlich innerhalb der letzten acht Jahre wohl mehr als irgend welcher anderen Pflanze von vielen Seiten Aufmerksamkeit geschenkt worden, in der Absicht vorgenommen, mich auf Grund meiner eigenen Erfahrung an diesen vielfach untersuchten Objecten in den Studien der Befruchtung bei den «Apetalen» richtig orientiren zu können. Ich habe meine Untersuchung des fraglichen Vorgangs bei der Wallnuss wegen ausserordentlicher Schwierigkeit des Objects (die männlichen Sexualkerne sind hier sehr winzig, und die Samenanlagen lassen sich mit keinem von den üblichen Mitteln genügend fixiren) einstweilen aufgegeben in der Hoffnung, auf dieselbe mit besserem Erfolge erst später zurückzukommen.

Es wurden kleine Stückchen der Fruchtknoten von *Fritillaria tenella* aus dem hiesigen botanischen Garten und von *Lilium Martagon*, das in der Umgebungen von Kiew wild wächst, hauptsächlich in die Flemming'sche Lösung eingelegt. Nach dem bekannten Flemming'schen Dreifärbungsverfahren wurden zahlreiche Schnittserienpräparate angefertigt. Die beiden Pflanzen wurden auch in vorgerückterer Jahreszeit mehrmals geprüft. Diese Prüfung zeigte, dass die Samen von *Fritillaria* sich eine Zeitlang ganz normal, d. h. unter Bildung eines normalen Embryo und

Физ.-Мат. стр. 259. 1

b

Fig. 1.2a,b Discovery of double fertilization. **a** Cover page of the journal in which Nawaschin's discovery of double fertilization was first published. **b** First page of the article describing double fertilization

of *Gloxinia hybrida* (Gesneriaceae), *Himantoglossum hircinum*, *Orchis latifolia* (Orchidaceae), and *Monotropa hypopitys* (Pyrolaceae). The most complete, illustrated details were provided on *M. hypopitys*, in which it was shown that one of the two male gametes conveyed by the pollen tube fused with the nucleus of the egg. At that time the male gametes were known as generative nuclei and it was uncertain whether these gametes were true cells or naked nuclei. However, the observation that a male gamete fused with the egg in the act of fertilization was contrary to a previous puzzling finding that this event was orchestrated by the diffusion of the cytoplasmic contents of the pollen tube (see Maheshwari 1950). Although Strasburger's work identified the embryo as the resulting product of fertilization, understanding of the fate of the second male gamete discharged by the pollen tube, and the source of origin of the endosperm (albumen), remained major hurdles in gaining a complete insight into the dynamics of fertilization in angiosperms.

1.1.1 Who Discovered Double Fertilization?

The breakthrough in the discovery of double fertilization occurred when S. Nawaschin in Russia showed that, in ovules of *Lilium martagon* and *Fritillaria tenella* (Liliaceae), both male gametes from the pollen tube penetrated the embryo sac; whereas

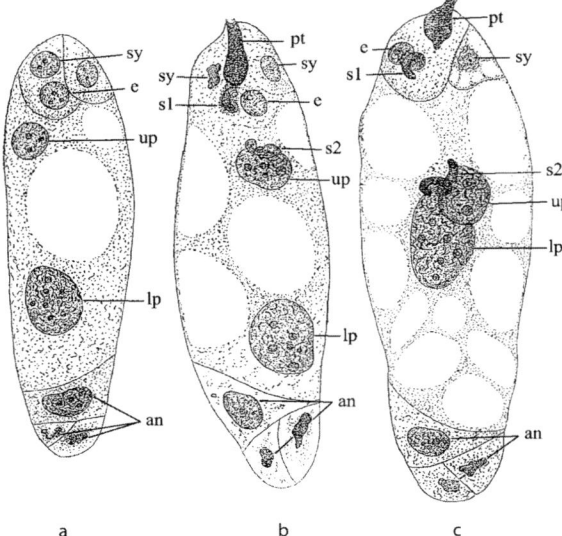

Fig. 1.3a–c Double fertilization in *Lilium martagon*. **a** Mature embryo sac showing the egg apparatus, consisting of the egg and synergids, antipodals, upper polar nucleus, and lower polar nucleus. **b** Mature embryo sac after discharge of male gametes from the pollen tube. The nucleus of one sperm has entered the egg and that of the second sperm is in contact with the upper polar nucleus. The nucleus of one of the synergids is disintegrating. **c** Union of one sperm with the egg nucleus and of the second sperm with the two polar nuclei. *an* Antipodals, *e* egg cell, *lp* lower polar nucleus, *pt* pollen tube, *s1* sperm that fuses with the egg, *s2* sperm that fuses with the polar nucleus, *sy* synergid, *up* upper polar nucleus. (Reprinted from Guignard 1899a)

one of them fused with the nucleus of the egg cell, the other fused with the polar fusion nucleus (at that time known as the definitive nucleus) floating in the central cell, initiating a second fertilization event (Nawaschin 1898, 1899). The results of this work were presented orally on 24 August 1898 to the botanical section of the "Naturforscherversammlung" held in Kiew, Russia (20–30 August 1898) and published as an abstract in the following year (Nawaschin 1899); the full paper appeared a few months after the meeting (Nawaschin 1898). Thus, reverent credit is due to Nawaschin for this legendary discovery of the two fusion events during fertilization in flowering plants (Fig. 1.2a,b). The phenomenon observed by Nawaschin was also independently confirmed in *L. martagon* and *Lilium pyrenaicum* by L. Guignard (1899a, 1899b) in France. The account of this investigation was communicated to the Academy of Sciences in Paris on 4 April 1899 and was published soon afterwards in its Report ("Comptes Rendus") (Guignard 1899a). Exactly the same paper, with a footnoted reference to the earlier paper with volume number and a middle page number, was also published in another journal in the same year (Guignard 1899b). The work described in these two papers, which included a reference to Nawaschin's 1899 abstract, was accompanied by a series of illustrations in the form of line drawings showing the two fusion events (Fig. 1.3a–c). Guignard's description and figures portrayed a precise two-step sequence of events involving the fusion of the second sperm with the upper polar nucleus, followed by integration of this fusion product into the lower polar nucleus. Within a few months of the publication of Guignard's papers, full confirmation of the startling discovery of fusion of the second sperm with the polar fusion nucleus came from a reexamination of previously prepared slides of fertilized ovules of *L. martagon* by E. Sargant in England (Sargant 1899). The coincident choice of ovules of species of *Lilium* and *Fritillaria* by investigators working in three European countries as the classic experimental system in these pioneering studies is not surprising because of the relatively large size of the embryo sac and its equally conspicuous nuclei as seen in microscopic preparations of ovules of these two genera. Indeed, because of this and other advantages, slides demonstrating embryo sac development in various species of *Lilium* and *Fritillaria* have been popular in the teaching of general plant biology; species of these genera have also been favored systems of subsequent investigators because embryo sac development in them appeared to be a simplified version of a complex series of nuclear fusions and divisions that did not have parallels in other plants studied (Maheshwari 1950). To designate the two fertilization events that occur at the inception of the sporophytic phase in flowering plants, Guignard (1899a, 1899b), in a seemingly visionary act, used the term 'double copulation' in the title of the first two papers and 'double fécondation' in later publications. Strasburger (1900) referred to the two fertilization events as 'doppelten Befruchtung' in the title of a paper, and nearly the same term ['die doppelte Befruchtung' and 'двойное оплодотворение' (in Russian)] appeared in the text of two papers by Nawaschin (1900a, 1900b). The term 'double fertilization' now in universal use was first employed in the title of a paper by Thomas (1900) and in the text of a paper by Sargant (1900). Putting to rest the prevalent assumption that the endosperm was generated by fusion of the two polar nuclei, the above-mentioned investigators also concluded correctly that the product of fusion of the second sperm with the polar fusion nucleus gives rise to the endosperm, typically constituted of cells with chromosomes of biparental origin from the coalescence of three nuclei. The discovery of double fertilization in the liliaceous species, and the confirmation of its occurrence in many other angiosperms, including both monocotyledons (monocots) and dicotyledons (eudicots), within a period of just over a year – for example, additional species within the Liliaceae such as *Fritillaria meleagris, Scilla bifolia, Lilium candidum, Tulipa celsiana, Tulipa gesneriana,* and *Tulipa sylvestris* (Guignard 1899c, 1900a, 1900b), *Narcissus poeticus* of the Amaryllidaceae (Guignard 1900a), and *Himantoglossum hircinum, Orchis latifolia, Orchis maculata,* and *Orchis mascula* of the Orchidaceae (Strasburger 1900) (all monocots), *Erigeron philadelphicus, Erigeron strigosa, Guizotia oleiflora, Helianthus annuus* (sunflower), *Heliopsis patula, Rudbeckia grandiflora, Rudbeckia laciniata, Rudbeckia speciosa, Silphium integrifolium, Silphium laciniatum, Silphium terebinthinaceum,* and *Spilanthes oleracea* of the Asteraceae (Guignard 1900a; Land 1900; Nawaschin 1900a, 1900b), *Hibiscus trionum* of the Malvaceae (Guignard 1900a), *Anemone nemorosa, Caltha palustris, Clematis viticella, Delphinium elatum, Helleborus foetidus, Nigella sativa,* and *Ranunculus flammula* of the Ranunculaceae (Guignard 1900a; Nawaschin 1900a, 1900b; Thomas 1900), *Reseda lutea* of the Resedaceae (Guignard 1900a), *Juglans* sp. of the Juglandaceae (Nawaschin 1900a, 1900b), and *Monotropa hypopitys* of the Pyrolaceae (Strasburger 1900) (all eudicots) – may be said to have ushered in twentieth century plant embryology, paving the way for what will surely go down as the golden age in the study of reproductive biology of flowering plants. Appropriately, the centennial of this discovery has been marked by the publication of several reviews on this topic (Jensen 1998; Erdelská and Dubová 2000; Faure 2001; Koul 2001; Friedman 2001b; Raghavan 2003b). Besides paying tribute to Nawaschin and Guignard, these articles show how their discovery has driven the field of plant embryology for more than a century, including most current research in this field.

1.1.2
Universality of Double Fertilization in Flowering Plants

The momentum created in the waning years of the nineteenth century to establish double fertilization as a ubiquitous feature in the reproductive biology of flowering plants was followed by a sustained effort in the first 2 years of the twentieth century leading to the discovery of this phenomenon in additional members of the Ranunculaceae (Guignard 1901c), Liliaceae (Ikeda 1902), Juglandaceae (Karsten 1902), and Pyrolaceae (Shibata 1902), as well as in plants belonging to Poaceae (Guignard 1901a), Najadaceae (Guignard 1901b), Solanaceae, Gentianaceae (Guignard 1901d), Asclepiadaceae (Frye 1902), Brassicaceae (Guignard 1902), and Ceratophyllaceae (Strasburger 1902). Guérin (1904), in a monograph devoted entirely to the topic of fertilization in seed-bearing plants, and Coulter and Chamberlain (1912) in their classic book on the *Morphology of Angiosperms*, refer to 16 families of angiosperms, encompassing about 40 genera and over 60 species definitely known to have a second fertilization event; these two publications surveyed the literature up to the end of 1902. From that time onwards, along with the presence of a reduced female gametophyte and embryo-nourishing endosperm, the occurrence of double fertilization was accepted as a general feature of the reproductive biology of angiosperms. Indeed, under this assumption, there were only occasional references to double fertilization in the numerous publications dating from the early 1900s to the present dealing with the variability and diversity of reproductive processes in flowering plants with special reference to their embryogenesis and endosperm development (Johansen 1950; Maheshwari 1950; Davis 1966; Johri et al. 1992). However, this period was notable for providing the first glimpses of electron microscopic details of double fertilization in several plants, including cotton (*Gossypium hirsutum*; Malvaceae; Jensen and Fisher 1967), maize (*Zea mays*; Poaceae; Diboll 1968; van Lammeren 1986), barley (*Hordeum vulgare*; Poaceae; Cass and Jensen 1970; Mogensen 1982, 1988), *Linum catharticum* (Linaceae; d'Alascio Deschamps 1974), spinach (*Spinacia oleracea*; Chenopodiaceae; Wilms 1981), *Plumbago zeylanica* (Plumbaginaceae; Russell 1982, 1983),

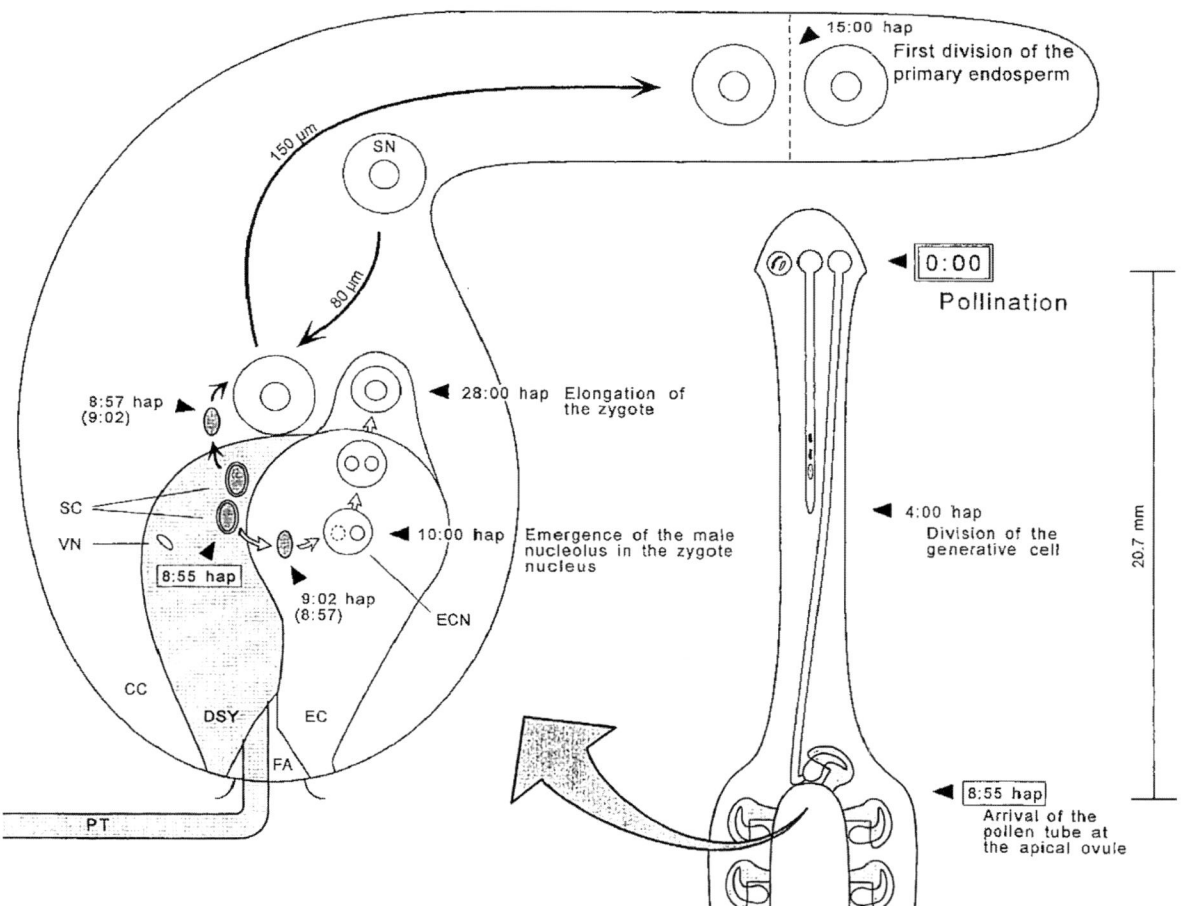

Fig. 1.4 A diagrammatic representation of the time course of double fertilization in *Torenia fournieri*. The time is indicated in hours after pollination (*hap*). Part of the carpel is shown on the right and the embryo sac of the apical ovule is on the left. *CC* Central cell, *DSY* degenerating synergid, *EC* egg cell, *ECN* egg cell nucleus, *FA* filiform apparatus, *PT* pollen tube, *SC* sperm cells, *SN* second polar nucleus, *VN* vegetative cell nucleus. (Reprinted from Higashiyama et al. 1997)

wheat (*Triticum aestivum*; Poaceae; You and Jensen 1985; Gao et al. 1992), *Triticale* (Poaceae; Hause and Schröder 1987), *Populus deltoides* (Salicaceae; Russell et al. 1990), and tobacco (*Nicotiana tabacum*; Solanaceae; Yu et al. 1994).

Almost all observations on double fertilization were made using fixed and/or fixed and sectioned materials. Over the years, complementary powerful insights into isolated aspects of double fertilization were provided by observations of living material of *Monotropa hypopitys* (Strasburger 1900), *Monotropa uniflora* (Shibata 1902), *Calanthe veitchii*, *Cypripedium insigne*, *Dendrobium nobile* (Orchidaceae; Poddubnaya-Arnoldi 1960), *Jasione montana* (Campanulaceae), *Galanthus nivalis* (Amaryllidaceae; Erdelská 1974, 1983), *Torenia fournieri* (Scrophulariaceae; Higashiyama et al. 1997), and *Arabidopsis* (Faure et al. 2002). It is believed that in *M. hypopitys* the male gametes find their way to the egg and the polar fusion nucleus by passively navigating between the cytoplasmic strands that criss-cross the embryo sac (Strasburger 1900). Cinematographic observations of ovules of *J. montana* and *G. nivalis* poised for double fertilization have provided data on the timing of movements of the two sperm in the central cell and on some hitherto unrecorded changes in size and shape of the embryo sac elements (Erdelská 1983). Because the naked embryo sac protrudes from the micropyle of the ovule, *T. fournieri* has proved an especially useful system for live monitoring of the fusion events of fertilization unhindered by the presence of ovular tissues (Fig. 1.4). Here the polar fusion nucleus engages in two targeted movements in the embryo sac. First is its slow migration from a region of the embryo sac to one side of the egg apparatus to await the arrival of the pollen tube with its cargo of male gametes. Second, after fertilization this nucleus is propelled from the vicinity of the egg apparatus to another specific site in the embryo sac (Higashiyama et al. 1997). These nuclear movements have raised wider questions about the involvement of specific signaling molecules during double fertilization, but their identity remains obscure. Using pollen grains from a transgenic line of *Arabidopsis* expressing the green fluorescent protein (GFP) fused with a pollen-specific promoter in the vegetative cell, Faure et al. (2002) have determined the precise time-course of the fertilization processes. Most importantly, this work has opened up the potential use of GFP, tagged to as yet unidentified sperm-cell- and embryo-sac-specific promoters, to follow labeled gametes during double fertilization in vivo without invasive manipulations.

1.2
Seed Development without Double Fertilization

One family of flowering plants whose members do not indulge in double fertilization is the Podostemaceae. Kapil (1970), beginning with relatively early studies, reviewed some of the problems in the embryology of members of the Podostemaceae, including the contradictory reports on the occurrence of double fertilization in members of this family. Compared with most other flowering plants, members of this family have a thalloid plant body that resembles an alga, lichen, or a liverwort. This, along with several other features in their vegetative and reproductive life, makes the Podostemaceae an extraordinary family of flowering plants (Mohan Ram and Sehgal 2001). The final configuration of the mature embryo sac in Podostemaceae studied from time to time initially influenced the reasons for attributing the absence of double fertilization to this family. Typically, the organized embryo sac is four-celled, consisting of a large egg cell and one or two small synergids constituting the egg apparatus, and a central cell harboring a polar nucleus or one or two antipodals. In some species with two synergids in the egg apparatus, the nucleus of the central cell has been shown to degenerate either before the pollen tube enters the embryo sac or before fertilization, or to survive as an antipodal (Battaglia 1971; Nagendran et al. 1976, 1980); in others in which the egg is flanked by only one synergid, the remaining two nuclei are designated as antipodals (Mukkada 1963, 1964; Arekal and Nagendran 1975). The implication is that the absence of a true polar nucleus in the embryo sac precludes fusion of the second male gamete initiating another fertilization event and formation of the endosperm. Understanding the reasons for the absence of double fertilization in this family is a real challenge because mechanical factors such as failure of the pollen tube to discharge the second sperm are probably also involved (Chopra and Mukkada 1966; Mukkada 1969). As double fertilization is a complex process requiring coordinated action of the component cells of the female gametophyte in concert with the male gametes, it is difficult to reconcile some of these observations with what may be actually happening, and hence more studies are required to understand the basis for the absence of double fertilization in the Podostemaceae; a great deal will be revealed by studying the widest possible selection of species.

Conclusive evidence of double fertilization is also lacking in most of the primitive angiosperms so far investigated. In spite of much research, views on the origin and early evolution of angiosperms have remained controversial, and it has not been possible to identify the earliest angiosperms from classifications based on morphological and physiological criteria and limited molecular systematic studies. Over a period of time, these studies designated groups such as Magnoliales, Ceratophyllaceae, and Chloranthaceae as candidates for the earliest angiosperms. However, a series of recent and concurrent investigations on angiosperm relationships inferred from phylogenetic analyses of DNA sequences that combined mitochondrial, chloroplast, ribosomal, and phytochrome genes have shown persuasively that the monotypic genus *Amborella trichopoda* (Amborellaceae), Nymphaeales (Nymphaeaceae and Cabombaceae), and the Illiciales-Trimeniaceae-Austrobaileyaceae complex (together known as the "ANITA" grade) are basal to the common ancestor of monocots and eudicots (Mathews and Donoghue 1999; Soltis et al. 1999; Qiu et al. 1999; Parkinson et al. 1999; The Angiosperm Phylogeny Group 2003). This conclusion was soon reinforced by molecular comparisons of additional chloroplast genes (Graham and Olmstead 2000). The current contenders for the earliest angiosperm lineages are Nymphaeales and Austrobaileyales (Illiciaceae, Schisandraceae, Trimeniaceae, and Austrobaileyaceae; Friedman et al. 2003). However, our knowledge of fertilization processes has not kept pace with the recognition of these new branches of angiosperm evolution, and it has not been definitely established that a representative selection of the earliest lineages of flowering plants identified by molecular phylogenetic analyses displays double fertilization. The limited contributions to the reproductive biology of basal angiosperms currently available pertain mostly to descriptive accounts of their floral morphology and comparative embryology (Friedman 2001a; Friedman and Floyd 2001). The closest that published studies in the comparative embryology of some basal angiosperm lineages such as *Illicium anisatum* (Illiciaceae; Hayashi 1963), *Brasenia schreberei* (Cabombaceae; Khanna 1965), and *Euryale ferox* and *Nymphaea stellata* (Nymphaeaceae; Khanna 1964, 1967) have come is to assume the existence of double fertilization and the formation of an endosperm, but without photographic or other convincing documentation. An exception is provided by studies showing that the embryo sac of *Nuphar polysepalum* (Nymphaeaceae) is typically four-celled, made up of an egg cell flanked by two synergids and a uninucleate central cell (Williams and Friedman 2002; Friedman and Williams 2003). Besides providing striking fluorescent micrographs of the fusion of the sperm nucleus with the haploid central cell nucleus, the authors of these reports have shown by DNA quantitation that the biparental endosperm generated by the second fusion event is diploid (see Plate 1, Fig. a–d). Two additional studies have followed the development of the endosperm from its single-celled origin in *A. trichopoda* and *Illicium floridanum*, but the ploidy level of the tissue has not been determined (Floyd and Friedman 2000, 2001). An investigation of female gametogenesis in *Kadsura japonica* (Schisandraceae) has revealed the development of a four-celled embryo sac, with a haploid central cell nucleus, with the clear implication of the origin of a diploid primary endosperm nucleus following double fertilization (Friedman et al. 2003). It will obviously be of great interest to establish unambiguously by refined microscopic methods the existence of double fertilization in other basal angiosperms, and to ascertain the ploidy level of the resulting endosperm to evaluate the evolutionary significance of this process and the origin of the embryo-nourishing tissue in flowering plants.

Despite the well-known advantages of sexual recombination in the transmission of hereditary characters, plants have also evolved various mechanisms for propagation of the progeny while remaining innocent of sex. In the context of double fertilization, the phenomenon known as apomixis leads to the formation of seeds enclosing a fertilization-independent embryo and, in some cases, an autonomously developing endosperm. Apomictic plants display prefertilization deviations from the normal sexual developmental program by aberrations in female meiosis to produce an unreduced diploid embryo sac enclosing an egg and polar fusion nucleus already endowed with a full complement of both male and female genomes (Ramachandran and Raghavan 1992; Koltunow et al. 2002). Whereas attempts to unravel the genetic control of apomixis in natural apomicts have not led to the isolation of genes involved in the process, mutational studies in the sexually reproducing *Arabidopsis* have provided new insights into the role of genes controlling certain steps in the cascade leading to an apomictic-type seed phenotype. Loss-of-function mutations in a cluster of genes now known as *FERTILIZATION-INDEPENDENT SEEDS2* (*FIS2*) (Chaudhury et al. 1997), *FERTILIZATION-INDEPENDENT ENDOSPERM* (*FIE*, allelic to *FIS3*) (Ohad et al. 1996, 1999; Luo et al. 1999), and *MEDEA* (*MEA*, allelic to *FIS1*, *F644*) (Ohad et al. 1996, 1999; Grossniklaus et al. 1998; Kiyosue et al. 1999; Luo et al. 1999) have been shown to initiate a substantial program of seed development resulting in the generation of a free-nuclear or a cellular endosperm, seed coat formation, and even partial embryogenesis in the absence of fertilization as in the case of some apomicts. Because embryo and endosperm development in the wild-type plants typically follows double fertilization, these genes have been justifiably assigned a role as suppressors of autonomous divisions in the prefertilization egg nucleus and polar fusion nucleus. As described in Chaps. 5 and 8, in addition to their ability to initiate partial embryo and endosperm developmental programs in the absence of fertilization, *fis2*, *fie*, and *mea* mutants (referred to as *fis* class mutants; Grossniklaus et al. 2001), and a few others identified later, display a maternal-effect seed abortion phenotype due to genomic imprinting.

1.3 A Case for Double Fertilization in Gymnosperms

It is well-known that in flowering plants the transformation of the ovule into a seed enclosing a diploid embryo and usually a triploid endosperm is based on the two fertilization events described above. Although gymnosperms share with angiosperms the seed habit, by the end of the last century only sporadic reports of the occurrence of double fertilization in gymnosperms had surfaced. Explicit evidence for a kind of double fertilization in the gymnosperm family Gnetales was first provided in *Ephedra nevadensis* (Friedman 1990a, 1990b) and *Ephedra trifurca* (Friedman 1991). Like most gymnosperms, in *E. trifurca* the egg is housed within the archegonium, which initially consists of a large central cell and a many-celled neck. The division of the nucleus of the central cell into an egg nucleus and a ventral canal nucleus sets the stage for fertilization (Land 1904; Friedman 1991). In contrast to the stray observations of previous investigators based on a limited number of sections of an ovule or of a few ovules of different species of *Ephedra*, Friedman (1990a, 1990b, 1991) has described in exquisite detail, supplemented with elegant light microphotographs and fluorescent micrographs, the odyssey of the two sperm nuclei entering the central cell from the pollen tube and their encounter with the egg cell nucleus and the ventral canal nucleus in *E. nevadensis* and *E. trifurca*. These observations, embracing serial sections of a large number of ovules of different ages, have provided indubitable proof of the occurrence of two fertilization events on a regular basis during sexual reproduction of *Ephedra*. In brief, after the two sperm nuclei generated within a single pollen tube migrate into the central cell, one sperm nucleus can be seen to move in a basal chalazal direction and fuse with the egg nucleus to initiate the first fertilization event. Contemporaneously, the ventral canal nucleus migrates to a deeper location within the central cell, where it fuses with the second sperm nucleus. Initial contact of the sperm nuclei with the egg nucleus and the ventral canal nucleus entails a characteristic invagination of the female nuclei (Fig. 1.5a–f). As in *Ephedra*, reports of previous investigations of fertilization in different species of *Gnetum*, another gnetalean genus, have also been contradictory, but the work of Carmichael and Friedman (1996) indicates the occurrence of a rudimentary process of double fertilization in *Gnetum gnemon*. Unlike in *Ephedra* and other gymnosperms, an archegonium housing the egg is lacking in *G. gnemon*; rather, at the time of fertilization, the female gametophyte appears as a large vacuolate cell with a thin parietal and a dense chalazal band of cytoplasm in which are embedded numerous free nuclei. Two fusion events occur

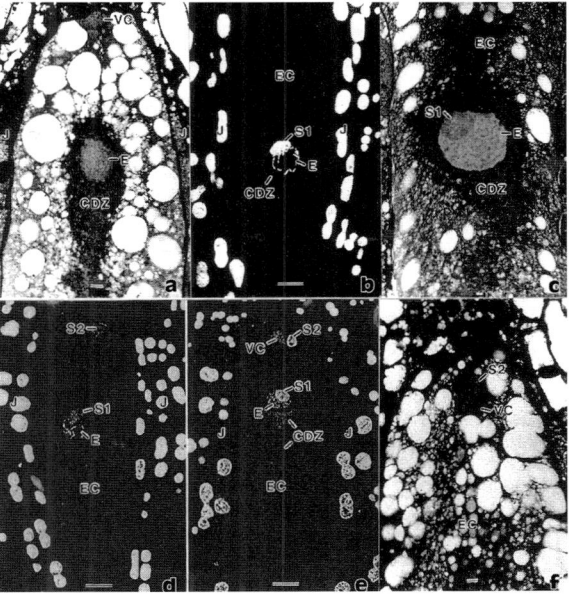

Fig. 1.5a–f Fertilization in *Ephedra trifurca*. **a** Section showing the micropylar end of the egg cell containing the ventral canal nucleus and egg nucleus prior to the entry of sperm nuclei into the egg cell. A cytoplasmically dense zone is prominent around the egg nucleus. **b** Fluorescence view of the first fertilization event in which a sperm nucleus has established contact with the egg nucleus and the two enjoined nuclei have migrated in the chalazal direction. **c** Section showing contact of the sperm nucleus with the egg nucleus. **d** Fluorescence view of the egg cell with one sperm nucleus in contact with the egg nucleus. The second sperm nucleus has not made contact with the ventral canal nucleus (not seen in the section). **e** Fluorescence view of the early stage of both fertilization events in which the first sperm nucleus and the egg nucleus, and the second sperm nucleus and the ventral canal nucleus have made contact. **f** A section showing an early stage of the second fertilization event. *CDZ* Cytoplasmically dense zone, *E* egg nucleus, *EC* egg cell, *J* jacket cells, *S1* sperm nucleus contacting the egg nucleus, *S2* sperm nucleus contacting the ventral canal nucleus, *VC* ventral canal nucleus. *Bars* **a**, **c**, **f** 10 µm; **b**, **d**, **e** 50 µm. (Reprinted from Friedman 1991)

when each of the two sperm nuclei discharged from a pollen tube fuses with a separate, undifferentiated female nucleus within the multinucleate female gametophyte. Thus, after more than a century of going their separate ways, angiosperms and gymnosperms appear to have come together in their fertilization episodes.

From a developmental perspective, it is not surprising that, among gymnosperms, the occurrence of two fertilization events billed as double fertilization was first described in Gnetales. Genera with membership in this family, which include *Ephedra*, *Gnetum*, and *Welwitschia*, collectively and separately possess several angiosperm features such as the presence of vessel elements in the xylem, the similarity of their strobili to some angiosperm inflorescences, reticulate leaf venation in *Gnetum*, and the lack of archegonia in *Gnetum* and *Welwitschia*. This has led to the view that double fertilization, hitherto considered as a defining feature of flowering plants, is evolutionarily homologous in both angiosperms and gymnosperms and may actually have evolved along parallel lines from a common ancestor of these phyla of seed plants (Friedman and Carmichael 1996; Friedman 1998). Acceptance of this hypothesis would lead to the assumption that development of the endosperm for the nurture of the nascent embryo through an intermediate stage is a significant component in the evolution of angiosperm reproductive tissues (Friedman 1990a, 1990b, 1992b, 1994, 1998). The fact that the second fertilization event in *E. trifurca* produces a zygote that yields additional embryos has been invoked in support of the view that the transitional stage in the evolution of the endosperm might be an extra embryo, modified to function as the endosperm to enhance the fitness of the sister embryo (Friedman 1995). Another factor to be reckoned with in discussions on endosperm origin is the endosperm bipolarity observed in several angiosperms, resulting in distinct micropylar and chalazal domains patterned after the anterio-posterior axis of developing angiosperm embryos (Floyd and Friedman 2000). Regarding evolutionary considerations on the origin of the embryo-nourishing tissue in seed plants, the formation of a diploid endosperm following double fertilization in the basal angiosperm *Nuphar polysepalum* alluded to earlier (Williams and Friedman 2002; Friedman and Williams 2003) might assign a strategic role as a homologue of the diploid endosperm to the diploid embryo generated by the second fertilization event in the gnetalean genera. This evolutionary scenario has been carried a step further by the assertion that, since it is thus a homologue of the embryo, the endosperm does not deserve to be designated as a tissue (Friedman 1994). Views of the evolution of double fertilization and endosperm in flowering plants have been aggressively promoted in several reviews (Friedman 1992a, 1994, 1998; Friedman and Carmichael 1996); these articles provide provocative reading on the subject.

Nevertheless, the hypothesis of evolutionary homology of double fertilization in gymnosperms and angiosperms has not gone unchallenged and remains controversial. A basic premise of this view, supported by cladistic analyses (Doyle and Donoghue 1986) and phylogenetic evaluation of molecular data (Hamby and Zimmer 1992), is that Gnetales are the closest extant relatives of angiosperms; consequently, a comparison of potential homologies in the reproductive features of these two groups of plants remains key to the hypothesis of the evolution of double fertilization from a common ancestor of Gnetales and angiosperms. However, evidence contrary to the hypothesis that Gnetales are sister to angiosperms has come from molecular phylogenetic analyses. Phylogenetic reconstructions to clarify the relationship between conifers, Gnetales, and angiosperms using different genes of the well-known MADS-box gene subfamilies that control floral organ identity as molecular markers have provided strong evidence for a closer affinity of Gnetales to conifers than to angiosperms (Winter et al. 1999). From an analysis of molecular data sets of sequences of mitochondrial small subunit ribosomal RNA (rRNA) and those of nuclear small subunit rRNA genes and the chloroplast *rbcL* (large subunit of Rubisco) gene of extant Lycophytes, ferns, Gymnosperms including Gnetales, and angiosperms, Chaw et al. (2000) have claimed that rather than sister to angiosperms, Gnetales are a monophyletic group with close affinity to conifers. Bowe et al. (2000) also find considerable support for gnetales/conifer grouping with the mitochondrial genes cytochrome-*c* subunit (*cox1*) and *atpA* (*atp1*, *ATPase I*), as well as with the chloroplast *rbcL* and nuclear 18S rDNA genes alone and with the two mitochondrial genes. As the relationship among major seed plant lineages continues to be debated, it appears unlikely that the last word has been written on the phylogenetic position of Gnetales among seed plants (Magallón and Sanderson 2002). The results of the studies reviewed above imply that the possible evolutionary relationship of double fertilization observed between Gnetales and angiosperms also remains unresolved and that, based on currently available data, double fertilization events might have originated in the two groups of seed plants independently rather than evolving once in a common ancestor. A close analogy between the two fertilization events in *E. trifurca*, *E. nevadensis*, and *G. gnemon* with double fertilization in angiosperms is also tenuous for other reasons. Unlike in angiosperms, where the second fusion product generates the nutritive tissue of the endosperm, the corresponding fusion nucleus in *E. trifurca* initially produces a zygote that subsequently embarks on a developmental program resulting in multiple cellular proembryos, very much like the zygote from the first fusion product (Friedman 1992a); the fate of the second fusion nucleus in *E. nevadensis* has not been established with certainty (Friedman 1990a, 1990b). Both fertilization products in *G. gnemon* also produce zygotes that evolve into identical proembryos (Carmichael and Friedman 1996). Moreover, the diploid genetic constitution of the second fusion product in the gymnosperm genera is identical to that of the normal zygote, in contrast to the triploid endosperm, which grows cohabitationally with a diploid embryo in angiosperms. Given the strong tendency for the second fusion product in *Ephedra* and *Gnetum* to give rise to multiple embryos, this author feels that the phenomenon of double fertilization observed in these two genera can probably be considered as another route to polyembryony, for which gymnosperms as a phylum are notorious. Like the supernumerary embryos produced in polyembryonic gymnosperms, derivatives of the second fertilization event in *E. trifurca* (Friedman 1992b), and possibly in *G. gnemon* (Carmichael and Friedman 1996), also tend to abort at early stages of development. Niklas (1997) has argued that double fertilization described in the gymnosperm genera is untenable by way of a strict definition of the concept because the term was coined to designate the interaction of two sperm cells with the egg cell and the central cell, respectively, giving rise to a diploid embryo and a triploid endosperm. In the light of the molecular phylogenetic studies assigning conifers a sister status to Gnetales, fresh approaches are clearly necessary to understand the evolution of double fertilization in seed-bearing plants. The implications of the new seed plant phylogenies in general, and those of the new angiosperm phylogenies in particular, on the evolutionary history of the embryo and endosperm resulting from double fertilization have been insightfully analyzed by Friedman and Floyd (2001).

1.4 Structural and Cytological Perspectives on Double Fertilization

The years following the discovery of double fertilization in angiosperms have seen not only a steady increase in the number of species showing this phenomenon, but also attempts to link it to other aspects of sexual reproduction in flowering plants and establish model systems to study the cell biology and nuclear cytology of the fusion events. The identification of discrete phases in the double fertilization episode, such as the arrival of the pollen tube in the embryo sac, release of sperm into the embryo sac, migration or alignment of the sperm nuclei with the female nuclei, and nuclear fusion, as proposed by Russell (1992), has proved to be valuable in this context. Uncovering the details of nuclear fusions that occur in the shrouded environment of the embryo sac to herald double fertilization has long daunted plant embryologists, but some inroads have been made toward this goal. Reviews by Faure (2001), Faure and Dumas (2001), and Lord and Russell (2002) have addressed the main issues involved in the cell biology of double fertilization, whereas Russell (1992) has reviewed most of the critical ultrastructural aspects of the process.

1.4.1 Cellular Nature of the Sperm and the Male Germ Unit

Electron microscopy has lifted the veil of secrecy that has obscured the structural details of the sperm of flowering plants, and its nature as a true cell is now hardly contestable. Aided by a precise knowledge of the location of male gametes born from the division of the generative cell in pollen tubes of germinating binucleate pollen grains of cotton, Jensen and Fisher (1968b) provided the first definitive evidence for the cellular nature of the sperm by examination of pollen tubes growing through the style. The surprising finding was that, even though the sperm has all the trappings of a true cell, it is a relatively simple and unspecialized cell surrounded by a distinct plasma membrane very much like a protoplast, designed solely to carry the genetic information contained in its prominent nucleus. Consequently, the cytoplasm of the sperm is reduced to a thin layer sparsely populated by dictyosomes, endoplasmic reticulum (ER), polysomes, vesicles, and organelles too unspecialized to be unambiguously identified in the electron microscope as either plastids or mitochondria. The structure-function relationship of cotton sperm formed in the pollen tube is also displayed by sperm cells present in ungerminated pollen grains of plants in which the generative cell divides before pollen germination and pollen tube growth (tricellular pollen). Besides the cytoplasmic organelles clearly identified in cotton sperm, sperm cells formed in the trinucleate pollen grains of sugar beet (*Beta vulgaris*; Chenopodiaceae) and barley have well-defined mitochondria and microtubules; the presence of microtubules suggests a role for cytoskeletal elements in sperm motility or changes in sperm shape (Hoefert 1969; Cass 1973). With minor variations in detail, electron microscopic studies of sperm cells of bicellular and tricellular pollen grains of many other angiosperms have essentially confirmed their undistinguished cytoplasmic structure. Although immunofluorescence studies have extended the evidence for the presence of microtubules in sperm cells, whether microfilaments form part of the sperm cytoskeleton has not been clearly established (see Raghavan 1997; Southworth and Russell 2001, for reviews). The simple organization of the sperm and the lack of any differences in structure between the two sperm carried by the pollen grain or pollen tube make them uniquely true protoplasts, yet also make it difficult to understand the multiple facets of double fertilization from the angle of sperm structure alone.

A limitation to the successful exploitation of angiosperm sperm cell cytology in later investigations of cytoplasmic inheritance, gametic transmission, and in vitro fertilization, has been the lack of information on the conformational relationship between sperm and other organelles of the pollen grain and pollen tube. Electron microscopic examination of sperm cells of *Plumbago zeylanica* provided exciting new insights allowing establishment of a model that assumes a physical association between the two sperm cells and between one sperm and the nucleus of the vegetative cell initially in the pollen grain and later in the pollen tube (Russell and Cass 1981; Russell 1984). In the pollen grain the two sperm cells are linked to each other by a common lateral wall with plasmodesmata. Whereas both sperm are enclosed in the inner plasma membrane of the vegetative cell, one of them is associated with the vegetative cell nucleus by a cytoplasmic extension that partially winds around the periphery of this nucleus and is to some extent ensheathed by its lobes. The three bodies travel as a package in the pollen tube, but the connection is lost following discharge of the pollen tube contents. Work on pollen grains of *P. zeylanica* also provided the basis for the profound conclusion that the two sperm cells born out of division of a generative cell, rather than being isomorphic, are different in external morphology, in size, and in organelle content. The important observation is that the sperm with the cytoplasmic extension by which it wraps around the vegetative cell nucleus is the larger of the two, is rich in mitochondria, and impoverished of plastids. In contrast, the small sperm that is free of contact with the vegetative cell nucleus is poor in mitochondria and rich in plastids. As shown by Singh et al. (2002), in addition to differences in the organelle complement, dimorphic sperm cells of *P. zeylanica* also display small differences in the situ expression of polyubiquitin genes. It thus appears that the fundamental problem of the identity of the sperm that fuses with the egg and of that which fuses with the polar fusion nucleus has been framed with observations that would allow sperm cells to respond with preestablished specificity in the double fertilization event. Following these studies, the focus has moved to the use of computer-aided three dimensional reconstructions of serial ultrathin sections of sperm-vegetative cell nucleus associations in pollen grains and pollen tubes of additional plants to show that the size differences between sperm cells and the presence within them of significantly different numbers of heritable organelles is not unique to *P. zeylanica* (Knox and Singh 1987; Mogensen 1992). Unequal distribution of DNA-containing organelles in the two sperm cells formed from a generative cell in *Erythrina crista-galli* (Fabaceae) also supports the concept of sperm dimorphism (Saito et al. 2000). Dimorphism in size or organelle content is not a general feature of populations of sperm cells isolated from maize pollen, although size dimorphism between two sperm cells of a pollen grain exists in a genetic line with supernumerary B chromosomes (Wagner et al. 1989; Faure et al. 2003).

Following the discovery of sperm dimorphism, the field was rapidly inundated with questions about the identity of the sperm that fuses with the egg and of that which fuses with the polar fusion nucleus. Counts of plastids of paternal origin in electron microscopic profiles of fertilized egg cells of *P. zeylanica* have brought a new twist to the study by showing that fusion of the small, plastid-rich, sperm with the egg is as frequent as fusion of the large mitochondrion-rich sperm with the polar fusion nucleus (Russell 1985). This suggests the occurrence of a putative recognition event at the gametic level, perhaps mediated by different cell surface molecules on sperm cells that recognize the egg and the polar fusion nucleus, respectively. The work of Xu et al. (1999) has led to the identification of a gene found to be expressed exclusively in the generative cell and sperm cells of *Lilium longiflorum* pollen. Immunolocalization of the protein product of this gene on the sperm cells has provided the impetus for further investigations into the role of this protein in sperm-egg cell recognition during fertilization. Lectin, a glycoprotein whose receptor sites have been localized on the surfaces of egg and sperm cells of certain plants, is another potential candidate molecule that might be involved in gamete interaction and fusion (Sun et al. 2002; Fang et al. 2003). Sequences from a complementary DNA (cDNA) library – made from isolated sperm cells of maize – predicted to encode plasma membrane-localized proteins might also be important in our quest to understand the molecular biology of egg-targeting determinants on the sperm (Engel et al. 2003). The flowering plant egg surface is probably specialized in many other ways, including the presence of cryptic recognition molecules and adaptations to prevent polyspermy, but these specializations have not yet been elucidated.

Because *P. zeylanica* lacks synergids in the embryo sac, the occurrence of preferential double fertilization mediated by sperm cells described in this species has not been accepted as a secure generalization applicable to other flowering plants in which sperm cells seemingly lack visible markers. Using isolated gametes from genetic lines of maize with and without supernumerary B chromosomes in an in vitro fertilization system (see Sect. 1.5), Faure et al. (2003) have shown that, irrespective of whether sperm cells are dimorphic or not, both sperm in a pollen grain possess the inherent ability to fuse with an egg cell. This finding not only challenges the notion of preferential double fertilization in angiosperms, but also leaves little compelling evidence to explain the basis for the preferential B chromosome transmission to the embryo rather than to the endosperm often observed in maize lines that harbor supernumerary chromosomes.

The discovery of sperm cells-vegetative cell nucleus packaging resulted in the introduction of the concept of male germ unit to designate this tripartite structure linking the two cells containing cytoplasmic and nuclear DNA of male heredity prior to fertilization. The emphasis placed on the male germ unit has led to the view that sperm cells and the nucleus of the vegetative cell function as a single transmitting unit for recognition and fusion with the female target cells during double fertilization (Dumas et al. 1984). The widespread occurrence of male germ unit associations among species that have been examined in detail attests to its role as a functional unit for male reproduction in flowering plants (Knox and Singh 1987; Russell 1997).

1.4.2
Pollen Tube Guidance and Sperm Entry into Embryo Sac

A combination of genetic, biochemical, and cell biological studies has provided much information about the multiple cues that guide the unidirectional growth of the pollen tube through the stigma and style to the ovary, ovule, and, in a final push, to the vicinity of the embryo sac through the micropyle. As an initial step, proteins of the extracellular matrix of the pollen grain interact with appropriate specificities with receptive proteins of the stigma to cause hydration of pollen grains and their subsequent germination (Mayfield et al. 2001). Lipids have been identified as members of a multicomponent complex of the stigma exudate considered essential for pollen tubes to penetrate the stigma following successful pollen germination (Wolters-Arts et al. 1998). The long distances that pollen tubes travel through the style before reaching the ovary have implicated a guidance system that involves multiple prompts acting in overlapping spatial and temporal frames, and includes chemical attractants and repellents and physical guidance, but it has remained a challenge to design studies that allow identification of the putative molecules or forces (Lush 1999). Jiang et al. (2005) have shown that the pectin methylesterase encoded by an *Arabidopsis* gene has an important bearing on the growth of pollen tubes through the transmitting tissue of the style, as a mutation in this gene causes a loss of enzyme activity and retardation of pollen tube growth. One area that is poorly understood is the identity of the signals that guide the pollen tube from the transmitting track of the style to the ovary and thence to the micropyle of the ovule to deliver sperm for successful double fertilization, but promising clues are provided by the emergence of the four-carbon amino acid, γ-aminobutyric acid (GABA), a neurotransmitter in animals, and nitrous oxide, another signal molecule in animal systems, as candidates in this navigation system (Palanivelu et al. 2003; Prado et al. 2004). Upon gaining entry into the ovule, the synergid becomes the cellular cue that guides the directional growth of the pollen tube into the embryo sac. Studies of the manner in which synergids of cotton respond to pollination have revealed a highly specific change: in unpollinated flowers, both synergids remain unchanged until the flower abscises, whereas, if the flower is pollinated, one of the synergids begins to degenerate within a few hours of pollination. It is not known how one of the synergids of an identical pair opts for suicide, but the growing pollen tube intrudes into the degenerating synergid through the filiform apparatus as if on cue, terminates its growth upon entry into this synergid, and discharges the baggage of sperm and some cytoplasm into the synergid (Jensen and Fisher 1968a). A definitive role for the synergid in nuclear fusions associated with double fertilization was established with the discovery that, besides cotton, in other plants such as maize (Diboll 1968), *Epidendrum scutella* (Orchidaceae; Cocucci and Jensen 1969), *Linum usitatissimum* (Vazart 1969), spinach (Wilms 1981), *Quercus gambelii* (Fagaceae; Mogensen 1972), wheat (You and Jensen 1985), sunflower (Yan et al. 1991), turnip (*Brassica campestris*; Brassicaceae; Sumner 1992), tobacco (Huang et al. 1993b), *Arabidopsis* (Christensen et al. 1997; Faure et al. 2002), and *Phaius tankervilliae* (Orchidaceae; Ye et al. 2002), the degenerating synergid becomes predisposed to facilitate entry of a pollen tube into the embryo sac. Variations of this scenario, such as degeneration of both synergids as a result of pollination or before the entry of the pollen tube into the embryo sac (van Went and Cresti 1988; Russell et al. 1990), or degeneration of only one synergid after pollen tube entry (Schulz and Jensen 1968; van Went 1970), have also been occasionally observed. Laser ablation of cells of the embryo sac of *Torenia fournieri* has identified the synergids as attractants of pollen tubes (Higashiyama et al. 2001). Implying a role for a combination of signals emanating from the ovule and the embryo sac, including the synergids in pollen tube guidance, it has been demonstrated that failure to attract pollen tubes is a way of life for ovules of several female gametophyte mutants of *Arabidopsis* harboring defective embryo sacs (Hülskamp et al. 1995; Ray et al. 1997; Shimizu and Okada 2000). However, some reports have cast doubts on the role of synergids in pollen tube attraction during double fertilization. In *P. zeylanica*, which lacks synergids, the contents of the pollen tube are discharged between the egg and the central cell near the chalazal end of the embryo sac (Russell 1982). As shown in other studies, ovules of certain *Arabidopsis* mutants that fail to undergo synergid degeneration nonetheless attract pollen tubes in the normal way without fertilizing the receptive egg cells (Drews and Yadegari 2002; Christensen et al. 2002; Huck et al. 2003; Rotman et al. 2003). In mutants designated as *feronia* (*fer*) and *sirène* (*srn*), wild-type pollen tubes that grow in bizarre ways in the mutant embryo sac are nonetheless prevented from discharging their cargo, implying a possible genetic regulation of this crucial step in double fertilization (Huck et al. 2003; Rotman et al. 2003). Analysis of these mutants has thus identified a new signaling process required for pollen tube reception by the female gametophyte, but not for pollen tube guidance to the vicinity of the latter.

Assuming that a healthy or a dead synergid provides a signal for pollen tube penetration into the embryo sac followed by arrest of growth of the pollen tube and discharge of its contents, the question arises: what is the nature of this signal? Suggestive of a key role for Ca^{2+} as a putative chemotropic attractant of the pollen tube is the finding that high concentrations of this ion in the synergid of wheat, pearl millet (*Pennisetum glaucum*; Poaceae), and tobacco may precede or follow its degeneration (Chaubal and Reger 1990, 1992; Huang and Russell 1992). A case has been made that, in *T. fournieri*, the synergids emit a diffusible signal that is species-specific and acts over a short range, in contrast to Ca^{2+}, which functions as a general attractant of pollen tubes over long distances; identification of the chemical nature of this molecule requires further work (Higashiyama et al. 2003). A close-range guidance cue that attracts pollen tubes of maize to the female gametophyte has been newly identified as a small protein with a predicted transmembrane domain, produced exclusively by the egg apparatus (Márton et al. 2005). As our understanding of pollen tube guidance into the embryo sac becomes sophisticated, it has become clear that even seemingly simple cellular attractions are intricately controlled.

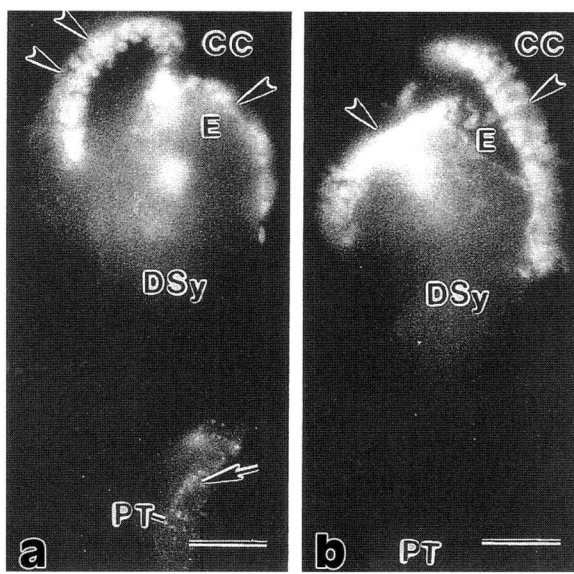

Fig. 1.6a,b Immunofluorecent localization of actin in the embryo sac of *Nicotiana tabacum*. **a** Actin aggregates forming 'coronas' (*arrowheads*) at the chalazal end of the degenerating synergid and in the interface between the egg and the central cell. A punctuate type of actin labeling (*arrow*) is seen in the terminal region of the pollen tube. **b** The same image in a different focal plane showing the coronas at the chalazal end of the degenerating synergid and in the interface between the egg and the central cell. *CC* Central cell, *DSy* degenerating synergid, *E* egg cell, *PT* pollen tube. *Bars* 10 µm. (Reprinted from Huang and Russell 1994)

1.4.3
Nuclear Fusions

Because of their lack of independent motility, the journey of sperm deposited in the degenerating synergid to align with the egg and the polar fusion nucleus is considered an arduous one, and some attention has been paid to the mechanism by which it is accomplished. It is well-established that only the sperm nuclei fuse with their female target cells; the rest of the pollen tube discharge and sperm cytoplasm remain trapped in the milieu of the synergid. As shown in Fig. 1.6 (a,b), two aggregates of actin filaments designated as 'coronas' that presumably guide the pathway of the male gametes in the initial step that brings together the compatible nuclei have been identified within the normally organized embryo sac of tobacco. One of the actin aggregates forms at the chalazal end of the degenerating synergid, extending from its middle lateral region to the region of the egg. The second band occurs in the interface between the egg and the central cell and extends from the side of the egg to the region of the polar nuclei (Huang and Russell 1994). As in tobacco, actin coronas have been identified in the embryo sacs of maize, *T. fournieri* (Huang and Sheridan 1998; Huang et al. 1999), and *Phaius tankervilliae* (Ye et al. 2002), charting the future pathway of the male gametes during fertilization. A view on the origin of actin coronas has been derived from studies that have taken advantage of the relative ease of microinjection of specific dyes that label the actin cytoskeleton during fertilization in living embryo sacs of *T. fournieri* (Fu et al. 2000). A striking change is found in the egg cell, where arrays of actin filaments present before pollination become fragmented into numerous patches after pollination. This actin, along with the actin present in the degenerating synergid and in the intercellular spaces between the egg and synergids, is presumed to contribute to the formation of the two coronas appearing during fertilization, but which disappear after fertilization (see Plate 1, Fig. e–i). These conformational changes in actin filaments in the egg apparatus before and after pollination, and of the coronas before and after fertilization, in addition to being interesting in their own right, have suggested a role for actin in the reception of the pollen tube and in the double fertilization process. However, actin constituties only one of the two principal proteins of a possible actomyosin-based sperm transport; in the absence of a clear demonstration of the presence of myosin on the surface of fertilization-prone sperm cells, a different type of regulatory machinery involving the actin coronas hauling sperm to their destination cannot be ruled out. A dense actin labeling at the boundary between the egg and the central cell constituting a single corona is also part of the cytoskeletal organization of the embryo sac of *P. zeylanica*; the observation that isolated sperm cells of *P. zeylanica* can effortlessly slide along actin bundles anchored in actively streaming cells of the alga *Nitella* has fueled speculation that sperm cells probably acquire soluble myosin from the pollen tube cytoplasm for locomotory purposes (Huang et al. 1993a; Zhang and Russell 1999).

Whereas fusion of the egg and sperm is a straightforward process, the order of fusion of the three nuclei during the second fertilization event always held a fascination for early investigators. Every possible order of their fusion, such as the sperm fusing with a diploid polar fusion nucleus, fusion of all three nuclei together, the sperm fusing with either polar nucleus or both, sperm fusing with the upper polar nucleus later joined by the lower polar nucleus, or the lower polar nucleus being the favorite to fuse with the sperm first, has been described in various plants (Coulter and Chamberlain 1912). The fine structure of karyogamy was first described in cotton by Jensen (1964) and later in a few additional eudicots such as spinach and *Petunia* (Solanaceae) and in monocots such as barley, wheat, and *Triticale* by other investigators (see Faure et al. 1993; Raghavan 1997, for review). In cotton, the mechanism of nuclear fusion involves the apposition and fusion of the outer membranes of the two nuclei, directly or via the ER, at numerous points, followed by fusion between the inner nuclear membranes forming bridges possibly entrapping some cytoplasm. Nuclear fusion is deemed complete when the bridges enlarge and coalesce, reversibly releasing any trapped cytoplasmic elements and providing a picture of the new nuclear membrane contributed by both nuclei (Jensen 1964). The central phenomenon of double fertilization is karyogamy, entailing the complete integration of the male chromatin into the egg nucleus. Regardless of whether gametic DNA synthesis occurs before or after karyogamy, synchrony in the phases of the cell cycle in each gamete has been proposed as being essential for successful fertilization and initiation of mitotic divisions in the zygote. Some early studies by Gerassimova-Navashina (1960) pointed to the importance of the cycling state of the male and female nuclei as a factor contributing to the variations observed in karyogamy. This investigator identified two types of karyogamy: the premitotic type, in which the sperm nucleus fuses immediately with the egg nucleus, after which the zygote nucleus passes through the cell cycle to complete the first zygotic division; and post-mitotic, in which the gametic nuclei enter the mitotic phase independently during the period of courtship, and fuse together during mitosis. Although only a few subsequent studies on the relationship between cell cycle, gamete differentiation, and fertilization have contributed to this generalization, its implications are profound because of the diverse patterns of gametogenesis observed in seed plants. Based on quantitative determination of the DNA content of nuclei of participating cells and of the fusion product, the works of several investigators have supported three mechanisms for karyogamy in seed plants, each of which is formally linked to a precise stage of the cell cycle in each gamete nucleus as proposed by Carmichael and Friedman (1995). An obvious, and probably the most common, mechanism – known as G_1 karyogamy – is one in which the gametic nuclei fuse immediately upon contact and the zygote nucleus subsequently passes through the S, G_2, and M phases of the cell cycle to prime the first division. The assumption here is that the male and female gametes remain in G_1 and contain a 1C (C = DNA quantity per haploid genome) amount of DNA at the time of karyogamy. As shown in maize and barley, nuclear fusion results in the formation of a zygote with a 2C amount of DNA, and the zygote passes through S phase prior to the first mitotic division (Mogensen and Holm 1995; Mogensen et al. 1995). In plants displaying S phase karyogamy, male and female gametes that initiate karyogamy also begin with 1C DNA. However, the gametic nuclei maintain courtship for a long period of time while they pass through S phase synchronously within the egg cytoplasm prior to completion of nuclear fusion, giving rise to a 4C zygote. Photometric data on the DNA content of gametic nuclei in the pollen grain, pollen tube, and in the process of fertilization, and of the zygote have confirmed the existence of

S phase karyogamy in *Ephedra trifurca* (Friedman 1991). In the third mechanism, known as G_2 karyogamy, the gametic nuclei, even before they enter into courtship, complete the S phase of the cell cycle independently, doubling their DNA, and then fuse with each other to form a zygote. Because DNA replication is completed in the male and female gametic nuclei before fusion, the zygote formed will have a 4C amount of DNA. Based on quantitation of cell cycle activity associated with sperm development, it has been inferred that *Arabidopsis* displays G_2 karyogamy, whereas the occurrence of G_2 karyogamy in *Gnetum gnemon* is supported by quantitation of DNA contents of both gametic nuclei as they pass through sexual maturation and form the zygote (Carmichael and Friedman 1995; Friedman 1999). The first example of bicellular pollen grains produced by a flowering plant showing G_2 fusion is in tobacco (Tian et al. 2005). The relationship between the cell cycle of gametes and karyogamy during fertilization in seed plants is illustrated in Fig. 1.7. One reason for the intense recent interest in the cell cycle activity of gametes of seed plants is to provide base-line data on the timing of zygotic DNA synthesis for genetic transformation experiments involving stable integration of novel genes at the time of fertilization.

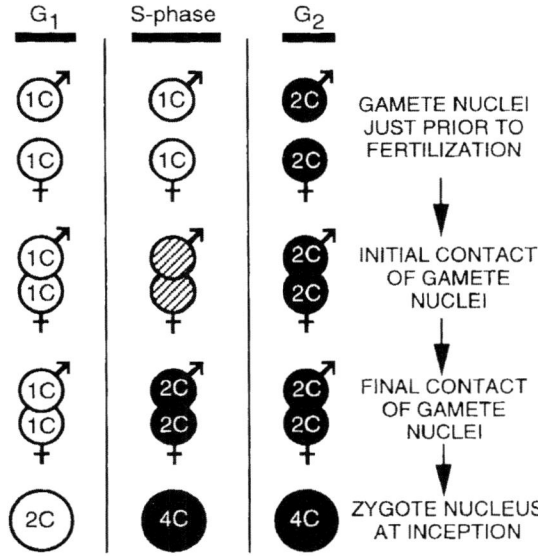

Fig. 1.7 Relationship between the cell cycle and gamete karyogamy in seed plants. *White circles* Nuclei in G_1 phase of the cell cycle, *hatched circles* nuclei in S phase of the cell cycle, *black circles* nuclei in G_2 phase of the cell cycle. (Reprinted from Carmichael and Friedman 1995)

1.5 In vitro Double Fertilization

In order to study the molecular interactions and other finer aspects of gamete fusion during double fertilization in flowering plants, in vitro fertilization with isolated, single gametes under defined conditions rightly deserves to be explored. A major impediment in developing an in vitro fertilization system as a productive encounter of a male and a female gamete – comparable to that routinely used in most animals and in some brown algae – has been the confinement of the sperm within the pollen grain, or in the pollen tube, and of the egg within the embryo sac. A step toward controlled fusion of the egg and sperm was achieved by culturing unpollinated receptive ovules of poppy (*Papaver somniferum*; Papaveraceae) in a nutrient medium and dusting them with viable pollen grains. Pollen germination, pollen tube entry into ovules, and double fertilization proceeded normally, as attested by the transformation of cultured ovules into seeds enclosing embryo and endosperm (Kanta et al. 1962). Subsequently, this technique, dubbed 'test-tube fertilization', was refined to obtain seeds from a variety of flowering plants, often only remotely related, by overcoming interspecific and intergeneric barriers and self-incompatibility (Zenkteler 1990).

Following on the heels of protoplast isolation and somatic hybridization in the 1970s and 1980s, studies were initiated to isolate sperm and embryo sacs from model eudicot and monocot plants. A common method employed for sperm isolation from mature tricellular pollen grains or tubes of bicellular pollen grains is to burst them in an osmoticum and to separate sperm from contaminants by density gradient centrifugation (Southworth 2001). Beginning with the isolation of embryo sacs from chemically fixed ovules by combining microdissection with treatment with pectinolytic and cellulolytic enzymes, the technique was further refined to isolate living, functional embryo sacs and constituent cells of the embryo sac such as the egg, synergids, and central cell (Cass and Laurie 2001). However, few developments in the 1990s created more of a stir among plant embryologists than the extraction of viable egg and sperm from ovules and pollen grains, respectively, of maize, and their use in developing a successful in vitro fertilization protocol;

reviews by Kranz and Kumlehn (1999) and Kranz (2001) describe the development of the field. The initial strategy used in this work is to select sperm cells individually from osmotically shocked pollen grains and egg cells microdissected from embryo sacs freed of maternal tissues by enzymatic treatment. Fusion is accomplished with the transfer of egg-sperm pairs to a microdrop of 0.55 M mannitol and subjecting them to bursts of electrical pulses. The fusion products, when nurtured in a microculture surrounded by a suspension of feeder cells of maize embryo origin as a nurse tissue, initially form a multicellular mass. This is later followed by the formation of bipolar embryos showing obvious similarities to stages in the in vivo embryogenic development of the zygote, culminating in the regeneration of a fertile plant from the fusion product (Kranz et al. 1991; Kranz and Lörz 1993). Efforts to achieve fusion of isolated gametes without electric current were rewarded when isolated egg and sperm cells of maize are reared in a medium containing a high concentration of $CaCl_2$ at pH 11 or in one containing mannitol and $CaCl_2$. On a general level, osmolality of the different media has been judged to play an important role in the success of egg-sperm fusions mediated by electrical pulses and by $CaCl_2$ (Faure et al. 1994; Kranz and Lörz 1994). Transmission electron microscopic study of the electrofusion products from isolated male and female gametes of maize showed that fusion of gametic nuclei occurs before zygotic mitosis, as is the case under in vivo conditions of the premitotic type of karyogamy (Faure et al. 1993). A later investigation demonstrated egg-sperm fusion in vitro in wheat leading to the formation of microcalluses (Kovács et al. 1995); research has also branched in other directions to demonstrate induction of a limited number of divisions in maize egg following fusion with sperm of other members of the Poaceae or when treated with a high concentration of the synthetic auxin, 2,4-dichlorophenoxyacetic acid (2,4-D; Kranz et al. 1995). Because fusion of the second sperm with the polar fusion nucleus in the central cell is a fundamental part of the double fertilization process, development of a successful in vitro egg-sperm fusion system in maize was followed by formulation of a procedure for the isolation of central cells and the fusion of sperm and polar fusion nucleus of maize. With the development of the in vitro crafted fusion nucleus (primary endosperm nucleus) into a typical

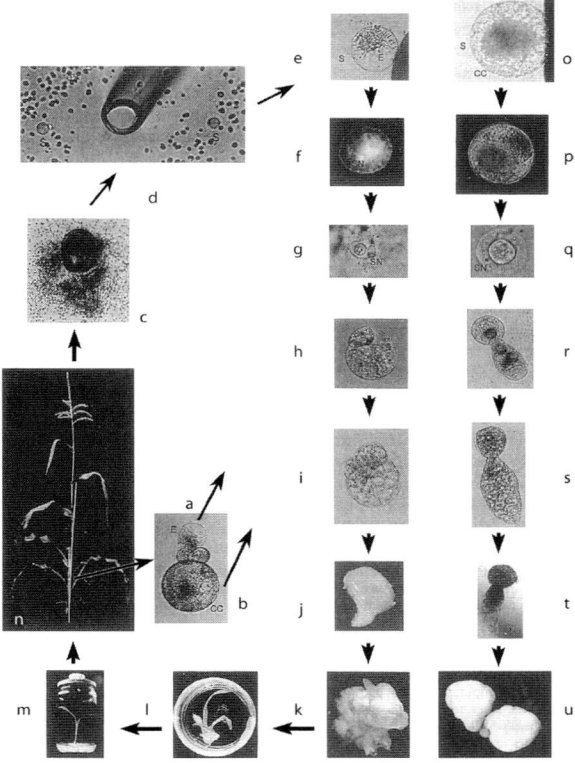

Fig. 1.8a–u A composite picture showing the sequential steps in in vitro double fertilization in maize. **a, b** Isolated egg apparatus (egg and two synergids) and central cell. **c** Pollen grains after bursting of the emerging pollen tube tip; released starch grains are visible. **d** Two isolated sperm cells before suction into the microcapillary. **e** Alignment of a sperm cell and an egg cell. **f** Fertilized egg cell stained with 4′6-diamidino-2-phenylindole (DAPI) showing strong fluorescence of the sperm nucleus inside. **g** Fluorescence of the sperm nucleus inside the isolated egg nucleus after DAPI-staining demonstrating karyogamy. **h** First division of the zygote. **i** Multicellular embryo, with small cells at one pole, and large, more vacuolated cells at the opposite pole. **j** Transition-phase embryo, with meristematic part and suspensor, 2 weeks after fertilization. **k** Callus of white and green tissues, 30 days after fertilization, showing emergence of a coleoptile from the white tissue. **l** Plantlet, 35 days after fertilization. **m** Another plantlet, 39 days after fertilization. **n** Regenerated fertile hybrid plant, 99 days after fertilization. **o** Alignment of a sperm cell and central cell. **p** DAPI-stained, fertilized central cell showing strong fluorescence of the sperm nucleus inside the central cell. **q** Karyogamy, demonstrated by the strong fluorescence of sperm nucleus inside the isolated primary endosperm nucleus after DAPI-staining. **r** Polarized primary endosperm cell at the syncytium stage, showing four nuclei after 1 day in culture. **s** Endosperm, 4 days after culture. **t** Endosperm, 5 days after culture. **u** Endosperm shown in **r** after 11 days in culture. *CC* Central cell, *E* egg cell, *S* sperm, *SN* sperm nucleus. (Reprinted from Kranz 2001)

endosperm tissue, it was demonstrated for the first time that, at least in maize, double fertilization can be accomplished in vitro. Here also, the osmolality of the different media used was found to play an important role; in particular, a shift from higher to lower osmolality between the isolation of the central cell, fusion, and culture of the fusion product was found to be advantageous (Kranz et al. 1998). The sequential steps in in vitro double fertilization in maize are illustrated in Fig. 1.8 (a–u). However, besides maize, tobacco remains the only other plant in which in vitro double fertilization has been accomplished (Tian and Russell 1997; Sun et al. 2000). Although these in vitro fertilization methods lack the elegant simplicity of sea urchins among animals and of *Fucus* among brown algae, we can reasonably expect that future developments will continue to improve on these protocols and lead to studies on the role of recognition molecules during egg-sperm interactions and the mechanism for the avoidance of polyspermy.

In the short period since the in vitro fertilization system was developed, it has proved to be extremely useful in identifying several early cytological and physiological events of fusion of egg and sperm cells in maize. These include the beginning of cell wall synthesis as early as 30 s after fusion (Kranz et al. 1995), a triggring of Ca^{2+} influx for sperm incorporation followed by an elevation of cytosolic Ca^{2+} to launch egg activation (Digonnet et al. 1997; Antoine et al. 2000, 2001), changes in membrane-bound Ca^{2+} and calmodulin levels (Tirlapur et al. 1995), expressional abundance of genes of calreticulun, a Ca^{2+}-binding protein, and of several novel ribosomal proteins at about 18 h (Dresselhaus et al. 1996, 1999b), down-regulation of transcripts of a gene encoding a translation initiation factor (Dresselhaus et al. 1999a), zygotic regulation of cell cycle genes (Sauter et al. 1998), switching off of expression of genes encoding defensin-related proteins after fertilization (Cordts et al. 2001), and changes in microtubule organization (Hoshino et al. 2004). Perhaps the most significant finding is the burst of free Ca^{2+} during egg-sperm fusion, as this has led to further analysis of the role of Ca^{2+} in the process by the demonstration that artificial elevation of cytosolic Ca^{2+} induced by Ca^{2+} ionophores mimics some signs of egg activation observed during in vitro fertilization in maize (such as establishment of a cellulosic cell wall; Antoine et al. 2000). However, further data are required to reveal whether the speculative scenario that Ca^{2+} elevation plays a role in egg activation during in vitro fertilization bears any relation to the actual situation in vivo.

Expression studies of stable integrates as well as microinjected foreign genes in egg cells and zygotes are being undertaken in several laboratories using in vitro fertilization and culture systems combined with transgene technology to decipher the molecular biology of double fertilization and early embryogenesis and endosperm formation in flowering plants (Scholten and Kranz 2001).

1.6 Double Fertilization and the Coming of Age of Plant Embryology

The fortuitous discovery of double fertilization led to a renaissance of interest in angiosperm reproductive biology, and generated a flow of new information on the development of the male and female gametophytes, embryo, and endosperm in a large number of species. This information served to connect the dots in the life cycle of flowering plants into a stunningly simple model of an alternation of generations between a gametophytic phase and a sporophytic phase. The resulting picture strengthened the idea that embryological processes lie at the interface of developmental pathways that initiate phase changes in the life cycle of plants. Included under the rubric of embryology in the angiosperm life cycle are the ensemble of changes involved in sporogenesis and gametogenesis in the anther and ovule, fertilization, and the development of the embryo and endosperm. In these investigations, the boundary between embryology, concerned with the developmental processes in reproductive biology beyond the strictly embryogenic phase, and embryogenesis, dealing with the development of the embryo from its single-celled beginning, was left somewhat vague and both terms were even used synonymously. In contrast to embryology, embryogenesis, in the sense used in this book, is concerned with the whole constellation of events following double fertilization and is regarded as the continuum of processes involved in the origin, growth, and orderly development of the zygote into a fully fledged embryo, and of the primary endosperm nucleus into the nutrient-rich tissue of the endosperm (Maheshwari 1950; Wardlaw 1955). Other terms such as general embryology and special and comparative embryology (recognizing broad divisions within angiosperm embryology), embryogeny (dealing with the dynamic aspects of embryo development under general embryology), and embryogenesis, embryotectonics, embryogenergy, and embryonomy (concerned with individual phases of embryogeny) were introduced by Johansen (1950), but none of them, except embryogenesis to describe the origin and development of the embryo, has caught on with plant biologists. No doubt, if real boundaries do not exist, then a rigorous practice of defining artificial boundaries may be misleading.

During most of the past century, embryogenesis has proved to be pivotal in the analysis of animal and plant development. It needs little emphasis to conclude that in all eukaryotes, embryogenesis is a phenomenon of great consequence as it leads in most cases to the formation of a functional adult organism endowed with full multicellularity, sexuality, and structure. Since animals and plants, despite their outward differences, share many common principles in ontogeny, embryos of representatives of both kingdoms have been studied to address basic developmental problems such as the induction of polarity, patterning of tissues and organs, gene function, and positional signals in cell specification. Study of animal embryogenesis crept forward at an uneven pace to come of age in the 1970s contemporaneous with the understanding of the role of genes in development, breaking of the genetic code, and advances in the techniques of molecular biology and genetic engineering, whereas after a period of neglect and stagnation, the coming of age of plant embryogenesis has been recent, and has been accomplished at an almost frightening speed (Raghavan 2000, 2001).

1.6.1
The Changing Scene

In the past, much of our basic knowledge of embryo and endosperm development in flowering plants has come from morphological, histological, cytological, and biochemical investigations of several species at one time or other over a long period. However, a few model systems have generated most of our current knowledge about the genetic and molecular biology of embryogenesis and endosperm development. When descriptive accounts of embryo development in angiosperms began to appear in the 1870s, a fairly clear picture of the structural organization of the male and female gametophytes was already available. The choice of Shepherd's purse (*Capsella bursa-pastoris*; Brassicaceae) for an important part of the first descriptive account of embryo development in angiosperms came as a gratifying surprise to later workers in the field, because this plant has received wide acceptance as a paradigm species to follow cleavage patterns of early-stage embryos and trace the ancestry of cells in a typical eudicot embryo (Hanstein 1870). For nearly 80 years following the work on *C. bursa-pastoris*, the field of descriptive embryology, involving not only the segmentation patterns of embryos but also events of micro- and megasporogenesis and gametogenesis and endosperm development in plants belonging to widely scattered families, emerged as a preeminent field of study. These investigations provided ample evidence of the diversity in the pattern of cell divisions during the development and organization of embryos of eudicots and monocots to suggest that cell lineages during embryo development are programmed by a blueprint characteristic of each species. They also led to an appreciation of the role of the suspensor and its bizarre haustorial outgrowths in anchoring the embryo and positioning it in relation to the endosperm and seed tissues, and of apomixis in short-circuiting the sexual pathway of reproduction in the formation of viable seeds (Maheshwari 1950; Wardlaw 1955). Embryological data were used in later investigations to identify for realignment doubtful genera and species delimited solely on the basis of vegetative characters and floral morphology. By providing a new level of information, these investigations opened up the field of comparative embryology for solving disputed taxonomic assignments of flowering plants (Herr 1984). However, given that nearly 300,000 species of flowering plants have been cataloged and identified, embryo and endosperm development has been adequately described in only a fraction of this number (Johri et al. 1992).

Beginning in the 1930s, advances made in the fields of plant physiology, biochemistry, and genetics, and refinements in the culture of plant organs, tissues, cells, and protoplasts under aseptic conditions, led to a swing of the spotlight in research in plant embryology from descriptive and comparative onto experimental. A focus area of research in experimental embryology was tissue culture involving the isolation and culture of embryos of different ages in defined media, first introduced by Hannig (1904). These studies provided, in their basic tenets, an invaluable guide to the type of nutrients necessary to grow embryos of different ages outside the environment of the ovule: early-stage embryos, bombarded as they are in the natural habitat of the embryo sac with nutrient substances present in the endosperm, require complex exogenously supplied metabolites to maximize their chances for survival and growth, whereas late-stage embryos, especially seed embryos can be nurtured to the stage of seedling plants in relatively simple media consisting of mineral salts and a carbon energy source such as sucrose. This confirmed what was suspected from other studies, i.e., that early-stage embryos are heterotrophic and depend on the nutrient materials present in the endosperm, whereas late-stage embryos are autotrophic and are able to synthesize the array of metabolites necessary for their growth (Raghavan 2003a).

To carry forward the concept of experimental embryology, tissue culture approaches have shown that single somatic cells and pollen grains of many angiosperms, and a few gymnosperms, can give rise to fertile plants, simulating stages strongly reminiscent of normal embryogenesis by processes known as somatic embryogenesis and pollen embryogenesis, respectively (see Thorpe and Stasolla 2001; Touraev et al. 2001, for reviews). This highlights the interesting fact that, whereas the zygote passes through cycles of growth and division to differentiate into an embryo, somatic cells and pollen grains follow dedifferentiative pathways to form embryo-like structures. The advantages inherent in the clonal multiplication of plants by somatic embryogenesis and in the production of isogenic haploids by anther and pollen culture techniques are enormous and are being exploited in horticultural and breeding practices.

1.6.2
Genetic and Molecular Studies of Embryogenesis and Endosperm Development

Although robust embryological investigations of very many additional species of flowering plants will be necessary to enlarge the database of wild-type relatives of cultivated crop plants, the need for this work was overshadowed by developments in molecular biology and genetics to study flowering plant embryology. This heralded the next frontier, the field of molecular embryology. Major insights into the genetic and molecular systems that underlie the progressive development of the embryo and endosperm of angiosperms have become possible by drawing largely on the experimental advantages of *Arabidopsis*, which has now entered the pantheon of plant model systems. Indeed, it is within this small plant, often called wall cress or mouse-ear cress, that our understanding of not only the genetic and molecular control of embryogenesis, but also of the whole spectrum of developmental episodes in the life of a flowering plant, is most advanced. *Arabidopsis* has nearly every characteristic that one could wish for in a model system, in particular, a short life-cycle of 4–5 weeks, coupled with the major molecular advantages of a small genome with a haploid DNA of 70,000 kb pairs in just five chromosomes and a low repetitive DNA content. The rich genetic potential of the plant has been further enhanced by the complete sequencing of its entire genome (The Arabidopsis Genome Initiative 2000).

In an assault on a gene of an organism, the weapon of choice is infliction of mutations. The genetic and molecular analysis of embryo development in *Arabidopsis* has been aided in large part by the systematic isolation and analysis of mutations that affect in an informative way virtually every aspect of embryo development from the morphology of the mature embryo down to the early-stage embryo generated by the first few rounds of division of the zygote. Most genetic screens have been phenotype-driven; once the mutant gene has been cloned, the real work begins in attempts to link gene action in the zygote and its immediate division products to progressive embryogenesis, dissect complex phenomena, and illuminate new issues. From these and other studies, which have thrust *Arabidopsis* to the forefront as a happy hunting ground for embryo developmental mutants, evidence has emerged that pattern formation, morphogenesis, and cytodifferentiation of the embryo are regulated independently by different sets of genes (Meinke 1994). Progress has also been made in isolating embryo-defective mutants from maize (Clark and Sheridan 1991), rice (*Oryza sativa*;Poaceae; Hong et al. 1995), and pea (*Pisum sativum*; Fabaceae; Liu et al. 1996) built on the rich genetic legacy of these plants. Studies of *Arabidopsis* mutants screened for endosperm development in unfertilized ovules have revealed that genesis of the endosperm by the division of the primary endosperm nucleus is regulated by a set of genes whose protein products act as gene silencers, and have led to renewed interest in the concept of genomic imprinting (Grossniklaus et al. 2001; Sørensen et al. 2001).

1.6.3
Problems and Prospects

Research on topics in the embryology of flowering plants, especially embryogenesis and endosperm development, seems to have attained a sufficient degree of sophistication to take its place among the most exciting and active areas of study in plant development, well ahead of other areas of plant reproductive biology and on a level nearly comparable to animal embryogenesis. The momentum of current research efforts using the whole modern bag of tricks of genetics and biotechnology has led to the creation of rice genetically engineered to make β-carotene in its endosperm cells. This rice, dubbed 'golden rice' because of its pale yellow color when polished as compared to pearly white ordinary rice, and its great humanitarian intent to improve the lives of millions who depend upon rice as a staple diet, has even caught the attention of the popular press (Nash 2000). Unfortunately, tinkering with genes of embryos of flowering plants has produced some dark clouds on the horizon, with profound social, ethical, philosophical, and economic implications. This is the development of the 'terminator technology' that does not allow embryos to grow into seedlings when seeds are sown for a crop in the next generation – an age-old practice of farmers who save their best seeds from one year's harvest for planting in a subsequent growing season. The technological feat dubbed 'suicide seeds' has also not escaped media scrutiny (Kluger 1999). It is something of an irony that a field of research that promises so much hope for economic breakthroughs, also portends disaster for farmers and even threatens to ruin the economy of some countries.

In addition to periodic reviews, symposium volumes, and multi-authored publications – too many to list here – that record the progress of research on selected topics in the reproductive biology of flowering plants covering the period from about 1940 to the present, books have been written to chronicle and synthesize in detail the accomplishments in descriptive and comparative (Coulter and Chamberlain 1912; Schnarf 1929; Johansen 1950; Maheshwari 1950; Davis 1966; Johri et al. 1992; Lersten 2004), experimental (Wardlaw 1955; Raghavan 1976, 1986), and molecular (Raghavan 1997) aspects of embryogenesis and endosperm development. Although there is thus no dearth of literature in the field, this book is intended to present an integrated picture of flowering plant embryogenesis and endosperm development, with emphasis on topics that have become the central focus for research in recent years. The chapters that follow can therefore be considered as state-of-the art accounts of published research to probe the developmental biology of the embryo and endosperm. Admittedly, much of this work is based on model systems such as *Arabidopsis* and maize, yet research on model systems might provide both an exciting and a fruitful background to those who want to keep abreast of developments that will roll forward in the next few years in the plant science scene relating to the themes considered in this book. It is likely that many genetic and molecular control systems active in post-fertilization events in flowering plants will have parallels in animal embryogenesis; a major thrust of future studies in this context lies in determining the extent of participation in angiosperm reproductive biology of genes and gene products implicated in the embryogenesis of animals.

1.7
Concluding Comments

Insight into the essential role of a second fertilization event in the formation of endosperm in angiosperm seeds came about as a result of the discovery of double fertilization. The age-old interest in improving crop efficiency, first with hybridization techniques and in later years with tissue culture-based and biotechnological approaches, may be traced to the discovery of double fertilization and our ability to intervene in the normal sexual reproductive processes of plants. Elucidation of the cell biology of double fertilization has been facilitated by investigations undertaken during the past 50 years, most importantly on the fine structural details of double fertilization and the role of synergids in the process, as well as on the development of techniques to isolate egg and sperm cells and their use in in vitro fertilization. After a long period of neglect on the evolutionary front, the discovery of two rudimentary fertilization events in certain gnetalean genera and the identification, based on molecular data, of new phylogenetic relationships of extant seed plants are beginning to reshape our thinking about the origin of double fertilization and the evolutionary equivalence of the two fertilization processes. These investigations have generated new interest in many fundamental questions about double fertilization, including the evolutionary history of the endosperm as an embryo-nourishing tissue, the nature of the signals activated at different times during the odyssey of the pollen tube, the mechanistic aspects of the movement of sperm cells to their female target cells, and the molecular basis of gene expression in the products of double fertilization. Thus, double fertilization has not only taken deep roots in the reproductive biology of flowering plants, but has also branched out in many unanticipated directions.

Acknowledgement. Parts of this chapter are from the Tansley review by the author entitled "Some reflections on double fertilization, from its discovery to the present" New Phytologist (2003) 159:565–583. Reproduced with permission of the New Phytologist Trust, © 2003.

REFERENCES

Antoine AF, Faure J-E, Cordeiro S, Dumas C, Rougier M, Feijó JA (2000) A calcium influx is triggered and propagates in the zygote as a wavefront during in vitro fertilization of flowering plants. Proc Natl Acad Sci USA 97:10643–10648

Antoine A-F, Faure J-E, Dumas C, Feijó JA (2001) Differential contribution of cytoplasmic C^{2+} and Ca^{2+} influx to gamete fusion and egg activation in maize. Nat Cell Biol 3:1120–1123

Arekal GD, Nagendran CR (1975) Embryo sac of *Hydrobryopsis sessilis* (Podostemaceae) – origin, organization and significance. Bot Not 128:332–338

Battaglia E (1971) The embryo sac of Podostemacae – an interpretation. Caryologia 24:403–420

Bowe LM, Coat G, dePamphilis CW (2000) Phylogeny of seed plants based on all three genomic compartments: extant gymnosperms are monophyletic and Gnetales' closest relatives are conifers. Proc Natl Acad Sci USA 97:4092–4097

Carmichael JS, Friedman WE (1995) Double fertilization in *Gnetum gnemon*: the relationship between the cell cycle and sexual reproduction. Plant Cell 7:1975–1988

Carmichael JS, Friedman WE (1996) Double fertilization in *Gnetum gnemon* (Gnetaceae): its bearing on the evolution of sexual reproduction within the Gnetales and the Anthophyte clade. Am J Bot 83:767–780

Cass DD (1973) An ultrastructural and Nomarski-interference study of the sperms of barley. Can J Bot 51:601–605

Cass DD, Jensen WA (1970) Fertilization in barley. Am J Bot 57:62–70

Cass DD, Laurie JD (2001) Embryo sac. Isolation and manipulation. In: Bhojwani SS, Soh WY (eds) Current trends in the embryology of angiosperms. Kluwer, Dordrecht, pp 89–100

Chaubal R, Reger BJ (1990) Relatively high calcium is localized in synergid cells of wheat ovaries. Sex Plant Reprod 3:98–102

Chaubal R, Reger BJ (1992) Calcium in the synergid cells and other regions of pearl millet ovaries. Sex Plant Reprod 5:34–46

Chaudhury AM, Ming L, Miller C, Craig S, Dennis ES, Peacock WJ (1997) Fertilization-independent seed development in *Arabidopsis thaliana*. Proc Natl Acad Sci USA 94:4223–4228

Chaw S-M, Parkinson CL, Cheng Y, Vincent TM, Palmer JD (2000) Seed plant phylogeny inferred from all three plant genomes: monophyly of extant gymnosperms and origin of Gnetales from Conifers. Proc Natl Acad Sci USA 97:4086–4091

Chopra RN, Mukkada AJ (1966) Gametogenesis and pseudo-embryo sac in *Indotristicha ramosissima* (Wight) van Royen. Phytomorphology 16:182–188

Christensen CA, King EJ, Jordan JR, Drews GN (1997) Megagametogenesis in *Arabidopsis* wild type and the *Gf* mutant. Sex Plant Reprod 10:49–64

Christensen CA, Gorsich SW, Brown RH, Jones LG, Brown J, Shaw JM, Drews GN (2002) Mitochondrial GFA2 is required for synergid cell death in *Arabidopsis*. Plant Cell 14:2215–2232

Clark JK, Sheridan WF (1991) Isolation and characterization of 51 *embryo-specific* mutations of maize. Plant Cell 3:935–951

Cocucci A, Jensen WA (1969) Orchid embryology: megametophyte of *Epidendrum scutella* following fertilization. Am J Bot 56:629–640

Cordts S, Bantin J, Wittich PE, Kranz E, Lörz H, Dresselhaus T (2001) *ZmES* genes encode peptides with structural homology to defensins and are specifically expressed in the female gametophyte of maize. Plant J 25:103–114

Coulter JM, Chamberlain CJ (1912) Morphology of angiosperms (Morphology of spermatophytes. Part II). Appleton, New York

d'Alascio Deschamps R (1974) Etude ultrastructurale de la double fécondation chez le *Linum catharticum* L. C R Acad Sci Paris 279D:263–265

Davis GL (1966) Systematic embryology of the angiosperms. Wiley, New York

Diboll AG (1968) Fine structural development of the megagametophyte of *Zea mays* following fertilization. Am J Bot 55:797–806

Digonnet C, Aldon D, Leduc N, Dumas C, Rougier M (1997) First evidence of a calcium transient in flowering plants at fertilization. Development 124:2867–2874

Doyle JA, Donoghue MJ (1986) Seed plant phylogeny and the origin of angiosperms: an experimental cladistic approach. Bot Rev 52:321–431

Dresselhaus T, Hagel C, Lörz H, Kranz E (1996) Isolation of a full-length cDNA encoding calreticulin from a PCR library of in vitro zygotes of maize. Plant Mol Biol 31:23–34

Dresselhaus T, Cordts S, Lörz H (1999a) A transcript encoding translation initiation factor eIF-5A is stored in unfertilized egg cells of maize. Plant Mol Biol 39:1063–1071

Dresselhaus T, Cordts S, Heuer S, Sauter M, Lörz H, Kranz E (1999b) Novel ribosomal genes from maize are differentially expressed in the zygotic and somatic cell cycles. Mol Gen Genet 261:416–427

Drews GN, Yadegari R (2002) Development and function of the angiosperm female gametophyte. Annu Rev Genet 36:99–124

Dumas C, Knox RB, McConchie CA, Russell SD (1984) Emerging physiological concepts in fertilization. What's New Plant Physiol 15:17–20

Engel ML, Chaboud A, Dumas C, McCormick S (2003) Sperm cells of *Zea mays* have a complex complement of mRNAs. Plant J 34:697–707

Erdelská O (1974) Contribution to the study of fertilization in the living embryo sac. In: Linskens HF (ed) Fertilization in higher plants. North-Holland, Amsterdam, pp 191–195

Erdelská O (1983) Microcinematographical investigation of the female gametophyte, fertilization and early embryo and endosperm development. In: Erdelská O (ed) Fertilization and embryogenesis in ovulated plants. VEDA, Bratislava, pp 49–54

Erdelská O, Dubová J (2000) Double fertilisation of angiosperms 1889–2000 (the origin and the significance of flowering plants double fertilisation). Biologia 55:311–319

Fang K-F, Sun M-X, Kranz E, Zhou C (2003) In vitro and in situ localization of concanavalin A and wheat germ agglutinin binding sites on the surface of female cells in *Torenia fournieri* L. Isr J Plant Sci 51:83–90

Faure J-E (2001) Double fertilization in flowering plants: discovery, study methods and mechanisms. C R Acad Sci Paris Sci de la vie 324:551–558

Faure J-E, Dumas C (2001) Fertilization in flowering plants. New approaches for an old story. Plant Physiol 125:102–104

Faure J-E, Mogensen HL, Dumas C, Lörz H, Kranz E (1993) Karyogamy after electrofusion of single egg and sperm cell protoplasts from maize: cytological evidence and time course. Plant Cell 5:747–755

Faure J-E, Digonnet C, Dumas C (1994) An in vitro system for adhesion and fusion of maize gametes. Science 263:1598–1600

Faure J-E, Rotman N, Fortune P, Dumas C (2002) Fertilization in *Arabidopsis thaliana* wild type: developmental stages and time course. Plant J 30:481–488

Faure J-E, Rusche ML, Thomas A, Keim P, Dumas C, Mogensen HL, Rougier M, Chaboud A (2003) Double fertilization in maize: the two male gametes from a pollen grain have the ability to fuse with egg cells. Plant J 33:1051–1062

Floyd SK, Friedman WE (2000) Evolution of endosperm developmental patterns among basal flowering plants. Int J Plant Sci 161:S57–S81

Floyd SK, Friedman WE (2001) Developmental evolution of endosperm in basal angiosperms: evidence from *Amborella* (Amborellaceae), *Nuphar* (Nymphaeaceae), and *Illicium* (Illiciaceae). Plant Syst Evol 228:153–169

Friedman WE (1990a) Double fertilization in *Ephedra*, a nonflowering seed plant: its bearing on the origin of angiosperms. Science 247:951–954

Friedman WE (1990b) Sexual reproduction in *Ephedra nevadensis* (Ephedraceae): further evidence of double fertilization in a nonflowering seed plant. Am J Bot 77:1582–1598

Friedman WE (1991) Double fertilization in *Ephedra trifurca*, a non-flowering seed plant: the relationship between fertilization events and the cell cycle. Protoplasma 165:106–120

Friedman WE (1992a) Double fertilization in nonflowering seed plants and its relevance to the origin of flowering plants. Int Rev Cytol 140:319–355

Friedman WE (1992b) Evidence of a pre-angiosperm origin of endosperm: implications for the evolution of flowering plants. Science 255:336–339

Friedman WE (1994) The evolution of embryogeny in seed plants and the developmental origin and early history of the endosperm. Am J Bot 81:1468–1486

Friedman WE (1995) Organismal duplication, inclusive fitness theory, and altruism: understanding the evolution of endosperm and the angiosperm reproductive syndrome. Proc Natl Acad Sci USA 92:3913–3917

Friedman WE (1998) The evolution of double fertilization and endosperm: an "historical" perspective. Sex Plant Reprod 11:6–16

Friedman WE (1999) Expression of the cell cycle in sperm of *Arabidopsis*: implications for understanding patterns of gametogenesis and fertilization in plants and other eukaryotes. Development 126:1065–1075

Friedman WE (2001a) Comparative embryology of basal angiosperms. Curr Opin Plant Biol 4:14–20

Friedman WE (2001b) Developmental and evolutionary hypotheses for the origin of double fertilization and endosperm. C R Acad Sci Paris Sci de la vie 324:59–567

Friedman WE, Carmichael JS (1996) Double fertilization in Gnetales: implications for understanding reproductive diversification among seed plants. Int J Plant Sci 157:S77–S94

Friedman WE, Floyd SK (2001) The origin of flowering plants and their reproductive biology – a tale of two phylogenies. Evolution 55:217–231

Friedman WE, Williams JH (2003) Modularity of the angiosperm female gametophyte and its bearing on the early evolution of endosperm in flowering plants. Evolution 57:216–230

Friedman WE, Gallup WN, Williams JH (2003) Female gametophyte development in *Kadsura*: implications for Schisandraceae, Austrobaileyales, and the early evolution of flowering plants. Int J Plant Sci 164:S293–S305

Frye TC (1902) A morphological study of certain Asclepiadaceae. Bot Gaz 34:389–413

Fu Y, Yuan M, Huang B-Q, Yang H-Y, Zee S-Y, O'Brien TP (2000) Changes in actin organization in the living egg apparatus of *Torenia fournieri* during fertilization. Sex Plant Reprod 12:315–322

Gao X, Francis D, Ormrod JC, Bennett MD (1992) An electron microscopic study of double fertilization in allohexaploid wheat *Triticum aestivum* L. Ann Bot 70:561–568

Gerassimova-Navashina H (1960) A contribution to the cytology of fertilization in flowering plants. Nucleus 3:111–120

Graham SW, Olmstead RG (2000) Utility of 17 chloroplast genes for inferring the phylogeny of the basal angiosperms. Am J Bot 87:1712–1730

Grossniklaus U, Vielle-Calzada J-P, Hoeppner MA, Gagliano WB (1998) Maternal control of embryogenesis by *MEDEA*, a polycomb group gene in *Arabidopsis*. Science 280:446–450

Grossniklaus U, Spillane C, Page DR, Köhler C (2001) Genomic imprinting and seed development: endosperm formation with and without sex. Curr Opin Plant Biol 4:1–27

Guérin P (1904) Les connaissances actuelles sur la fécondation chez les phanérogames. Joanin, Paris

Guignard L (1899a) Sur les anthérozoïdes et la double copulation sexuelle chez les végétaux angiospermes. C R Acad Sci Paris 128:864–871

Guignard L (1899b) Sur les anthérozoïdes et la double copulation sexuelle chez les végétaux angiospermes. Rev Gén Bot 11:129–135

Guignard L (1899c) Les découvertes récentes sur la fécondation chez les végétaux angiospermes. In: Cinquantenaire de la Société de Biologie, vol Jubilaire. Masson, Paris, pp 189–198

Guignard L (1900a) Nouvelles recherches sur la double fécondation chez les végétaux angiospermes. C R Acad Sci Paris 131:153–160

Guignard L (1900b) L'appareil sexuel et la double fécondation dans les tulipes. Ann Sci Nat Bot Ser 8, 11:365–387

Guignard L (1901a) La double fécondation dans le maïs. J Bot 15:37–50

Guignard L (1901b) La double fécondation dans le *Naias major*. J Bot 15: 205–213

Guignard L (1901c) La double fécondation chez les Renonculacées. J Bot 15: 394–408

Guignard L (1901d) Sur la double fécondation chez les Solanées et les Gentianées. C R Acad Sci Paris 133:1268–1272

Guignard L (1902) La double fécondation chez les Crucifères. J Bot 16:361–368

Hamby RK, Zimmer EA (1992) Ribosomal RNA as a phylogenetic tool in plant systematics. In: Soltis PS, Soltis DE, Doyle JJ (eds) Molecular systematics of plants. Chapman and Hall, New York, pp 50–91

Hannig E (1904) Zur Physiologie pflanzlicher Embryonen. I. Ueber die Cultur von Cruciferen-Embryonen ausserhalb des Embryosacks. Bot Zeit 62:45–80

Hanstein J (1870) Die Entwicklung des Keimes der Monokotylen und Dikotylen. Botanische Abhandlungen aus dem Gebiet der Morphologie und Physiologie, I. Marcus, Bonn

Hause G, Schröder M-B (1987) Reproduction in *Triticale*. 2. Karyogamy. Protoplasma 139:100–104

Hayashi Y (1963) The embryology of the family Magnoliaceae sens. lat. I. Megasporogenesis, female gametophyte and embryogeny in *Illicium anisatum* L. Sci Rep Tôhoku Univ Ser IV 29:27–33

Herr JM Jr (1984) Embryology and taxonomy. In: Johri BM (ed) Embryology of angiosperms. Springer, Berlin Heidelberg New York, pp 647–696

Higashiyama T, Kuroiwa H, Kawano S, Kuroiwa T (1997) Kinetics of double fertilization in *Torenia fournieri* based on direct observations of the naked embryo sac. Planta 203:101–110

Higashiyama T, Yabe S, Sasaki N, Nishimura Y, Miyagishima S, Kuroiwa H, Kuroiwa T (2001) Pollen tube attraction by the synergid cell. Science 293:1480–1483

Higashiyama T, Kuroiwa H, Kuroiwa T (2003) Pollen-tube guidance: beacons from the female gametophyte. Curr Opin Plant Biol 6:36–41

Hoefert LL (1969) Fine structure of sperm cells in pollen grains of *Beta*. Protoplasma 68:237–240

Hong S-K, Aoki T, Kitano H, Satoh H, Nagato Y (1995) Phenotypic diversity of 188 rice embryo mutants. Dev Genet 16:298–310

Hoshino Y, Scholten S, von Wiegen P, Lörz H, Kranz E (2004) Fertilization-induced changes in the microtubular architecture of the maize egg cell and zygote – an immunocytochemical approach adapted to single cells. Sex Plant Reprod 17:89–95

Huang B-Q, Russell SD (1992) Synergid degeneration in *Nicotiana*: a quantitative, fluorochromatic and chlorotetracycline study. Sex Plant Reprod 5:151–155

Huang B-Q, Russell SD (1994) Fertilization in *Nicotiana tabacum*: cytoskeletal modifications in the embryo sac during synergid degeneration. Planta 194:200–214

Huang, B-Q, Sheridan WF (1998) Actin coronas in normal and *indeterminate gametophyte1* embryo sacs of maize. Sex Plant Reprod 11:257–264

Huang B-Q, Pierson ES, Russell SD, Tiezzi A, Cresti M (1993a) Cytoskeletal organisation and modification during pollen tube arrival, gamete delivery and fertilisation in *Plumbago zeylanica*. Zygote 1:143–154

Huang B-Q, Strout GW, Russell SD (1993b) Fertilization in *Nicotiana tabacum*: ultrastructural organization of propane-jet-frozen embryo sacs in vivo. Planta 191:256–264

Huang B-Q, Fu Y, Zee SY, Hepler PK (1999) Three-dimensional organization and dynamic changes of the actin cytoskeleton in embryo sacs of *Zea mays* and *Torenia fournieri*. Protoplasma 209:105–119

Huck N, Moore JM, Federer M, Grossniklaus U (2003) The *Arabidopsis* mutant *feronia* disrupts the female gametophytic control of pollen tube reception. Development 130:2149–2159

Hülskamp M, Schneitz K, Pruitt RE (1995) Genetic evidence for a long-range activity that directs pollen tube guidance in *Arabidopsis*. Plant Cell 7:57–64

Ikeda T (1902) Studies in the physiological functions of antipodals and the phenomena of fertilization in Liliaceae. I. *Tricyrtis hirta*. Bull Coll Agric Tôkyô Imp Univ 5:41–72

Jensen WA (1964) Observations on the fusion of nuclei in plants. J Cell Biol 23:669–672

Jensen WA (1998) Double fertilization: a personal view. Sex Plant Reprod 11:1–5

Jensen WA, Fisher DB (1967) Cotton embryogenesis: double fertilization. Phytomorphology 17:261–269

Jensen WA, Fisher DB (1968a) Cotton embryogenesis: the entrance and discharge of the pollen tube in the embryo sac. Planta 78:158–183

Jensen WA, Fisher DB (1968b) Cotton embryogenesis: the sperm. Protoplasma 65:277–286

Jiang L, Yang S-L, Xie L-F, Puah CS, Zhang X-Q, Yang W-C, Sundaresan V, Ye D (2005) *VANGUARD1* encodes a pectin methylesterase that enhances pollen tube growth in the *Arabidopsis* style and transmitting tract. Plant Cell 17:584–595

Johansen DA (1950) Plant embryology. Embryology of the spermatophyta. Chronica Botanica, Waltham

Johnson MA, Preuss D (2002) Plotting a course: multiple signals guide pollen tubes to their targets. Dev Cell 2:273–281

Johri BM, Ambegaokar KB, Srivastava PS (1992) Comparative embryology of angiosperms, vol 1 and 2. Springer, Berlin Heidelberg New York

Kanta K, Ranga Swamy NS, Maheshwari P (1962) Test-tube fertilization in a flowering plant. Nature 194:1214–1217

Kapil RN (1970) Podostemaceae. Bull Indian Natl Sci Acad 41:104–109

Karsten G (1902) Ueber die Entwickelung der weiblichen Blüthen bei einigen Juglandaceen. Flora 90:316–333

Khanna P (1964) Morphological and embryological studies in Nymphaeaceae I. *Euryale ferox* Salisb. Proc Indian Acad Sci 59B:237–243

Khanna P (1965) Morphological and embryological studies in Nymphaeaceae II. *Brasenia schreberei* Gmel and *Nelumbo nucifera* Gaertn. Aust J Bot 13:379–387

Khanna P (1967) Morphological and embryological studies in Nymphaeaceae III. *Victoria cruziana* D'Orb, and *Nymphaea stellata* Willd. Bot Mag Tokyo 80:305–312

Kiyosue T, Ohad N, Yadegari R, Hannon M, Dinneny J, Wells D, Katz A, Margossian L, Harada JJ, Goldberg RB, Fischer RL (1999) Control of fertilization-independent endosperm development by the *MEDEA* polycomb gene in *Arabidopsis*. Proc Natl Acad Sci USA 96:4186–4191

Kluger J (1999) The suicide seeds. Time 153(4):44–45

Knox RB, Singh MB (1987) New perspectives in pollen biology and fertilization. Ann Bot 60[Suppl 4]:15–37

Koltunow AM, Vivian-Smith A, Tucker MR, Paech N (2002) The central role of the ovule in apomixis and parthenocarpy. In: O'Neill SD, Roberts JA (eds) Plant reproduction. Annual Plant Reviews vol 6. Sheffield Academic Press, Sheffield, pp 221–256

Koul AK (2001) Double fertilization: changing frontiers. In: Rangaswamy NS (ed) Phytomorphology Golden Jubilee issue 2001: Trends in plant sciences. International Society of Plant Morphologists, Delhi, pp 237–250

Kovács M, Barnabás B, Kranz E (1995) Electro-fused isolated wheat (*Triticum aestivum* L.) gametes develop into multicellular structures. Plant Cell Rep 15:178–180

Kranz E (2001) In vitro fertilization. In: Bhojwani SS, Soh WY (eds) Current trends in the embryology of angiosperms. Kluwer, Dordrecht, pp 143–166

Kranz E, Kumlehn J (1999) Angiosperm fertilisation, embryo and endosperm development in vitro. Plant Sci 142:183–197

Kranz E, Lörz H (1993) In vitro fertilization with isolated, single gametes results in zygotic embryogenesis and fertile maize plants. Plant Cell 5:739–746

Kranz E, Lörz H (1994) In vitro fertilisation of maize by single egg and sperm cell protoplast fusion mediated by high calcium and high pH. Zygote 2:125–128

Kranz E, Bautor J, Lörz H (1991) In vitro fertilization of single, isolated gametes of maize mediated by electrofusion. Sex Plant Reprod 4:12–16

Kranz E, von Wiegen P, Lörz H (1995) Early cytological events after induction of cell division in egg cells and zygote development following in vitro fertilization with angiosperm gametes. Plant J 8:9–23

Kranz E, von Wiegen P, Quader H, Lörz H (1998) Endosperm development after fusion of isolated, single maize sperm and central cells in vitro. Plant Cell 10:511–524

Land WJG (1900) Double fertilization in Compositae. Bot Gaz 30:252–260

Land WJG (1904) Spermatogenesis and oogenesis in *Ephedra trifurca*. Bot Gaz 38:1–18

Lersten NR (2004) Flowering plant embryology. Blackwell, Ames, IA

Liu C-M, Johnson S, Hedley CL, Wang TL (1996) The generation of a legume embryo: morphological and cellular defects in pea mutants. In: Wang TL, Cuming A (eds) Embryogenesis. The generation of a plant. Bios, Oxford, pp 191–213

Lord EM, Russell SD (2002) The mechanisms of pollination and fertilization in plants. Annu Rev Cell Dev Biol 18:81–105

Luo M, Bilodeau P, Koltunow A, Dennis ES, Peacock WJ, Chaudhury AM (1999) Genes controlling fertilization-independent seed development in *Arabidopsis thaliana*. Proc Natl Acad Sci USA 96:296–301

Lush WM (1999) Whither chemotropism and pollen tube guidance? Trends Plant Sci 4:413–418

Magallón S, Sanderson MJ (2002) Relationships among seed plants inferred from highly conserved genes: sorting conflicting phylogenetic signals among ancient lineages. Am J Bot 89:1991–2006

Maheshwari P (1950) An introduction to the embryology of angiosperms. McGraw-Hill, New York.

Márton ML, Cordts S, Broadhvest J, Dresselhaus T (2005) Micropylar pollen tube guidance by egg apparatus 1 of maize. Science 307:573–576

Mathews S, Donoghue MJ (1999) The root of angiosperm phylogeny inferred from duplicate phytochrome genes. Science 286:947–950

Mayfield JA, Fiebig A, Johnstone SE, Preuss D (2001) Gene families from the *Arabidopsis thaliana* pollen coat proteome. Science 292:2482–2485

Meinke DW (1994) Seed development in *Arabidopsis thaliana*. In: Meyerowitz EM, Somerville CR (eds) *Arabidopsis*. Cold Spring Harbor Laboratory Press, Cold Spring Harbor, New York, pp 253–295

Mogensen HL (1972) Fine structure and composition of the egg apparatus before and after fertilization in *Quercus gambelii*: the functional ovule. Am J Bot 59:931–941

Mogensen HL (1982) Double fertilization in barley and the cytological explanation for haploid embryo formation, embryoless caryopses, and ovule abortion. Carlsberg Res Commun 47:313–354

Mogensen HL (1988) Exclusion of male mitochondria and plastids during syngamy in barley as a basis for maternal inheritance. Proc Natl Acad Sci USA 85:2594–2597

Mogensen HL (1992) The male germ unit: concept, composition, and significance. Int Rev Cytol 140:129–147

Mogensen HL, Holm PB (1995) Dynamics of nuclear DNA quantities during zygote development in barley. Plant Cell 7:487–494

Mogensen HL, Leduc N, Matthys-Rochon E, Dumas C (1995) Nuclear DNA amounts in the egg and zygote of maize (*Zea mays* L.). Planta 197:641–645

Mohan Ram HY, Sehgal A (2001) Biology of Indian Podostemaceae. In: Rangaswamy NS (ed) Phytomorphology Golden Jubilee issue 2001: Trends in plant sciences. International Society of Plant Morphologists, Delhi, pp 365–391

Mukkada AJ (1963) Some observations on the embryology of *Dicraea stylosa* Wight. In: Plant embryology – a symposium. Council of Scientific & Industrial Research, New Delhi, pp 139–145

Mukkada AJ (1964) An addition to the bisporic embryo sacs – the *Dicraea* type. New Phytol 63:289–292

Mukkada AJ (1969) Some aspects of the morphology, embryology and biology of *Terniola zeylanica* (Gardner) Tulasne. New Phytol 68:1145–1158

Nagendran CR, Subramanyam K, Arekal GD (1976) Development of the female gametophyte in *Hydrobryum griffithii* (Podostemaceae). Ann Bot 40:511–513

Nagendran CR, Anand VV, Arekal GD (1980) The embryo sac of *Podostemum subulatus* (Podostemaceae) – a reinvestigation. Plant Syst Evol 134:121–125

Nash JM (2000) Grains of hope. Time 156(5):39–46

Nawaschin S (1898) Resultate einer Revision der Befruchtungsvorgänge bei *Lilium martagon* und *Fritillaria tenella*. Bull Acad Imp Sci St-Pétersbourg Ser 5, 9:377–382

Nawaschin S (1899) Neuen Beobachtungen über Befruchtung bei *Fritillaria tenella* und *Lilium martagon*. Bot Centralbl 77:62

Nawaschin S (1900a) Ueber die Befruchtungsvorgänge bei einigen Dicotyledoneen. Ber Dtsch Bot Ges 18:224–230

Nawaschin S (1900b) On fertilization in Compositae and Orchidaceae. Bull Acad Imp Sci St- Pétersbourg Ser 5, 13:335–340

Niklas KJ (1997) The evolutionary biology of plants. University of Chicago Press, Chicago, IL.

Ohad N, Margossian L, Hsu Y, Williams C, Repetti P, Fischer RL (1996) A mutation that allows endosperm development without fertilization. Proc Natl Acad Sci USA 93:5319–5324

Ohad N, Yadegari R, Margossian L, Hannon M, Michaeli D, Harada JJ, Goldberg RB, Fischer RL (1999) Mutations in *FIE*, a WD polycomb group gene, allow endosperm development without fertilization. Plant Cell 11:407–415

Palanivelu R, Brass L, Edlund AF, Preuss D (2003) Pollen tube growth and guidance is regulated by *POP2*, an *Arabidopsis* gene that controls GABA levels. Cell 114:47–59

Parkinson CL, Adams KL, Palmer JD (1999) Multigene analyses identify the three earliest lineages of extant flowering plants. Curr Biol 9:1485–1488

Poddubnaya-Arnoldi VA (1960) Studies of fertilization in the living material of some angiosperms. Phytomorphology 10:185–198

Prado AM, Porterfield DM, Feijó JA (2004) Nitrous oxide is involved in growth regulation and re-orientation of pollen tubes. Development 131:2707–2714

Qiu Y-L, Lee J, Bernasconi-Quadroni F, Soltis DE, Soltis PS, Zanis M, Zimmer EA, Chen Z, Savolainen V, Chase MW (1999) The earliest angiosperms: evidence from mitochondrial, plastid and nuclear genomes. Nature 402:404–407

Raghavan V (1976) Experimental embryogenesis in vascular plants. Academic Press, London

Raghavan V (1986) Embryogenesis in angiosperms. A developmental and experimental study. Cambridge University Press, New York

Raghavan V (1997) Molecular embryology of flowering plants. Cambridge University Press, New York

Raghavan V (2000) Embryogenesis at the crossroads – some perspectives on over a century of plant embryo research. Acta Biol Cracov Ser Bot 42:31–38

Raghavan V (2001) The coming of age of plant embryology. Curr Sci 80:244–251

Raghavan V (2003a) One hundred years of zygotic embryo culture investigations. In Vitro Cell Dev Biol – Plant 39:437–442

Raghavan V (2003b) Some reflections on double fertilization, from its discovery to the present. New Phytol 159:565–583

Ramachandran C, Raghavan V (1992) Apomixis in distant hybridization. In: Kalloo G, Chowdhury JB (eds) Distant hybridization in crop plants. Springer, Berlin Heidelberg New York, pp 106–121

Ray S, Park S-S, Ray A (1997) Pollen tube guidance by the female gametophyte. Development 124:2489–2498

Rotman N, Rozier F, Boavida, L, Dumas C, Berger F, Faure J-E (2003) Female control of male gamete delivery during fertilization in *Arabidopsis thaliana*. Curr Biol 13:432–436

Russell SD (1982) Fertilization in *Plumbago zeylanica*: entry and discharge of the pollen tube in the embryo sac. Can J Bot 60:2219–2230

Russell SD (1983) Fertilization in *Plumbago zeylanica*: gametic fusion and the fate of the male cytoplasm. Am J Bot 70:416–434

Russell SD (1984) Ultrastructure of the sperm of *Plumbago zeylanica* II. Quantitative cytology and three-dimensional organization. Planta 162:385–391

Russell SD (1985) Preferential fertilization in *Plumbago*: ultrastructural evidence for gamete-level recognition in an angiosperm. Proc Natl Acad Sci USA 82:6129–6132

Russell SD (1992) Double fertilization. Int Rev Cytol 140:357–388

Russell SD (1997) Male germ unit. In: Batygina TB (ed) Embryology of flowering plants. Terminology and concepts, vol 2. Seed. World & Family-95, St. Petersburg, pp 127–135

Russell SD, Cass DD (1981) Ultrastructure of the sperms of *Plumbago zeylanica* 1. Cytology and association with the vegetative nucleus. Protoplasma 107:85–107

Russell SD, Rougier M, Dumas C (1990) Organization of the early post-fertilization megagametophyte of *Populus deltoides*. Ultrastructure and implications for male cytoplasmic transmission. Protoplasma 155:153–165

Saito C, Nagata N, Sakai A, Mori K, Kuroiwa H, Kuroiwa T (2000) Unequal distribution of DNA-containing organelles in generative and sperm cells of *Erythrina cristagalli* (Fabaceae). Sex Plant Reprod 12:296–301

Sargant E (1899) On the presence of two vermiform nuclei in the fertilised embryo-sac of *Lilium martagon*. Proc R Soc London 65:163–165

Sargant E (1900) Recent work on the results of fertilization in angiosperms. Ann Bot 14:689–712

Sauter M, von Wiegen P, Lörz H, Kranz E (1998) Cell cycle regulatory genes from maize are differentially controlled during fertilization and first embryonic cell division. Sex Plant Reprod 11:41–48

Schnarf K (1929) Embryologie der Angiospermen. Handbuch der Pflanzenanatomie II. Abteilung 2. Teil: Archegoniaten. Bd X/2. Borntraeger-verlag, Berlin

Scholten S, Kranz E (2001) In vitro fertilization and expression of transgenes in gametes and zygotes. Sex Plant Reprod 14:35–40

Schulz R, Jensen WA (1968) *Capsella* embryogenesis: the synergids before and after fertilization. Am J Bot 55:541–552

Shibata K (1902) Die Doppelbefruchtung bei *Monotropa uniflora* L. Flora 90:61–66

Shimizu KK, Okada K (2000) Attractive and repulsive interactions between female and male gametophytes in *Arabidopsis* pollen tube guidance. Development 127:4511–4518

Singh MB, Xu H, Bhalla PL, Zhang Z, Swoboda I, Russell SD (2002) Developmental expression of polyubiquitin genes and distribution of ubiquinated proteins in generative and sperm cells. Sex Plant Reprod 14:325–329

Soltis PS, Soltis DE, Chase MW (1999) Angiosperm phylogeny inferred from multiple genes as a tool for comparative biology. Nature 402:402–404

Sørensen MB, Chaudhury AM, Robert H, Bancharel E, Berger F (2001) Polycomb group genes control pattern formation in plant seed. Curr Biol 11:277–281

Southworth D (2001) Sperm and generative cell. Isolation and manipulation. In: Bhojwani SS, Soh WY (eds) Current trends in the embryology of angiosperms. Kluwer, Dordrecht, pp 17–32

Southworth D, Russell SD (2001) Male gametogenesis. Development and structure of sperm. In: Bhojwani SS, Soh WY (eds) Current trends in the embryology of angiosperms. Kluwer, Dordrecht, pp 1–16

Strasburger E (1884) Neue Untersuchungen über den Befruchtungsvorgang bei den Phanerogamen als Grundlage für eine Theorie der Zeugung. Fischer-verlag, Jena

Strasburger E (1900) Einige Bemerkungen zur Frage nach der "doppelten Befruchtung" bei den Angiospermen. Bot Zeit 58:293–316

Strasburger E (1902) Ein Beitrag zur Kenntniss von *Ceratophyllum submersum* und phylogenetische Erörterungen. Jahrb Wiss Bot 37:477–526

Sumner MJ (1992) Embryology of *Brassica campestris*: the entrance and discharge of the pollen tube in the synergid and the formation of the zygote. Can J Bot 70:1577–1590

Sun M-X, Moscatelli A, Yang H-Y, Cresti M (2000) In vitro double fertilization in *Nicotiana tabacum* (L.): fusion behavior and gamete interaction by video-enhanced microscopy. Sex Plant Reprod 12:267–275

Sun M-X, Kranz E, Moscatelli A, Yang H-Y, Lörz H, Cresti M (2002) A reliable protocol for direct detection of lectin binding sites on the plasma membrane of a single living sperm cell in maize. Sex Plant Reprod 15:53–55

The Angiosperm Phylogeny Group (2003) An update of the angiosperm phylogeny group classification for the orders and families of flowering plants: APG II. Bot J Linn Soc 141:399–436

The Arabidopsis Genome Initiative (2000) Analysis of the genome sequence of the flowering plant *Arabidopsis thaliana*. Nature 408:796–815

Thomas EN (1900) Double fertilization in a dicotyledon – *Caltha palustris*. Ann Bot 14:527–535

Thorpe TA, Stasolla C (2001) Somatic embryogenesis. In: Bhojwani SS, Soh WY (eds) Current trends in the embryology of angiosperms. Kluwer, Dordrecht, pp 279–336

Tian HQ, Russell SD (1997) Micromanipulation of male and female gametes of *Nicotiana tabacum*: II. Preliminary attempts for in vitro fertilization and egg cell culture. Plant Cell Rep 16:657–661

Tian HQ, Yuan T, Russell SD (2005) Relationship between double fertilization and the cell cycle in male and female gametes of tobacco. Sex Plant Reprod 17:243–252

Tirlapur UK, Kranz E, Cresti M (1995) Characterisation of isolated egg cells, in vitro fusion products and zygotes of *Zea mays* L. using the technique of image analysis and confocal laser scanning microscopy. Zygote 3:57–64

Touraev A, Pfosser M, Heberle-Bors E (2001) The microspore: a haploid multipurpose cell. Adv Bot Res 35:53–109

van Lammeren AAM (1986) A comparative ultrastructural study of the megagametophytes in two strains of *Zea mays* L. before and after fertilization. Agric Univ Wageningen Papers 86-1:1–37

van Went JL (1970) The ultrastructure of the fertilized embryo sac of *Petunia*. Acta Bot Neerl 19:468–480

van Went J, Cresti M (1988) Pre-fertilization degeneration of both synergids in *Brassica campestris* ovules. Sex Plant Reprod 1:208–216

Vazart J (1969) Organisation et ultrastructure du sac embryonnaire du lin (*Linum usitatissimum* L.). Rev Cytol Biol Vég 32:227–232

Wagner VT, Dumas C, Mogensen HL (1989) Morphometric analysis of isolated *Zea mays* sperm. J Cell Sci 93:179–184

Wardlaw CW (1955) Embryogenesis in plants. Methuen, London

Williams JH, Friedman WE (2002) Identification of diploid endosperm in an early angiosperm lineage. Nature 415:522–526

Wilms HJ (1981) Pollen tube penetration and fertilization in spinach. Acta Bot Neerl 30:101–122

Winter K-U, Becker A, Münster T, Kim JT, Saedler H, Theissen G (1999) MADS-box genes reveal that gnetophytes are more closely related to conifers than to flowering plants. Proc Natl Acad Sci USA 96:7342–7347

Wolters-Arts M, Lush WM, Mariani C (1998) Lipids are required for directional pollen-tube growth. Nature 392:818–821

Xu H, Swoboda I, Bhalla PL, Singh MB (1999) Male gametic cell-specific gene expression in flowering plants. Proc Natl Acad Sci USA 96:2554–2558

Yan H, Yang H-Y, Jensen WA (1991) Ultrastructure of the developing embryo sac of sunflower (*Helianthus annuus*) before and after fertilization. Can J Bot 69:191–202

Ye X-L, Yeung EC, Zee S-Y (2002) Sperm movement during double fertilization of a flowering plant, *Phaius tankervilliae*. Planta 215:60–66

You R, Jensen WA (1985) Ultrastructural observations of the mature megagametophyte and the fertilization in wheat (*Triticum aestivum*). Can J Bot 63:163–178

Yu H-S, Huang B-Q, Russell SD (1994) Transmission of male cytoplasm during fertilization in *Nicotiana tabacum*. Sex Plant Reprod 7:313–323

Zenkteler M (1990) In vitro fertilization and wide hybridization in higher plants. Crit Rev Plant Sci 9:267–279

Zhang Z, Russell SD (1999) Sperm cell surface characteristics of *Plumbago zeylanica* L. in relation to transport in the embryo sac. Planta 208:539–544

2 Establishment of the Embryo Body Plan – A Reassessment of Cell Lineage and Cell Fate

It is hardly necessary to state that embryonic cells arise by simple successive bipartitionings of the zygote and its derivative cells. Their theoretical process of generation may be simply described in homely language, since there are no precise scientific terms that can be employed: the first two cells are sisters and, at the same time, daughters of the zygote. In the second cell generation the cells number four, two of which are sisters and first cousins of the other two; all are granddaughters of the zygote. At the third cell generation there are eight cells, all of which are great-granddaughters of the zygote; they form two groups of first cousins once removed and, in each of these groups, two sisters, as in the preceding cell generation, are first cousins of two other sisters.

D.A. Johansen 1950

2.1 Organization of the Egg and Zygote   30
2.2 From the Zygote to the Embryo   34
2.2.1 A Model of Embryogenesis in Eudicots   35
2.2.2 A Model of Embryogenesis in Monocots   41
2.2.3 Are Embryonic Organs and Tissues Lineage-restricted Compartments?   43
2.2.4 Abnormal Embryo Types   44
2.3 Physiological Considerations of Embryogenesis   45
2.3.1 A Role for Auxin Polar Transport in Embryogenesis   46
2.3.2 Embryo Nutrition   47
2.3.3 Embryo Culture Investigations   48
2.4 Concluding Comments   51
References   52

As discussed in the previous chapter, our present understanding of the complex processes involved in the formation of embryos in flowering plants unifies the first description of embryo development in certain eudicots and monocots by Hanstein in 1870 with discoveries of syngamy by Strasburger in 1884 and of double fertilization by Nawaschin in 1898. Early studies were chiefly concerned with understanding the sequence and planes of divisions of the zygote and its cellular descendants, and the basic structural configuration of the primary tissues and organs of developing embryos, with interest gradually shifting to ultrastructural and physiological investigations of embryogenesis. In contrast to animals, which complete their morphogenetic events including the formation of most adult organs during embryogenesis, the embryogenic phase of flowering plants lays out the primary body plan, leaving major ontogenetic events to occur post-embryogenically by the activity of the shoot and root apical meristems; hence, flowering plants are said to be in a state of continuing embryogenesis.

This chapter provides overviews of the establishment of the structural and functional body plan of embryos of representative eudicot and monocot species, followed by a consideration of the nutrition of developing embryos. The representatives have been chosen as models to illustrate in an uncomplicated way the successive divisions of the zygote to form the embryo, and in recognition of their relatively widespread use in experimental investigations. The genetic and molecular basis for cellular pattern formation in early-stage embryos is the subject of the next chapter. Two reviews germane to the topics discussed in this chapter are by Yadegari and Goldberg (1997) and Perez-Grau (2002).

2.1
Organization of the Egg and Zygote

Fertilization is the trigger that transforms the egg into a zygote, which subsequently cleaves into cells that become part of the embryo and ultimately contribute to the body plan of the mature plant. After years of focus on the orchestrated division patterns of the zygote, other aspects of this cell, such as polarity and organelle disposition, have begun to unravel their secrets. Whereas the zygote possesses all the structural and functional qualities of a typical plant cell, it is also highly differentiated as a storehouse of developmental information in anticipation of a complex division program. A comparison of the ultrastructural profile of the zygote with that of the egg indicates that fertilization results in considerable metabolic turmoil. Although egg cells of flowering plants have a similar overall organization characterized by inherent polarity due to a tapered, basal micropylar end attached to the embryo sac wall, and a broad, unattached, terminal chalazal end, polarization of the cytoplasm resulting in ultrastructural differences between the terminal and basal parts of the egg is a hallmark of many plants. For example, superimposed upon the predetermined polarity of egg cells of cotton (Jensen 1963, 1965), *Capsella bursa-pastoris* (Schulz and Jensen 1968), turnip (Sumner and van Caeseele 1989), and *Arabidopsis* (Mansfield et al. 1991) is a large vacuole toward the micropylar end, with the cytoplasmic organelles including the nucleus displaced toward the chalazal end. The total amount of cytoplasm present in the egg cells of these plants is sparse and is spread in a thin strip surrounding the vacuole except near the nucleus. Plastids, mitochondria, and dictyosomes are randomly and parsimoniously distributed in the egg cytoplasm of cotton, turnip, and *C. bursa-pastoris*. Although the egg cytoplasm of turnip and *Arabidopsis* has also a low chloroplast count, a large number of undifferentiated starch-containing plastids (amyloplasts) surround the nucleus at the chalazal end, leading to the suggestion that egg cells may serve as a sink for carbohydrates prior to fertilization. Strands of ER are relatively abundant in the egg of cotton, where they seem to partially surround the plastids, mitochondria, and dictyosomes. Occasional strands of ER also appear unique in having an internal network of tubes probably formed by the invagination of the inner membrane. By contrast, egg cells of turnip and *C. bursa-pastoris* have very little ER, which occurs in the form of short, randomly oriented strands. Eggs of both cotton and *C. bursa-pastoris* also contain liberal supplies of ribosomes that exist predominantly as monosomes. Polarity in the egg of maize is conferred by the presence of vacuoles of various sizes at the chalazal end, whereas the cytoplasm, along with the nucleus and numerous abnormal mitochondria, is confined to the micropylar end. A close structural relationship of the egg to its maturation stage is underlined by the observation that, whereas the immature egg is small and nonvacuolate, the mature egg is large with a proportionately conspicuous apical vacuole (van Lammeren 1986; Mól et al. 2000). From the functional point of view, the ultrastructural simplicity of the mature egg cells of the species considered here, in particular the comparative poverty of their cytoplasm, tends to suggest that the angiosperm egg is an inactive cell whose metabolism is at a low ebb. Since synergids are intimately aligned with the egg in the egg apparatus, metabolic quiescence of the egg is often compensated by the presence of metabolically active synergids (Jensen 1965). A different situation is observed in the egg of *Plumbago zeylanica*, whose embryo sac lacks synergids. The major ultrastructural features of the egg cell are the elaboration of wall ingrowths at the micropylar end corresponding to the filiform apparatus normally found in synergids, and the presence of a metabolically active cytoplasm with a large number of relatively well-developed mitochondria and dictyosomes as well as ER studded with polysomes (Cass and Karas 1974). Here, the egg not only plays its genetically ordained role as the female gamete, but probably also performs synergid functions.

The chalazal end of egg cells of many angiosperms examined in the electron microscope has been found to be attenuated – a feature achieved in large measure by a decreasing amount of organized cell wall material. In most cases, this is manifest by the presence of wall material around the micropylar half of the cell, the chalazal half being covered by just the plasma membrane, or by the deposition of patches of wall material doting the chalazal part of the egg, or by the wall disappearing from the chalazal part of the egg with maturity (Jensen 1965; Folsom and Peterson 1984; Yan et al. 1991). Possibly, the naked, or partially naked, chalazal part of the egg facilitates entry of the sperm for fertilization and absorption of nutrients from the central cell. Whether signals from the stigma, style, or the central cell at the time of pollination are involved in the differential accumulation of wall material on the egg surface is not known.

As will become clear later, the developmental pattern of the embryo is determined by the structural and molecular polar cues embedded in the unfertilized egg. Consequently, the mechanism that establishes egg polarity by maternal information can be expected to exercise an overriding effect on future development, although the mechanism itself raises questions that have no answers. For over 100 years, eggs and zygotes of brown algae belonging to the family Fucales, especially species of *Fucus* and its close relative *Silvetia* (renamed from *Pelvetia*), have been used in investigations on polarity, and the relevance of these studies to formation of the apicobasal axis in embryos of flowering plants has been emphasized (Quatrano and Shaw 1997). The unfertilized egg cell of fucoid algae is apolar, but polarity is established soon after fertilization by a two-stage process, namely, the initial formation of a reversible axis and its later fixation. The first visible sign of polarity fixation is the appearance, near the spot of sperm entry on the egg, of a pear-shaped protuberance that grows into a rhizoid. Next, the zygote is cleaved in a plane at right angles to the emerging rhizoid, cutting off a small rhizoid cell and a large thallus cell. However, a number of external cues, such as application of unilateral light, temperature, osmotic and ionic gradients, and the auxin, indoleacetic acid (IAA), among others, override fixation of the growth axis by sperm entry. The most commonly used external factor in experimental investigations is unilateral illumination, which causes emergence of the rhizoid from the shaded side of the zygote as the first morphological expression of polarity. Fixation of polarity occurs when the labile axis formed can no longer be reoriented by the direction of the external stimulus (Jaffe 1968; Kropf 1992). One of the earliest recruits, which still remains as a viable candidate, to an expanding group of diverse factors that lead to formation of the polar axis in fucoid zygotes is a localized Ca^{2+} influx at the site of the future rhizoid; this has led to the suggestion that an essential link in the polarization process may be a cellular component that binds strongly to Ca^{2+} (Robinson and Jaffe 1975; Nuccitelli 1978). The role of Ca^{2+} as one of the earliest detectable polar phenomena has been reinforced by the observation that, in *Fucus* zygotes, fluorescently labeled dihydropyridine (which probably labels calcium channels) binds to a specific plasma membrane-localized receptor in the region of high Ca^{2+} concentration that predicts the site of rhizoid growth (Shaw and Quatrano 1996). Actin-depolymerizing agents such as cytochalasins B or D and latrunculin, the actin-stabilizing agent jasplakinolide, and the fluorescent probe rhodamine phalloidin have been used to show that actin networks play a pivotal role as an early marker of zygote polarity at the presumptive rhizoid pole. In general, it was found that disruption of the actin cytoskeleton inhibits subsequent polar growth of the rhizoid and that, in contrast to the uniform distribution of actin in the early stage of development of the zygote, formation/fixation of the polar axis is associated with a redistribution to the rhizoid site of dynamic actin generated by depolymerization within existing arrays and by polymerization of new arrays (Kropf et al. 1989; Alessa and Kropf 1999; Hable et al. 2003). Changes in actin organization resulting in the formation of an 'actin patch' are followed by the polarized vesicle secretion at this site necessary for axis stabilization (Hable and Kropf 1998). Because vesicle secretion is Ca^{2+}-dependent, it is tempting to conclude that actin filaments recruit Ca^{2+} channel proteins to aid in vesicle secretion at the rhizoid pole (Kropf et al. 1999). In an attempt to decipher the transduction of environmental gradients in the initiation of polarity in *Fucus* zygotes, Corellou et al. (2000) have implicated phosphorylation cascades involving tyrosine phosphorylation in polar axis formation in the zygote and in stable actin localization at the rhizoid site. A series of experiments using auxin efflux inhibitors such as naphthylphthalamic acid (NPA) have indicated interactions between actin and auxin transport in the development of po-

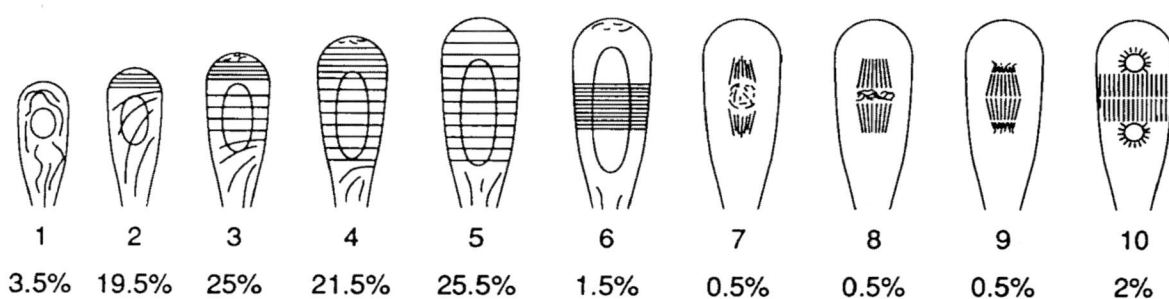

Fig. 2.1 Diagrams showing changes in the distribution of microtubules during development and division of the zygote of *Arabidopsis*. The formation of the preprophase band is shown in 6. The frequency of each stage as a percentage of 200 zygotes examined is indicated. (Reprinted from Webb and Gunning 1991)

larity in *Fucus* zygotes (Sun et al. 2004), but sorting out the interactions involved is not likely to be so simple. Although common threads connect the pattern of early division of algal and angiosperm zygotes, polarity remains an intriguing problem, and a unifying theme about the mechanism controlling the initiation and stabilization of the polar axis in unfertilized eggs is yet to emerge.

The initial responses of the angiosperm egg to fertilization do not follow well-ordered patterns but, in various species investigated, involve changes in size, laying down of wall material at the chalazal end, and overhauling of the cytoplasmic contents. Cotton provides a good model, showing that before the zygote nucleus divides, there is a dramatic decrease in cell size to nearly one-half of the volume of the egg. This is accompanied by a decrease in size of the vacuole soon after fertilization, presumably due to the loss of water into the central cell, continuing even after the cell size ceases to decrease (Jensen 1963, 1968). Zygote shrinkage has also been observed in *Hibiscus costatus, H. costatus-aculeatus, H. costatus-furcellatus* (Ashley 1972), tobacco (Mogensen and Suthar 1979), and turnip (Sumner 1992). In contrast, within a few hours of fertilization, the egg cell of *Arabidopsis* executes a nearly three-fold elongation along the apicobasal axis. The prime mover of this event is also the vacuole, whose reorganization involves replacement of the large micropylar vacuole of the egg by numerous small vacuoles, which finally coalesce to a large vacuole filling most of the volume of the zygote (Mansfield and Briarty 1991; Jürgens and Mayer 1994). As shown diagrammatically in Fig. 2.1, zygote elongation in *Arabidopsis* coincides with a gradual change in the configuration of the array of microtubules from a perpendicular to a transverse cortical alignment, predominantly in a subapical band (Webb and Gunning 1991). This implies that cortical microtubules provide the force for elongation of the zygote and that this activity is largely restricted to the apical region of the cell. Changes in microtubule organization resulting in the replacement of sparsely scattered cortical microtubules in the egg cytoplasm by dense strands radiating from the nucleus also occur during in vitro fertilization of isolated egg cells of maize (Hoshino et al. 2004). In some plants it has been shown that laying down of wall materials at the chalazal end of the egg as an early post-fertilization event involves increased activity of the cortical microtubules as well as of dictyosome vesicles in the cytoplasm; interesting questions regarding the regulatory mechanisms and biological roles of the newly formed wall are raised by this observation (Jensen 1968; Schulz and Jensen 1968; Yan et al. 1991; Sumner 1992). In *Torenia fournieri*, there is evidence for a change in the symplastic traffic of solutes between the egg and the central cell as a consequence of fertilization. Microinjection of water-soluble molecular tracers into the embryo sac showed that, in contrast to the unhindered passage of tracers in the 3–10 kDa range from the central cell to the egg before fertilization, there is a complete cessation of movement of solutes of all sizes into the zygote (Han et al. 2000). It is possible, but as yet unproven, that in the microcosm of the embryo sac, isolation of the zygote by a cell wall, and symplastic prevention of cell-to-cell communication promotes subsequent division unhindered by the influence of a primary endosperm nucleus of a different genotype.

Other fertilization-related events contribute to

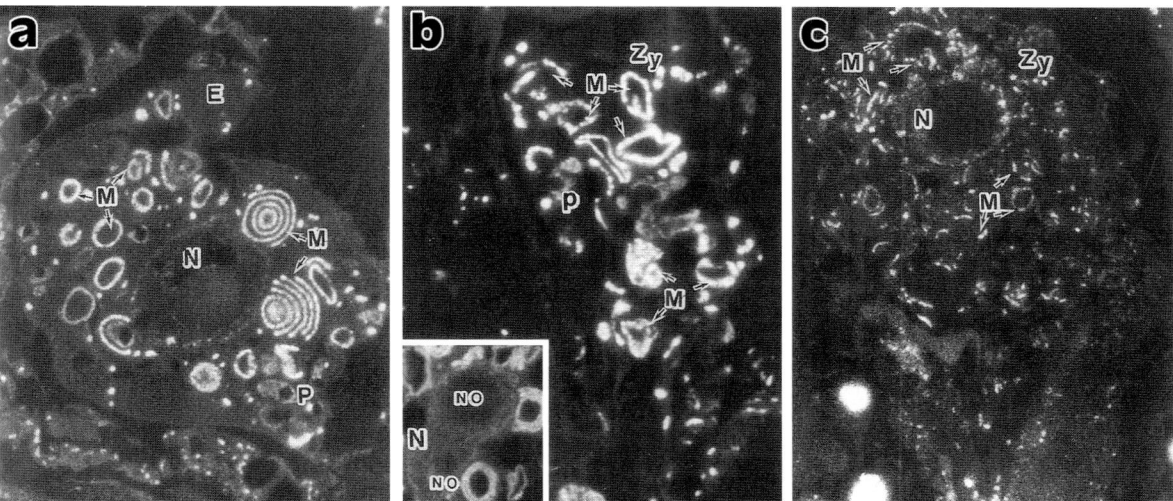

Fig. 2.2a–c Fluorescent micrographs of sections of the egg and zygote of *Pelargonium zonale* showing mitochondrial profile. **a** Egg cell with centrally placed nucleus and concentric, half-concentric, and ring-shaped mitochondria and plastids. **b** Zygote (15 h after pollination), in which mitochondria appear as crushed. *Inset* shows fusion of the egg and sperm nuclei. **c** Zygote (20 h after pollination) at the beginning of the first division, with mitochondria appearing as fine rods or small rings. *E* Egg cell, *M* mitochondria, *N* nucleus, *No* nucleolus, *P* plastid, *Zy* zygote. *Bars* 5 μm (*bar* in **b** applies also to **c**). (Reprinted from Kuroiwa et al. 2002)

accentuation of the inherent polarity of the egg and to an increase in the metabolic state of the zygote. Implicated in the further development of polarity following fertilization of the egg in cotton are ER, ribosomes, plastids, and mitochondria, which gather around the nucleus at the chalazal end. A consistent feature of this change is that the organelles complete their migration from the periphery of the cell to their new positions within 24 h after fertilization (Jensen 1968). A complete reversal of the organelle traffic underpinning zygote establishment occurs in *Papaver nudicaule* and *Zea mays*, resulting in the migration of the nucleus and cytoplasmic contents from their pre-fertilization micropylar locations to the chalazal pole (Olson and Cass 1981; van Lammeren 1986). The aggregation of ribosomes into polysomes, generation of new ribosome populations, increase in the number of lipid bodies, mitochondria, and dictyosomes, change in ER from rough to smooth or tubular type, and increase in RNA and protein contents observed in zygotes of various plants are consistent with the scenario associated with increased metabolic activity of this cell (Jensen 1968; Schulz and Jensen 1968; Cocucci and Jensen 1969; Mogensen 1972; Mansfield and Briarty 1991; Sumner 1992).

Pelargonium zonale (Geraniaceae) is a relative newcomer to the ranks of plants favored for ultrastructural investigations of the egg-to-zygote transformation. Following fertilization, changes are confined mainly to the mitochondria, which appear in sections of the egg cell as stacks of cup-shaped rings consisting of several concentric or half-concentric circles. It has been estimated that 6–9 h elapse between pollination and gametic fusion, and that the zygote does not divide until 20–24 h after pollination. As shown in Fig. 2.2 (a–c), by 15 h the ring-shaped mitochondria of the egg separate into single cups and appear crushed to form oval or rod-shaped structures. Their rod-shaped configuration is further accentuated by 20 h after pollination and before division of the zygote. These changes in mitochondrial morphology associated with fertilization are also accompanied by a substantial decrease in the amount of mitochondrial DNA in the zygote (Kuroiwa et al. 2002).

There is now compelling evidence to show that fertilization activates a cascade of changes in the egg. Formation of a highly polarized zygote not only ensures the fidelity of subsequent divisions in the embryogenic pathway, but also allows for the formation of phenotypically and functionally different cells. Despite the limited accessibility of the angiosperm egg to experimental manipulation, sorting out the developmental interactions that operate after fertilization is of importance in deciphering the genetic and molecular switches that are turned on or off to augur the changes described.

2.2
From the Zygote to the Embryo

Within the zygote lies the potential to form an entire plant, a feat that is accomplished by extensive changes in form in defined and dramatic ways, and by the progressive change of an undifferentiated cell to a mass of differentiated cells. Common cellular activities involved in the transformation of the zygote into the embryo are division, expansion, maturation, and differentiation; these are terminated by the formation of meristems and embryonic organs. In a very young embryo, all cells divide faithfully to produce a new generation of daughter cells. However, during progressive embryogenesis, cell divisions are restricted to certain parts of the embryo, predictable by their position in the cell lineage, to produce specialized cells, tissues, and organs. Unfortunately, in the still-unfolding molecular biology of embryogenesis, not much is known about the mechanisms that restrict functional activities of cells in the developing embryo.

The zygotic genome becomes fully activated, potentiating the zygote to divide within a few hours of fertilization, but division does not occur until after the endosperm nucleus has generated a syncytium of free nuclei (Maheshwari 1950). The first division of the zygote is almost invariably asymmetric and transverse to its long axis, cutting off a large, vacuolate, basal cell toward the micropylar end and a small, densely cytoplasmic, terminal cell toward the chalazal end. The embryo proper is derived from the terminal cell with varying degrees of contribution from the basal cell; however, a common scenario is one in which the basal cell wholly or partially forms a suspensor, which anchors the embryo to the embryo sac wall and probably functions as a conduit for nutrients. Obviously, the fates of the terminal and basal cells are reflected in the afore-mentioned polarity of the zygote. The plane of the subsequent division of the terminal cell and the extent of contribution of the basal cell to the formation of the embryo proper have been linked together to provide the framework for a widely used classification of embryo development types in flowering plants; the basis of this classification has changed little since it was introduced more than 50 years ago (Johansen 1950; Maheshwari 1950). In this classification, embryo development types are separated into two major groups: in one group, the first division of the terminal cell of the two-celled embryo is longitudinal; in the second group, the division is transverse. A small number of plants in which the first division of the zygote itself is oblique or vertical have been lumped into a third group. Within the first two groups, different embryo segmentation types are recognized and identified to customize the division sequence for specific families, and are designated by the name of the family in which many examples of the type are found or in which the type was first described. These include the Crucifer (or Onagrad) and Asterad types in the first group, Chenopodiad, Solanad, and Caryophyllad types in the second group, and Piperad type in the third group. However, some families show great diversity in their embryo development, and more than one type of development represented by one or two genera in each case is not uncommon within any one family. A case in point is Fabaceae, in which as many as four (Asterad, Caryophyllad, Crucifer, and Solanad) of the six possible types and many variations of the basic types of embryo development are known to exist in the subfamily Papilionoideae alone (Prakash 1987). The following account of the major types of embryo development has been summarized from a review by Natesh and Rau (1984), to which reference is made for additional details. Embryogenesis in representative plants of each type is described in books by Maheshwari (1950) and Johri et al. (1992).

Although the embryo is almost wholly generated by the terminal cell in the Crucifer type, in widely investigated plants such as *Capsella bursa-pastoris* and *Arabidopsis*, part of the embryonic root meristem and root cap are derived from the division products of the suspensor cell closest to the embryo proper known as the hypophysis. On balance, the contribution of the basal cell to formation of the mature embryo in these two species does not appear to be insignificant. Whereas embryo development in several genera of Onagraceae and Brassicaceae displays the hallmarks of the Crucifer type, the ranks of this type have also been swollen by additions of isolated genera from such widely dispartate families as Bignoniaceae, Fabaceae, Lamiaceae, Lythraceae, Ranunculaceae, Rutaceae, and Scrophulariaceae, among eudicots, and Juncaceae and Liliaceae, among monocots. The defining feature of the Asterad type is that derivatives of both terminal and basal cells blend indistinguishably to adopt an embryonic fate and form the mature embryo. Indeed, there is a nice division of labor between the terminal and basal cells in formation of the embryo, the former giving rise to the cotyledons and stem tip and the latter to the hypocotyl, root cortex, and root cap. Although the basal cell thus makes a substantial contribution to the crafting of the embryo, not all descendants of this cell are incorporated into the final product; a small number of cells assume the characteristics of a suspensor. The family Asteraceae claims many genera displaying the Asterad type of embryo development; variations of the Asterad type are also represented in a few genera of Geraniaceae, Lamiaceae, Oxalidaceae, Polygonaceae, Rosaceae, Urticaceae (eudicots), Liliacee, and Poaceae (monocots).

The Chenopodiad, Solanad, and Caryophyllad types of embryo development included in the second group have infused the field of descriptive embryogenesis with new questions about the precise contribution of descendants of the basal cell to the fabrication of the mature embryo. The distinguishing feature of these three types of embryo development is the transverse division of the terminal cell; however, in the Chenopodiad and Solanad types, it is not unusual for the basal cell also to divide transversely and, together with the terminal cell, form a linear strand of four cells. Since most of the hypocotyl, root cap, and root cortex of the mature Chenopodiad embryo is generated by the division of the basal cell, this type of embryo development shares significant similarity with the Asterad type. Besides Chenopodiaceae, other families with reported cases of Chenopodiad-type embryo development are Amaranthaceae and Polemoniaceae. In the Solanad type, described in members of the Solanaceae, Hydnoraceae, Linaceae, Papaveraceae, and Rubiaceae, descendants of the terminal cell give rise to most of the mature embryo; the root epidermis and suspensor are born out of division of the basal cell. Failure of the basal cell to divide gives the terminal cell an exclusive role in the formation of the Caryophyllad type of embryo. A suspensor is not a regular feature of the Caryophyllad embryo and, if one is present, it is also derived from the terminal cell. Systematic examination has resulted in the identification of the basic Caryophyllad type in Caryophyllaceae and its variations in members of eudicot families such as Crassulaceae, Droseraceae, Fabaceae, Fumariaceae, Haloragaceae, Portulaceae, and Pyrolaceae, and monocot families such as Alismataceae, Araceae, Potamogetonaceae, and Zannichelliaceae. Finally, clear evidence for the initial division of the zygote in an oblique or vertical plane reported mostly in a few eudicot families such as Balanophoraceae, Dipsacaceae, Loranthaceae, Piperaceae, and Santalaceae, constituting the Piperad type, was slow to be recognized, but this division pathway of embryogenesis is sufficiently different from the other two groups to be placed in a third group.

2.2.1
A Model of Embryogenesis in Eudicots

The above classification encompasses embryos of both eudicots and monocots and, indeed, it is generally accepted that plants included in the two divisions of flowering plants initially exhibit a largely identical and orderly series of embryogenic divisions. Our understanding of the challenging problem of how the fertilized egg gives rise to a diverse array of organs and tissues constituting the embryo is based on careful documentation of division sequences of the zygote in a broad spectrum of both eudicots and monocots. Although *Capsella bursa-pastoris* has served for many years as a text-book example of embryogenesis in eudicots, recent elucidation of the precise embryo division patterns in the related *Arabidopsis* had considerable symbolic significance in opening up powerful genetic and molecular approaches to the study of embryogenesis. As an introduction to embryogenesis in a representative eudicot, it is therefore appropriate to describe the cellular details during progressive divisions of the zygote of *Arabidopsis*; there are common threads connecting embryogenic division sequences in *C. bursa-pastoris* and *Arabidopsis*, which are, no doubt, anticipated due to their membership in the same family. Illustrated accounts of embryo development in *Arabidopsis* at the light microscopic and electron microscopic levels are given by Jürgens and Mayer (1994) and Mansfield and Briarty (1991), respectively.

The first division of the *Arabidopsis* zygote is unequal and gives rise to a small terminal cell and a large basal cell, constituting a two-celled embryo. This division is marked by the concentration of microtubules in a discrete band girdling the nucleus as a preprophase band; the appreance of the preprophase band of microtubules marking the future cell plate bisecting the zygote is surprising since a similar alignment of these cytoskeletal elements is suppressed during divisions between megasporogenesis and fertilization (see Fig. 2.1; Webb and Gunning 1991). It was mentioned earlier that growth of the zygote is associated with a specific arrangement of microtubules. Mutants of *Arabidopsis* designated as *tonneau* (*ton1* and *ton2*), which are unable to form the interphase and preprophase band of microtubules in dividing embryo cells predictably provoke irregular cell expansion and inability to align the division planes in cells, yet the regenerated phenotypes readjust their subsequent development to produce tissues and organs in the correct spatial positions. This observation is of particular significance: by linking the mutant phenotype to abnormalities in the cytoskeleton, it negates the notion that genes affecting polarized cell expansion and division plane alignment are necessary for spatial positioning of tissues and organs during embryogenesis (Traas et al. 1995). Indeed, this message would not have been discernible or even imaginable without the experimental analysis of mutants.

A major developmental decision appears to be made during the first division of the zygote, as descendants of the terminal cell become the organogenetic part of the embryo (embryo proper), whereas cells derived from the basal cell form the suspensor. Various ultrastructural and histochemical changes have served as markers for the differing functional potentials of the terminal and basal cells. For example, the terminal cell of the two-celled embryo of cotton acquires a dense, organelle-enriched cytoplasm with a high concentration of RNA, whereas the organelle profile and macromolecule concentration remains unaffected in the basal cell (Jensen 1963). To underscore the difference in cell fates at the molecular level, transcripts of the *Arabidopsis thaliana MERISTEM L1 LAYER* (*ATML1*) gene, encoding a homeodomain, are found to accumulate in the terminal, but not in the basal cell born out of the division of the *Arabidopsis* zygote (Liu et al. 1996). Haecker et al. (2004) have identified a novel family of genes named *WOX* (for *WUSCHEL*-related homeobox) as potential regulators of the apicobasal body axis of the embryo of *Arabidopsis* beginning with the unfertilized egg. Transcripts of the *WOX2* and *WOX8* genes are co-expressed in the egg and zygote, but following the first division of the zygote they segregate to the apical and basal cells, respectively. A general theme that has emerged from this work is that redistribution of transcripts establishes the identities of the two cells as the first step in the initiation of the apicobasal body axis, using developmental and cell growth strategies. Cytologically, the divergence in fate of these two cells is highlighted by the occurrence of longitudinal divisions in the terminal cell during the first two rounds, and several transverse divisions in the basal cell. Another variable that contributes to the divergence in cell fate is the unlimited division potential of the terminal cell compared to the limited number of cells generated in the basal cell. Much research will be required to decipher the mechanism that determines the cell fate of the two-celled embryo, although it might involve the polarization of the zygote alluded to earlier.

The subsequent divisions of the basal cell to form the suspensor and of the terminal cell to form the embryo proper are now fairly well understood. The basal cell divides first, and it does so once or occasionally twice transversely. The cell closest to the terminal cell, designated as the suspensor cell, undergoes additional transverse divisions to form a filament of seven to nine cells connected to one another by end-wall plasmodesmata. These cells are terminated at the micropylar end of the embryo sac by the enlarged basal cell. The entire filamentous structure, including the basal cell is known as the suspensor. A role in the specification of suspensor cells has been assigned for the *Arabidopsis* homolog of a protein kinase gene related to *SHAGGY* that encodes a serine/threonine protein kinase involved in the regulation of cell fate and/or pattern formation in *Drosophila*. Transcripts of this gene, designated as *Arabidopsis shaggy*-related protein kinase *etha* (*ASKη*) are found exclusively in the suspensor (excluding the hypophysis) derived from the basal cell of the two-celled embryo and not in the derivatives of the terminal cell (Dornelas et al. 1999). Weterings et al. (2001) have identified two mRNAs that accumulate preferentially within the suspensor of four-celled and older embryos of *Phaseolus coccineus* (Fabaceae) but not in the cells of the embryo proper (see Plate 1, Fig. j–l). A reporter gene coupled to the promoter of a *P. coccineus* gene introduced into tobacco plants is also found to be transcribed exclusively in the basal region and in the suspensor cells of preglobular embryos of transgenic plants. Thus, mRNAs whose genes or protein products are as yet uncharacterized are unveiled as versatile markers of embryogenesis by specifying the basal cells at the molecular level soon after the first zygotic division. As will be described in a later chapter, the basal cell and other cells of the suspensor display specific ultrastructural features to facilitate absorption and transport of solutes from the surrounding endosperm.

In *Arabidopsis*, the terminal cell divides longitudinally when the zygote has produced three to four cells. An additional longitudinal division in each of the two daughter cells of the terminal cell (quadrant stage) followed by a transverse division produces an octant-stage embryo, comprising an upper and lower tier of four cells each. At this stage, the suspensor cell might have sired four or five cells (Mansfield and Briarty 1991; Jürgens and Mayer 1994). Histological techniques combined with clonal analysis have shown that the fate of the different cell groups is already fixed in the octant-stage embryo, with the caveat that derivatives of more than one group may be integrated into specific organs of the mature embryo (see Plate 3, Fig. l). Thus, the upper tier of cells is destined to form exclusively the shoot apex and most of the cotyledons; the lower tier, in addition to providing derivatives to part of the cotyledons, generates the hypocotyl, radicle, and most of the root meristem; the central part of the root cap known as the columella and the remainder of the root apical meristem including the quiescent center are derived from the hypophysis (Dolan et al. 1993; Scheres et al. 1994). The apicobasal pattern of the future plant, built up by the reiterative action of the meristems or stem-cell systems in the shoot and root apices, is thus established in the octant-stage embryo. A single round of tangential divisions separating eight peripheral cells from a core of eight inner cells heralds the next phase of development of the embryo. In the 16-celled embryo, the eight external cells form the protoderm or the precursor cells of the epidermis, and the eight internal cells, organized into an upper and a lower tier of four cells each, differentiate into the procambium and ground meristem (precursors of the vascular and ground tissues of the cortex, respectively). The tangential divisions initiate the formation of the radial pattern elements made up of concentric tissue layers first seen in the basal part of the embryo. If these divisions are disrupted by mutations, as in the *fist* mutant of *Arabidopsis*, the embryos formed lack a histologically distinct protoderm and procambium (Dunn et al. 1997). The function of the protoderm on the embryo is enhanced by the formation of a waxy layer of cuticle. Tanaka et al. (2001) have identified an *Arabidopsis* gene, *ABNORMAL LEAF SHAPE1* (*ALE1*) encoding a subtilisin-like serine protease involved in cuticle differentiation, and have shown that only a rudimentary cuticle covers the protoderm of embryos of *ale1* mutants. Differentiation of the inner cells of the embryo is characterized by elongation and a subsequent anticlinal division. In another mutant designated as *auxin-binding protein1* (*abp1*), these cells fail to elongate or divide in the globular-stage embryo, which consequently does not make the transition to the next heart-shaped stage. Since the ABP1 protein is a well-known auxin receptor mediating auxin-induced cell elongation, it has been suggested that this protein has a role in embryo axialization via organized cell elongation and cell divisions (Chen et al. 2001).

Evidence that the Kunitz trypsin inhibitor (KTi) mRNA is an early marker of the root pole of the globular embryo of soybean (*Glycine max*; Fabaceae) has been obtained by in situ hybridization of sections of embryos using a cloned *KTi* gene. *KTi* transcripts are found to persist in the axial cells of the ground meristem of developing embryos to signal progressive establishment of the apicobasal axis (Perez-Grau and Goldberg 1989). The *Arabidopsis* gene *POLARIS* (*PLS*) has been found to be a specific marker for the root tip of wild-type embryos and for embryos of some mutants that fail to regenerate a normal root meristem. Gene expression in the mutants appears to be influenced by the biochemical differentiation of cells as root meristem in a position-dependent manner (Topping and Lindsey 1997). Apicobasal and radial pattern formation in *Arabidopsis* embryos is reflected in the expression of transcripts of the *A. thaliana LIPID TRANSFER*

PROTEIN1 (*AtLTP1*) gene, encoding a lipid transfer protein, and of the *ATML1* gene. That pattern formation is mediated in part by the position-specific expression of the *AtLTP1* gene is inferred from its initial expression in the protoderm of globular and older embryos and later expression in the cotyledons and upper end of the hypocotyl (Vroemen et al. 1996). Although *ATML1* mRNA is expressed uniformly in all cells of the globular embryo, specificity of the gene as a marker for radial pattern-forming elements is vividly seen from the disappearance of transcripts from the inner cells of the 16-celled embryo and the restriction of transcripts to the protoderm cells and to the epidermal cells of later-stage embryos (Lu et al. 1996; Sessions et al. 1999). Probably, different combinations of marker genes mediate in the region- and cell layer-specific interpretation of some basic positional information during embryogenesis.

In the 16-celled stage of the *Arabidopsis* embryo, the hypophysis is formed by a transverse division of the uppermost suspensor cell. At the same time, another round of divisions in the derivatives of the terminal cell of the two-celled embryo produces a globular embryo consisting of an epidermis and a central core, each of 16 cells. The first division of the hypophysis yields a small, lens-shaped, upper cell that abuts the lower end of the globular embryo, and a large lower cell that contacts laterally with the embryo epidermis and, at its basal end, with the uppermost suspensor cell (see Plate 4, Fig. a). The suspensor has now attained its genetically permissible number of cells and, apparently having fulfilled their function, the cells gradually begin to lose connection with one another and from the embryo and disintegrate. The globular stage of the embryo is completed by approximately three additional rounds of divisions, mostly of the inner core of cells (Mansfield and Briarty 1991). The end of the globular stage also signifies a change from radial to bilateral symmetry of the embryo, which initially flattens and attains a transient triangular or early heart-shaped stage. Emerging from the triangular stage, the embryo expands laterally by cell divisions to forecast the imminent formation of a pair of cotyledons and assumes the heart-shaped stage at the same time as the two hypophyseal cells divide twice vertically to form two layers of four cells each. The generic term 'proembryo' is used to refer to the above stages of embryo development, which precede initiation of cotyledons. Further growth of cotyledons and elongation of the embryo axis, which occur during the heart-shaped stage, are accompanied by the appearance of meristems from which the future seedling organs are derived. For example, the shoot apical meristem is organized in the depression between cotyledons and appears as a mound of rapidly dividing cells. At the opposite end of the embryo, cells of the lower hypophyseal layer divide horizontally to produce four superimposed layers of four cells each. By further divisions, these cells become the root cap columella; cells of the lateral root cap and root epidermis generated by accompanying periclinal divisions of cells adjacent to the hypophyseal cells contribute additional derivatives to the generation of the root apex from the lower tier of cells of the octant embryo and hypophysis (see Plate 4, Fig. a). The formation of the root apex is complete upon incorporation into the root apical meristem of the four upper hypophyseal layer of cells as the quiescent center (Jürgens and Mayer 1994).

Plasmodesmata are rampant in the globular embryo of *Arabidopsis*, suggesting that the entire embryo might represent a single symplastic domain (Mansfield and Briarty 1991). This was confirmed by using a fluorescent tracer loading method that showed that the torpedo-shaped embryo represented a critical transition point at which embryos ceased to traffic large tracers that had traveled freely through cells of earlier stage embryos (Kim et al. 2002). Use of autofluorescent markers of two different sizes (27 kDa and 54 kDa) to monitor intercellular trafficking through plasmodesmata of embryos of different developmental stages has now confirmed that there is indeed a difference in plasmodesmatal function between cells of embryos of early and late stages of development. Whereas proteins of both sizes move freely throughout the early heart-shaped embryo symplast, as the embryo matures into the torpedo-shaped stage not only does the permeability of the 54 kDa marker decrease, the movement of the marker is also restricted to the region around the shoot apical meristem (Kim et al. 2005). Although fluorescent tracer methodology has not been around for a long time in plant embryology, its use to show that plasmodesmatal openings in younger embryos are more dilated and more dynamic than those of older embryos illustrates the power of this technique.

Using a fluorescent reporter gene construct to monitor the dynamics of auxin distribution in individual cells, Friml et al. (2003) have provided a conceptual framework for cell-to-cell auxin transport to account for the development of the root apical meristem and apicobasal patterning in the *Arabidopsis* embryo, beginning with the single-celled zygote. In this model, much or all of the zygote is activated so that auxin accumulates in the small terminal cell born out of the asymmetric division of the zygote specifying the apical pole. There is an increase in signal intensity in the developing embryo proper up to the globular stage, compared to the virtual absence of signal in the suspensor. The auxin gradient is however reversed in the globular embryo, as auxin begins to accumulate in the uppermost suspensor cells, specifying the basal pole of the embryo. At later stages of embryogenesis, additional auxin signals appear at the tips of cotyledons. Evidence suggests that the products of at least four *PIN-FORMED* (*PIN*) genes, most importantly *PIN1* and *PIN7*, function as regulators of auxin efflux. Correlation of defects in apicobasal patterning with spatial patterns of auxin distribution and defects in embryos of *pin* mutants and with subcellular localization of PIN proteins in developing wild-type embryos has provided the strongest evidence in support of the proposed model. What determines the apical-basal shift in the localization of PIN protein to direct the auxin flow in the few-celled embryos is an important question. One view has assigned this function to the protein kinase encoded by the *PINOID* (*PID*) gene; this idea is supported by the observation that overexpression of the *PID* gene in early-stage embryos targets the PIN protein apically, whereas a low level of the gene retargets PIN protein basally (Friml et al. 2004).

The heart-shaped stage of the *Arabidopsis* embryo is followed by the torpedo-shaped stage, when further elongation of cotyledons and hypocotyl, as well as extension of vascular tissues carved out from the inner core of cells, occur. Although the embryo continues to increase in size and exhibit further changes in shape and organizational complexity as it goes through the bent-cotyledon and mature stages, the basic body plan of a shoot-root axis becomes unmistakably clear in the torpedo-shaped embryo. The striking morphological feature of the bent-cotyledon stage embryo is the curvature of cotyledons toward the hypocotyl; at the mature stage, because of space restrictions within the ovule, tips of cotyledons come to lie opposite the root pole. It has been estimated that a mature embryo of *Arabidopsis* has 15,000–20,000 cells and, under favorable growth conditions, it takes about 9 days from the time of fertilization to the mature embryo stage. The main tissues formed in the embryo are the protoderm, cortex, endodermis (innermost layer of the root cortex), pericycle (the outermost layer of cells of the vascular cylinder of the root), and xylem and phloem (differentiated from the procambium; Mansfield and Briarty 1991; Jürgens and Mayer 1994). The stages in embryogenesis described above are easily seen in sections of *Arabidopsis* ovules of different ages; some of these are illustrated in Plate 2, Fig. a–h and Plate 3, Fig. i–k.

Traditionally, the cotyledons are considered to be the first formative organs of the embryo produced by the shoot apical meristem, giving them the status of derivatives of the latter. The new era of genetic and molecular studies of *Arabidopsis* embryos has suggested that establishment of the shoot apical meristem follows the outgrowth of cotyledons (Long and Barton 1998; Mayer and Jürgens 1998). Analysis of the fate map of the embryonic shoot apical region has also led to the view that specification of cotyledons is a necessary ground state for the formation of the shoot apical meristem (Woodrick et al. 2000). The development of cotyledons on the incipient embryo revolves around the question of how certain cells in a homogeneous population are positioned to respond to specific signals and differentiate. The patterns of expression of transcripts of *ASYMMETRIC LEAVES1* (*AS1*) and *AINTEGUMENTA* (*ANT*) genes in the two flanking regions of the globular embryo representing the sites of future cotyledons and later in the cotyledons of torpedo-shaped embryo of *Arabidopsis* are consistent with their having a general role as early markers of cotyledon initiation (Elliott et al. 1996; Byrne et al. 2000). Another gene that has been assigned an important role in cotyledon identity in the embryo is *LEAFY COTYLEDON* (*LEC*). Although trichomes constitute a leaf trait in *Arabidopsis*, Meinke (1992) showed that cotyledons of the *lec1* mutant have trichomes on their adaxial surface and an internal anatomy intermediate between that of a cotyledon and a leaf. The implication of this finding is that the *LEC1* gene is required for cotyledon identity as the absence of the gene product causes cotyledons to revert partially to leaf-like organs. A successful genetic approach has shown that, in contrast to the moderate disruption of cotyledon symmetry inflicted by mutations in the *PIN1* and *PID* genes, the *pin1/pid* double mutant completely lacks cotyledons. However, cotyledon formation is not a trait normally associated with *PIN1* and *PID* genes, as elimination of the activity of the *SHOOT MERISTEMLESS (STM)* gene, among others, reinstates cotyledon symmetry to the double mutant (Furutani et al. 2004).

Based on the account given in this section, one can conceptually divide embryogenesis in *Arabidopsis* into an early phase, when all cells of the embryo engage in divisions to generate a population of new cells, and a later phase during which divisions are restricted to cells of certain embryonic regions to produce tissues and organs; during the second phase, when the embryo grows mostly by cell elongation, the cotyledons are programmed to accumulate storage products. The apicobasal and radial patterns of the embryo are established during the first phase; as will be described in the next chapter, sensitive genetic screens have begun to unravel the molecular components of the pattern-forming system. Although the morphogenetic control mechanisms active during embryogenesis remain largely unexplored, tentative beginnings made to characterize shape mutants of *Arabidopsis* embryos and seedlings have implications for the future genetic dissection of embryo morphogenesis. The first shape mutants, defined by their abnormal embryos and seedlings, are *fass (fs), knopf (knf), mickey (mic)*, and *radially swollen1 (rsw1)*. Abnormalities such as round and irregularly spaced cells in *fs*, densely stacked epidermal cells of the hypocotyl in *knf*, and bloated epidermal cells of the hypocotyl in *mic* and *rsw* mutants, also abound (Mayer et al. 1991; Gillmor et al. 2002). Mutations in the *FS, HYDRA (HYD)*, and *PASTICCINO (PAS)* genes not only cause abnormalities in embryo or seedling morphology, but also inflict lesions in their apicobasal or radial patterns (Torres-Ruiz and Jürgens 1994; Topping et al. 1997; Faure et al. 1998). The mutant phenotypes suggest that the correct shape of the embryo depends upon the frequency and plane of division of cells executed during the first phase of development. In another approach to highlight the link between cell division and embryo shape, the frequency of cell divisions during embryogenesis was found to be reduced in *Arabidopsis* transformed by the introduction of a dominant-negative mutant of a cell division cyclin (*CDC*) gene. Embryos of these transgenic lines displayed a range of distortions in their apicobasal pattern and, upon germination, produced seedlings with phyllotactically misplaced, distorted leaves (Hemerly et al. 2000). The mechanisms that underlie the asymmetric allocation of cell fate during the first zygotic divisions are also being addressed by analysis of *Arabidopsis* embryo-defective mutants disrupted in cell cycle control, described in Chap. 5. As described in Chap. 4, mutant analysis has provided new insights into the embryogenic potential of the suspensor cells and the possible interactions that occur between cells of the embryo and suspensor.

ACCUMULATION OF STORAGE RESERVES

Anatomical and ultrastructural investigations have generated a complex picture of cellular changes associated with the period of cell elongation resulting in the accumulation of storage reserves in embryos of eudicots. During this period, cells of the cotyledons and embryo axis typically interpret their biochemical environment by changing course from a program of synthesis of house-keeping proteins to one concerned with the synthesis of storage proteins, starch, and lipids. The most bountiful storage reserves that accumulate in embryos are proteins. The use of cotyledons of embryos of agronomically important crops such as peas, peanuts, beans, and lentils, among others, for human and animal feed has been one of the main driving forces behind the convergence of plant anatomists, biochemists, and molecular biologists to probe the structural, biochemical, and molecular aspects of storage reserve synthesis and accumulation in embryos of eudicots. As will be described in a later chapter, a similar parallel effort has gone into the study of storage reserves of the endosperm of cereal grains.

In the cotyledons of *Arabidopsis* embryos, many steps in the deposition of storage reserves take place during a relatively short period of 72 h beginning about 6 days after flowering, in parallel with cell elongation (Mansfield and Briarty 1992). During this period, the chloroplasts appear swollen due to

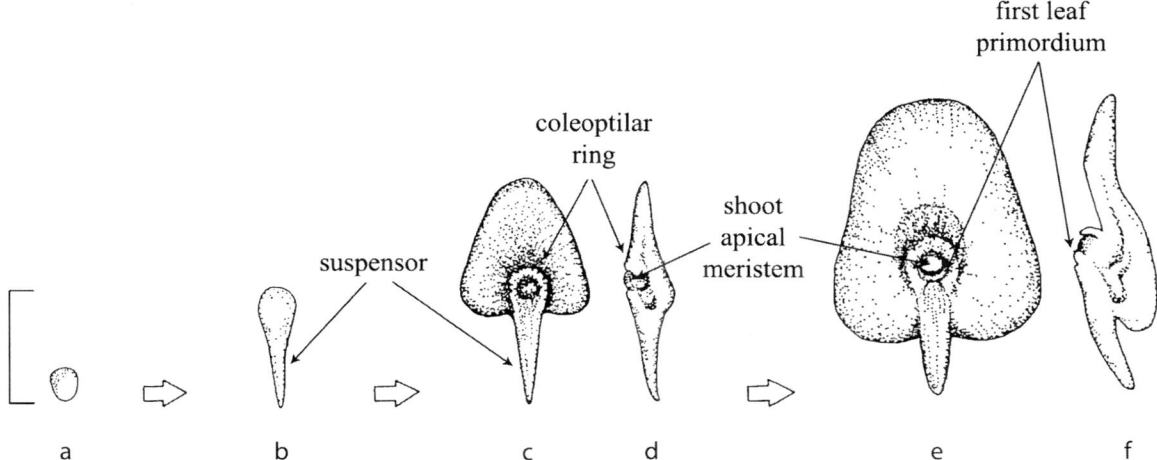

Fig. 2.3a–f Diagrams of early embryogenesis in maize. **a** Face view of proembryo. **b** Face view of transition-stage embryo. **c** Face view of coleoptilar-stage embryo. **d** Schematic section of coleoptilar-stage embryo. **e** Face view of embryo with first leaf primordium beginning to grow upward from the lower side of the shoot apical meristem. **f** Schematic section of the type of embryo in **e**. *Large arrows* indicate transition from one stage to the next. *Bar* 0.5 mm. (Reprinted with modifications from Sheridan and Clark 1994)

the presence of starch bodies that literally push the thylakoids to the periphery. Initially, storage proteins accumulate in the vacuoles as small clumps of homogeneous, densely staining material attached to the tonoplast and, eventually, the whole vacuole is filled with proteins. Judging from the abundance of ER and dictyosomes during the period of storage protein accumulation, these organelles appear to be involved in protein synthesis and transport to the vacuoles. Storage lipid is deposited initially in small spherical bodies partially surrounded by cisterna of rough ER close to the cell wall but later appears dispersed throughout the cytoplasm. The question of whether ER is responsible for lipid synthesis remains unsolved. It is remarkable, but perhaps not entirely surprising, that details of the involvement of various cytoplasmic organelles in the synthesis and accumulation of storage reserves in *Arabidopsis* embryos are no different from those described in embryos of other plants (see Raghavan 1997, for review).

The major groups of embryo storage proteins are the globulins (soluble in salt solutions) and albumins (mostly soluble in water). The two components of globulins, which differ in molecular mass and sedimentation coefficients are the 7S and 12S proteins. Albumins generally migrate with a 2S sedimentation coefficient. Four genes encoding 12S globulins and five genes encoding 2S albumins are present in the embryos of *Arabidopsis* (Pang et al. 1988; van der Klei et al. 1993). Work on storage protein synthesis and accumulation in embryos of several leguminous plants has been strongly complemented by studies showing that the period of embryo growth by cell expansion when storage reserves accumulate coincides with continued DNA synthesis in the cells of cotyledons by endoreduplication – a process in which nuclei go through several successive rounds of DNA synthesis unaccompanied by mitosis (see Chap. 7). It has not, however, been established that storage protein genes are selectively amplified in the cells of cotyledons, and thus much work remains to be done to determine the mechanism by which storage reserves accumulate during embryogenesis (see Raghavan 1997, for review).

2.2.2
A Model of Embryogenesis in Monocots

Because of the presence of a single cotyledon, embryos of monocots present a picture strikingly different from that of eudicots. There is now general agreement that development of the embryo up to the octant stage is almost identical in monocots and eudicots, and that in the former both the shoot apex and cotyledon share a common origin in the terminal cell of a three-celled proembryo (see Raghavan and Sharma 1995, for review). However, in their ontogeny and mature structure, embryos of the large monocot family of grasses, Poaceae, do not have much in common with other monocots. This is exemplified by an account of embryogenesis in *Zea mays* (Fig. 2.3a–f). As in *Arabidopsis*, an asymmetric division producing a small terminal cell and a large basal cell is the hallmark of the zygote in *Z. mays*, but subsequent divisions are variable. One or two longitudinal divisions in the terminal cell are followed by further irregular divisions in the daughter cells as well as in the basal cell to produce a club-shaped embryo-suspensor complex in about 5 days after pollination. Consistent with the polarity of the zygote, the upper part formed by the descendants of the terminal cell generates the embryo proper, whereas descendants of the basal cell form the suspensor (Randolph 1936; Clark 1996). Based on the localization in the suspensor of transcripts of the transcription factor-encoding homeobox (a conserved DNA sequence motif) gene *Z. mays OUTER CELL LAYER3* (*ZmOCL3*) from an early stage of embryogenesis, this gene can be considered as a suspensor marker (Ingram et al. 2000). Coincidentally, the embryo proper, which is radially symmetrical at this stage, contains cells that are small and dense compared to the large and vacuolated cells of the suspensor. Compared to the suspensor, embryo cells, beginning at an early stage, are enriched for transcripts of *Z. mays Homeobox* (*ZmHox1* and *ZmHox2*) genes, which encode a different class of transcription factors (Klinge and Werr 1995). Differentiation of the protoderm as a distinct layer of homogeneous cells, followed by the formation of new cells in the embryo proper and elongation of the suspensor, moves the embryo into the transition stage. Given the expression of transcripts of the *ZmOCL1* gene in the protoderm layer throughout maize embryogenesis, this gene has been identified with a commitment for epidermis specification. Since the gene is also expressed in the emerging shoot and root apical meristems, it might additionally be considered to function in apico-basal pattern formation in the embryo (Ingram et al. 1999). Protoderm-specific expression of *ZmOCL4* and *ZmOCL5* is initially confined to the top of the globular mass of cells of the embryo proper, shifting subsequently to its adaxial and abaxial regions, respectively, but global alterations in the expression of these two genes have not been observed during later stages of embryogenesis (Ingram et al. 2000). In developing embryos of rice, transcripts of a gene called *Roc1* (for *rice outermost cell-specific1*) are expressed in the outermost cells shortly after fertilization, much earlier than differentiation of the protoderm, suggesting that expression of the gene may be dependent on positional information of cells of the embryo (Ito et al. 2002).

About 7 days after pollination, the maize embryo assumes a bilateral symmetry as it undergoes morphogenesis by the initial ramification of the two embryonic structures, the scutellum and embryo axis. Whereas the axially and laterally growing scutellum (considered by some morphologists as equivalent to the single cotyledon) contributes to the bulk of the bilaterally symmetrical embryo, initiation of two groups of cells within the embryo axis foreshadows the future shoot and root apices. A small elevation on the anterior face of the embryo axis demarcates the shoot apex, whereas the root apex arises endogenously as a dark-staining region. The first sign of the coleoptile, the sheathing structure around the shoot apical meristem and the embryonic leaves, is the formation of a bulging ring of cells (coleoptilar ring) on the face of the scutellum encircling the shoot apex (coleoptilar stage). This is followed by the appearance of the first leaf primordium on the surface of the shoot apical meristem. Considerable expansion and growth of embryonic organs occurs during this period and extends into the maturation period. It is during this latter period that the root apex becomes ensheathed by the coleorhiza that originates during the transition stage by division of cells in the lower part of the embryo axis. The final morphogenetic event before embryo maturation is the formation of the mesocotyl, lying between nodes of the coleoptile and scutellum, as an internode. The mature embryo is formed in about 45 days after pollination and reaches a length of 7–10 mm; by this time it would also have generated five or six leaf primordia (see Plate 4, Fig. b). The leaf primordia are thus considered as products of embryogenic events rather than of post-germination development as in eudicots (Randolph 1936; Sheridan and Clark 1994; Klinge and Werr 1995; Clark 1996; Elster et al. 2000). Consistent with the fact that the scutellum and coleoptile do not form parts of the adult maize plant, DNA fragmentation, characteristic of programmed cell death (pcd), has been detected by terminal deoxyribonucleotidyl transferase (TdT)-mediated dUPT-fluorescein nick end-labeling (TUNEL) and by genomic DNA laddering in the cells of these organs as they are specified during embryogenesis (Giuliani et al. 2002). Along with the presence of the scutellum, coleoptile, coleorhiza, mesocotyl, and epiblast (a flap on one side of the coleorhiza, absent in maize embryo, but present in embryos of many other members of the Poaceae), the grass embryo represents one of those rare examples in which no counterparts to these organs are found in embryos of other angiosperms. However, these organs have functional similarities with organs described in the different genera and species of some eudicots, suggesting that mechanisms regulating their development have been evolutionarily conserved.

A case has been made to do away with the stages of embryogenesis in angiosperms likened to the stages in animal embryogenesis and to adopt a model based on initial histogenetic processes, such as the formation of the protoderm and cortical and vascular precursor cells, and on fundamental developmental processes, such as the origin of the shoot and root apical meristems and tissue differentiation, rather than shape changes (Kaplan and Cooke 1997), but the idea has not fulfilled early expectations of acceptability and has not caught on.

2.2.3
Are Embryonic Organs and Tissues Lineage-restricted Compartments?

The formation of the embryo requires that descendants of each cell generated during early stages of segmentation of the zygote find their appropriate pathway among other cells searching for their own destinations. The developmental patterns of embryos of most eudicots and some monocots seem to involve an orderly series of transverse and longitudinal divisions to make it possible to assign by histological observations the ancestry of a tissue or organ of the mature embryo to a particular cell, or group of cells, of the early division phase embryo. Indeed, the invariance of cell lineage might indicate that there is a link between the pattern of cell divisions throughout embryogenesis and the ultimate fate of each cell. This view of cell lineage relationship inferred from the stereotyped segmentation of the zygote formed the basis of the classification of embryo developmental types considered earlier and even led to the formulation of laws of embryogenesis to define each genus or species based on the fundamental organization of its embryo (Crété 1963).

The wisdom of applying such laws to embrace embryo development in the widely disparate division of flowering plants became questionable, relegating the laws of embryogenesis to no more than a historical footnote.

As an adjunct to conventional histological studies, clonal analysis has been used to construct fate maps of embryos of *Arabidopsis* and maize. These studies, in which the distribution of genetically marked cells of early division phase embryos is followed in seedlings or adult plants, have revealed that the acquisition of cell fate is less lineage-dependent than would be predicted by histological analysis. In the strategy used for *Arabidopsis*, excision of the maize transposable element *Activator* (*Ac*) from a transgenic marker made with a reporter gene construct [cauliflower mosaic virus (CaMV) 35S promoter-β-*GLUCURONIDASE* (*GUS*) gene] linked through the transposon is used to mark cell clones. As the reporter gene is expressed in sectors of the seedling constituted of cell progenies in which *Ac* excision occurs, sectors marking gene expression can be detected histochemically by a characteristic blue stain. This method has beautifully confirmed the predictions of cell lineage studies that the hypocotyl, an intermediate zone near the upper boundary of the root, the radicle, and the root apical meristem have their origin in cells produced by the basal two cell layers derived from the lower tier of cells of the octant embryo. This is not the entire story, however: strong evidence that there are no restricted lineages that result in the progressive allocation of cells to specify the root and hypocotyl has come from the observation that the sector boundaries spanning these adjacent organs of the seedling intrude into each other, thus violating the clonal boundaries set up in the embryo (Scheres et al. 1994). In another investigation, a fate-mapping of genetic chimeras induced during the first division of the terminal cell of the two-celled embryo of *Arabidopsis* showed that there are no lineage restrictions on the daughter cells of this division in the formation of seedling organs (Saulsberry et al. 2002). Analysis of the shape mutant *fs* has also raised doubts about the notion of lineage-restricted cell compartments during embryogenesis. Embryos of this mutant, whose cells are disorganized and irregularly enlarged with misaligned division planes beginning with the first zygotic division, produce misshapen seedlings that nonetheless have root, stem, and leaves in their correct places (Torres-Ruiz and Jürgens 1994). This is consistent with the thesis that seedling morphogenesis is not coupled to the production of lineage-derived cells.

Experiments in which the fate of cells of early-stage maize embryos was followed after X-irradiation of developing ears heterozygous for cell marker mutations that affect pigmentation of mature embryos and seedlings also provide good evidence to indicate that the cell lineage of the embryo is variable and somewhat indeterminate. Using a maize stock that produces sectors in both the scutellum and seedling, it was found that marking cells after the first longitudinal division of the terminal cell of the two-celled embryo by irradiation yields a few kernels in which the sectors occupy more than half of the scutellum overlapping with the embryo axis and others in which the sectors occupy less than half of the scutellum. Obviously, this would not have been the case if the developmental potential within a lineage is restricted before the first longitudinal division of the terminal cell; it also appears unlikely from these results that derivatives of this division contribute equally to the growth of the embryo (Poethig et al. 1986). While these studies have undoubtedly raised questions about the importance of lineage relationships in specifying embryonic organs, they do not show whether any far-reaching positional information is necessary for the allocation of prospective cell fates for organ formation.

2.2.4
Abnormal Embryo Types

It is clear by now that mature embryos of both eudicots and monocots attain a basic organization consisting of a bipolar axis terminated at each pole by an apical meristem, with one or two cotyledons attached at a node below the shoot apical meristem; the point of attachment of the cotyledon(s) separates the embryonic axis into a lower hypocotyl-root region and an upper epicotyl-plumule region. However, in plants belonging to about 15 eudicot families and 5 monocot families, embryogenesis does not proceed to completion in seeds, which thus harbor underdeveloped or reduced embryos at the time of shedding. Balanophoraceae, Ranunculaceae, Scrophulariaceae, Orobanchaceae, and Pyrolaceae among eudicots, and Eriocaulaceae, Orchidaceae, Burmanniaceae, and Pandanaceae among monocots, are noteworthy in this respect (Natesh and Rau 1984). Classic examples of seeds with reduced embryos are *Monotropa uniflora*, in which the embryo embedded in the endosperm consists of no more

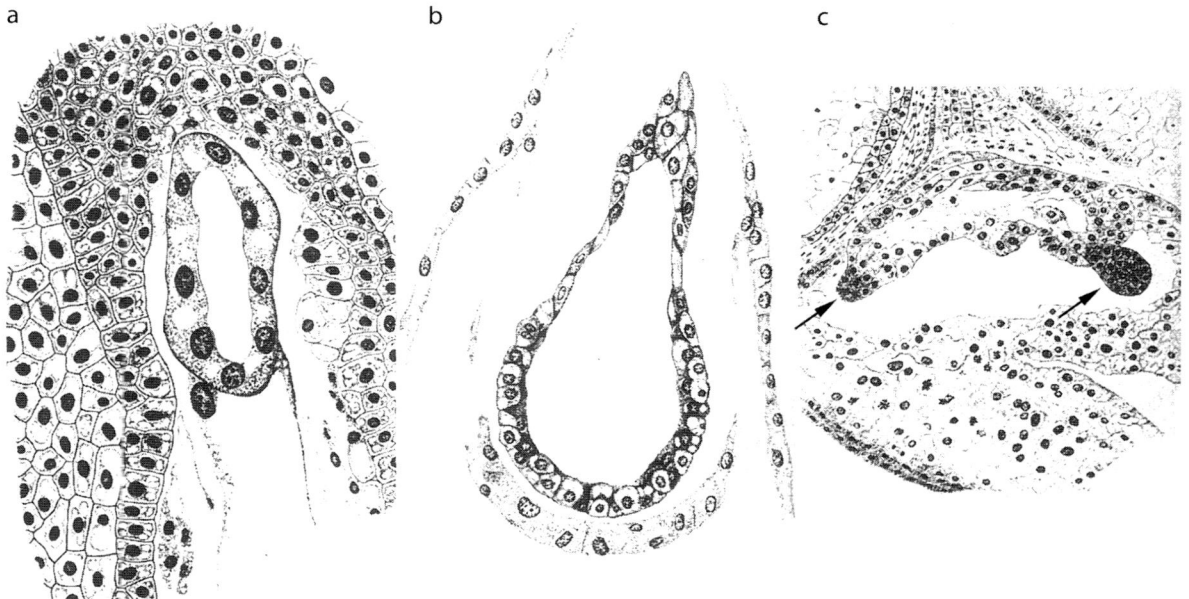

Fig. 2.4a–c Embryogenesis in *Paeonia*. **a** Free-nuclear proembryo in *Paeonia wittmanniana*. **b** Organization of multicellular proembryo in *Paeonia anomala*. **c** Multicellular proembryo of *P. anomala* showing embryo primordia (*arrows*). (Reprinted from Yakovlev and Yoffe 1957)

than two cells separated by a transverse wall (Olson 1980), and *Burmannia pusilla* (Burmanniaceae) in which a four-celled embryo, surrounded by a robust endosperm, is the norm (Arekal and Ramaswamy 1973). Seeds of various orchids and of root-parasites such as *Orobanche aegyptiaca* and *Cistanche tubulosa* (both Orobanchaceae) harbor globular embryos lacking differentiated organs. Despite the lack of morphological differentiation, a distinction is possible between the chalazal and micropylar ends of the embryo of the orchid *Spathoglottis plicata*, as the former is occupied by cells slightly smaller in size than those at the micropylar end (Raghavan and Goh 1994). Similarly, in both *O. aegyptiaca* and *C. tubulosa*, cells at one end of the embryo identified as the radicular pole are small, while those at the opposite plumular pole are large and vacuolated (Rangaswamy 1967). Mature seeds of other root parasites such as *Alectra vogelii* and *Striga gesnerioides* (both Scrophulariaceae) contain heart-shaped embryos with two rudimentary cotyledons and a radicular pole (Okonkwo and Raghavan 1982). On the other hand, reduced embryos enclosed in seeds of *Eriocaulon robusto-brownianum* and *E. xeranthemum* (Eriocaulaceae) show some differential cell division activity resulting in a relatively quiescent region representing the site of epicotyl differentiation and an actively dividing sector for cotyledon initiation (Ramaswamy et al. 1981; Ramaswamy and Arekal 1982). These examples reveal the occurrence of developmental blocks at programmed stages of embryogenesis, but the causal factors that lead to arrest of embryo development require further study.

Embryogenesis in *Paeonia* (Paeoniaceae) seems to be dramatically different from that of the eudicot and monocot models, although it has a fundamental similarity to the process in Gymnosperms. The pattern of embryogenesis in this genus was first revealed by Yakovlev and Yoffe (1957) in *P. anomala*, *P. moutan*, and *P. wittmanniana*. The uniqueness of embryogenesis in the species investigated is that the proembryo is forced to spend part of its existence as a coenocyte due to failure of wall formation following the first few rounds of division of the zygote. The free nuclei later migrate to the periphery of the coenocyte where wall formation takes place to form a multicellular proembryo. Soon after, several groups of marginal cells begin to emerge as embryo primordia, but only one outlives the others and differentiates into a eudicot-type embryo (Fig. 2.4a–c). Although a later work claimed that the first division of the zygote results in wall formation and that the coenocyte is actually a greatly hypertrophied suspensor (Murgai 1959), subsequent investigations of additional species (Cave et al. 1961; Matthiessen 1962; Moscov 1964; Carniel 1967; Mu and Wang 1985) have made most embryologists feel comfortable with the unique themes in the early division patterns of embryos of *Paeonia*. A free nuclear stage culminating in polyembryony is typical of embryogenesis in gymnosperms and thus it remains a formal possibility that the occurrence of coenocyte and incipient polyembryony in *Paeonia* representing the angiosperm equivalents of gymnosperm traits might indicate parallel evolution.

2.3 Physiological Considerations of Embryogenesis

Embryogenesis in flowering plants does not take place in a structural vacuum, although such an impression is conveyed by photographs and diagrams that focus on the embryo alone to the total obliteration of the milieu surrounding it. As a result, the effects of the chemical constituents of the endosperm and of the nutrient substances supplied by the sporophytic parts of the plant in nourishing the growing embryo have not been fully appreciated. Moreover, merely discussing the nutritional aspects of the endosperm and of substances originating in the mother plant becomes a rather unrealistic exercise if the biological significance of the hormonal and signaling molecules from these sources in the morphogenesis of the embryo is not considered. Despite the fact that the archetypal plant hormone, auxin, is known to mediate a wide array of plant growth effects, only recently have the effects of auxin gradients and auxin transport been implicated in the establishment of polarity and patterning along the apicobasal axis of embryos to provide a springboard for further exploration. On the other hand, although there are still many mysteries concerning embryo nutrition, structural studies and investigations based on the culture of excised embryos have been reasonably successful in identifying the sources and nature of nutrients necessary for continued growth and differentiation of embryos. The following account is thus limited to these aspects of the cell physiology and nutrition of developing embryos.

2.3.1
A Role for Auxin Polar Transport in Embryogenesis

Several indirect lines of evidence have implicated auxin action in various stages of embryo development. Fry and Wangermann (1976) suggested that the morphological polarity of the embryo might be determined by initiation of polarized auxin transport in the apolar globular embryo. In an important extension of this view, a series of studies using tissue culture approaches assigned auxin a major role in providing positional cues within the apical dome of early-stage embryos for the differentiation of certain organs. For example, basipetal polar auxin transport might be involved in directing localized cell divisions in the globular embryo preparatory to the outgrowth of cotyledons. The basis for this suggestion is the observation that treatment of cultured globular embryos of *Brassica juncea* with auxin transport inhibitors such as *p*-chlorophenoxyisobutyric acid, *trans*-cinnamic acid, 9-hydroxyfluorene-9-carboxylic acid, NPA, and 2,3,5-triiodobenzoic acid (TIBA) leads to the formation of varying proportions of aberrant morphologies, including a ring-like structure around the incipient shoot apex, akin to fused cotyledons, instead of the two separate cotyledons that establish bilateral symmetry. Assuming that auxin is synthesized in the apical cells of the globular embryo, failure of auxin transport to the two sides of the embryo to form cotyledons could, in principle, lead to auxin diffusion into a ring of cells below the apex and induce their division into the cotyledonary tissue. Since the effect is specific to the globular embryo, it appears that auxin signaling is an integral part of the mechanism that directs localized cell divisions in the embryo to form cotyledons (Liu et al. 1993b; Hadfi et al. 1998). Work with embryos of different ages of *B. juncea* has also shown that maintenance of the correct auxin gradient in the growing embryo is essential for normal morphogenesis, as overwhelming the gradient by exogenous auxin application leads to a broad range of developmental aberrations (Hadfi et al. 1998). Using a promising preparative microtechnique, Ribnicky et al. (2002) have demonstrated that the globular to heart-shaped transition of carrot (*Daucus carota*; Umbelliferae) embryos is associated with an impressive auxin surge.

Interestingly, a small percentage of embryos of the *Arabidopsis* mutant *pin1*, which has a defect in polar auxin transport (Okada et al. 1991) resembles embryos of *B. juncea* treated with auxin transport inhibitors; inhibitor-treated embryos also mimic in varying degrees phenotypes of embryos of other *Arabidopsis* mutants such as *bodenlos* (*bdl*), *gnom* (*gn*), and *monopteros* (*mp*), suggesting a link between the function of the mutated genes and auxin transport and perception (Steinmann et al. 1999; Jürgens 2001). These findings were as striking as they were unexpected. Molecular analysis has shown that *BDL* and *MP* encode auxin-response proteins with opposite effects, whereas the protein product of the *GN* gene has been implicated in vesicle trafficking required for recycling of auxin transport components (Hardtke and Berleth 1998; Hamann et al. 2002; Geldner et al. 2003). Another *Arabidopsis* mutant with a modest defect in polar auxin transport in the stem and inflorescence axis and in cotyledon initiation in the embryo is *pid*. The *PID* gene has been cloned and found to encode a serine-threonine protein kinase. The timing of radial to bilateral transition of the embryo is found to correlate well with the expression of this gene in the flanks of the globular embryo representing potential sites of cotyledon initiation (Christensen et al. 2000). Observations such as the formation of phenocopies of embryos of the *Arabidopsis* shape mutant *fs* by culture of growing heart-shaped wild-type embryos in a medium containing the synthetic auxin naphthaleneacetic acid (NAA) and the presence of much lower levels of free auxin in wild-type seedlings than in mutant seedlings have implicated the *FS* gene as a negative regulator of auxin levels during embryogenesis (Fisher et al. 1996).

As described in Sect. 2.2.2, delineation of the scutellum and the embryo axis signifies the shift from radial to bilateral symmetry in gramineam embryos. In experiments with cultured wheat embryos, it was found that flooding the medium with exogenous auxins like 2,4-D and 2,4,5-trichlorophenoxyacetic acid perpetuates radial growth without the attainment of bilateral symmetry. Auxin seems to be required for the establishment of a normal embryonic symmetry at the globular and early transition stages, as both stages are affected by manipulation of auxin levels. Although TIBA alters the position of the shoot apical meristem in relation to the scutellum in cultured embryos without basically affecting their transition from radial to bilateral symmetry, other auxin transport inhibitors, such as NPA and quercitin, induce profound morphogenetic changes resulting in the differentiation of multiple shoot and root meristems, coleoptiles, and scutella (Fischer and Neuhaus 1996; Fischer et al. 1997). These results support the notion that differentiation of the scutellum and embryo axis on graminean embryos is determined by polar auxin transport. When the distribution of the photoaffinity agent, labeled azidoindoleacetic acid ($[^3H]$,5-N_3IAA) was followed to deduce the auxin transport pathway in embryos of wheat, a shift from radial to bilateral symmetry was found to be correlated with a redistribution of the label in the embryo; a change in the distribution of the label is also observed in morphologically abnormal embryos produced by treatment with auxin polar transport inhibitors. Based on these results, a model for auxin distribution and transport in bilaterally symmetrical wheat embryos envisages unidirectional polar transport of auxin toward the scutellum, and bidirectional transport toward both the scutellum and the shoot apical meristem (Fischer-Iglesias et al. 2001). On the whole, the role of auxin in the radial-bilateral transition of the embryo is an area we are sure to hear more about in the future.

2.3.2
Embryo Nutrition

There is considerable indirect evidence to support the view that continued growth of the zygote through progressive embryogenesis depends upon an uninterrupted supply of nutrients from the milieu of the embryo sac, the endosperm, cells of the ovule surrounding the embryo sac, and from parts of the mother plant. Although a mass of free nuclei or cells produced by division of the primary endosperm nucleus are usually in place in the central cell of the embryo sac before the zygote begins to divide in the embryogenic pathway, there is scant evidence to support the dependency of the zygote on the initial products of division of the endosperm nucleus as a nutrient source. Rather, structural modifications such as haustorial outgrowths of the embryo sac and plasmodesmata, as well as the legendary transfer cell-type wall projections in the central cell described in several plants, seem to implicate the female gametophyte in the large-scale absorption of nutrients from neighboring cells. It is likely that these modifications of the female gametophyte are carried forward unchanged after fertilization, and that they have some primary consequences for nutrition of the zygote. Wall projections presumably involved in the nutrition of the zygote have been shown to originate from the inner wall of the embryo sac at the micropylar end close to the zygote in cotton (Schulz and Jensen 1977), soybean (Tilton et al. 1984; Chamberlin et al. 1993), and *Arabidopsis* (Mansfield and Briarty 1991). Consistent with the role of wall ingrowths in the absorption of nutrients, autoradiography of the fate of ^{14}C-labeled photosynthates in the ovule of soybean has shown that, at the zygote stage, the label is concentrated at the wall projections of the embryo sac and in the hypostase – a group of nucellar cells at the chalazal end of the embryo sac. These observations have led to the suggestion that the zygote obtains its nutrients through two major transport pathways, one from the outer integument to the base of the zygote and the central cell, and the other through the hypostase to the central cell (Chamberlin et al. 1993).

The suspensor has figured in a number of investigations related to the transfer of nutrients from the endosperm to the embryo. Most of these studies have focused on the plasma membrane-lined invaginations from the outer wall of suspensor cells that project into the endosperm or from the inner walls of suspensor cells (Schulz and Jensen 1969, 1974; Yeung and Clutter 1979; Hu et al. 1983). Such ultrastructural modifications of the suspensor cell walls, akin to transfer cells, have been conjured up to support the role of the suspensor in nutrient exchange between the endosperm and the growing embryo. That the suspensor acts as a conduit for metabolites to the developing embryo has also become evident from studies showing that the administration of radioactively labeled sucrose or putrescine (a polyamine) to pods or isolated embryos of *Phaseolus coccineus* and *P. vulgaris* leads to the uptake, translocation, and accumulation of much of the radioactivity in the suspensor (Yeung 1980; Nagl 1990). Additional evidence in support of a role for the endosperm in the nutrition of embryos is described in Chap. 7.

Morphological and anatomical studies of developing embryos of certain plants have provided evidence for the possible utilization of materials from cells of the ovule surrounding the embryo sac. For example, development of elaborate haustorial structures from the suspensor, which come in contact with cells of the integument, nucellus, and placenta, is common in plants belonging to Rubiaceae, Fabaceae, Orchidaceae, and Trapaceae, whose seeds also lack a well-developed endosperm (see Raghavan 1976, for review). It is difficult to find structural evidence for the supply of nutrients from the vegetative parts of the plant to the growing embryo, although it is known that seeds that accumulate large quantities of storage reserves in the embryo or endosperm act as powerful sinks for metabolites from other parts of the plant. It is commonly observed that in developing seeds of pea and other legumes, most of the endosperm is consumed by the time storage protein synthesis is initiated in the cotyledons. However, storage protein synthesis coincides with a marked increase in the amount of vascular tissues in the ovule and funiculus, and in the number of phloem transfer cells in the latter, although it is uncertain whether nutrients transported from the vegetative parts of the plant through the vascular tissues contribute to the nutrition of the embryo or are converted into storage products (Hardham 1976). An investigation that traced the fate of labeled photosynthates in developing soybean seeds has suggested that materials transported through the vascular system into the micropylar, chalazal, and lateral poles of the embryo sac are used for the nurture of globular- to heart-shaped-embryos (Chamberlin et al. 1993).

In summary, it is clear that developing embryos employ strategies involving whole plant physiology, as well as structural and ultrastructural modifications, for their nurture. Admittedly, much work remains to be done to demonstrate in a straightforward way the specific nutrients utilized by embryos of different ages.

2.3.3
Embryo Culture Investigations

In attempts to gain some insight into the nature of the nutrient requirements for continued growth of embryos, considerable attention has been focused on the artificial culture of embryos outside the environment of the ovule under aseptic conditions in media of known chemical composition. A generalization that has emerged from embryo culture studies is that medium requirements of cultured embryos are dependent upon their age at excision, a rule-of-thumb being that whereas mature seed embryos require a relatively simple medium, immature embryos and proembryos require increasingly complex media.

SEED EMBRYO CULTURE

In general, cultured seed embryos grow into plantlets when supplied with a limited diet consisting of inorganic salts, and a carbon energy source such as sucrose. The large size of the embryo enclosed in the seeds of many plants, and the ease with which they can be isolated from seeds, have fostered a great tradition in plant physiology of using isolated seed embryos in metabolic and physiological studies, especially those relating to carbohydrate and nitrogen nutrition and the effects of plant hormones. It is the general experience that growth and survival of cultured embryos is markedly enhanced by supplementation of the medium with sucrose as a carbon energy source and rarely has any other carbohydrate been as successful. At the morphogenetic level, addition of sucrose to the medium enhances the growth of the root and shoot primordia of cultured embryos (Raghavan 1980). Studies on the nitrogen nutrition of embryos have led to three conclusions of general interest: (1) embryos are able to grow moderately well in a medium utilizing nitrates or ammonium salts as the sole source of nitrogen; (2) the amide glutamine is an efficient source of nitrogen for the growth of embryos of a number of plants (Rijven 1952, 1956; Paris et al. 1953; Matsubara 1964; Kost et al. 1992); (3) mutual antagonism and synergism exist between different amino acids in the growth of embryos, as first reported by Sanders and Burkholder (1948), who found that a mixture of 20 amino acids in the proportion in which they occur in casein hydrolyzate is as effective as the latter in promoting the growth of pre- and early heart-shaped embryos of *Datura stramonium* and *D. innoxia* (Solanaceae). The reality of an interaction between amino acids became evident when it was found that favorable effects of the complete mixture containing both beneficial and inhibitory amino acids on the growth of *D. stramonium* embryos are not reproduced by the beneficial compounds alone. In another avenue followed in these investigations, negative interactions between individual amino acids and their alleviation by other amino acids in the same biosynthetic pathways have been noted in the growth of cultured embryos of oat (*Avena sativa*; Poaceae; Harris 1956), barley (Miflin 1969), wheat (Bright et al. 1978), and maize (Green and Donovan 1980), for possible selection of feed-back sensitive mutants.

The effects of the three major groups of plant hormones, namely, auxins, gibberellins, and cytokinins, on the growth of seed embryos have been extensively studied and it can be stated as a secure generalization that the principal organs of cultured embryos respond to auxins, gibberellins, and cytokinins in a way quite similar to the corresponding organs of seedlings (see Raghavan 1980; Raghavan and Srivastava 1982, for reviews). For example, culture of mature embryos of *Capsella bursa-pastoris* in a range of concentrations of IAA leads to promotion of growth of the radicle at low concentrations, inhibition of growth of the root, hypocotyl, and shoot at intermediate concentrations, and induction of callus growth at high concentrations. A range of concentrations of gibberellic acid (GA) is found to promote hypocotyl and root elongation, whereas kinetin (a cytokinin) generally suppresses root growth, but promotes leaf expansion and callus growth (Raghavan and Torrey 1964). A role for the endosperm in supplying hormonal substances for embryo growth is implied in studies of Nyman et al. (1986, 1987) who found that seed embryos of *Colocasia esculenta* (Araceae) develop into plantlets when cultured in the presence of the endosperm including an intact aleurone layer, and that NAA and 6-dimethylaminopurine (a cytokinin) can substitute for the endosperm effect.

CULTURE OF IMMATURE EMBRYOS AND PRECOCIOUS GERMINATION

In contrast to cultured mature embryos, which develop into normal seedlings, cultured immature embryos skip the later stages of embryogenesis and evolve into weak seedlings; to acknowledge the similarity of the process in some respects to normal germination, the process is known as precocious germination. Embryo culture investigations have shown that it is possible to control precocious germination by manipulation of the medium composition by provision of high osmotic pressure (Ziebur et al. 1950), or by addition of abscisic acid (ABA; Norstog 1972), and by changes in the environmental conditions of culture such as provision for high light intensities, moderately high temperatures, and reduced oxygen tension (Norstog and Klein 1972). The effect of ABA in curtailing precocious germination of embryos has been confirmed in many plants, raising the possibility that this hormone is a natural factor that suppresses precocious germination during embryogenesis *in planta*. Some biochemical aspects of precocious germination in the context of germination and embryo dormancy are considered in Chaps. 5 and 6, respectively.

PROEMBRYO CULTURE

Nutritional requirements for growth in culture of proembryos are more exacting than those found necessary for growth of mature and immature embryos. With different species, successful culture of proembryos has been possible by the use of nutrients of endospermic origin, by modifications of the physical conditions of culture, by application of hormonal and organic additives, and by manipulations of the suspensor. In a few model systems, zygotes enclosed in cultured ovules, created by in vitro fertilization of isolated gametes, isolated after fertilization *in planta*, and enclosed in the embryo sac after fertilization, have also been cultured and reared into plants.

The use of nutritionally rich substances of endospermic origin, such as the liquid endosperm of coconut (*Cocos nucifera*; Arecaceae) known as coconut milk or coconut water, for the culture of proembryos is traced to the work of van Overbeek et al. (1942). It was found that, although it was possible to grow to plantlet stage heart-shaped and torpedo-shaped embryos of *Datura stramonium* in an inorganic nutrient medium enriched with a mixture of vitamins and assorted organic substances, smaller embryos failed to grow in this medium or grew feebly before callusing. The clue to the recrudescence of growth in the cultured proembryos lay in the supplementation of the medium with nonautoclaved coconut water. In a later work, a hormonal factor, designated as 'embryo factor', which promoted growth of embryos at very low dilution, was isolated from coconut water (van Overbeek et al. 1944). These studies spawned several successful attempts to culture proembryos of other plants by the use of extracts of endospermic or nonendospermic origin, but none of these extracts have come close to coconut water in terms of their efficacy or wider use in plant tissue culture. Support for the view that growth promotion of cultured proembryos by endosperm extracts is mediated by specific chemical components came from the successful attempt to substitute for the requirement for coconut water in the growth promotion of undifferentiated barley embryos by a phosphate-enriched mineral salt medium fortified by several amino acids with alanine and glutamine as major nitrogen sources (Norstog and Smith 1963).

Under most growing conditions, the amorphous liquid endosperm in which proembryos are constantly bathed has a low (negative) osmotic potential that substantially decreases (becomes more negative) as the embryo matures (Ryczkowski 1960; Smith 1973; Yeung and Brown 1982). This observation suggests that the osmotic pressure of the liquid endosperm might play a role in promoting growth of proembryos in vivo and, by extrapolation, also in vitro. In line with this view, it was found that the requirement for coconut water for the successful growth of proembryos of *D. stramonium* could be met by supplementing a mineral salt medium with 8–12% sucrose or with 2% sucrose plus enough mannitol to be isotonic with 8–12% sucrose (Rietsema et al 1953). Increasing the osmotic concentration of the culture medium by the addition of high concentrations of sucrose or mannitol led to the successful culture of proembryos of other plants such as *Capsella bursa-pastoris* (Rijven 1952), *D. tatula* (Matsubara 1964), *Linum usitatissimum* (Pretová 1974), and *Triticum aestivum* (Fischer and Neuhaus 1995). It is not clear how a high osmolality of the medium promotes growth of proembryos, although its effectiveness might support a possible mechanism that controls the traffic of metabolites and inorganic ions into embryo cells starved of these components.

Two technical modifications of the culture system have eliminated the need to gradually reduce the osmolality of the medium during growth of embryos without their sequential transfer from one medium to another. For the culture of embryos of *C. bursa-pastoris* as small as 50 μm in length, this is achieved by using two media of different osmolalities solidified in juxtaposition in a Petri dish. During initial growth of the embryos, the high osmolality of the medium is gradually reduced by diffusion of water from the medium of low osmolality (Monnier 1976). Continued growth and differentiation of 8- to 36-celled proembryos of *Brassica juncea* is obtained in a culture system composed of two agar layers, with the top layer having a higher osmolality than the bottom layer. Embryos are embedded in the top layer, the osmolality of which decreases during culture (Liu et al. 1993a).

The promotion of growth of proembryos by hormonal additives to the medium is illustrated in studies on the culture of progressively smaller embryos of *C. bursa-pastoris*. Although heart-shaped and older embryos have been routinely cultured in an inorganic liquid medium of high osmolality secured by the addition of 12–18% sucrose (Rijven 1952), later work opened up the feasibility of growing heart-shaped embryos in an agar-solidified mineral salt medium supplemented with 2% sucrose. Growth of still smaller embryos, down to about 55 μm long, was secured by fortifying this medium with a balanced mixture of IAA, kinetin, and adenine sulfate (Raghavan and Torrey 1963). A requirement for kinetin in inducing growth of proembryos of *L. usitatissimum* (Pretová 1986), and for zeatin (a cytokinin) or benzylaminopurine for proembryos of maize (Matthys-Rochon et al. 1998), has also been reported.

A role for the suspensor in embryo nutrition implied from morphological and cytological studies has been strengthened by investigations in which the growth of the embryo severed of its connection with the suspensor was followed. This approach has been used to show that continued growth of embryos of *Eruca sativa* (Brassicaceae; Corsi 1972) and *Phaseolus coccineus* (Yeung and Sussex 1979) is more enhanced in the presence of an attached suspensor than in its absence. Indeed, growth of proembryos of *P. coccineus* is promoted even by the presence of a detached suspensor kept in close proximity in the medium (Yeung and Sussex 1979). Other experiments on *P. coccineus*, described in Chap. 4, involving supplementation of the medium with growth hormones, and determination of the growth hormone levels in the embryo and suspensor cells at specific stages of development have provided indirect evidence to show that the presumed suspensor function is due to the production of gibberellins and cytokinins. The results also underscore the role of hormonal gradients from the suspensor in promoting proembryo growth.

With many plants, isolating proembryos from the ovule remains a stumbling block in their successful culture. Some insights into the growth requirements of proembryos and even of zygotes of certain plants that have hitherto defied attempts at excision and culture have been obtained by an alternative method of ovule culture. Although growth in culture of isolated ovules of *Papaver somniferum* containing the zygote or the two-celled proembryo in a mineral salt medium containing 5% sucrose was sporadic, growth of the nascent sporophyte ensued when the medium was supplemented with casein hydrolyzate, yeast extract, or kinetin (Maheshwari 1958; Maheshwari and Lal 1961). Following this early success, growth of the enclosed zygote or proembryo into a full-term embryo has been induced in cultured ovules of *Zephyranthes* (Liliaceae; Kapoor 1959), cotton (Stewart and Hsu 1977), *C. bursa-pastoris* (Lagriffol and Monnier 1985), barley (Töpfer and Steinbiss 1985; Holm et al. 1995), wheat (Zenkteler and Nitzsche 1985; Comeau et al. 1992) and *Arabidopsis* (Sauer and Friml 2004). Ovules of cotton enclosing the zygote were successfully cultured to the mature embryo stage by supplementing a mineral salt medium with low concentrations of IAA, kinetin, GA, and 15 mM NH_4^+. Whereas the addition of hormones enabled ovules to attain their normal size, NH_4^+ promoted growth and differentiation of the zygote in the embryogenic pathway (Stewart and Hsu 1977).

Successful attempts to induce growth of the zygote of maize in vitro began with ovary culture (Schel and Kieft 1986), followed by culture of the zygote-containing embryo sac surrounded by the nucellus with or without a block of the endosperm (van Lammeren 1988; Campenot et al. 1992; Leduc et al. 1995). Initial culture of embryo sacs enclosing the zygote with subsequent transfer of embryos to a medium of different composition proved suitable for regenerating maize plants by in vitro division of the zygote (Mól et al. 1993, 1995). Using enzyme digestion and microdissection, zygotes isolated from embryo sacs of maize have been induced to form fertile plants simulating typical stages of in vivo embryogenesis. An unexpected finding was that a nurse tissue of embryogenic pollen grains of barley was necessary for induction of continued growth and division of explanted maize zygotes (Leduc et al. 1996). The same nurse tissue culture system was used to induce growth of zygotes extruded from embryo sacs of barley (Holm et al. 1994) and wheat (Kumlehn et al. 1998). As noted in Chap. 1, growth of maize eggs fertilized in vitro was possible in the presence of a nurse tissue system originating from maize embryo. Thus, in spite of the technical success achieved in isolating the zygote from the embryo sac, formulation of a defined medium for its successful growth into an embryo is yet to be perfected. Identification of the nutritional requirements for successful culture of embryos of different stages beginning with the zygote will permit conclusions to be drawn about the biosynthetic pathways activated during progressive embryogenesis in flowering plants.

2.4
Concluding Comments

Long the steady fare of plant embryologists, anatomical characterization of progressive development of the embryo from the zygote laid the foundation for our understanding of how mature embryos of flowering plants are put together by a succession of cell divisions and formation of organized domains or organs. Focused investigations on species that are amenable to genetic and molecular analysis are now beginning to provide insights into the heart of the genetic and molecular mechanisms of polarity of the zygote and its first division that modulate the establishment of the final body plan of the embryo. The intervening years have also seen great progress in understanding the structural and physiological basis for the continued growth and development of the embryo. While much needs to be done to unravel the complexities associated with organization of the embryo body plan, how cell identities are generated in early-stage embryos, and how cell fate becomes fixed in late-stage embryos, are probably the most challenging questions remaining.

REFERENCES

Alessa L, Kropf DL (1999) F-Actin marks the rhizoid pole in living *Pelvetia compressa* zygotes. Development 126:201–209

Arekal GD, Ramaswamy SN (1973) Embryology of *Burmannia pusilla* (Wall. ex Miers) THW. and its taxonomic status. Beitr Biol Pflanz 49:35–45

Ashley T (1972) Zygote shrinkage and subsequent development in some *Hibiscus* hybrids. Planta 108:303–317

Bright SWJ, Wood EA, Miflin BJ (1978) The effect of aspartate-derived amino acids (lysine, threonine, methionine) on the growth of excided embryos of wheat and barley. Planta 139:113–117

Byrne ME, Barley R, Curtis M, Arroyo JM, Dunham M, Hudson A, Martienssen RA (2000) *Asymmetric leaves1* mediates leaf patterning and stem cell function in *Arabidopsis*. Nature 408:967–971

Campenot MK, Zhang G, Cutler AJ, Cass DD (1992) *Zea mays* embryo sacs in culture. I. Plant regeneration from 1 day after pollination embryos. Am J Bot 79:1368–1373

Carniel K (1967) Über die Embryobildung in der Gattung *Paeonia*. Oesterr Bot Z 114:4–19

Cass D, Karas I (1974) Ultrastructural organization of the egg of *Plumbago zeylanica*. Protoplasma 81:49–62

Cave MS, Arnott HJ, Cook SA (1961) Embryogeny in the California peonies with reference to their taxonomic position. Am J Bot 48:397–404

Chamberlin MA, Horner HT, Palmer RG (1993) Nutrition of the ovule, embryo sac, and young embryo in soybean: an anatomical and autoradiographic study. Can J Bot 71:1153–1168

Chen J-G, Ullah H, Young JC, Sussman MR, Jones AM (2001) ABP1 is required for organized cell elongation and division in *Arabidopsis* embryogenesis. Genes Dev 15:902–911

Christensen SK, Dagenais N, Chory J, Weigel D (2000) Regulation of auxin response by the protein kinase PINOID. Cell 100:469–478

Clark JK (1996) Maize embryogenesis mutants. In: Wang TL, Cuming A (eds) Embryogenesis. The generation of a plant. Bios, Oxford, pp 89–112

Cocucci A, Jensen WA (1969) Orchid embryology: megagametophyte of *Epidendrum scutella* following fertilization. Am J Bot 56:629–640

Comeau A, Nadeau P, Plourde A, Simard R, Maës O, Kelly S, Harper L, Lettre J, Landry B, St-Pierre C-A (1992) Media for the in ovulo culture of proembryos of wheat and wheat-derived interspecific hybrids or haploids. Plant Sci 81:117–125

Corellou F, Potin P, Brownless C, Kloareg B, Bouget F-Y (2000) Inhibition of the establishment of zygotic polarity by protein tyrosine kinase inhibitors leads to an alteration of embryo pattern in *Fucus*. Dev Biol 219:165–182

Corsi G (1972) The suspensor of *Eruca sativa* Miller (Cruciferae) during embryogenesis in vitro. G Bot Ital 106:41–54

Crété P (1963) Embryo. In: Maheshwari P (ed) Recent advances in the embryology of angiosperms. International Society of Plant Morphologists, Delhi, India, pp 171–220

Dolan L, Janmaat K, Willemsen V, Linstead P, Poethig S, Roberts K, Scheres B (1993) Cellular organisation of the *Arabidopsis thaliana* root. Development 119:71–84

Dornelas MC, Wittich P, von Recklinghausen I, van Lammeren A, Kreis M (1999) Characterization of three novel members of the *Arabidopsis SHAGGY*-related protein kinase (*ASK*) multigene family. Plant Mol Biol 39:137–147

Dunn SM, Drews GN, Fischer R L, Harada JJ, Goldberg RB, Koltunow AM (1997) *fist*: an *Arabidopsis* mutant with altered cell division planes and radial pattern disruption during embryogenesis. Sex Plant Reprod 10:358–367

Elliott RC, Betzner AS, Huttner E, Oakes MP, Tucker WQJ, Gerentes D, Perez P, Smyth DR (1996) *AINTEGUMENTA*, an *APETALA2*-like gene of Arabidopsis with pleiotropic roles in ovule development and floral organ growth. Plant Cell 8:155–168

Elster R, Bommert P, Sheridan WF, Werr W (2000) Analysis of four embryo-specific mutants in *Zea mays* reveals that incomplete radial organization of the proembryo interferes with subsequent development. Dev Genes Evol 210:300–310

Faure J-D, Vittorioso P, Santoni V, Fraisier V, Prinsen E, Barlier I, van Onckelen H, Caboche M, Bellini C (1998) The *PASTICCINO* genes of *Arabidopsis thaliana* are involved in the control of cell division and differentiation. Development 125:909–918

Fischer C, Neuhaus G (1995) In vitro development of globular zygotic wheat embryos. Plant Cell Rep 15:186–191

Fischer C, Neuhaus G (1996) Influence of auxin on the establishment of bilateral symmetry in monocots. Plant J 9:659–669

Fischer C, Speth V, Fleig-Eberenz S, Neuhaus G (1997) Induction of zygotic polyembryos in wheat: influence of auxin polar transport. Plant Cell 9:1767–1780

Fischer-Iglesias, C, Sundberg B, Neuhaus G, Jones AM (2001) Auxin distribution and transport during embryonic pattern formation in wheat. Plant J 26:115–129

Fisher RH, Barton MK, Cohen JD, Cooke TJ (1996) Hormonal studies of *fass*, an *Arabidopsis* mutant that is altered in organ elongation. Plant Physiol 110:1109–1121

Folsom MW, Peterson CM (1984) Ultrastructural aspects of the mature embryo sac of soybean, *Glycine max* (L.) Merr. Bot Gaz 145:1–10

Friml J, Vieten A, Sauer M, Weijers D, Schwarz H, Hamann T, Offringa R, Jürgens G (2003) Efflux-dependent auxin gradients establish the apical-basal axis of *Arabidopsis*. Nature 426:147–153

Friml J, Yang X, Michniewicz M, Weijers D, Quint A, Tietz O, Benjamins R, Ouwerkerk PBF, Ljung K, Sandberg G, Hooykaas PJJ, Palme K, Offringa R (2004) A PINOID-dependent binary switch in apical-basal PIN polar targeting directs auxin efflux. Science 306:862–865

Fry SC, Wangermann E (1976) Polar transport of auxin through embryos. New Phytol 77:313–317

Furutani M, Vernoux T, Traas J, Kato T, Tasaka M, Aida M (2004) *PIN-FORMED1* and *PINOID* regulate boundary formation and cotyledon development in *Arabidopsis* embryogenesis. Development 131:5021–5030

Geldner N, Anders N, Wolters H, Keicher J, Kornberger W, Muller P, Delbarre A, Ueda T, Nakano A, Jürgens G (2003) The *Arabidopsis* GNOM ARF-GEF mediates endosomal recycling, auxin transport, and auxin-dependent plant growth. Cell 112:219–230

Gillmor CS, Poindexter P, Lorieau J, Palcic MM, Somerville C (2002) α-Glucosidase is required for cellulose biosynthesis and morphogenesis in *Arabidopsis*. J Cell Biol 156:1003–1013

Giuliani C, Consonni G, Gavazzi G, Colombo M, Dolfini S (2002) Programmed cell death during embryogenesis in maize. Ann Bot 90:287–292

Green CE, Donovan CM (1980) Effect of aspartate-derived amino acids and aminoethyl-cysteine on growth of excised mature embryos of maize. Crop Sci 20:358–362

Hable WE, Kropf DL (1998) Roles of secretion and the cytoskeleton in cell adhesion and polarity establishment in *Pelvetia compressa* zygotes. Dev Biol 198:45–56

Hable WE, Miller NR, Kropf DL (2003) Polarity establishment requires dynamic actin in fucoid zygotes. Protoplasma 221:193–204

Hadfi K, Speth V, Neuhaus G (1998) Auxin-induced developmental patterns in *Brassica juncea* embryos. Development 125:879–887

Haecker A, Groß-Hardt R, Geiges B, Sarkar A, Breuninger H, Hermann M, Laux T (2004) Expression dynamics of *WOX* genes mark cell fate decisions during early embryonic patterning in *Arabidopsis thaliana*. Development 131:657–668

Hamann T, Benkova E, Bäurle I, Kientz M, Jürgens G (2002) The *Arabidopsis* BODENLOS gene encodes an auxin response protein inhibiting MONOPTEROS-mediated embryo patterning. Genes Dev 16:1610–1615

Han Y-Z, Huang B-Q, Zee S-Y, Yuan M (2000) Symplastic communication between the central cell and the egg apparatus cells in the embryo sac of *Torenia fournieri* Lind. before and during fertilization. Planta 211:158–162

Hanstein J (1870) Die Entwicklung des Keimes der Monokotylen und Dikotylen. Botanische Abhandlungen aus dem Gebiet der Morphologie und Physiologie, I. Marcus, Bonn

Hardham AR (1976) Structural aspects of the pathways of nutrient flow to the developing embryo and cotyledons of *Pisum sativum* L. Aust J Bot 24:711–721

Hardtke CS, Berleth T (1998) The *Arabidopsis* gene *MONOPTEROS* encodes a transcription factor mediating embryo axis formation and vascular development. EMBO J 17:1405–1411

Harris GP (1956) Amino acids as sources of nitrogen for the growth of isolated oat embryos. New Phytol 55:53–268

Hemerly AS, Ferreira PCG, van Montagu M, Engler G, Inzé D (2000) Cell division events are essential for embryo patterning and morphogenesis: studies on dominant-negative *cdc2aAt* mutants of *Arabidopsis*. Plant J 23:123–130

Holm PB, Knudsen S, Mouritzen P, Negri D, Olsen FL, Roué C (1994) Regeneration of fertile barley plants from mechanically isolated protoplasts of the fertilized egg. Plant Cell 6:531–543

Holm PB, Knudsen S, Mouritzen P, Negri D, Olsen FL, Roué C (1995) Regeneration of the barley zygote in ovule culture. Sex Plant Reprod 8:49–59

Hoshino Y, Scholten S, von Wiegen P, Lörz H, Kranz E (2004) Fertilization-induced changes in the microtubular architecture of the maize egg cell and zygote – an immunocytochemical approach adapted to single cells. Sex Plant Reprod 17:89–95

Hu S, Zhu C, Zee SY (1983) Transfer cells in suspensor and endosperm during early embryogeny of *Vigna sinensis*. Acta Bot Sin 25:1–7

Ingram GC, Magnard J-L, Vergne P, Dumas C, Rogowsky PM (1999) *ZmOCL1*, an HDGL2 family homeobox gene, is expressed in the outer cell layer throughout maize development. Plant Mol Biol 40:343–354

Ingram GC, Boisnard-Lorig C, Dumas C, Rogowsky PM (2000) Expression patterns of genes encoding HD-ZipIV homeo domain proteins define specific domains in maize embryos and meristems. Plant J 22:401–414

Ito M, Sentoku N, Nishimura A, Hong S-K, Sato Y, Matsuoka M (2002) Position dependent expression of *GL2*-type homeobox gene, *Roc1*: significance for protoderm differentiation and radial pattern formation in early rice embryogenesis. Plant J 29:497–507

Jaffe LF (1968) Localization in the developing *Fucus* egg and the general role of localizing currents. Adv Morphog 7:295–328

Jensen WA (1963) Cell development during plant embryogenesis. In: Meristems and differentiation. Brookhaven Symp Biol 16:179–202

Jensen WA (1965) The ultrastructure and composition of the egg and central cell of cotton. Am J Bot 52:781–797

Jensen WA (1968) Cotton embryogenesis: the zygote. Planta 79:346–366

Johansen DA (1950) Plant embryology. Embryogeny of the spermatophyta. Chronica Botanica, Waltham, MA

Johri BM, Ambegaokar KB, Srivastava PS (1992) Comparative embryology of angiosperms, vol 1 and 2. Springer, Berlin Heidelberg New York

Jürgens G (2001) Apical-basal pattern formation in *Arabidopsis* embryogenesis. EMBO J 20:3609–3616

Jürgens G, Mayer U (1994) *Arabidopsis*. In: Bard J (ed) Embryos. Color atlas of development. Wolfe, London, pp 7–21

Kaplan DR, Cooke TJ (1997) Fundamental concepts in the embryogenesis of dicotyledons: a morphological interpretation of embryo mutants. Plant Cell 9:1903–1919

Kapoor M (1959) Influence of growth substances on the ovules of *Zephyranthes*. Phytomorphology 9:313–315

Kim I, Hempel FD, Sha K, Pfluger J, Zambryski PC (2002) Identification of a developmental transition in plasmodesmatal function during embryogenesis in *Arabidopsis thaliana*. Development 129:1261–1272

Kim I, Cho E, Crawford K, Hempel FD, Zambryski PC (2005) Cell-to-cell movement of GFP during embryogenesis and early seedling development in *Arabidopsis*. Proc Natl Acad Sci USA 102:2227–2131

Klinge B, Werr W (1995) Transcription of the *Zea mays* homeobox (*ZmHox*) genes is activated early in embryogenesis and restricted to meristems of the maize plant. Dev Genet 16:349–357

Kost B, Potrykus I, Neuhaus G (1992) Regeneration of fertile plants from excised immature zygotic embryos of *Arabidopsis thaliana*. Plant Cell Rep 12:50–54

Kropf DL (1992) Establishment and expression of cellular polarity in fucoid zygotes. Microbiol Rev 56:316–339

Kropf DL, Berge SK, Quatrano RS (1989) Actin localization during *Fucus* embryogenesis. Plant Cell 1:191–200

Kropf DL, Bisgrove SR, Hable WE (1999) Establishing a growth axis in fucoid algae. Trends Plant Sci 4:490–494

Kumlehn J, Lörz H, Kranz E (1998) Differentiation of isolated wheat zygotes into embryos and normal plants. Planta 205:327–333

Kuroiwa H, Nishimura Y, Higashiyama T, Kuroiwa T (2002) *Pelargonium* embryogenesis: cytological investigations of organelles in early embryogenesis from the egg to the two-celled embryo. Sex Plant Reprod 15:1–12

Lagriffol J, Monnier M (1985) Effects of endosperm and placenta on development of *Capsella* embryos in ovules cultivated in vitro. J Plant Physiol 118:127–137

Leduc N, Matthys-Rochon E, Dumas C (1995) Deleterious effect of minimal enzymatic treatments on the development of isolated maize embryo sacs in culture. Sex Plant Reprod 8:313–317

Leduc N, Matthys-Rochon E, Rougier M, Mogensen L, Holm P, Magnard J-L, Dumas C (1996) Isolated maize zygotes mimic in vivo embryonic development and express microinjected genes when cultured in vitro. Dev Biol 177:190–203

Liu C, Xu Z, Chua N-H (1993a) Proembryo culture: in vitro development of early globular-stage zygotic embryos from *Brassica juncea*. Plant J 3:291–300

Liu C, Xu Z, Chua N-H (1993b) Auxin polar transport is essential for the establishment of bilateral symmetry during early plant embryogenesis. Plant Cell 5:621–630

Long JA, Barton MK (1998) The development of apical embryonic pattern in *Arabidopsis*. Development 125:3027–3035

Lu P, Porat R, Nadeau JA, O'Neill SD (1996) Identification of a meristem L1 layer-specific gene in *Arabidopsis* that is expressed during embryonic pattern formation and defines a new class of homeobox genes. Plant Cell 8:2155–2168

Maheshwari N (1958) In vitro culture of excised ovules of *Papaver somniferum*. Science 127:342

Maheshwari N, Lal M (1961) In vitro culture of excised ovules of *Papaver somniferum* L. Phytomorphology 11:307–314

Maheshwari P (1950) An introduction to the embryology of angiosperms. McGraw-Hill, New York

Mansfield SG, Briarty LG (1991) Early embryogenesis in *Arabidopsis thaliana*. II. The developing embryo. Can J Bot 69:461–476

Mansfield SG, Briarty LG (1992) Cotyledon cell development in *Arabidopsis thaliana* during reserve deposition. Can J Bot 70:151–164

Mansfield SG, Briarty LG, Erni S (1991) Early embryogenesis in *Arabidopsis thaliana*. I. The mature embryo sac. Can J Bot 69:447–460

Matsubara S (1964) Effect of nitrogen compounds on the growth of isolated young embryos of *Datura*. Bot Mag Tokyo 77:253–259

Matthiessen Å (1962) A contribution to the embryogeny of *Paeonia*. Acta Hortic Berg 20:57–61

Matthys-Rochon E, Piola F, le Deunff E, Mól R, Dumas C (1998) In vitro development of maize immature embryos: a tool for embryogenesis analysis. J Exp Bot 49:839–845

Mayer U, Jürgens G (1998) Pattern formation in plant embryogenesis: a reassessment. Semin Cell Dev Biol 9:187–193

Mayer U, Torres Ruiz RA, Berleth TE, Miséra S, Jürgens G (1991) Mutations affecting body organization in the *Arabidopsis* embryo. Nature 353:402–407

Meinke DW (1992) A homeotic mutant of *Arabidopsis thaliana* with leafy cotyledons. Science 258:1647–1650

Miflin BJ (1969) The inhibitory effects of various amino acids on the growth of barley seedlings. J Exp Bot 20:810–819

Mogensen HL (1972) Fine structure and composition of the egg apparatus before and after fertilization in *Quercus gambelii*: the functional ovule. Am J Bot 59: 931–941

Mogensen HL, Suthar HK (1979) Ultrastructure of the egg apparatus of *Nicotiana tabacum* (Solanaceae) before and after fertilization. Bot Gaz 140:168–179

Mól R, Matthys-Rochon E, Dumas C (1993) In-vitro culture of fertilized embryo sacs of maize: zygotes and two-celled proembryos can develop into plants. Planta 189:213–217

Mól R, Matthys-Rochon E, Dumas C (1995) Embryogenesis and plant regeneration from maize zygotes by in vitro culture of fertilized embryo sacs. Plant Cell Rep 14:743–747

Mól R, Idzikowska K, Dumas C, Matthys-Rochon E (2000) Late steps of egg cell differentiation are accelerated by pollination in *Zea mays* L. Planta 210:749–757

Monnier M (1976) Culture in vitro de l'embryon immature de *Capsella bursa-pastoris* Moench. Rev Cytol Biol Vég 39:1–120

Moscov IV (1964) On the development of the embryo in several species of *Paeonia*. Bot Zhu 49:887–894

Mu X, Wang F (1985) The early development of embryo and endosperm of *Paeonia lactiflora*. Acta Bot Sin 27:7–12

Murgai P (1959) The development of the embryo in *Paeonia*. – A reinvestigation. Phytomorphology 9:275–277

Nagl W (1990) Translocation of putrescine in the ovule, suspensor and embryo of *Phaseolus coccineus*. J Plant Physiol 136:587–591

Natesh S, Rao MA (1984) The embryo. In: Johri BM (ed) Embryology of angiosperms. Springer, Berlin Heidelberg New York, pp 377–443

Nawaschin S (1898) Resultate einer Revision der Befruchtungsvorgänge bei *Lilium martagon* und *Fritillaria tenella*. Bull Acad Imp Sci St-Pétersbourg Ser 5, 9:377–382

Norstog K (1972) Factors relating to precocious germination in cultured barley embryos. Phytomorphology 22:134–139

Norstog K, Klein RM (1972) Development of cultured barley embryos. II. Precocious germination and dormancy. Can J Bot 50:1887–1894

Norstog K, Smith JE (1963) Culture of small barley embryos on defined media. Science 142:1655–1656

Nuccitelli R (1978) Oöplasmic segregation and secretion in the *Pelvetia* egg is accompanied by a membrane-generated electrical current. Dev Biol 62:13–33

Nyman LP, Webb EL, Gu Z, Arditti J (1986) Structure and in vitro growth of zygotic embryos of taro (*Colocasia esculenta* var. *antiquorum*). Ann Bot 57:623–630

Nyman LP, Webb EL, Gu Z, Arditti J (1987) Effects of growth regulators and glutamine on in vitro development of zygotic embryos of taro (*Colocasia esculenta* var. *antiquorum*). Ann Bot 59:517–523

Okada K, Ueda J, Komaki MK, Bell CJ, Shimura Y (1991) Requirement of the auxin polar transport system in early stages of *Arabidopsis* floral bud formation. Plant Cell 3:677–684

Okonkwo SNC, Raghavan V (1982) Studies on the germination of seeds of the root parasites, *Alectra vogelii* and *Striga gesnerioides*. I. Anatomical changes in the embryos. Am J Bot 69:1636–1645

Olson AR (1980) Seed morphology of *Monotropa uniflora* L. (Ericaceae). Am J Bot 67:968–974

Olson AR, Cass DD (1981) Changes in megagametophyte structure in *Papaver nudicaule* L. (Papaveraceae) following in vitro placental pollination. Am J Bot 68:1333–1341

Pang PP, Pruitt RE, Meyerowitz EM (1988) Molecular cloning, genomic organization, expression and evolution of 12S seed storage protein genes of *Arabidopsis thaliana*. Plant Mol Biol 11:805–820

Paris D, Rietsema J, Satina S, Blakeslee AF (1953) Effect of amino acids, especially aspartic and glutamic acid and their amides, on the growth of *Datura stramonium* embryos in vitro. Proc Natl Acad Sci USA 39:1205–1212

Perez-Grau L (2002) Plant embryogenesis – the cellular design of a plant. In: O'Neill SD, Roberts JA (eds) Plant reproduction, Annual Plant Reviews vol 6. Sheffield Academic Press, Sheffield, pp 154–192

Perez-Grau L, Goldberg RB (1989) Soybean seed protein genes are regulated spatially during embryogenesis. Plant Cell 1:1095–1109

Poethig RS, Coe EH Jr, Johri MM (1986) Cell lineage patterns in maize embryogenesis: a clonal analysis. Dev Biol 117:392–404

Prakash N (1987) Embryology of the Leguminosae. In: Stirton CH (ed) Advances in legume systematics, part 3. Royal Botanic Gardens, Kew, pp 241–278

Preťová A (1974) The influence of the osmotic potential of the cultivation medium on the development of excised flax embryos. Biol Plant 16:14–20

Preťová A (1986) Growth of zygotic flax embryos in vitro and influence of kinetin. Plant Cell Rep 3:210–211

Quatrano RS, Shaw SL (1997) Role of the cell wall in the determination of cell polarity and the plane of cell division in *Fucus* embryos. Trends Plant Sci 2:15–21

Raghavan V (1976) Experimental embryogenesis in vascular plants. Academic Press, London

Raghavan V (1980) Embryo culture. Int Rev Cytol Suppl 11B:209–240

Raghavan V (1997) Molecular embryology of flowering plants. Cambridge University Press, New York

Raghavan V, Goh CJ (1994) DNA synthesis and mRNA accumulation during germination of embryos of the orchid *Spathoglottis plicata*. Protoplasma 183:137–147

Raghavan V, Sharma KK (1995) Zygotic embryogenesis in gymnosperms and angiosperms. In: Thorpe TA (ed) In vitro embryogenesis in plants. Kluwer, Dordrecht, pp 73–115

Raghavan V, Srivastava PS (1982) Embryo culture. In: Johri BM (ed) Experimental embryology of vascular plants. Springer Heidelberg New York, Berlin, pp 195–230

Raghavan V, Torrey JG (1963) Growth and morphogenesis of globular and older embryos of *Capsella* in culture. Am J Bot 50:540–551

Raghavan V, Torrey JG (1964) Effects of certain growth substances on the growth and morphogenesis of immature embryos of *Capsella* in culture. Plant Physiol 39:691–699

Ramaswamy SN, Arekal GD (1982) Embryology of *Eriocaulon xeranthemum* Mart. (Eriocaulaceae). Acta Bot Neerl 31:41–54

Ramaswamy SN, Swamy BGL, Govindappa DA (1981) From zygote to seedling in *Eriocaulon robusto-brownianum* Ruhl. (Eriocaulaceae). Beitr Biol Pflanz 55:179–188

Randolph LF (1936) Developmental morphology of the caryopsis in maize. J Agric Res 53:881–916

Rangaswamy NS (1967) Morphogenesis of seed germination in angiosperms. Phytomorphology 17:477–487

Ribnicky DM, Cohen JD, Hu W-S, Cooke TJ (2002) An auxin surge following fertilization in carrots: a mechanism for regulating plant totipotency. Planta 214:505–509

Rietsema J, Satina S, Blakeslee AF (1953) The effect of sucrose on the growth of *Datura stramonium* embryos in vitro. Am J Bot 40:538–545

Rijven AHGC (1952) In vitro studies on the embryos of *Capsella bursa-pastoris*. Acta Bot Neerl 1:157–200

Rijven AHGC (1956) Glutamine and asparagines as nitrogen sources for the growth of plant embryos in vitro: a comparative study of 12 species Aust J Biol Sci 9:511–527

Robinson KR, Jaffe LF (1975) Polarizing Fucoid eggs drive a calcium current through themselves. Science 187:70–72

Ryczkowski M (1960) Changes of the osmotic value during the development of the ovule. Planta 55:343–356

Sanders ME, Burkholder PR (1948) Influence of amino acids on growth of *Datura* embryos in culture. Proc Natl Acad Sci USA 34:516–526

Sauer M, Friml J (2004) In vitro culture of *Arabidopsis* embryos within their ovules. Plant J 40:835–843

Saulsberry A, Martin PR, O'Brien T, Sieburth LE, Pickett FB (2002) The induced sector *Arabidopsis* apical embryonic fate map. Development 129:3403–3410

Schel JHN, Kieft H (1986) An ultrastructural study of embryo and endosperm development during in vitro culture of maize ovaries (*Zea mays*). Can J Bot 64:2227–2238

Scheres B, Wolkenfelt H, Willemsen V, Terlouw M, Lawson E, Dean C, Weisbeek P (1994) Embryonic origin of the *Arabidopsis* primary root and root meristems. Development 120:2475–2487

Schulz P, Jensen WA (1969) *Capsella* embryogenesis: the suspensor and the basal cell. Protoplasma 67:139–163

Schulz P, Jensen WA (1974) *Capsella* embryogenesis: the development of the free nuclear endosperm. Protoplasma 80:183–205

Schulz P, Jensen WA (1977) Cotton embryogenesis: the early development of the free nuclear endosperm. Am J Bot 64:384–394

Schulz R, Jensen WA (1968) *Capsella* embryogenesis: the egg, zygote, and young embryo. Am J Bot 55:807–819

Sessions A, Weigel D, Yanofsky MF (1999) The *Arabidopsis thaliana* MERISTEM LAYER 1 promoter specifies epidermal expression in meristems and young primordia. Plant J 20:259–263

Shaw SL, Quatrano RS (1996) Polar localization of a dihydropyridine receptor on living *Fucus* zygotes. J Cell Sci 109:335–342

Sheridan WF, Clark JK (1994) Fertilization and embryogeny in maize. In: Freeling M, Walbot V (eds) The maize handbook. Springer, New York Berlin Heidelberg, pp 3–10

Smith JG (1973) Embryo development in *Phaseolus vulgaris* II. Analysis of selected inorganic ions, ammonia, organic acids, amino acids, and sugars in the endosperm liquid. Plant Physiol 51:454–458

Steinmann T, Geldner N, Grebe M, Mangold S, Jackson CL, Paris S, Gälweiler L, Palme K, Jürgens G (1999) Coordinated polar localization of auxin efflux carrier PIN1 by GNOM ARF GEF. Science 286:316–318

Stewart JM, Hsu CL (1977) In-ovulo embryo culture and seedling development of cotton (*Gossypium hirsutum* L.). Planta 137:113–117

Strasburger E (1884) Neue Untersuchungen über den Befruchtungsvorgang bei den Phanerogamen als Grundlage für eine Theorie der Zeugung. Fischer-verlag, Jena

Sumner MJ (1992) Embryology of *Brassica campestris*: the entrance and discharge of the pollen tube in the synergid and the formation of the zygote. Can J Bot 70:1577–1590

Sumner MJ, van Caeseele L (1989) The ultrastructure and cytochemistry of the egg apparatus of *Brassica campestris*. Can J Bot 67:177–190

Sun H, Basu S, Brady SR, Luciano RL, Muday GK (2004) Interactions between auxin transport and the actin cytoskeleton in developmental polarity of *Fucus distichus* embryos in response to light and gravity. Plant Physiol 135:266–278

Tanaka H, Onouchi H, Kondo M, Hara-Nishimura I, Nishimura M, Machida C, Machida Y (2001) A subtilisin-like serine protease is required for epidermal surface formation in *Arabidopsis* embryos and juvenile plants. Development 128:4681–4689

Tilton VR, Wilcox LW, Palmer RG (1984) Postfertilization wandlabrinthe formation and function in the central cell of soybean, *Glycine max* (L.) Merr. (Leguminosae). Bot Gaz 145:334–339

Töpfer R, Steinbiss H-H (1985) Plant regeneration from cultured fertilized barley ovules. Plant Sci 41:49–54

Topping J, Lindsey K (1997) Promoter trap markers differentiate structural and positional components of polar development in *Arabdiopsis*. Plant Cell 9:1713–1725

Topping JF, May VJ, Muskett PR, Lindsey K (1997) Mutations in the *HYDRA1* gene of *Arabidopsis* perturb cell shape and sirupt embryonic and seedling morphogenesis. Development 124:4415–4424

Torres-Ruiz RA, Jürgens G (1994) Mutations in the FASS gene uncouple pattern formation and morphogenesis in *Arabidopsis* development. Development 120:2967–2978

Traas J, Bellini C, Nacry P, Kronenberger J, Bouchez D, Caboche M (1995) Normal differentiation patterns in plants lacking microtubular preprophase bands. Nature 375:676–677

van der Klei H, van Damme J, Casteels P, Krebbers E (1993) A fifth 2S albumin isoform is present in *Arabidopsis thaliana*. Plant Physiol 101:1415–1416

van Lammeren AAM (1986) A comparative ultrastructural study of the megagametophytes in two strains of *Zea mays* L. before and after fertilization. Agric Univ Wageningen Papers 86(1):1–37

van Lammeren AAM (1988) Observations on the structural development of immature maize embryos (*Zea mays* L.) during in vitro culture in the presence or absence of 2,4-D. Acta Bot Neerl 37:49–61

van Overbeek J, Conklin ME, Blakeslee AF (1942) Cultivation in vitro of small *Datura* embryos. Am J Bot 29:472–477

van Overbeek J, Siu R, Haagen-Smit AJ (1944) Factors affecting the growth of *Datura* embryos in vitro. Am J Bot 31:219–224

Vroemen CW, Langeveld S, Mayer U, Ripper G, Jürgens G, van Kammen A, de Vries SC (1996) Pattern formation in the *Arabidopsis* embryo revealed by position-specific lipid transfer protein gene expression. Plant Cell 8:783–791

Webb MC, Gunning BES (1991) The microtubular cytoskeleton during development of the zygote, proembryo and free-nuclear endosperm in *Arabidopsis thaliana* (L.) Heynh. Planta 184:187–195

Weterings K, Apuya NR, Bi Y, Fischer RL, Harada JJ, Goldberg RB (2001). Regional localization of suspensor mRNAs during early embryo development. Plant Cell 13:2409–2425

Woodrick R, Martin PR, Birman I, Pickett FB (2000) The *Arabidopsis* embryonic shoot fate map. Development 127:813–820

Yadegari R, Goldberg RB (1997) Embryogenesis in dicotyledonous plants. In: Larkins BA, Vasil IK (eds) Cellular and molecular biology of plant seed development. Kluwer, Dordrecht, pp 3–52

Yakovlev MS, Yoffe MD (1957) On some peculiar features in the embryogeny of *Paeonia* L. Phytomorphology 7:74–82

Yan H, Yang H-Y, Jensen WA (1991) Ultrastructure of the developing embryo sac of sunflower (*Helianthus annuus*) before and after fertilization. Can J Bot 69:191–202

Yeung EC (1980) Embryogeny of *Phaseolus*: the role of the suspensor. Z Pflanzenphysiol 96:17–28

Yeung EC, Brown DCW (1982) The osmotic environment of developing embryos of *Phaseolus vulgaris*. Z Pflanzenphysiol 106:149–156

Yeung EC, Clutter ME (1979) Embryogeny of *Phaseolus coccineus*: the ultrastructure and development of the suspensor. Can J Bot 57:120–136

Yeung EC, Sussex IM (1979) Embryogeny of *Phaseolus coccineus*: the suspensor and the growth of the embryoproper in vitro. Z Pflanzenphysiol 91:423–433

Zenkteler M, Nitzsche W (1985) In vitro culture of ovules of *Triticum aestivum* at early stages of embryogenesis. Plant Cell Rep 4:168–171

Ziebur NK, Brink RA, Graf LH, Stahmann MA (1950) The effect of casein hydrolysate on the growth in vitro of immature *Hordeum* embryos. Am J Bot 37:144–148

3 Pattern Formation in Embryos – Interpretation of Positional Information

The term 'pattern formation' is being used at present to describe practically everything that happens in development, and is thus robbed of much precise meaning. In addition, it retains some woolly, premodern connotations that have little to do with the nature of the molecular mechanisms that explain how morphogenetic domains are divided into territories of prospective fate. However, semantics aside, the last several years have brought forth a very interesting, compelling and general concept of at least the initial phases of regional morphogenesis. Results from several well-studied systems show that the morphogenesis of specific structures that are composed of various substructures begins with the regional expression of sets of transcription factors, such that each region defines the cells that will produce a working part of the structure (transient or ultimate). Thus the sole function of this initial process appears to be to install different regulatory states in the territories from which the different parts will develop.

E.H. Davidson 1994

3.1   Initiation and Maintenance of Embryo Meristems   58
3.1.1 Shoot Apical Meristem   58
3.1.2 Root Apical Meristem   66
3.2   Genetic and Molecular Control of Embryo Pattern Formation   69
3.2.1 Apicobasal Patterning of the Embryo   69
3.2.2 Radial Patterning of the Embryo   72
3.3   Concluding Comments   75
      References   76

Considerable uncertainty prevails amongst developmental biologists in defining at some sophisticated level the concept of pattern formation – the processes by which the developing embryo specifies the spatial and temporal program of cellular activities (morphogenetic domains) that generate well-ordered embryonic structures, but excluding differentiation processes involved in the structural gene expression determining the final product of each morphogenetic domain. Despite the fact that animal cells undergo considerable rearrangement and migration, sculpturing of the shape of the animal's adult body along the axis of symmetry, and establishment of the organ rudiments and their differentiation into functional organs are accomplished during embryogenesis. By contrast, in most flowering plants, the rudiments of the body plan of the mature plant are carved out during early embryogenesis by two distinct and largely independent processes – one defining the apicobasal pattern and the other the radial pattern – and are elaborated during late embryogenesis. As pointed out in Chap. 2, in *Arabidopsis*, the first indication of apicobasal pattern surfaces in the octant-stage embryo, whose upper tier of cells is destined to form exclusively the shoot apex and most of the cotyledons. Part of the cotyledon structure, the hypocotyl, the radicle, and most of the root meristem are derived from the lower tier of cells whereas the hypophysis contributes to the central portion of the root cap and the remainder of the root apical meristem comprising the quiescent center. The next round of divisions of the octant-stage embryo, which is tangential, sets off its radial pattern made up of concentric layers of the epidermis, ground tissues, and vascular tissues. The main organ systems of the adult flowering plant are, however, initiated at the seedling stage from two groups of initial cells – the shoot apical meristem and the root apical meristem – superimposed upon opposite ends of the bipolar embryo. Thus, unlike in animals, the adult plant is vastly different from the mature embryo from which it is derived; since the shoot and root apical meristems assume primary roles in organizing the basic body plan of the plant throughout its life beginning with the seedling stage, concepts bearing on pattern formation in animal embryos do not necessarily apply to the patterning of plant embryos.

Genetic and molecular approaches have been employed in recent years to investigate the mechanisms that initiate development of the shoot and root apical meristems and establishment of the apicobasal and radial patterning in embryos of flowering plants. These studies have shown that the genome of *Arabidopsis* harbors an amazingly large and diverse set of genes whose mutation can lead to havoc in the patterning of embryos and in the organization of the meristems. Underscoring the significance of specific cell division patterns in the crafting of an embryo, most such mutations have been traced back to defects in the early stages of embryogenic divisions. A new era of multifaceted analyses extending to the cloning of the genes, analysis of their expression patterns, and characterization of their protein products is beginning to usher in new insights into the mechanisms that control cell identities during embryo patterning processes. This chapter summarizes the contribution of information derived from mutant analysis and gene cloning to our understanding of the organization, specification of cell fate, and pattern formation in flowering plant embryos and their meristems. Parts of this story are covered in reviews by, among others, Laux and Jürgens (1997), Scheres and Heidstra (1999), Raghavan (2000a), Jürgens (2001), Berleth and Chatfield (2002), Casson and Lindsey (2003), Laux et al. (2004), and Willemsen and Scheres (2004), and in books by Raghavan (1997, 2000b) and Howell (1998).

3.1 Initiation and Maintenance of Embryo Meristems

In analytical studies of plant embryogenesis, special attention is given to the shoot and root apical meristems for two reasons. First, these meristems, which are groups of undifferentiated cells confined to the tips of growing stems and roots, are delimited early in embryogenesis and play a central role in establishing the apicobasal body plan of the embryo. Second, the shoot and root apical meristems have the innate ability for self-renewal, producing new cells and tissues throughout the life of the plant and hence, empowered with substantial longevity and replicative potential, groups of cells in these meristems are considered analogous to the pluripotent stem cell population of animals. In a sense, the shoot and root apical meristems can be considered immortal, not because the cells comprising them live for ever, but because they are continuously being replaced by new cells. No doubt, many of the cellular interactions that maintain the meristems in functional mode throughout the life of the plant are initiated in the embryo and hence explaining why the shoot and root apical meristems appear where they do, and function in the way they do, could very well explain the apicobasal body plan of the plant.

3.1.1 Shoot Apical Meristem

In *Arabidopsis* the lineage of the shoot apical meristem can be traced to cells in the apical half of the globular embryo, although the meristem itself becomes first visible later in the torpedo-shaped embryo. However, it is close to the mature stage of the embryo that the shoot apical meristem appears as a distinct histological entity organized into the outer tunica and the inner corpus layers of cells (Barton and Poethig 1993). Investigations of experimentally induced periclinal chimeras in *Datura* sp. have led to the identification, based on differences in chromosome number, nuclear size, and cell size, of separate lineages of cells both in the shoot apical meristem and its derivatives in the primary plant body, and have shown that cells of the meristem that give rise to the tunica and corpus are arranged in three clonally discrete layers. The tunica is derived from the two outer layers (L1 and L2), which undergo predominantly anticlinal divisions, whereas cells of the inner L3 layer, constituting a group of cells rather than a single cell layer, display mostly periclinal divisions to produce the major portion of the internal tissues (Satina et al. 1940). During postembryonic development, the shoot apical meristem begins to divide and cut off leaf primordia by evoking regulatory components that coordinate the rate of cell division and frequency of organ initiation. To account for its dual functions of self-perpetuation and the ability to form lateral appendages, the shoot apical meristem has also been descriptively compartmentalized into a cluster of indeterminate cells positioned at its summit known as the central zone, providing a reservoir of self-renewing cells akin to stem cells and a surrounding flanking region – the peripheral zone – involved in the generation of lateral organs. These zones, which include cells from all three layers, differ in their cell division rates, with divisions being more frequent in the peripheral zone than in the central zone. A column of large vacuolate cells referred to as the rib zone has also been identified in the deeper layers of the meristem beneath the central zone (Steeves and Sussex 1989; Lyndon 1998). Although the ultimate number of initial cells that function as stem cells in the shoot apical meristem has not been precisely defined, clonal analysis of certain angiosperms and gymnosperms has delineated one to three cells as initial cells in each of the three outer layers in the central zone of the meristem (Stewart and Dermen 1970).

Since cells of the shoot and root apical meristems are interconnected by plasmodesmata, it has long been assumed that coordination of morphogenetic events within the meristems, as well as meristem maintenance, requires some kind of intercellular signaling. Two independent investigations tracking the movement of membrane-impermeable fluorescent tracer dyes in the shoot apical meristem of different plants have demonstrated that zones and layers of the meristem form separate symplastic fields. In one study, Rinne and van der Schoot (1998) found that a tracer microinjected into the outer layers of the shoot apical meristem of birch (*Betula pubescens*; Betulaceae) seedlings is localized in the tunica but not in the central zone, thus demonstrating a potential restriction of symplastic diffusion of signaling molecules between the central and peripheral fields. Suggestive of alterations in symplastic signaling pathways in the shoot apical meristem during seedling development, exposure of seedlings to dormancy-inducing short days, was found to lead to a breakdown of these fields into symplastically isolated cells incapable of any further dye trafficking. Exclusion of tracer uptake by the central cells of the corpus was observed in the shoot apical meristem of *Arabidopsis* loaded with a dye via cut leaves (Gisel et al. 1999). It is easy to imagine how exchange of signals within and between symplastic fields could influence morphogenetic events in the meristem, thereby conferring unique patterns of gene expression in these fields.

The specification and maintenance of the shoot apical meristem in the embryo and in the seedling plant has been shown to involve members of a fascinating family of genes, the homeobox genes. These genes, which gained prominence as determinants of segmental and cellular identity and regionalization in animal embryos, encode an evolutionarily conserved 64-amino acid, DNA-binding sequence known as homeodomain. The basic molecular structure of homeodomain is a string of three α-helices to form a 'helix-turn-helix' motif. By analogy with their DNA-binding role in animal embryogenesis, homeodomains, which have turned up as products of several plant genes, apparently function as transcription factors controlling certain facets of the embryonic and post-embryonic development of plants. The first plant homeobox gene, *KNOTTED1* (*KN1*), to be fished out and cloned was from maize; dominant mutations in this gene cause anarchic growth of the leaf, resulting in the formation of irregular patches of tissue known as knots. The *KN1* gene was also the first gene to be clearly linked as a molecular marker to the prepatterning information in the shoot apical meristem delimited during embryogenesis. Transcripts of the *KN1* gene as well as its protein product are initially detected in a handful of precursor cells in the bilaterally symmetrical maize embryo where the shoot apical meristem is ordained to appear (see Plate 5, Fig. a–i). The continued expression of *KN1* in the shoot apical meristem formed from these cells during progressive embryogenesis and during post-embryonic development of maize has supported a role for this gene in maintenance of the meristem in an undifferentiated state (Jackson et al. 1994; Smith et al. 1995). The establishment of the shoot apical meristem in rice embryo is also foreshadowed by the appearance

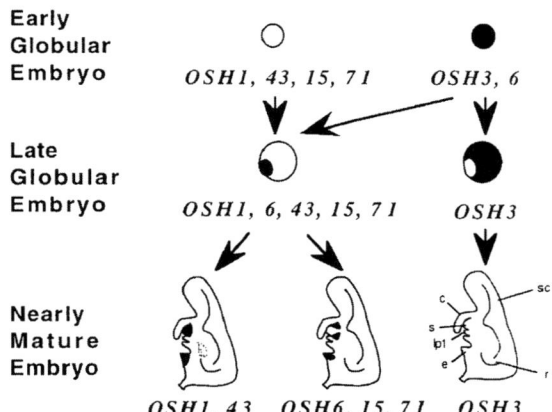

Fig. 3.1 Diagrammatic representation of expression patterns of rice *OSH* genes (*shaded black*) during embryogenesis in rice. *c* Coleoptile, *e* epiblast, *lp1* first leaf primordium, *r* radicle, *s* shoot apical meristem, *sc* scutellum. (Reprinted from Sentoku et al. 1999)

of transcripts of *KN1*-type homeobox genes cloned from embryos and shoot apices of rice. Detailed studies have yielded a picture of different sets of homeobox genes controlling the specification and development of the embryonic shoot apical meristem in rice (Sato et al. 1996, 1998; Ito et al. 1999; Postma-Haarsma et al. 1999; Sentoku et al. 1999). As shown in Fig. 3.1, from a battery of six rice homeobox genes whose expression in developing embryos was monitored by in situ hybridization, genes *ORYZA SATIVA HOMEOBOX1* [*OSH1*, also designated as *Oryza sativa KNOTTED1-like* (*OsKn1*)], *OSH6*, *OSH15*, *OSH43*, and *OSH71* are considered to promote meristem development, whereas the *OSH3* gene provides basic positional cues to generate and allocate prospective cell fates. This conclusion is based on the observation that although the *OSH3* gene is uniformly expressed in the early globular embryo, its expression is suppressed in the region of the developing shoot apical meristem beginning in the late globular embryo; in contrast, expression of the other genes is confined to the area of the developing shoot apical meristem preceding its delimitation (Sentoku et al. 1999). Two other *OsKn* gene family members, namely, *OsKn2* (allelic to *OSH71*) and *OsKn3* (allelic to *OSH15*) and novel genes encoding a *KN1*-like homeodomain belonging to the *HOS* family (for *HOMEOBOX GENE OF Oryza sativa*) are also expressed at the site of the shoot apical meristem, suggesting their involvement in the positioning of the meristem (Postma-Haarsma et al. 1999, 2002; Ito et al. 1999).

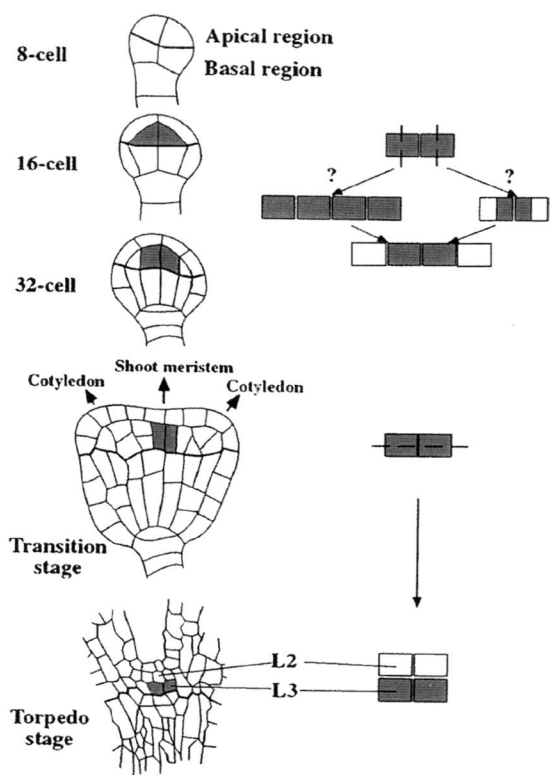

Fig. 3.2 Dynamics of *WUS* gene expression pattern during early embryogenesis in *Arabidopsis*. *Left* Diagrammatic representations of expression of gene transcripts (*shaded*) beginning in the four subepidermal cells in the apical region of the 16-celled embryo. Although these cells divide longitudinally, only the central daughter cells continue *WUS* gene expression as the embryo progresses into the torpedo-shaped stage. *Right* Diagrams of divisions of cells that lead to the formation of new cells expressing the *WUS* gene in the shoot apical meristem in the torpedo-shaped embryo. As shown in the top diagram, two mechanisms to explain how the central daughter cells sustain *WUS* gene expression have been proposed. The middle diagram shows that *WUS*-expressing cells eventually divide horizontally (*dashed lines*) setting up the L2 and L3 cell layers. (Reprinted from Mayer et al. 1998)

Previous studies that elucidated the structural aspects of the shoot apical meristem in flowering plants have now been coupled with genetic and molecular analyses to sketch its progression from a subset of undistinguished cells in the globular embryo through a continuum of developmental states and to identify the signaling pathways that coordinate the meristem behavior necessary to provide a balance between stem cell maintenance and initiation of organ differentiation. Much of this work has been done in *Arabidopsis*, in which more than a dozen mutations that affect the organization of the embryonic shoot apical meristem in subtle ways have been identified (Jürgens 2001; Sharma and Fletcher 2002; Carles and Fletcher 2003; Bäurle and Laux 2003). Based on analysis of expression of wild-type genes as molecular markers, the prevailing view is that a stepwise appearance of characteristic transcriptional domains beginning at an early stage of embryogenesis collectively defines initiation, and subsequent maintenance, of the shoot apical meristem. Mutants designated as *wuschel* (*wus*), *stm*, and *clavata* (*clv*) along with a few others have served as useful paradigms to illustrate the critical need for specific genes in determining shoot apical meristem cell fate and its maintenance in the developing embryo and seedling plant. Mutation in the *WUS* gene incorrectly specifies the embryonic shoot apical meristem, which consequently displays defective organization and attains, at best, a flat contour in both the mature embryo and the seedling. The phenotypic effects of the mutation are most dramatically seen in the seedling, in which defective meristems are initiated repeatedly only to terminate in aberrant flat structures. The *WUS* gene is also probably the only molecular marker that identifies the shoot apical meristem at the earliest recognizable stage of embryogenesis; as seen in Fig. 3.2, transcripts of the gene are first detected in the four inner cells of the apical region of the 16-celled wild-type *Arabidopsis* embryo long before the shoot apical meristem is histologically incarnated. These transcripts gradually become limited to a group of cells in the lower part of the central zone and above the rib zone. This pattern of *WUS* gene expression is maintained essentially unchanged during most of embryogenesis, and is even continued during vegetative development. Based on the pattern of transcript expression narrowed down to a small group of cells underneath the two or three outermost layers of cells in the central zone of wild-type embryonic and postembryonic shoot apical meristems, it has been hypothesized that stem cell identity in the meristem is specified by the *WUS* gene in the underlying group of cells (considered as the stem cell organizer), which in turn let the outer layers of cells know that they are to become pluripotent stem cells (Mayer et al. 1998). The remarkable observation of Gallois et al. (2004) showing that ectopic expression of *WUS* gene in the root meristem of *Arabidopsis* prompts the neighboring cells to participate in the making of a shoot testifies to the function of this gene in establishing stem cells as well as in specifying their shoot

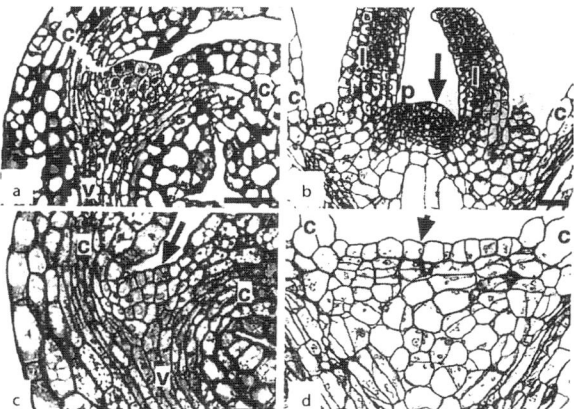

Fig. 3.3a–d Sections of shoot apices of mature embryos and seedlings of wild-type and *zll* mutants of *Arabidopsis*. **a, b** Shoot apical meristems (*arrows*) of wild-type embryo and seedling, respectively, showing small, densely stained cells. **c, d** Shoot apical meristems (*arrows*) of *zll* mutant embryo and seedling, respectively, showing large, lightly stained cells. *c* Cotyledon, *l* leaf primordium, *p* emerging leaf primordium, *v* vascular tissue. *Bars* 20 μm. (Reprinted from Moussian et al. 1998)

identity. The predicted protein product of the *WUS* gene shows homology to homeodomain sequences, consistent with its postulated role as a transcription factor in recognizing specific DNA fragments, but how this is translated into signal transduction is part of the larger problem of understanding the mode of functioning of homeodomains in general (Laux et al. 1996; Mayer et al. 1998).

Mutations in *STM* and the related *ZWILL* [*ZLL*; also represented by the *PINHEAD* (*PNH*) allele] genes are systematically correlated with a disrupted shoot apical meristem in otherwise morphologically normal embryos and seedlings; severe alleles of the *stm* mutant cause absence of the shoot apical meristem in the embryo even though other organs, such as cotyledons, hypocotyl, and root, are formed. An extreme defect observed in *zll* mutant embryos and seedlings is a flat apex composed of cells that lack features of a meristem, although *zll* mutant phenotypes exhibit variability ranging from the presence of an enlarged shoot apical meristem to its complete absence (Fig. 3.3a–d). As initiation of the meristem is blocked close to the torpedo-shaped stage of mutant embryos, no leaves develop at the junction between cotyledons in the seedling phenotypes of *stm* and *zll* mutants (Barton and Poethig 1993; Endrizzi et al. 1996; Moussian et al. 1998; Lynn et al. 1999). The amino acid sequence of the *STM* gene shows a remarkable degree of homology to the homeodomain of the *KN1* gene of maize, implying its role as a transcriptional regulator in shoot apical meristem development. Expression of the *STM* gene first becomes evident in a few cells slightly displaced from the center of the globular embryo, later spreading to both sides of the embryo; from the heart-shaped stage onwards, expression is detected in the shoot apical meristem as a continuous band between cotyledons (Long et al. 1996; Long and Barton 1998). However, beyond confirming that the *STM* gene is required for shoot apical meristem formation during embryogenesis and for the continued functioning of the meristem during the vegetative growth of the plant, the expression pattern of gene transcripts in wild-type embryos has not provided a clear picture of the function of this gene. One view is that the gene is required in the central region of the meristem to inhibit differentiation of stem cells, whereas in the peripheral region it functions to prevent organ outgrowth (Long and Barton 1998). The ultimate fate of the central and peripheral domains of the shoot apical meristem, the former constituting the stem cells and the latter serving as the presumptive sites of cotyledons and leaf primordia, is also reflected in the spatio-temporal patterns of expression of transcripts of other molecular markers in these regions of the shoot apical meristem of developing embryos. A particularly interesting example is provided by the *AS1* gene, which encodes a MYB-domain transcription factor (MYB is a recognition site in the genome identified with myeloblastosis-associated viruses). As mentioned in Chap. 2, transcripts of the *AS1* gene are expressed in the peripheral domains of developing embryos corresponding to the cotyledons. That this gene has another pivotal function in shoot apical meristem organization became evident in *stm/as1* double mutants in which the meristem appears normal without suffering the defects of the *stm* mutant. Given that the typical phenotype of the *stm* mutant is suppressed by the loss of activity of the *AS1* gene, one model that has been proposed is that the *STM* gene maintains the undifferentiated state of cells in the central region of the shoot apical meristem by negatively regulating the *AS1* gene (Byrne et al. 2000). By preventing the differentiation of cells of the central region of the shoot apical meristem as stem cells, the *ZLL* gene appears to play a role similar to that of the *STM* gene. Because the *ZLL* gene encodes a relatively novel protein with undefined functions, it is not far-fetched to assume that this protein may serve as a determinant of the stem cell fate of the shoot apical meristem by signaling positional information, or in the translational regulation of a specific mRNA subset required for the development of the meristem (Moussian et al. 1998; Lynn et al. 1999). A genetic analysis of *zll/wus* and *zll/clv* double mutants has led to the suggestion that stability of the shoot apical meristem in *Arabidopsis* is driven by *ZLL* gene activity, which ensures the availability of the required critical cell number (Moussian et al. 2003).

Since gross defects in the organization of the shoot apical meristem of the *stm* mutant are more or less similar to those displayed by the *wus* mutant, some attention has been paid to determining how *STM* and *WUS* gene functions are integrated in the formation of the meristem. A variety of experiments have suggested that the *WUS* and *STM* genes play independent and complementary roles involving stem cell specification by the former and suppression of differentiation of stem cells in the embryo apex by the latter (Gallois et al. 2002; Lenhard et al. 2002).

The *CUP-SHAPED COTYLEDON* genes (*CUC1–CUC3*) have also facilitated fine-tuning of our understanding of shoot apical meristem initiation in *Arabidopsis* embryos. Although the effect of mutational inactivation of *CUC1* and *CUC2* genes singly is very weak, double mutations not only impair development of the embryonic shoot apical meristem, but also cause defects in the separation of cotyledons in embryos and of sepals and stamens in flowers (Aida et al. 1997; Vroemen et al. 2003). In addressing the question of how *CUC* genes control shoot apical meristem formation and cotyledon separation, a comparison of the expression patterns of *CUC1–CUC3* and *STM* gene transcripts in the wild-type *Arabidopsis* embryos has proved informative. Suggestive of a role for *CUC1* and *CUC3* in the initiation of the embryonic shoot apical meristem, transcripts of these genes are detected earlier than those of *CUC2* and *STM* genes in a few cells of the globular embryo predicted to form the shoot apical meristem. However, there is an overlap of expression of the four genes in the early stages of embryogenesis as expression of *CUC2* and *STM* genes follows closely on the heels of that of *CUC1* and

CUC3 in the same domain of the globular embryo. Later, in the bent-cotyledon stage embryo, coincident with the bulging of the developing shoot apical meristem, *CUC* gene transcripts bypass the center of the meristem and appear in a region surrounding it. This is consistent with the assumed role of these genes in cotyledon separation by repressing bulging in the boundary between the cotyledons, but contrasts with the expression of the *STM* gene, which becomes restricted to the site of the presumptive shoot apical meristem. Indicative of a requirement for *CUC* genes in the separation of the shoot apical meristem and cotyledons, *CUC* transcripts are also expressed in the boundary between these parts of the developing embryo. Based on the observation that the *STM* gene is not expressed in *cuc1/cuc2* double mutants, it is believed that *CUC* genes probably promote transcription of the *STM* gene in the embryonic shoot apical meristem of *Arabidopsis* (Aida et al. 1999; Takada et al. 2001; Vroemen et al. 2003). Through their control of auxin distribution, two other genes, *MP* and *PIN1*, have also been implicated in the regulation of expression of *CUC* genes in patterning of the shoot apical meristem and cotyledon separation in the *Arabidopsis* embryo (Aida et al. 2002). In the current picture, the main players in the molecular organization of the shoot apical meristem in *Arabidopsis* are the *WUS* gene, whose expression marks meristem initiation in the globular embryo, followed by the *CUC* genes, and finally, the *STM* gene, whose expression is activated by the *CUC1* gene (Sharma and Fletcher 2002).

As alluded to earlier, central to the functioning of the shoot apical meristem close to immortality is the maintenance of a reservoir of stem cells that are available for ongoing organogenesis throughout the life of the plant. Although this is a problem to be reckoned with during the post-embryonic development of the plant, the stage is set during early embryogenesis by the activation of a meristem signal transduction pathway. The *CLV* family of genes, represented by *CLV1*, *CLV2*, and *CLV3*, has been assigned the primary role of promoting stem cells to make the transition into the differentiated state, so that they become recruited into potential leaf or flower primordia at the periphery of the meristem; this view is supported by the observation that mutations in the *CLV* loci result not only in delayed organ initiation at the shoot apical meristem, but also in a disproportionate increase in size of the meristem due to the accumulation of surplus stem cells. For example, the shoot apical meristem of *clv3* mutant embryos is visibly larger than wild type embryos due to the accumulation of excess stem cells, and, in plants poised to flower, the meristem attains an increase in volume of almost 1,000-fold (Clark et al. 1993, 1995; Kayes and Clark 1998). Increases in size of the embryonic and post-embryonic shoot apical meristems of *mgoun* (*mgo1* and *mgo2*) mutants are of smaller magnitude than those of *clv* mutants (Laufs et al. 1998). Although there are no abnormalities in the development of the shoot apical meristem in the *poltergeist* (*pol*) mutant, double mutant analyses have revealed that the *pol* mutation suppresses the phenotypes of all *clv* mutant alleles and enhances the effect of the *wus* mutation on meristem development; the *POL* gene probably promotes the undifferentiated state of cells in the center of the meristem (Yu et al. 2000). The *POL* gene encodes a protein phosphatase 2C (PP2C) with a nuclear-localization motif. Because the gene is expressed not only in the shoot and floral meristems of *Arabidopsis*, but also in many other tissues of the plant, it is justifiably considered as a common regulator of multiple signaling pathways (Yu et al. 2003). Despite the fact that mutations in *EXTRA COTYLEDON1* (*XTC1*) and *XTC2* genes perturb the globular to heart-shaped transition of embryos, the shoot apical meristem becomes visibly large in mature mutant embryos, suggesting that development of the embryo proper and that of the shoot apical meristem are independently regulated (Conway and Poethig 1997).

Genetic and molecular studies of interactions between the *CLV*, *STM*, and *WUS* genes have provided new information about the regulatory pathways that oversee the integrity of the shoot apical meristem and coordinate the rates of cell division in the stem cell population, and the timing of organ formation in the meristem. An important component of this pathway, which provides the all-important balance between proliferation of stem cells and their differentiation at the shoot apical meristem, is controlled by the *CLV* genes, which restrict the stem cell population and are in turn subject to negative regulation by the stem cell-promoting pathway. How this is accomplished remains a mystery but some pointers have been gained as a result of cloning of the three

CLV genes, analyzing their expression patterns, identifying their protein products, and various other transgenic and biochemical approaches. The *CLV1* gene encodes a putative receptor-like kinase, which might provide critical extracellular signaling cues through activation by a peptide ligand. Indeed, the deduced amino acid sequence of the *CLV3* gene as a low molecular mass secreted protein, and the expression of transcripts of *CLV1* and *CLV3* genes first detected in a packet of cells between the developing cotyledons of heart-shaped embryos, and post-embryonically in a limited zone of cells in the summit of the shoot apical meristem, have pointed to the CLV1 and CLV3 proteins as a candidate receptor and ligand pair in the signaling pathway that extends between the three layers in the central zone of the shoot apical meristem; however, uncertainty still surrounds the role in the signal processing complex of the receptor-like protein encoded by the *CLV2* gene (Clark et al. 1997; Fletcher et al. 1999; Jeong et al. 1999). Other related investigations have suggested a role for the CLV3 protein alone as a ligand secreted from stem cells that binds to the CLV1-CLV2 receptor complex to activate signaling events in the meristem (Rojo et al. 2002; Lenhard and Laux 2003), and for CLV2 and CLV3 proteins in the assembly of an active CLV1 protein signaling complex by interaction with a kinase-associated protein phosphatase (KAPP) and a Rho-like GTPase (Rop) protein (Trotochaud et al. 1999). The interaction of a CLV1 fusion protein with KAPP in vitro and in vivo, and analysis of transgenic plants expressing the *KAPP* gene, have indicated that the latter may function as a negative regulator of the *CLV1* signal transduction pathway in the development of the shoot apical meristem (Williams et al. 1997; Stone et al. 1998). Overexpression of the *CLV3* gene in transgenic *Arabidopsis* has made it possible to assign a definitive role for this gene in negatively regulating the stem cell-promoting pathway, and in signaling through a CLV1/CLV2 receptor kinase complex. Transgenes overexpressing the *CLV3* gene are unable to generate a continuous supply of stem cells in the shoot apical meristem, as seen by their inability to produce lateral organs regularly during the vegetative and reproductive phases in the life of the plant. It was also found that overexpression of the *CLV3* gene in a *clv1* or *clv2* mutant background leads to the appearance of a typical *clv* mutant phenotype. This could occur only if a functional *CLV1/CLV2* gene complex is required for *CLV3* gene action (Brandt et al. 2000).

Of particular interest is the mechanism by which *STM* and *WUS* genes act antagonistically to *CLV* genes, yet their combined effects yield a functional meristem. Whereas the *STM* gene promotes the establishment and maintenance of the shoot apical meristem and *CLV1* and *CLV3* loci repress meristem proliferation, formation of the embryonic shoot apical meristem is not restored in double mutants carrying a weak *stm* allele and a *clv1* or *clv3* allele. This shows *STM* gene activity is required for meristem proliferation when *CLV* genes are knocked out, and that suppression of meristem development in the *stm* mutant depends on *CLV* gene activity. These results suggest competing, but closely related functions for *STM* and *CLV* genes on a common cellular target in regulating the development of the shoot apical meristem as it undergoes proliferation and organogenesis during the life of the plant beginning with the embryo (Clark et al. 1996). On the other hand, embryo apices of *wus/clv1* and *wus/clv3* double mutants are indistinguishable from those of *wus* single mutants; given the opposite phenotypes of *wus* and single *clv* mutants, this indicates that *CLV* genes act by negatively regulating the *WUS* gene. This view is also supported by the ectopic *WUS* gene transcript expression observed in an enlarged domain of *clv* mutant embryos and by the absence of *WUS* gene transcripts in the arrested shoot apical meristem of transgenic plants overexpressing the *CLV3* gene (Brandt et al. 2000; Schoof et al. 2000). Thus, a sufficiently diverse array of interactions between a set of three major gene products involved in the maintenance of the shoot apical meristem in *Arabidopsis* seems to exist. The functional analysis of another gene, *SHEPHERD* (*SHD*), shows that its protein product might interact with the CLV complex, perhaps by activating the CLV1/CLV2 receptor and/or CLV3 ligand. The basis for this view is the observation that although the *shd* mutation inflicts the same defects in the shoot apical meristem of *Arabidopsis* as weak and intermediate *clv* alleles, *wus/clv* and *wus/shd* double mutants have phenotypes identical to those of *wus* single mutants. Because the *SHD* gene encodes an ortholog of the mammalian GRP94 chaperone protein, the latter is believed to interfere in the proper conformation of the CLV protein (Ishiguro et al. 2002).

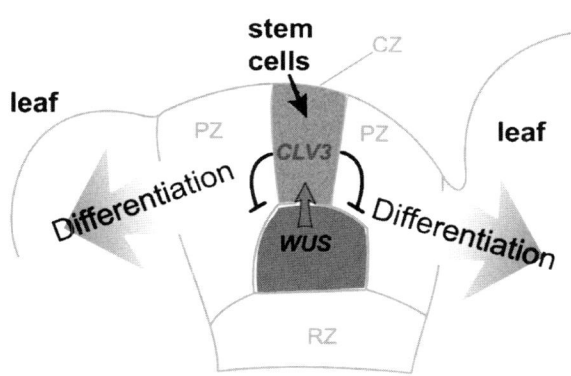

Fig. 3.4 A model illustrating the interaction between *CLV3* and *WUS* genes to maintain shoot meristem identity in the *Arabidopsis* embryo. An organizing center requiring *WUS* gene activity specifies stem cell identity by signaling (*small arrow*). The stem cells restrict the range of *WUS* gene expression by *CLV3* signaling. Cells that have passed the boundary defined by *CLV3* function become the founder cells for leaves which differentiate at the flanks of the shoot apical meristem. *CZ* Central zone, *PZ* peripheral zone, *RZ* rib zone. (Reprinted from Schoof et al. 2000)

Since *WUS* gene expression appears necessary for the enlarged phenotype of the shoot apical meristem in *clv* mutants, a model has been proposed incorporating an interaction of *CLV3* and *WUS* genes to keep the meristem in a dynamic equilibrium between a rising stem cell population at its summit and a population of cells displaced toward its periphery as organ primordia (Fig. 3.4). The core of this proposal is that *WUS* gene expression confers stem cell identity in the shoot apical meristem, which in turn promotes *CLV3* gene expression. The secreted protein product of the *CLV3* gene subsequently interacts with the CLV1/CLV2 receptor complex to complete the signaling cascade. In reality, when *CLV3* activates the signal transduction pathway, this scenario predicts a corresponding down-regulation of *WUS* gene expression to maintain a group of permanent stem cells in the meristem throughout plant life. This model has the hallmarks of a regulatory feedback loop between the stem cells and the stem cell organizing center in the shoot apical meristem mediated by the activities of two sets of genes (Schoof et al. 2000; Waites and Simon 2000). Using complementary strategies, Lenhard and Laux (2003) have provided additional molecular insights by showing that CLV3 protein functions as a mobile intercellular signal by repressing *WUS* gene transcription in the stem cells, and by spreading laterally in a controlled fashion by binding to CLV1 protein. Given that the availability of the *CLV3* gene is critical for regulation of the *WUS* gene to control the size of the stem cell population in the shoot apical meristem, a role for some other gene to regulate *CLV3* expression becomes obvious. As shown by the work of Brandt et al. (2002), unexpectedly, this gene turns out to be none other than *STM*. Results of an investigation on the role of the newly identified *CORONA* (*CNA*) gene in stem cell specification in the shoot apical meristem of *Arabidopsis* seedlings are in conflict with the above model based essentailly on *WUS* gene expression in the *clv* mutant background (Green et al. 2005).

In their failure to produce the shoot apical meristem, the *no apical meristem* (*nam*) mutant embryos and seedlings of *Petunia* (Souer et al. 1996) and the *shootless* (*shl*) mutant embryos of rice (Satoh et al. 1999) resemble their counterparts in *Arabidopsis*. Based on anatomical observations of developing embryos and in situ hybridization analysis with gene transcripts, the *NAM* gene is considered to be associated with determination of the position of the shoot apical meristem and leaf primordia. The protein encoded by the *NAM* gene, which has a highly conserved N-terminal domain, belongs to a family that includes several other *Arabidopsis* proteins, including the proteins encoded by the *CUC* genes, which are now considered as transcription factors (Souer et al. 1996; Aida et al. 1997; Taoka et al. 2004). Of the four *SHL* (*SHL1–SHL4*) genes studied in detailed as indispensable for initiation of the shoot apical meristem in embryos of rice, *SHL1* and *SHL2* appear unusual as they are required for both initiation and maintenance of the meristem (Satoh et al. 2003). The evidence that supports this notion is the observation that in weak alleles of *shl1* and *shl2* mutants, the shoot apical meristem is gradually consumed by leaf primordia at the same time as the indeterminate cells of the meristem that express *OSH1* gene transcripts are reduced in number. Embryos of a *shootmeristemless* (*sml*) phenotype described in maize display a disrupted shoot apical meristem coupled with the failure of scutellum elongation and coleoptile differentiation. Genetic analysis has attributed the failure of the shoot apical meristem to the synergistic interaction of mutations at the *sml* and an unlinked *distorted growth* (*dgr*) loci (Pilu et al. 2002).

In summary, although it is apparent that only a

few genes involved in the specification and functioning of the shoot apical meristem have been identified, even in its unfinished state, the available information has provided tantalizing glimpses into the mechanism that keeps the meristem in a functional mode during the life of the plant.

3.1.2
Root Apical Meristem

Unlike the shoot apical meristem, the root apical meristem is bipolar in nature, cutting off cells both distally and proximally. The distal derivatives form the root cap, whereas the proximal derivatives differentiate into the mature tissues of the root. Furthermore, the presence of a group of cells in the root apical meristem known as the quiescent center, which divide rather infrequently or not at all, presents this meristem with an unusual cytological feature. The root apical meristem has a distinct advantage over the shoot apical meristem for cell lineage studies because the continuous, monotonous files of cells formed can be traced to repeated transverse divisions of layers of meristematic cells. Despite their differences, one recurrent theme is that the shoot and root apical meristems of flowering plants maintain the same basic structural innovation of layered cells responsible for the elaboration of the above-ground and underground architecture of the plant, respectively. Analysis of transgenic tomato (*Lycopersicon esculentum*; Solanaceae) carrying maize transposable elements has identified a single gene, *DEFECTIVE EMBRYO AND MERISTEMS* (*DEM*) that functions in the maintenance of both shoot and root apical meristems. Consistent with the purported role of this gene, embryos and seedlings of *dem* mutants are found to lack organized shoot apical and root apical meristems. *DEM* gene encodes a novel 72 kDa protein; whether this protein is an essential link to ensure that some basic aspect of regulation of the shoot and root apical meristems follows the same signaling pathway remains unclear (Keddie et al. 1998).

The embryonic root apical meristem of *Arabidopsis* is composed of four sets of initials or stem cells (see Plate 6, Fig. a–e). Of these, a distal plate of 12 initial cells (columella initials) and a block of 4 central cells that comprise the quiescent center have their origin in the hypophysis, whereas cells derived from the embryo proper form the core initials of the remaining root tissues such as the epidermis, cortex, endodermis, pericycle, vascular tissues, and lateral root cap. The columella initials are constituted of a peripheral ring of eight cells surrounding an inner core of four cells and periclinal divisions of the columella initials give rise to the columella or cells of the central region of the root cap. The lateral root cap cells that envelop the columella and the root epidermal cells are derived by periclinal divisions of a ring of about 16 initial cells (epidermal-lateral root cap initials) that surround the peripheral columella initials. The cortical-endodermal cells can be traced to a ring of eight initials (cortex-endodermal initials) encircling the central cells that undergo periclinal divisions to generate the endodermis on the inside and the cortex on the outside. Finally, the cells of the vascular cylinder and the pericycle are derived from a proximal plate of initials (pericycle-vascular initials) located above the central cells (Dolan et al. 1993, 1994).

Clonal analysis of the *Arabidopsis* seedling root has provided strong hints that derivatives from the central cells give rise to the pericycle, columella and the vascular cylinder; in this sense, the central cells can be considered to act as stem cells or as a stem cell reservoir analogous to the central zone of the shoot apical meristem. In a different sense, this observation emphasizes the importance of positional information in the determination of cell fate in the root apical meristem (Kidner et al. 2000). In support of the role of the quiescent center as stem cells, it has been shown that a *WUS*-type gene isolated from rice is specifically expressed in a few cells located in the basal region of the developing embryo of rice prior to the differentiation of the radicle, and that the localized expression is continued into the quiescent center during growth of the radicle (Kamiya et al. 2003a).

Although the phenomenon of mitotic quiescence in a group of cells in the root apical meristem has stimulated great interest, no insights have been generated into the mechanism by which cells are forced to lapse into quiescence. The model of origin of the quiescent center from the hypophysis has been supported by in situ hybridization experiments using ^{3}H-polyuridylic acid [^{3}H-poly(U)] as a probe to monitor the development of the quiescent center during embryogenesis in *Capsella bursa-pasto-*

ris. In early-stage embryos, autoradiographic silver grains are localized in all cells of the presumptive root apex except in the hypophysis. As the inner cell formed by the transverse division of the hypophysis cuts off cells proximally, these cells remain characteristically unlabeled in contrast to the labeled cells of the rest of the embryo. As would be expected of the quiescent center, the same group of cells of the seedling root of *C. bursa-pastoris* that fail to bind ^{3}H-poly(U) do not incorporate ^{3}H-thymidine (Raghavan 1990). The origin of the embryonic root from two clonal groups of cells derived from the small terminal cell (which gives rise to the embryo proper) and the large basal cell (from which the hypophysis is formed) of the zygote accords with the view that certain poorly understood cell-cell interactions coordinate this process. This view has been highlighted by monitoring the changes in cell fate evoked by interference with signal transduction from specific cells of the root meristem of *Arabidopsis* seedlings. A typical example of this approach involves laser ablation of a single cell of the four-celled quiescent center, hence forcing a decision by the columella initial adjacent to the destroyed cell to either divide and produce another cell of its own kind or differentiate into a mature columella cell. The columella initial chooses the latter option and differentiates into a mature columella cell by the accumulation of starch. Laser damage to a cell of the quiescent center also alters the fate of the cortical-endodermal initial in contact with the damaged cell, promoting its differentiation into cortical and endodermal cells by an asymmetric periclinal division instead of the usual anticlinal division to produce another cortical-endodermal initial cell. In contrast, the initials in contact with the surviving cells of the quiescent center undergo their normal divisions. These observations on the role of cellular interactions in the formation of the root apical meristem provide indirect evidence indicating that the developmental fate of cells of the meristem, including the stem cell properties of initials, is intimately determined by contact-dependent signals from cells of the quiescent center (van den Berg et al. 1997). It was also found that laser killing of the complete quiescent center prompts cells of the proximal vascular tissue to occupy the position of the dead cells, and it has been claimed that, in their new position, the vascular tissue cells can reform the cells of the quiescent center (van den Berg et al. 1995). Another experiment revealed that ablation of the cortex-endodermal initials results in a response in the adjacent pericycle cells, which attempt to restore the initial pattern: the available space of the ablated cells is occupied by pericycle cells, which divide and generate a file of cortical cells and endodermis in accordance with their new positions. Similarly, the neighboring cortical cells invade the space previously occupied by the ablated epidermal-lateral root cap initials and function as progenitors of the epidermis and root cap cells. Thus, pericycle and cortical cells are not predetermined to become the pericycle or the cortex. The importance of a signaling cascade by which cells interpret their positional information to differentiate into appropriate cell types in the root apical meristem is evident from these results, as switches in cell specification occur in spite of the largely invariant cell lineages of the meristem (van den Berg et al. 1995). The root apical meristem of *Arabidopsis* presents itself as the very epitome of a group of cells of which some differentiate according to their position relative to other cells, which are themselves differentiating in different ways.

The sum of the above investigations indicates that stem cells actually surround the quiescent center and, in this sense, by maintaining the stem cell fate in the surrounding cells, the quiescent center can, as suggested by Kidner et al. (2000), be appropriately considered as the root apical meristem-equivalent of the stem cell organizer identified in the shoot apical meristem. One corollary of this ostensible similarity in meristem maintenance in the shoot and root is the need to explore the regulatory pathways of the stem cell population in the root apical meristem and narrow down the candidate receptor proteins in both meristems. With this in mind, Casamitjana-Martinez et al. (2003) have shown that overexpression of a *CLV*-like gene (*CLE19*) in *Arabidopsis* roots restricts the size of the root apical meristem, resulting in loss of meristematic cells. Since this occurs without interfering with the quiescent center and stem cell maintenance, it appears that, in contrast to the CLV pathway in the shoot apical meristem, the CLE protein might target differentiating cells rather than the quiescent center. In extending this work, it was found that that two novel genes, *SUPPRESSOR OF LLP1* (LIGAND LIKE PROTEIN1)

(*SOL1*) and *SOL2* completely suppress the meristem defect inflicted by the *CLE19* phenotype. This fascinating twist to the story is notable because the *SOL1* gene encodes a putative Zn^{2+}-carboxypeptidase that is expressed in both shoots and roots and is thus a potential component of the CLV signaling pathway. Although it appears that, at the very least, a CLV-like pathway is required for meristem maintenance in the root, there is a long way ahead before the complexities of this pathway are fully revealed.

Histochemical localization of free auxin in individual cells of the root apical meristem of *Arabidopsis* seedlings by means of a reporter gene under the control of an auxin-responsive regulatory sequence has revealed that, after ablation of the quiescent center, an auxin peak is established in the vascular tissue cells that become respecified as the quiescent center. From this work, which suggests that cellular organization of the root is modulated by localized auxin concentrations, it seems likely that auxin is one member of a chain of molecules that conveys and registers positional information affecting the organization of the quiescent center (Sabatini et al. 1999). According to Friml et al. (2002), the *A. thaliana PIN4* (*AtPIN4*) gene, which encodes a member of the putative auxin efflux carriers, is an important mediator of patterning of the embryonic root. The basis for this conclusion is the finding that an apical shift in the auxin response maximum visualized by a reporter gene construct in early-stage *Atpin4* mutant embryos is correlated with premature and abnormal cell divisions in the hypophysis, quiescent center, and columella precursors. Compared to expression of a quiescent center marker confined to just four cells of the wild-type embryo, the mutant embryo expresses this marker in a much broader domain in the root apex. These results favor an important role for the AtPIN4 protein in channeling auxin through the root meristem, which evidently leads to correct patterning at the root apex. Two new additions to the list of genes that mediate in the specification of the quiescent center in the *Arabidopsis* embryonic root meristem are the transcription factor-encoding *PLETHORA1* (*PLT1*) and *PLT2*. These genes are also transcribed in response to auxin accumulation in the distal part of the root meristem, and induce varying degrees of formation of root tissues in developing embryos, including root primordia complete with a quiescent center (Aida et al. 2004). According to the most recent view, *PLT* gene expression is not straightforward – nature has come up with its own solution to keep this gene in check. This is accomplished by restricting the action of the *PLT* gene by a network of five *PIN* genes (*PIN1*–*PIN4* and *PIN7*) to define the stem cell region of the root apex. The interaction is completed when the auxin flux necessary for the patterning of the root apex occurs by the action of *PLT* genes controlling *PIN* gene transcription (Blilou et al. 2005).

Earlier studies of *Arabidopsis* mutants had revealed that determination of cell fate in the root apical meristem is under the control of additional genes that are activated during early stages of embryogenesis. The *hobbit* (*hbt*) mutant was isolated by virtue of its failure to form a functional root meristem, in particular, a recognizable quiescent center and a columella in the root cap, both derived from the hypophysis, and lateral root cap cells. The first deviation from the wild-type embryo observed in the mutant is the occurrence of a vertical division in the hypophysis at the four- or eight-celled stage of the embryo instead of a transverse division. This defect, which persists throughout embryogenesis as evidence of further atypical divisions in the hypophysis, suggests that differentiation of the quiescent center and columella requires *HBT* gene activity (Willemsen et al. 1998). However, identification of the protein product of the *HBT* gene, which encodes a subunit of the anaphase-promoting-complex (APC) involved in regulating cell cycle progression, provides no clues to its specific role in patterning of the root apical meristem (Blilou et al. 2002). In the *bdl* mutant, which also fails to form the primary root meristem, embryos deviate from normal development as early as the two-celled stage, in which the terminal daughter cell undergoes a horizontal instead of a vertical division. This is also accompanied by failure of the potential hypophysis to form the quiescent center and columella. Transcripts of the *BDL* gene, which encodes an auxin-response protein, are expressed in the cells of the embryo proper but not in the hypophysis (Hamann et al. 2002). Overall, the results from laser surgery, mutant analysis, and in situ localization studies have provided a fruitful framework involving signal transduction for establishing the root apical meristem during the early stages of embryogenesis. The model is based on initial signaling from the progeny of the apical daughter cell of the zygote to fix the fate of the uppermost cell of the suspensor as the hypophysis, and a later signal from the quiescent center to the adjacent cells derived from the apical daughter cell of the zygote to maintain their fate as undifferentiated root meristem initials. The identification of the *BDL* gene product as an auxin-response factor was an exciting step indicating that the signal relayed from the apical cell of the zygote might be auxin (Hamann et al. 1999, 2002). If cells of the developing embryo can detect differences in their neighbors, and respond to these differences, it is not far-fetched to suppose that such a mechanism could be used to generate the intricate cellular pattern in the highly dynamic meristem of the embryonic root.

3.2 Genetic and Molecular Control of Embryo Pattern Formation

In concert with cell lineage studies, identification of genes – and their protein products – that are activated during pattern formation in developing embryos is an important step in analyzing the mechanism underlying the allocation of cell fate along pattern-forming pathways. The work of Mayer et al. (1991) generated the first new wave of interest in the genetic control of apicobasal and radial patterning elements in embryos of flowering plants, leading to phenotype-based gene discovery. The investigation involved screening zygotic mutations of *Arabidopsis* that delete specific cell lineages in the embryos and cause chaos in the formation of identifiable embryonic regions in the seedlings. The basic strategy behind this approach is the assumption that mutations in pattern-forming genes allow embryogenesis to proceed to completion, but inflict diagnostic defects in the body organization of the seedlings; the origin of the defect could then be traced back to deviations in the division patterns of the wild-type embryo. The information generated by this work and several other studies that followed in subsequent years is that obvious differences in the apicobasal and radial patterns in the embryo axis of mutants are reflected in lesions in the different classes of genes that generate these patterns with exquisite precision and fidelity. These investigations also revealed an advantageous note that genes involved in pattern formation in plant embryos may number only a few rather than hundreds.

3.2.1 Apicobasal Patterning of the Embryo

The apicobasal axis of a seedling plant is divided into four major component parts: the shoot, cotyledons, hypocotyl, and root. With a little overlap, four mutant classes that give phenotypes with defects in the apical (shoot apical meristem and cotyledons), central (hypocotyl), basal (hypocotyl and root), and terminal (shoot and root apical meristems) parts define genes that control the apicobasal organization of the embryo. Evidence that the shoot apical meristem and cotyledons are specified mostly, if not exclusively, by a single gene has come from analysis of *gurke* (*gk*) mutants. Cotyledons are lacking in phenotypes of all *gk* alleles, but the most striking phenotype is seen in a few strong alleles in which the entire shoot apex, cotyledons, and part of the hypocotyl are obliterated. Ontogenetic studies showed that *gk* phenotypes result from the failure of the organized cell divisions in the globular/heart-shaped embryo that initiate cotyledons, rather than from the death of cells already formed (Torres-Ruiz et al. 1996). Identification of the *GK* gene product as acetyl-CoA carboxylase, which catalyzes malonyl-CoA synthesis, has led to the view that metabolites related to malonyl-CoA are probably required to specify the apical part of the embryo (Baud et al. 2004; Kajiwara et al. 2004). A variety of phenotypes affecting the shoot have been described in seedlings carrying the temperature-sensitive *topless1* (*tpl1*) mutation; perhaps the most striking phenotype is seen at a high temperature, where the embryo forms an apical root in place of the shoot apical meristem and cotyledons. As a definitive indication of the transformation of the shoot into a root primordium, the mutational change is associated with concomitant expression of root marker genes in the apical pole of the embryo (Long et al. 2002).

In *pas* mutants, the hypocotyl remains short and thick, thus preventing curvature of the embryo in the seed. Mutant embryos are also characterized by defects in cotyledon initiation (Faure et al. 1998). Similar to the *GK* gene, a *PAS* gene has been shown to encode an acetyl-CoA carboxylase (Baud et al.

2004). A gene with the spatially restricted task of hypocotyl specification is *FACKEL* (*FK*), as mutations in this gene give rise to seedlings in which the cotyledons are directly attached to a short root. Seedling phenotypes of some *fk* mutants that delete the central portion of the axis harbor other defects such as the formation of abnormal cotyledons and the presence of more than two cotyledons. The mutant phenotype has been traced to cytokinetic defects such as enlarged cells, random orientation of cell divisions, and incomplete cell walls beginning with globular stage embryos, resulting in disorganized misshapen embryos (Schrick et al. 2000). Simultaneous investigations undertaken in two different laboratories have shown that *fk* is a sterol biosynthetic mutant, and that the *FK* gene encodes a protein belonging to the sterol reductase family. Biochemical analysis showing that the *fk* mutation causes a lesion in the pathway of C-14 sterol reductase synthesis has given some insight into how the biochemical defect is translated into phenotypic abnormalities (Schrick et al. 2000; Jang et al. 2000). Other *fk*-like phenotypes displaying short hypocotyl and root and malformed cotyledons in the seedlings, abnormal embryo development, and defects in the sterol biosynthetic pathway are *sterol methyl transferase1* (*smt1*; Diener et al. 2000), *cephalopod* (*cph*; Schrick et al. 2002), *hyd1*, and *hyd2* (Topping et al. 1997; Schrick et al. 2002; Souter et al. 2002). The results of these studies are compatible with the notion that a novel sterol signaling pathway is involved in the cell-cell communication necessary for organized cell growth during apicobasal pattern formation in the embryo (Schrick et al. 2000, 2002; Jang et al. 2000). As integral membrane components, and as biosynthetic precursors of various steroid hormones, sterol molecules are known to play a role in animal embryogenesis, but are somewhat new to the field of plant embryo development, whatever their function might be.

Mutants with impaired capacity in the production of both hypocotyl and root are *mp*, *bdl*, and *auxin-resistant6* (*axr6*). The former two mutants were isolated based on their severely distorted seedling phenotypes, which end basally in a conical structure, whereas the *axr6* mutant appeared as a heterozygote based on its resistance to 2,4-D (Berleth and Jürgens 1993; Hamann et al. 1999; Hobbie et al. 2000). Features that characteristically distinguish mutant embryos from the wild-type are confined to a narrow developmental window between the octant, or even the two-celled stage, and the heart-shaped stage. Although the octant stage of the *mp* mutant embryo consists of four tiers of cells compared to two in the wild-type, the descendants of the lower tier(s) fail to undergo the stereotyped division pattern and oriented expansion that produce the elongate cell files characteristic of the wild-type embryo as it phases out into the heart-shaped stage. Additionally, unlike the hypophysis of the wild-type embryo, the hypophysis in the *mp* mutant forms a central column of cells by transverse divisions. Consistent with the failure of the lower tier(s) of cells of the heart-shaped embryo to produce elongate files of cells, the vascular system of the mutant seedling is impaired, probably due to inhibition of polar auxin transport (Berleth and Jürgens 1993). The existence of cellular interactions for differentiation evident in this last observation implies the provision for signaling mechanisms. This view has led to the assignment of multiple roles for the *MP* gene as a signal transducer for establishing an axis (axialization) in the embryo, development of the vascular system, and polar auxin transport (Przemeck et al. 1996). With somewhat similar aberrant division patterns, embryos of *bdl* and *axr6* mutants also fail to form the apical and basal cell files that give rise to the hypocotyl and root, although in both mutants the first defective division occurs as early as the two-celled embryo stage (Hamann et al. 1999; Hobbie et al. 2000). As described in Chap. 2, unexpected support for a role for auxin transport in embryo pattern formation has come from pharmacological experiments using auxins and auxin transport inhibitors on cultured embryos of *Brassica juncea* (Liu et al. 1993; Hadfi et al. 1998).

The *MP* gene has been cloned and the predicted protein product is similar to AUXIN RESPONSE FACTOR 1, a transcriptional regulator that binds auxin-responsive promoter elements and is thought to mediate responses to auxin. The developmental profile of *MP* gene activity revealed by in situ hybridization shows that expression of this gene is very dynamic, with transcripts initially expressed in broad regions of the embryo later becoming confined to the vascular tissues, in harmony with the existence of possible genetic interactions or molecular signaling between the apical and basal parts of the embryo axis in the establishment of continuous files of vascular tissues (Hardtke and Berleth 1998). The *MP* gene could thus harbor a considerable amount of information on the coordination of vascular- and embryo body-pattern formation, but further work will be required to support this important notion. As mentioned above, the protein product of the *BDL* gene is an auxin response protein. Hellmann et al. (2003) have shown that the *AXR6* gene encodes the SCF [for SKP1 (SUPPRESSOR OF KINETOCHORE PROTEINS1)/CDC53 (or CULLIN), *F*-box protein] subunit of CULLIN 1 (CUL1) protein. Since an important effect of auxin is to promote degradation of the short-lived auxin response proteins by the action of ubiquitin protein ligase SCF, the results imply that the embryonic *axr6* phenotype is due to a defect in SCF function. The convergence of the work on the cloning of *MP*, *BDL*, *HBT*, and *AXR6* genes has thus brought auxin into the fold of molecular embryogenesis as a major factor in the determination of the basal pole of the embryo and in the maintenance of cellular organization in the embryonic root.

The gene with the mutant phenotype most easily interpretable as causing defects in both apical parts of the *Arabidopsis* seedling is *GN*, also represented by the *EMBRYO-DEFECTIVE* (*EMB30*) locus. With failure of formation of both shoot and root apical meristems, mutant *gn* seedlings appear mostly cone-shaped with reduced root and cotyledons or, in strong alleles, as an undifferentiated mass of tissue with no apparent apicobasal axis. At the physiological level, developmental abnormalities have been attributed to defective establishment of polarity to direct auxin flow in cells along the apicobasal axis of the embryo. Cytologically, defects in the mutant lines have been precisely traced to the zygote, whose first division is deflected to produce two nearly equal cells, rather than two asymmetrical cells. The very first division of the apical cell born of partitioning of the zygote occurs in various planes, setting the stage for subsequent formation of the defective embryo by random divisions of daughter cells formed, without any contribution from the basal cell including the hypophysis. These observations support the view that the cellular targets of the *GN* gene are the asymmetric division of the zygote and the precise pattern of divisions of the hypophysis to form part of the root meristem. Since *mp/gn* double mutants have a *gn* phenotype and thus do not show the hypocotyl deletion characteristic of the *mp* mutant, *MP* gene function appears dependent upon prior *GN* gene action (Mayer et al. 1993; Steinmann et al. 1999). The GN protein shows partial sequence homology to yeast proteins YEC2, of unknown function, and SEC7 [a member of a new family of ADP-ribosylation factor (ARF) nucleotide exchange factors], that facilitates intracellular transport mediated by Golgi bodies (Shevell et al. 1994; Busch et al. 1996). Of particular interest will be the physiological and molecular mechanisms by which the *GN* gene product functions during early embryogenesis. A likely scenario involves streamlining of polar auxin transport in the embryo by regulating the dictyosome vesicle trafficking required for localizing auxin efflux carriers along the route (Steinmann et al. 1999). Suggestive of a dimerization reaction involved in the GN protein function in vesicular trafficking, a direct interaction between identical domains within the SEC7 subunit of the GN protein has also been demonstrated (Grebe et al. 2000). These observations have been integrated into a common model of action of the *GN* gene in other aspects of the mutant phenotype such as disorganization of the vascular tissues and the defects in auxin-induced lateral root formation seen during post-embryonic development of *Arabidopsis* (Geldner et al. 2004). Another model suggesting that the *emb30* mutation might cause defects in cell wall architecture is indicative of a broader function for the GN protein (Shevell et al. 2000).

In summary, it appears that in the partitioning of the embryo axis, the patterning genes might act in either a hierarchical or a combinatorial fashion. In the hierarchical scenario, *GK*, *FK*, and *MP* are considered to represent a cascade of genes that are deployed to specify the apical, central, and basal parts of the embryo, respectively. Before the hierarchical tier is activated, it is assumed that the apicobasal polarity of the zygote is initiated by the action of the *GN* gene. The pairs of complementary genes that might account for these same regions of the embryo are *GK* and *GN* for the apical, *FK* and *MP* for the central, and *GN* and *MP* for the basal regions (Mayer et al. 1991). Although the hierarchical pattern seems to have gained the upper ground based on the results of *gn/mp* double mutants, the combinatorial pattern cannot be ruled out. Because the humble auxin molecule has been implicated in a wide range of developmental effects in plants, it is not surprising that the experiments described in this chapter have elevated auxin to the status of a model molecule required for the maintenance of cellular organization in the embryonic root.

Apicobasal embryo pattern mutants with abnormalities in root and cotyledon development, including the formation of a single cotyledon, have also been isolated from pea (Johnson et al. 1994; Liu et al. 1995, 1999). One such mutant, designated as *cytokinesis-defective* (*cyd*), has an uneven surface with reduced cotyledons. An intriguing feature of the cells of mutant cotyledons is the failure of cytokinesis during division, resulting in multinucleate cells. Consistent with this observation, the cytokinetic defect is traced to the absence, or only partial formation, of cell plate, generating cell wall stubs (Johnson et al. 1994; Liu et al. 1995). Our current understanding of the role of the *CYD* gene in embryo pattern formation is that it is probably indirect, by extending the cell plate.

3.2.2
Radial Patterning of the Embryo

After the apicobasal axis of the embryo is established, the shoot and root apical meristems delimit the three sets of primary meristems of the protoderm, ground meristem, and procambium, which subsequently differentiate into the main tissues of the embryo axis to provide the stereotypical radial pattern of the embryo. Genetic screens have revealed that mutations in *Arabidopsis* specifically cause defects in tissue differentiation or deletions of specific cell layers in the embryonic organs; analysis of these mutants has provided genetic and molecular insights into the processes that determine the radial patterning of the tissues during embryogenesis. Two of the most informative and thoroughly investigated mutants of the group in which imperfections in cell differentiation can be traced to early-stage embryos are *knolle* (*kn*) and *keule* (*keu*). The severity of mutations varies in the seedling phenotypes, which appear mostly as round or tuber-shaped structures with a rough epidermis and lacking functional meristems in *kn* alleles, and as elongate axis topped by reduced cotyledons in *keu* alleles. Embryogenesis in the *kn* mutant is relatively normal up to the globular stage when the division delimiting the protoderm layer and the central core of cells goes awry and, consequently, embryos lack a well-defined epidermal layer. Interestingly, the seedling phenotype of the *keu* mutant uses a different strategy to regulate the radial pattern. Here the protoderm layer is detectable in some embryos, whereas in others it may be absent or incompletely formed and when present, the cells of the protoderm are abnormally swollen and irregularly arranged around a normal complement of inner cells. The cellular effects of both mutations are perhaps most vividly illustrated in the octant-stage and heart-shaped embryos during divisions that are generally anarchic, resulting in the formation of large multinucleate cells with gapped or incomplete crosswalls (Fig. 3.5a–f). These defects in the execution of cytokinesis may be construed to influence cell differentiation and embryo shape (Mayer et al. 1991; Assaad et al. 1996; Lukowitz et al. 1996; Sørensen et al. 2002). Thus, pattern defects in these mutants can be attributed indirectly to defects in cytokinesis because vesicles transported to the equator of the dividing cell do not fuse to form the cell plate. Several other *Arabidopsis* mutants to be described later in Chap. 5, along with *kn* and *keu*, which display defects in the orientation of the plane of division and in the execution of cytokinesis, are designated as cytokinesis-defective mutants. Since most cytokinesis-defective mutants are seedling lethal and do not cause any derangement of the radial pattern of embryos, they are not considered further here (see Nacry et al. 2000; Söllner et al. 2002). Suggestive of a possible link between the molecular functions of the *KN* gene and cytokinetic defects during embryogenesis resulting in the accumulation of unfused vesicles at the site of the cell plate, the predicted protein product of this gene is found to have similarity to syntaxins, members of a protein family known as SNARE (for soluble N-ethylmaleimide-sensitive factor attachment protein receptors). The SNARE complexes have been assigned important roles in membrane fusion events and in diverse vesicle trafficking pathways in eukaryotic cells, although the extent to which they contribute to the specificity of these processes is not fully determined. Immunofluorescence localization of the KN protein in dividing cells of embryos of *Arabidopsis* has implied participation of this protein in vesicle fusion preparatory to cell plate formation during cytokinesis, rather than in vesicle transport (Lukowitz et al. 1996; Lauber et al. 1997). The timing of KN protein accumulation confined specifically to the M phase of the cell cycle has been judged critical for membrane fusion and cell plate formation: in transgenic *Arabidopsis* lines overexpressing the *KN* gene under control of the CaMV 35S promoter, the protein is localized at the plasma membrane in nondividing cells, in contrast to its accumulation in the nascent cell plate of dividing cells (Völker et al. 2001). Although the KN-syntaxin specificity of cytokinesis in *Arabidopsis* embryos has been established by showing that most KN-related syntaxins do not rescue the *kn* mutant in a transgenic setting, how it is that only the KN protein facilitates vesicle fusion at the site of the future cell plate is unclear (Müller et al. 2003).

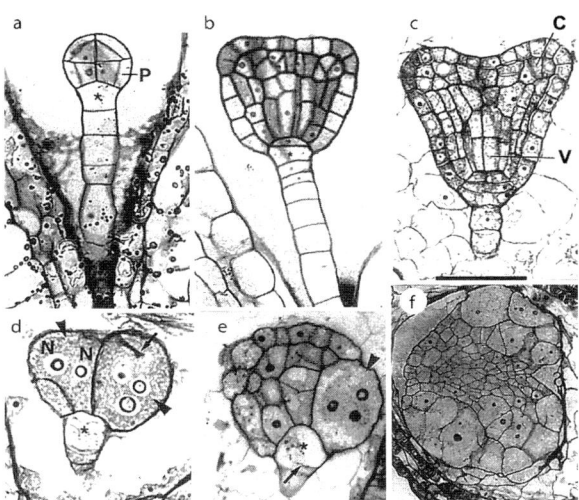

Fig. 3.5a–f Cytokinetic defects in embryos of the *keu* mutant of *Arabidopsis*. **a–c** Wild-type embryos. **a** Sixteen-celled embryo consisting of eight epidermal precursor cells and an inner core of eight cells. *Asterisk* Uppermost suspensor cell. **b** Triangular-stage embryo. *Asterisks* Two hypophyseal cells. **c** Heart-shaped embryo. **d–f** Mutant embryos corresponding to wild-type embryos. **d** Note the multinucleate cells (*arrowheads*) and an incomplete crosswall (*arrow*) in the mutant embryo corresponding to the 16-celled wild-type embryo. **e** Mutant embryo corresponding to the wild-type triangular-stage embryo. The epidermis is interrupted by a large multinucleate cell (*arrowhead*). Note the incomplete crosswall (*arrow*) in the uppermost suspensor cells (*asterisk*). **f** Mutant embryo corresponding to the heart-shaped wild-type embryo. The entire epidermis consists of large cells. *C* Cotyledon, *N* nuclei, *P* epidermal precursor cells, *V* vascular tissue. *Bar* in **c** 50 µm (applies also to **b**, **e**, **f**). **a** and **d** are at the same magnification. (Reprinted from Assaad et al. 1996)

The demonstration that *kn/keu* double mutants exhibit a much more profound defect in cytokinesis in the embryos, resulting in their death as multinucleate single cells, than either single mutant has strengthened the view that the two genes function independently of each other. Since the double mutants are different from *kn* null mutants, the results might reflect a genetic interaction in vivo between *KN* and *KEU* genes in membrane trafficking to the cell plate (Waizenegger et al. 2000). Molecular evidence in support of this conclusion comes from the identification of the *KEU* gene product as a member of the Sec1 family, and the demonstration that KEU protein binds to KN protein in in vitro binding assays (Assaad et al. 2001). In a further attempt to identify other proteins involved in vesicle fusion at the site of the cell plate in cooperation with KN and KEU proteins, a biochemical and reverse genetic approach has implicated a vesicle trafficking gene, an *Arabidopsis* homolog of the *SNAP25* type *SNARE* called *SNAP33*, interacting with the KN syntaxin and KEU Sec1 homologue during cell plate formation in *Arabidopsis* embryos (Heese et al. 2001).

The primary meristems produced, produced by the root apical meristem of the *Arabidopsis* embryo differentiate into concentrically arranged single layers of cells of the epidermis, cortex, endodermis, and pericycle, and a central mass of cells of the vascular cylinder constituting the radial pattern of the radicle (Scheres et al. 1995). Epidermal cells in the post-embryonic root of *Arabidopsis* differentiate either as hair or non-hair cells in a distinct position-dependent pattern. Characterization of mutants with defects in the specification of epidermal cell types in seedling roots has identified genes such as *TRANSPARENT TESTA GLABRA* (*TTG*; Galway et al. 1994), *GLABRA2* (*GL2*; Masucci et al. 1996), and *WEREWOLF* (*WER*; Lee and Schiefelbein 1999) as positive regulators of non-hair cells and *CAPRICE* (*CPC*) as a positive regulator of hair cells (Wada et al. 1997). Other studies now indicate that the root hair patterning mechanism is initiated in developing embryos as early as the heart-shaped stage and is subsequently refined by genetic interactions during embryogenesis and post-embryogenesis (Berger et al. 1998; Lin and Schiefelbein 2001; Costa and Dolan 2003).

Genes such as *SHORT ROOT* (*SHR*), *SCARE-*

CROW (*SCR*), *PINOCCHIO* (*PIC*), *FS*, *GOLLUM* (*GLM*), *SCHIZORIZA* (*SCZ*), and *WOODEN LEG* (*WOL*), which are uncovered by their root phenotype, are the most crucial to the fashioning of the radial pattern of embryos of *Arabidopsis*, as mutations in these genes result in the replacement of certain cell layers by others in the seedling root and hypocotyl (Benfey et al. 1993; Scheres et al. 1995; Mylona et al. 2002). The closely investigated *shr* mutant, as the name implies, has a shorter root than the wild-type seedling due to retardation of growth. The developmental problem arises during formation of the endodermis, which is found wanting in the mutant root as indicated by the absence of cells decorated with the Casparian strip – a marker of endodermis. The notion that the *SHR* gene plays a role in both cell division and endodermis specification fits in with this observation. The defect in root anatomy has been traced to a failure of the formative division in the cortical-endodermal initials of the torpedo-shaped embryo that generates the cortex and endodermal cell lineages; instead, a single layer with cortical cell attributes results. Like the *shr* mutant, *scr* and *pic* mutants have only a single layer of cells between the epidermis and pericycle due to interference with the periclinal division of the cortical-endodermal initials, but cells of the surviving layer in the mutant roots have mixed structural and molecular identities of both endodermis and cortex. The *scr* and *pic* mutations thus show that the identity of the cortex and endodermal cell is specified independently of the division of the cortical-endodermal initial cells, and that the mutated genes have a role in regulating this division but not in the differentiation of the tissue layers (Benfey et al. 1993; Scheres et al. 1995).

The core of *SHR* and *SCR* genes is a small family of putative transcription factors that they encode. The products of these two genes, as well as those of the *GIBBERELLIN-INSENSITIVE* (*GAI*) and *REPRESSOR OF GA* (*RGA*) genes involved in the gibberellin signal transduction pathway, show high structural and sequence similarity and have been designated by the acronym GRAS (Pysh et al. 1999). Indicative of the role of the GRAS protein as a key regulator of the asymmetric division of the cortex-endodermal initials, transcripts of the *SCR* gene are detected during *Arabidopsis* embryo development in the initial cells before their asymmetric division, and in the endodermal cells born of this division. In a developmental sequence beginning with the first evidence for *SCR* involvement in radial patterning seen in the ground tissue and hypophysis of the globular embryo, gene expression shifts to the precursor of the central cells formed by the division of the hypophysis, and appears in the presumptive initials of the cortex and endodermis in the hypocotyl region of the torpedo-shaped embryo before settling finally in the endodermis of the mature stage embryo (see Plate 7, Fig. a–r). As evident from these figures, the pattern of expression of GFP driven by the *SCR* gene promoter in embryos of transgenic *Arabidopsis* plants is very similar to that of *SCR* transcripts. Moreover, both the root and hypocotyl regions of developing embryos display identical patterns of *SCR* gene expression, supporting the view that a common mechanism might be involved in the radial patterning of the cortical-endodermal layers in these embryonic organs (Di Laurenzio et al. 1996; Wysocka-Diller et al. 2000). Evidence that the SCR protein is required for specification of the quiescent center and maintenance of the surrounding cells as stem cells has also been presented (Sabatini et al. 2003). In contrast to the expression of the *SCR* gene described above, *SHR* gene transcripts are expressed in the procambial cells of early stage embryos, and the vascular cylinder and pericycle of mature embryos; no expression is observed in the cortex, endodermis, or the corresponding precursor cell layer. One possible explanation advanced for the lack of expression of the *SHR* gene in the ground tissue is that the gene controls radial patterning of the embryo in a non-cell-autonomous manner. In support of this view, it has been shown that ectopic expression of the *SHR* gene under the control of the *SCR* gene promoter produces an increased number of radial layers of cells with endodermal features in the roots of transgenic plants (Helariutta et al. 2000). The ability of the SHR protein to diffuse from the cells of the vascular tissue to the endodermis, as shown by Nakajima et al. (2001) could also be cited as evidence in support of the non-cell-autonomous manner of *SHR* gene action. According to these authors, SHR protein expressed in the vascular cylinder of the root moves into the neighboring cells to induce expression of the *SCR* gene; the latter, in turn specifies the endodermis.

The *fs* mutant was initially identified as a shape mutant (Chap. 2). Although *fs* mutant roots have additional layers of cells in the cortex, the bloated cortex does not come at the expense of another radial pattern layer in the root. Developmental analysis has revealed that the frequency and orientation of cell divisions in the developing embryo are affected without interfering with the concentric arrangement of the respective tissue layers in the root. This begs the question whether a regularity of cell division is crucial in embryonic pattern determination (Torres-Ruiz and Jürgens 1994). An increased number of periclinal divisions in the cells of the ground tissue initiated during embryogenesis likewise contributes to the defective radial patterning in roots of *scz* mutants. Suggestive of a role for other genes in the development of the ground tissue, roots of *scz/shr* and *scz/scr* double mutants are found to develop only a single layer of cells of the ground tissue (Mylona et al. 2002). The organization of vascular bundles is an appealing target for the action of the *GLM* and *WOL* genes, and anatomical studies of the mutant roots support this notion. The vascular tissues and pericycle cannot easily be identified in the center of the mutant *glm* root, whereas the *wol* mutant root contains fewer cells, especially of the phloem, in the vascular system. Double mutants incorporating the *shr* and *wol* mutant traits show additive phenotypes indicating that, with some functional overlap, two independent mechanisms produce defects in the endodermis and vascular tissues of the root (Scheres et al. 1995). The predicted *WOL* gene product belongs to a small protein family possessing the hallmarks of a two-component signal transducer. Expression of *WOL* gene transcripts in the precursors of the vascular tissue beginning in the globular embryo indicates that this gene probably functions as a receptor molecule in the control of the asymmetric cell division of the vascular initials (Mähönen et al. 2000).

The radial pattern genes so far identified play a role in the development of features that distinguish each concentric layer of the embryo. These genes probably remain active throughout the later life of the plant, serving as a reference system to propagate the positional information elaborated during early embryogenesis. Both the apicobasal and radial pattern mutants described above are phenotypes of disrupted zygotic genes – those activated after fertilization. Despite the fact that increasing evidence suggests a critical role for maternal effect genes – genes activated in the egg cytoplasm rather than in the zygote – in animal embryogenesis, very few maternal genes affecting plant development are known. The identity of a maternal gene controlling the apicobasal pattern of the embryo has emerged from genetic analysis of embryo development in the *short integument* (*sin1*) mutant of *Arabidopsis*. Whereas embryos of homozygous mutant plants developed within the embryo sac of a heterozygous mutant maternal sporophyte are normal in every respect, defects confined mainly to the cotyledons are frequently observed in embryos of mutants developed within a homozygous mutant maternal sporophyte (Ray et al. 1996). This implies that the *SIN1* gene product might influence pattern formation of the developing embryo by the production of a signaling factor from the tissues of the ovule. Another pattern forming gene, *GN*, discussed above, also appears to be affected by transcription from a maternally inherited allele (Vielle-Calzada et al. 2000).

3.3
Concluding Comments

Most of the work reviewed in this chapter can be attributed to the increasing pace in the study of plant embryo development using genetic and molecular techniques, but would not have been possible without the availability of good, robust information on the ontogeny of embryos in model eudicots and monocots, and on the role of plant hormones in plant growth and development. While seedling pattern mutants of *Arabidopsis* so far isolated and characterized have provided a fruitful framework within which to link genes and pattern-forming events during embryogenesis, it will be important to integrate genetic data with selective marker gene expression profiles in specific cells to gain a deeper understanding of the basis for the morphogenetic organization of the embryo along the apicobasal and radial axes. An additional challenge is to determine how the initial cell specification events in the developing embryo are linked to position-dependent information mediated through cellular interactions to generate the body organization of the embryo. Despite the fact that *Arabidopsis* has served as an excellent model to show how gene action is involved in establishing embryonic patterns, it seems likely that embryos of other angiosperms will have some secrets of their own. Although mutational studies in rice are beginning to acquire a rich tradition in providing special insights into the genetic control of embryogenesis in a monocot, the difficulty of isolating pattern mutants without interference by embryo lethality has led to the characterization of only a single radial pattern mutant from an arsenal of more than 200 mutants identified in this system (Kamiya et al. 2003b). Thus, studies on other angiosperms harboring virtues of the type abundantly displayed by *Arabidopsis* are necessary to illuminate the fundamental genetic and molecular biology of embryo body pattern formation – a topic that, as emphasized at the beginning of this chapter, is of paramount importance to plant and animal biologists.

REFERENCES

Aida M, Ishida T, Fukaki H, Fujisawa H, Tasaka M (1997) Genes involved in organ separation in *Arabidopsis*: an analysis of the *cup-shaped cotyledon* mutant. Plant Cell 9:841–857

Aida M, Ishida T, Tasaka M (1999) Shoot apical meristem and cotyledon formation during *Arabidopsis* embryogenesis: interaction among the *CUP-SHAPED COTYLEDON* and *SHOOT MERISTEMLESS* genes. Development 126:1563–1570

Aida M, Vernoux T, Furutani M, Traas J, Tasaka M (2002) Roles of *PIN-FORMED1* and *MONOPTEROS* in pattern formation of the apical region of the *Arabidopsis* embryo. Development 129:3965–3974

Aida M, Beis D, Heidstra R, Willemsen V, Blilou I, Galinha C, Nussaume L, Noh Y-S, Amasino R, Scheres B (2004) The *PLETHORA* genes mediate patterning of the *Arabidopsis* root stem cell niche. Cell 119:109–120

Assaad FF, Mayer U, Wanner G, Jürgens G (1996) The *KEULE* gene is involved in cytokinesis in *Arabidopsis*. Mol Gen Genet 253:267–277

Assaad FF, Huet Y, Mayer U, Jürgens G (2001) The cytokinesis gene *KEULE* encodes a Sec1 protein that binds the syntaxin KNOLLE. J Cell Biol 152:531–543

Barton MK, Poethig RS (1993) Formation of the shoot apical meristem in *Arabidopsis thaliana*: an analysis of development in the wild type and in the *shoot meristemless* mutant. Development 119:823–831

Baud S, Bellec Y, Miquel M, Bellini C, Caboche M, Lepiniec L, Faure J-D, Rochat C (2004) *gurke* and *pasticcino3* mutants affected in embryo development are impaired in acetyl-CoA carboxylase. EMBO Rep 5:515–520

Bäurle I, Laux T (2003) Apical meristems: the plant's fountain of youth. Bioessays 25:961–970

Benfey PN, Linstead PJ, Roberts K, Schiefelbein JW, Hauser M-T, Aeschbacher RA (1993) Root development in *Arabidopsis*: four mutants with dramatically altered root morphogenesis. Development 119:57–70

Berger F, Haseloff J, Schiefelbein J, Dolan L (1998) Positional information in root epidermis is defined during embryogenesis and acts in domains with strict boundaries. Curr Biol 8:421–430

Berleth T, Chatfield S (2002) Embryogenesis: pattern formation from a single cell. In: Somerville CR, Meyerowitz EM (eds) The *Arabidopsis* book. American Society of Plant Biologists, Rockville, MD, doi/10.1199/tab 0051, http://www.aspb.org/publications/arabidopsis/

Berleth T, Jürgens G (1993) The role of the *monopteros* gene in organising the basal body region of the *Arabidopsis* embryo. Development 118:575–587

Blilou I, Frugier F, Folmer S, Serralbo O, Willemsen V, Wolkenfelt H, Eloy NB, Ferreira PCG, Weisbeek P, Scheres B (2002) The *Arabidopsis HOBBIT* gene encodes a CDC27 homolog that links the plant cell cycle to progression of cell differentiation. Genes Dev 16:2566–2575

Blilou I, Xu J, Wildwater M, Willemsen V, Paponov I, Friml J, Heldstra R, Aida M, Palme K, Scheres B (2005) The PIN auxin efflux facilitator network controls growth and patterning in *Arabidopsis* roots. Nature 433:39–44

Brandt U, Fletcher JC, Hobe M, Meyerowitz EM, Simon R (2000) Dependence of stem cell fate in *Arabidopsis* on a feedback loop regulated by *CLV3* activity. Science 289:617–619

Brandt U, Grünewald M, Hobe M, Simon R (2002) Regulation of CLV3 expression by two homeobox genes in *Arabidopsis*. Plant Physiol 129:565–575

Busch M, Mayer U, Jürgens G (1996) Molecular analysis of the *Arabidopsis* pattern formation gene *GNOM*: gene structure and intragenic complementation. Mol Gen Genet 250:681–691

Byrne ME, Barley R, Curtis M, Arroyo JM, Dunham M, Hudson A, Mertienssen RA (2000) *Asymmetric leaves1* mediates leaf patterning and stem cell function in *Arabidopsis*. Nature 408:967–971

Carles CC, Fletcher JC (2003) Shoot apical meristem maintenance: the art of a dynamic balance. Trends Plant Sci 8:394–401

Casamitjana-Martinez E, Hofhuis HF, Xu J, Liu C-M, Heidstra R, Scheres B (2003) Root-specific *CLE19* overexpression and the *sol1/2* suppressors implicate a CLV-like pathway in the control of *Arabidopsis* root meristem maintenance. Curr Biol 13:1435–1441

Casson SA, Lindsey K (2003) Genes and signaling in root development. New Phytol 158:11–38

Clark SE, Running MP, Meyerowtiz EM (1993) *CLAVATA1*, a regulator of meristem and flower development in *Arabidopsis*. Development 119:397–418

Clark SE, Running MP, Meyerowitz EM (1995) *CLAVATA3* is a specific regulator of shoot and floral meristem development affecting the same processes as *CLAVATA1*. Development 121:2057–2067

Clark SE, Jacobsen SE, Levin JZ, Meyerowitz EM (1996) The *CLAVATA* and *SHOOT MERISTEMLESS* loci competitively regulate meristem activity in *Arabidopsis*. Development 122:1567–1575

Clark SE, Williams RW, Meyerowitz EM (1997) The *CLAVATA1* gene encodes a putative receptor kinase that controls shoot and floral meristem size in *Arabidopsis*. Cell 89:575–585

Conway LJ, Poethig RS (1997) Mutations of *Arabidopsis thaliana* that transform leaves into cotyledons. Proc Natl Acad Sci USA 94:10209–10214

Costa S, Dolan L (2003) Epidermal patterning genes are active during embryogenesis in *Arabidopsis*. Development 130:2893–2901

Davidson EH (1994) Molecular biology of embryonic development: how far have we come in the last ten years? Bioessays 16:603–615

Diener AC, Li H, Zhou W, Whoriskey WJ, Nes WD, Fink GR (2000) *STEROL METHYL TRANSFERASE 1* controls the level of cholesterol in plants. Plant Cell 12:853–870

Di Laurenzio L, Wysocka-Diller J, Malamy JE, Pysh L, Helariutta Y, Freshour G, Hahn MG, Feldmann KA, Benfey PN (1996) The *SCARECROW* gene regulates an asymmetric cell division that is essential for generating the radial organization of the *Arabidopsis* root. Cell 86:423–433

Dolan L, Janmaat K, Willemsen V, Linstead P, Poethig S, Roberts K, Scheres B (1993) Cellular organisation of the *Arabidopsis thaliana* root. Development 119:71–84

Dolan L, Duckett CM, Grierson C, Linstead P, Schneider K, Lawson E, Dean C, Poethig S, Roberts K (1994) Clonal relationships and cell patterning in the root epidermis of *Arabidopsis*. Development 120:2465–2474

Endrizzi K, Moussian B, Haecker A, Levin JZ, Laux T (1996) The *SHOOT MERISTEMLESS* gene is required for maintenance of undifferentiated cells in *Arabidopsis* shoot and floral meristems and acts at a different regulatory level than the meristem genes *WUSCHEL* and *ZWILLE*. Plant J 10:967–979

Faure J-D, Vittorioso P, Santoni V, Fraisier V, Prinsen E, Barlier I, van Onckelen H, Caboche M, Bellini C (1998) The *PASTICCINO* genes of *Arabidopsis thaliana* are involved in the control of cell division and differentiation. Development 125:909–918

Fletcher JC, Brand U, Running MP, Simon R, Meyerowitz EM (1999) Signaling of cell fate decisions by *CLAVATA3* in *Arabidopsis* shoot meristems. Science 283:1911–1914

Friml J, Benkova E, Blilou I, Wisniewska J, Hamann T, Ljung K, Woody S, Sandberg G, Scheres B, Jürgens G, Palme K (2002) AtPIN4 mediates sink-driven auxin gradients and root patterning in *Arabidopsis*. Cell 108:661–673

Gallois J-L, Woodward C, Reddy GV, Sablowski R (2002) Combined SHOOT MERISTEMLESS and WUSCHEL trigger ectopic organogenesis in *Arabidopsis*. Development 129:3207–3217

Gallois J-L, Nora FR, Mizukami Y, Sablowski R (2004) WUSCHEL induces shoot stem cell activity and developmental plasticity in the root meristem. Genes Dev 18:375–380

Galway ME, Masucci JD, Lloyd AM, Walbot V, Davis RW, Schiefelbein JW (1994) The *TTG* gene is required to specify epidermal cell fate and cell patterning in the *Arabidopsis* root. Dev Biol 166:740–754

Geldner N, Richter S, Vieten A, Marquardt S, Torres-Ruiz RA, Mayer U, Jürgens G (2004) Partial loss-of-function alleles reveal a role for *GNOM* in auxin transport-related, post-embryonic development of *Arabidopsis*. Development 131:389–400

Gisel A, Barella S, Hempel FD, Zambryski PC (1999) Temporal and spatial regulation of symplastic trafficking during development in *Arabidopsis thaliana* apices. Development 126:1879–1889

Grebe M, Gadea J, Steinmann T, Kientz M, Rahfeld J-U, Salchert K, Koncz K, Jürgens G (2000) A conserved domain of the *Arabidopsis* GNOM protein mediates subunit interaction and cyclophilin 5 binding. Plant Cell 12:343–356

Green KA, Priggs MJ, Katzman RB, Clark SE (2005) *CORONA*, a member of the class III homeodomain leucine zipper gene family in *Arabidopsis*, regulates stem cell specification and organogenesis. Plant Cell 17:691–704

Hadfi K, Speth V, Neuhaus G (1998) Auxin-induced developmental patterns in *Brassica juncea* embryos Development 125:879–887

Hamann T, Mayer U, Jürgens G (1999) The auxin-insensitive *bodenlos* mutation affects primary root formation and apical-basal patterning in the *Arabidopsis* embryo. Development 126:1387–1395

Hamann T, Benkova E, Bäurle I, Kientz M, Jürgens G (2002) The *Arabidopsis BODENLOS* gene encodes an auxin response protein inhibiting MONOPTEROS-mediated embryo patterning. Genes Dev 16:1610–1615

Hardtke CS, Berleth T (1998) The *Arabidopsis* gene *MONOPTEROS* encodes a transcription factor mediating embryo axis formation and vascular development. EMBO J 17:1405–1411

Heese M, Gansel X, Sticher L, Wick P, Grebe M, Granier F, Jürgens G (2001) Functional characterization of the KNOLLE-interacting t-SNARE AtSNAP33 and its role in plant cytokinesis. J Cell Biol 155:239–249

Helariutta Y, Fukaki H, Wysocka-Diller J, Nakajima K, Jung J, Sena G, Hauser M-T, Benfey PN (2000) The *SHORT-ROOT* gene controls radial patterning of the *Arabidopsis* root through radial signaling. Cell 101:555–567

Hellmann H, Hobbie L, Chapman A, Dharmasiri S, Dharmasiri N, del Pozo C, Reinhardt D, Estelle M (2003) *Arabidopsis AXR6* encodes CUL1 implicating SCF E3 ligases in auxin regulation of embryogenesis. EMBO J 22:3314–3325

Hobbie L, McGovern M, Hurwitz LR, Pierro A, Liu NY, Bandyopadhyay A, Estelle M (2000) The *axr6* mutants of *Arabidopsis thaliana* define a gene involved in auxin response and early development. Development 127:23–32

Howell SH (1998) Molecular genetics of plant development. Cambridge University Press, New York

Ishiguro S, Watanabe Y, Ito N, Nonaka H, Takeda N, Sakai T, Kanaya H, Okada K (2002) SHEPHERD is the *Arabidopsis* GRP94 responsible for the formation of functional CLAVATA proteins. EMBO J 21:898–908

Ito Y, Eiguchi M, Kurata N (1999) Expression of novel homeobox genes in early embryogenesis in rice. Biochim Biophys Acta 1444:445–450

Jackson D, Veit B, Hake S (1994) Expression of maize *KNOTTED1* related homeobox genes in the shoot apical meristem predicts patterns of morphogenesis in the vegetative shoot. Development 120:405–413

Jang J-C, Fujioka S, Tasaka M, Seto H, Takatsuto S, Ishii A, Aida M, Yoshida S, Sheen J (2000) A critical role of sterols in embryonic patterning and meristem programming revealed by the *fackel* mutants of *Arabidopsis thaliana*. Genes Dev 14:1485–1497

Jeong S, Trotochaud AE, Clark SE (1999) The *Arabidopsis CLAVATA2* gene encodes a receptor-like protein required for the stability of the *CLAVATA1* receptor-like kinase. Plant Cell 11:1925–1933

Johnson S, Liu C-M, Hedley CL, Wang TL (1994) An analysis of seed development in *Pisum sativum* XVIII. The isoation of mutants defective in embryo development. J Exp Bot 45:1503–1511

Jürgens G (2001) Apical-basal pattern formation in *Arabidopsis* embryogenesis. EMBO J 20:3609–3616

Kajiwara T, Furutani M, Hibara K, Tasaka M (2004) The *GURKE* gene encoding an acetyl-CoA carboxylase is required for partitioning the embryo apex into three subregions in *Arabidopsis*. Plant Cell Physiol 45:1122–1128

Kamiya N, Nagasaki H, Morikami A, Sato Y, Matsuoka M (2003a) Isolation and characterization of a rice *WUSCHEL*-type homeobox gene that is specifically expressed in the central cells of a quiescent center in the root apical meristem. Plant J 35:429–441

Kamiya N, Nishimura A, Sentoku N, Takabe E, Nagato Y, Kitano H, Matsuoka M (2003b) Rice *globular embryo 4* (*gle4*) mutant is defective in radial pattern formation during embryogenesis. Plant Cell Physiol 44:875–883

Kayes JM, Clark SE (1998) *CLAVATA2*, a regulator of meristem and organ development in *Arabidopsis*. Development 125:3843–3851

Keddie JS, Carroll BJ, Thomas CM, Reyes MEC, Klimyuk V, Holtan H, Gruissem W, Jones JDG (1998) Transposon tagging of the *defective embryo and meristems* gene of tomato. Plant Cell 10:877–887

Kidner C, Sundaresan V, Roberts K, Dolan L (2000) Clonal analysis of the *Arabidopsis* root confirms that position, not lineage, determines cell fate. Planta 211:191–199

Lauber MH, Waizenegger I, Steinmann T, Schwarz H, Mayer U, Hwang I, Lukowitz W, Jürgens G (1997) The *Arabidopsis* KNOLLE protein is a cytokinesis-specific syntaxin. J Cell Biol 139:1485–1493

Laufs P, Dockx J, Kronenberger J, Traas J (1998) *MGOUN1* and *MGOUN2*: two genes required for primordium initiation at the shoot apical and floral meristems in *Arabidopsis thaliana*. Development 125:1253–1260

Laux T, Jürgens G (1997) Embryogenesis: a new start in life. Plant Cell 9:989–1000

Laux T, Mayer KFX, Berger J, Jürgens G (1996) The *WUSCHEL* gene is required for shoot and floral meristem integrity in *Arabidopsis*. Development 122:87–96

Laux T, Würschum T, Breuninger H (2004) Genetic regulation of embryonic pattern formation. Plant Cell 16:S190–S202

Lee MM, Schiefelbein JW (1999) WEREWOLF, a MYB-related protein in *Arabidopsis*, is a position-dependent regulator of epidermal cell patterning. Cell 99:473–483

Lenhard M, Laux, T (2003) Stem cell homeostasis in the *Arabidopsis* shoot meristem is regulated by intercellular movement of CLAVATA3 and its sequestration by CLAVATA1. Development 130:3163–173

Lenhard M, Jürgens G, Laux T (2002) The *WUSCHEL* and *SHOOTMERISTEMLESS* genes fulfil complementary roles in *Arabidopsis* shoot meristem regulation. Development 129:3195–3206

Lin Y, Schiefelbein JW (2001) Embryonic control of epidermal cell patterning in the root and hypocotyl of *Arabidopsis*. Development 128:3697–3705

Liu C, Xu Z-H, Chua N-H (1993) Auxin polar transport is essential for the establishment of bilateral symmetry during early plant embryogenesis. Plant Cell 5:621–630

Liu C-M, Johnson S, Wang TL (1995) *Cyd*, a mutant of pea that alters embryo morphology is defective in cytokinesis. Dev Genet 16:321–331

Liu C-M, Johnson S, di Gregorio S, Wang TL (1999) *single cotyledon* (*sic*) mutants of pea and their significance in understanding plant embryo development. Dev Genet 25:11–22

Long JA, Barton MK (1998) The development of apical embryonic pattern in *Arabidopsis*. Development 125:3027–3035

Long JA, Moan EI, Medford JI, Barton MK (1996) A member of the KNOTTED class of homeodomain proteins encoded by the *STM* gene of *Arabidopsis*. Nature 379:66–69

Long JA, Woody S, Poethig S, Meyerowitz EM, Barton MK (2002) Transformation of shoots into roots in *Arabidopsis* embryos mutant at the *TOPLESS* locus. Development 129:2297–2306

Lukowitz W, Mayer U, Jürgens G (1996) Cytokinesis in the *Arabidopsis* embryo involves the syntaxin-related KNOLLE gene product. Cell 84:61–71

Lyndon RF (1998) The shoot apical meristem. Its growth and development. Cambridge University Press, New York

Lynn K, Fernandez A, Aida M, Sedbrook J, Tasaka M, Masson P, Barton MK (1999) The *PINHEAD/ZWILLE* gene acts pleiotropically in *Arabidopsis* development and has overlapping functions with the *ARGONAUTE1* gene. Development 126:469–481

Mähönen AP, Bonke M, Kauppinen L, Riikonen M, Benfey PN, Helariutta Y (2000) A novel two-component hybrid molecule regulates vascular morphogenesis of the *Arabidopsis* root. Genes Dev 14:2938–2943

Masucci JD, Rerie WG, Foreman DR, Zhang M, Galway ME, Marks MD, Schiefelbein JW (1996) The homeobox gene *GLABRA2* is required for position-dependent cell differentiation in the root epidermis of *Arabidopsis thaliana*. Development 122:1253–1260

Mayer KFX, Schoof H, Haecker A, Lenhard M, Jürgens G, Laux T (1998) Role of *WUSCHEL* in regulating stem cell fate in the *Arabidopsis* shoot meristem. Cell 95:805–815

Mayer U, Torres Ruiz R, Berleth T, Miséra S, Jürgens G (1991) Mutations affecting body organization in the *Arabidopsis* embryo. Nature 353:402–407

Mayer U, Büttner G, Jürgens G (1993) Apical-basal pattern formation in the *Arabidopsis* embryo: studies on the role of the *gnom* gene. Development 117:149–162

Moussian B, Schoof H, Haecker A, Jürgens G, Laux T (1998) Role of the *ZWILLE* gene in the regulation of central shoot meristem cell fate during *Arabidopsis* embryogenesis. EMBO J 17:1799–1809

Moussian B, Haecker A, Laux T (2003) *ZWILLE* buffers meristem stability in *Arabidopsis thaliana*. Dev Genes Evol 213:534–540

Müller I, Wagner W, Völker A, Schellmann S, Nacry P, Küttner F, Schwarz-Sommer S, Mayer U, Jürgens G (2003) Syntaxin specificity of cytokinesis in *Arabidopsis*. Nat Cell Biol 5:531–534

Mylona P, Linstead P, Martienssen R, Dolan L (2002) *SCHIZORIZA* controls an asymmetric cell division and restricts epidermal identity in the *Arabidopsis* root. Development 129:4327–4334

Nacry P, Mayer U, Jürgens G (2000) Genetic dissection of cytokinesis. Plant Mol Biol 43:719–733

Nakajima K, Sena G, Nawy T, Benfey PN (2001) Intercellular movement of the putative transcription factor SHR in root patterning. Nature 413:307–311

Pilu R, Consonni G, Busti E, MacCabe AP, Giulini A, Dolfini S, Gavazzi G (2002) Mutations in two independent genes lead to suppression of the shoot apical meristem in maize. Plant Physiol 128:502–511

Postma-Haarsma AD, Verwoert IIGS, Stronk OP, Koster J, Lamers GEM, Hoge JHC, Meijer AH (1999) Characterization of the KNOX class homeobox genes Oskn2 and Oskn3 identified in a collection of cDNA libraries covering the early stages of rice embryogenesis. Plant Mol Biol 39:257–271

Postma-Haarsma AD, Rueb S, Scarpella E, den Besten W, Hoge HC, Meijer AH (2002) Developmental regulation and downstream effects of the knox class homeobox genes Oskn2 and Oskn3 from rice. Plant Mol Biol 48:423–441

Przemeck GKH, Mattsson J, Hardtke CS, Sung ZR, Berleth T (1996) Studies on the role of the Arabidopsis gene MONOPTEROS in vascular development and plant cell axialization. Planta 200:229–237

Pysh LD, Wysocka-Diller JW, Camilleri C, Bouchez D, Benfey PN (1999) The GRAS gene family in Arabidopsis: sequence characterization and basic expression analysis of SCARECROW-LIKE genes. Plant J 18:111–119

Raghavan V (1990). Origin of the quiescent center in the root of Capsella bursa-pastoris (L.) Medik. Planta 181:62–70

Raghavan V (1997) Molecular embryology of flowering plants. Cambridge University Press, New York

Raghavan V (2000a) Pattern formation in angiosperm embryos. Botanica 50:33–47

Raghavan V (2000b) Developmental biology of flowering plants. Springer, New York Berlin Heidelberg

Ray S, Golden T, Ray A (1996) Maternal effects of the short integument mutation on embryo development in Arabidopsis. Dev Biol 180:365–369

Rinne PLH, van der Schoot C (1998) Symplastic fields in the tunica of the shoot apical meristem coordinate morphogenetic events. Development 125:1477–1485

Rojo E, Sharma VK, Kovaleva V, Raikhel NV, Fletcher JC (2002) CLV3 is localized to the extracellular space, where it activates the Arabidopsis CLAVATA stem cell signaling pathway. Plant Cell 14:969–977

Sabatini S, Heidstra R, Wildwater M, Scheres B (2003) SCARECROW is involved in positioning the stem cell niche in the Arabidopsis root meristem. Genes Dev 17:354–358

Sabatini S, Beis D, Wolkenfelt H, Murfett J, Guilfoyle T, Malamy J, Benfey P, Leyser O, Bechtold N, Weisbeek P, Scheres B (1999) An auxin-dependent distal organizer of pattern and polarity in the Arabidopsis root. Cell 99:463–472

Satina S, Blakeslee AF, Avery AG (1940) Demonstration of three germ layers in the shoot apex of Datura by means of polyploidy in periclinal chimeras. Am J Bot 25:895–905

Sato Y, Hong S-K, Tagiri A, Kitano H, Yamamoto N, Nagato Y, Matsuoka M (1996) A rice homeobox gene, OSH1, is expressed before organ differentiation in a specific region during early embryogenesis. Proc Natl Acad Sci USA 93:8117–8122

Sato Y, Sentoku N, Nagato Y, Matsuoka M (1998) Isolation and characterization of a rice homeobox gene, OSH15. Plant Mol Biol 38:983–998

Satoh N, Hong S-K, Nishimura A, Matsuoka M, Kitano H, Nagato Y (1999) Initiation of shoot apical meristem in rice: characterization of four SHOOTLESS genes. Development 126:3629–3636

Satoh N, Itoh J-I, Nagato Y (2003) The SHOOTLESS2 and SHOOTLESS1 genes are involved in both initiation and maintenance of the shoot apical meristem through regulating the number of indeterminate cells. Genetics 164:335–346

Scheres B, Heidstra R (1999) Digging out roots: pattern formation, cell division, and morphogenesis in plants. Curr Top Dev Biol 45:207–247

Scheres B, Di Laurenzio L, Willemsen V, Hauser M-T, Janmaat K, Weisbeek P, Benfey PN (1995) Mutations affecting the radial organisation of the Arabidopsis root display specific defects throughout the embryonic axis. Development 121:53–62

Schoof H, Lenhard M, Haecker A, Mayer KFX, Jürgens G, Laux T (2000) The stem cell population of Arabidopsis shoot meristems is maintained by a regulatory loop between the CLAVATA and WUSCHEL genes. Cell 100:635–644

Schrick K, Mayer U, Horrichs A, Kuhnt C, Bellini C, Dangl J, Schmidt J, Jürgens G (2000) FACKEL is a sterol C-14 reductase required for organized cell division and expansion in Arabidopsis embryogenesis. Genes Dev 14:1471–1484

Schrick K, Mayer U, Martin G, Bellini C, Kuhnt C, Schmidt J, Jürgens G (2002) Interactions between sterol biosynthesis genes in embryonic development of Arabidopsis. Plant J 31:61–73

Sentoku N, Sato Y, Kurata N, Ito Y, Kitano H, Matsuoka M (1999) Regional expression of the rice KN1-type homeobox gene family during embryo, shoot, and flower development. Plant Cell 11:1651–1663

Sharma VK, Fletcher JC (2002) Maintenance of shoot and floral meristem cell proliferation and fate. Plant Physiol 129:31–39

Shevell DE, Leu WM, Gillmor CS, Xia G, Feldmann KA, Chua N-H (1994) EMB30 is essential for normal cell division, cell expansion, and cell adhesion in Arabidopsis and encodes a protein that has similarity to Sec7. Cell 77:1051–1062

Shevell DE, Kunkel T, Chua N-H (2000) Cell wall alterations in the Arabidopsis EMB30 mutant. Plant Cell 12:2047–2059

Smith LG, Jackson D, Hake S (1995) Expression of knotted1 marks shoot meristem formation during maize embryogenesis. Dev Genet 16:344–348

Söllner R, Glässer G, Wanner G, Somerville CR, Jürgens, Assaad FF (2002) Cytokinesis-defective mutants of Arabidopsis. Plant Physiol 129:678–690

Sørensen MB, Mayer U, Lukowitz W, Robert H, Chambrier P, Jürgens G, Somerville C, Lepiniec L, Berger F (2002) Cellularisation in the endosperm of Arabidopsis thaliana is coupled to mitosis and shares multiple components with cytokinesis. Development 129:5567–5576

Souer E, van Houwelingen A, Kloos D, Mol J, Koes R (1996) The no apical meristem gene of Petunia is required for pattern formation in embryos and flowers and is expressed at meristem and primordia boundaries. Cell 85:159–170

Souter M, Topping J, Pullen M, Friml J, Palme K, Hackett R, Grierson D, Lindsey K (2002) *hydra* mutants of *Arabidopsis* are defective in sterol profiles and auxin and ethylene signaling. Plant Cell 14:1017–1031

Steeves TA, Sussex IM (1989) Patterns in plant development, 2nd edn. Cambridge University Press, New York

Steinmann T, Geldner N, Grebe M, Mangold S, Jackson CL, Paris S, Gälweiler L, Palme K, Jürgens G (1999) Coordinated polar localization of auxin efflux carrier PIN1 by GNOM ARF GEF. Science 286:316–318

Stewart RN, Dermen H (1970) Determination of number and mitotic activity of shoot apical initial cells by analysis of mericlinal chimeras. Am J Bot 57:816–826

Stone JM, Trotochaud AE, Walker JC, Clark SE (1998) Control of meristem development by CLAVATA1 receptor kinase and kinase-associated protein phosphatase interactions. Plant Physiol 117:1217–1225

Takada S, Hibara K, Ishida T, Tasaka M (2001) The *CUP-SHAPED COTYLEDON1* gene of *Arabidopsis* regulates shoot apical meristem formation. Development 128:1127–1135

Taoka K, Yanagimoto Y, Daimon Y, Hibara K, Aida M, Tasaka M (2004) The NAC domain mediates functional specificity of CUP-SHAPED COTYLEDON proteins. Plant J 40:462–473

Topping JF, May VJ, Muskett PR, Lindsey K (1997) Mutations in the *HYDRA1* gene of *Arabidopsis* perturb cell shape and disrupt embryonic and seedling morphogenesis. Development 124:4415–4424

Torres-Ruiz RA, Jürgens G (1994) Mutations in the *FASS* gene uncouple pattern formation and morphogenesis in *Arabidopsis* development. Development 120:2967–2978

Torres-Ruiz RA, Lohner A, Jürgens G (1996) The *GURKE* gene is required for normal organization of the apical region in the *Arabidopsis* embryo. Plant J 10:1005–1016

Trotochaud AE, Hao T, Wu G, Yang Z, Clark SE (1999) The CLAVATA1 receptor-like kinase requires CLAVATA3 for its assembly into a signaling complex that includes KAPP and a Rho-related protein. Plant Cell 11:393–405

van den Berg C, Willemsen V, Hage W, Weisbeek P, Scheres B (1995) Cell fate in the *Arabidopsis* root meristem determined by directional signalling. Nature 378:62–65

van den Berg C, Willemsen V, Hendriks G, Weisbeek P, Scheres B (1997) Short-range control of cell differentiation in the *Arabidopsis* root meristem. Nature 390:287–289

Vielle-Calzada J-P, Baskar R, Grossniklaus U (2000) Delayed activation of the paternal genome during seed development. Nature 404:91–94

Völker A, Stierhof Y-D, Jürgens G (2001) Cell cycle-independent expression of the *Arabidopsis* cytokinesis-specific syntaxin KNOLLE results in mistargeting to the plasma membrane and is not sufficient for cytokinesis. J Cell Biol 114:3001–3012

Vroemen CW, Mordhorst AP, Albrecht C, Kwaaitaal MACJ, de Vries SC (2003) The *CUP-SHAPED COTYLEDON3* gene is required for boundary and shoot meristem formation in *Arabidopsis*. Plant Cell 15:1563–1577

Wada T, Tachibana T, Shimura Y, Okada K (1997) Epidermal cell differentiation in *Arabidopsis* determined by a Myb homolog, *CPC*. Science 277:1113–1116

Waites R, Simon R (2000) Signaling cell fate in plant meristems: three clubs on one tousle. Cell 103:835–838

Waizenegger I, Lukowitz W, Assaad F, Schwarz H, Jürgens G, Mayer U (2000) The *Arabidopsis* KNOLLE and KEULE genes interact to promote vesicle fusion during cytokinesis. Curr Biol 10:1371–1374

Willemsen V, Scheres B (2004) Mechanisms of pattern formation in plant embryogenesis. Annu Rev Genet 38:587–614

Willemsen V, Wolkenfelt H, de Vrieze G, Weisbeek P, Scheres B (1998) The *HOBBIT* gene is required for formation of the root meristem in the *Arabidopsis* embryo. Development 125:521–531

Williams RW, Wilson JM, Meyerowitz EM (1997) A possible role for kinase-associated protein phosphatase in the *Arabidopsis* CLAVATA1 signaling pathway. Proc Natl Acad Sci USA 94:10467–10472

Wysocka-Diller JW, Helariutta Y, Fukaki H, Malamy JE, Benfey PN (2000) Molecular analysis of SCARECROW function reveals a radial patterning mechanism common to root and shoot. Development 127:595–603

Yu LP, Simon EJ, Trotochaud AE, Clark SE (2000) *POLTERGEIST* functions to regulate meristem development downstream of the *CLAVATA* loci. Development 127:1661–1670

Yu LP, Miller AK, Clark SE (2003) *POLTERGEIST* encodes a protein phosphatase 2C that regulates CLAVATA pathways controlling stem identity at *Arabidopsis* shoot and flower meristems. Curr Biol 13:179–188

4 Life and Times of the Suspensor – Cell Signaling between the Embryo and Suspensor

When the comparative morphologist asserts that the suspensor is a primitive organ found in the embryogeny of some modern survivors of ancient stocks, he has probably seized upon a truth, even though he cannot account for the mechanism involved: it is a particular feature of the hereditary constitution which happens to become manifest in the embryogeny. In those genera and families in which a suspensor is known, it occurs with a high degree of regularity. In the contemporary view, the suspensor could be regarded as a gene-determined organ in the inception of which, perhaps, only a few genes are involved; and the genic action leading to its formation would be seen in the differential protoplasmic changes which take place in the elongating zygote.

C.W. Wardlaw 1955

4.1 Morphological and Physiological Considerations 82
4.1.1 Subcellular Morphology of the Suspensor 85
4.1.2 Nuclear Cytology of the Suspensor 88
4.1.3 Functional Physiology of the Suspensor 90
4.1.4 Developmental Physiology and Programmed Death of Suspensor Cells 92
4.2 Genetic Control of Suspensor Form 94
4.3 Concluding Comments 97
References 97

In the numerous accounts of embryology of flowering plants, the suspensor is considered as a product of the first division of the zygote. Since the suspensor appears as an attachment to the embryo in some pteridophytes, the presence of suspensor is viewed as a primitive feature retained on the angiosperm embryo (Wardlaw 1955). With a few exceptions, in most angiosperms, soon after the suspensor is fully formed its cells begin to degenerate without contributing derivatives to the formation of the embryo; however, despite its short lifespan, the suspensor has received more than its share of attention in the embryology literature. Early investigations of the suspensor were focused on its origin and morphology in relation to the embryo proper developing within the ovule. Indeed, to the plant embryologist, one of the enduring images of the suspensor is the structure first described in the 1870s in *Capsella bursa-pastoris* as a string of cells attached to the organogenetic part of the embryo (Hanstein 1870). Based on early morphological investigations of suspensors, and the bizarre forms of their haustoria observed in many plants, it has been accepted as common wisdom that the main function of the suspensor is to anchor the embryo and position it in relation to the nutritionally rich endosperm and ovular tissues. Later cytological and physiological investigations have forged the general impression over the years that suspensor cells play a direct role in the nutrition of the embryo, perhaps by absorbing and transmitting food materials from the surrounding tissues to the embryo and by functioning as a temporary storage station for reserve products. This and other features, such as the short life span of the suspensor, its embryonic origin, and the high degree of endoreduplication and polyteny in its cells, have prompted a comparison between the suspensor and the trophoblast, an ectoderm layer that functions in the nutrition and implantation of animal embryos (Nagl 1974). In recent years, genetic studies have provided insights into the fundamental cell signaling mechanisms between the suspensor and embryo involved in maintaining suspensor cell identity. As a consequence of this multifaceted approach, it has been established that rather than functioning mechanistically in physical support, the suspensor has evolved into a remarkable structure that interacts with the embryo in a complex way and whose activities are entwined with the growth of the embryo. It is therefore no wonder that the suspensor has persisted in flowering plants as an almost ubiquitous attachment to the embryo.

This chapter focuses on the morphological, physiological, and cytological features of suspensors of representative angiosperms and the genetic control of suspensor form in *Arabidopsis* – the model plant in which the establishment of suspensor cell identity is understood in most detail. The principal findings covering these topics are reviewed by Yeung and Meinke (1993), Schwartz et al. (1997), and Raghavan (2001).

4.1
Morphological and Physiological Considerations

As extensive observations on various plants have now shown, the suspensor attains remarkably diverse morphological forms ranging from vesicular single-celled to elaborate multicellular structures. These variations may be confined to genera and species of different families, and members of even single families may show a fair amount of diversity in suspensor morphology. How true this really is we shall consider below with selected examples from four families.

As shown in Fig. 4.1 (a–d), certain genera of the Orchidaceae, such as *Bulbophyllum*, *Calanthe*, *Coelogyne*, *Dendrobium*, *Peristeria*, *Phaius*, and *Spathoglottis*, are noted for their single-celled suspensor, which becomes elongate, vesicular, or tubular and persists even after germination of the seed (Swamy 1949; Poddubnaya-Arnoldi 1967; Ye et al. 1997). Although enlargement of the suspensor cell can be accounted for by its vacuolation, as shown in *Phaius tankervilliae*, this is also accompanied by striking changes in the configuration of the cytoskeletal elements: actin filaments and microtubules. When the newly cut off suspensor cell elongates into the endosperm cavity, microtubules form a cortical array at its chalazal end, whereas actin filaments assume a central location with some concentration at the micropylar end of the cell. At the time of enlargement of the suspensor cell due to vacuolation, actin filaments appear in the cortical region while microtubules form a perinuclear array with extensions towards both poles of the cell. Further elongation of the suspensor through the micropyle leads to the reappearance of cortical microtubules without any change in the location of actin filaments (Ye et al. 1997).

Unusually long, at times coiled or twisted, uniseriate or biseriate suspensors are common in some members of the Loranthaceae. In *Macrosolen cochinchinensis*, the suspensor remains biseriate throughout, although it becomes multiseriate close to its attachment to the embryo (Fig. 4.1e). In *Peraxilla tetrapetala*, the suspensor becomes multiseriate and fleshy (Fig. 4.1f). The length attained by the suspensor in the Loranthaceae is not surprising since, in different members of the family, the tip of the embryo sac harboring the egg apparatus extends to varying heights into the style, even reaching the stigma; following fertilization, the embryo is thrust into the vicinity of the endosperm by the long suspensor (Maheshwari and Singh 1952; Prakash, 1960).

A morphologically simple, filamentous, multicellular suspensor is the type described in *Capsella bursa-pastoris*, *Arabidopsis*, and *Diplotaxis erucoides* – all members of the Brassicaceae (Fig. 4.1g,h). In the octant stage embryo of *C. bursa-pastoris*, the suspensor has a single file of six cells, which increases to about ten cells in the globular-stage embryo. It is terminated by a large basal cell at the micropylar end and is connected to the embryo proper through the hypophysis at the chalazal end (Schulz and Jensen 1969). Excluding the basal cell and the hypophysis, suspensors of *Arabidopsis* (Mansfield and Briarty 1991; Jürgens and Mayer 1994) and *D. erucoides* (Simoncioli 1974) have three-to-eight and six-to-eight cells, respectively.

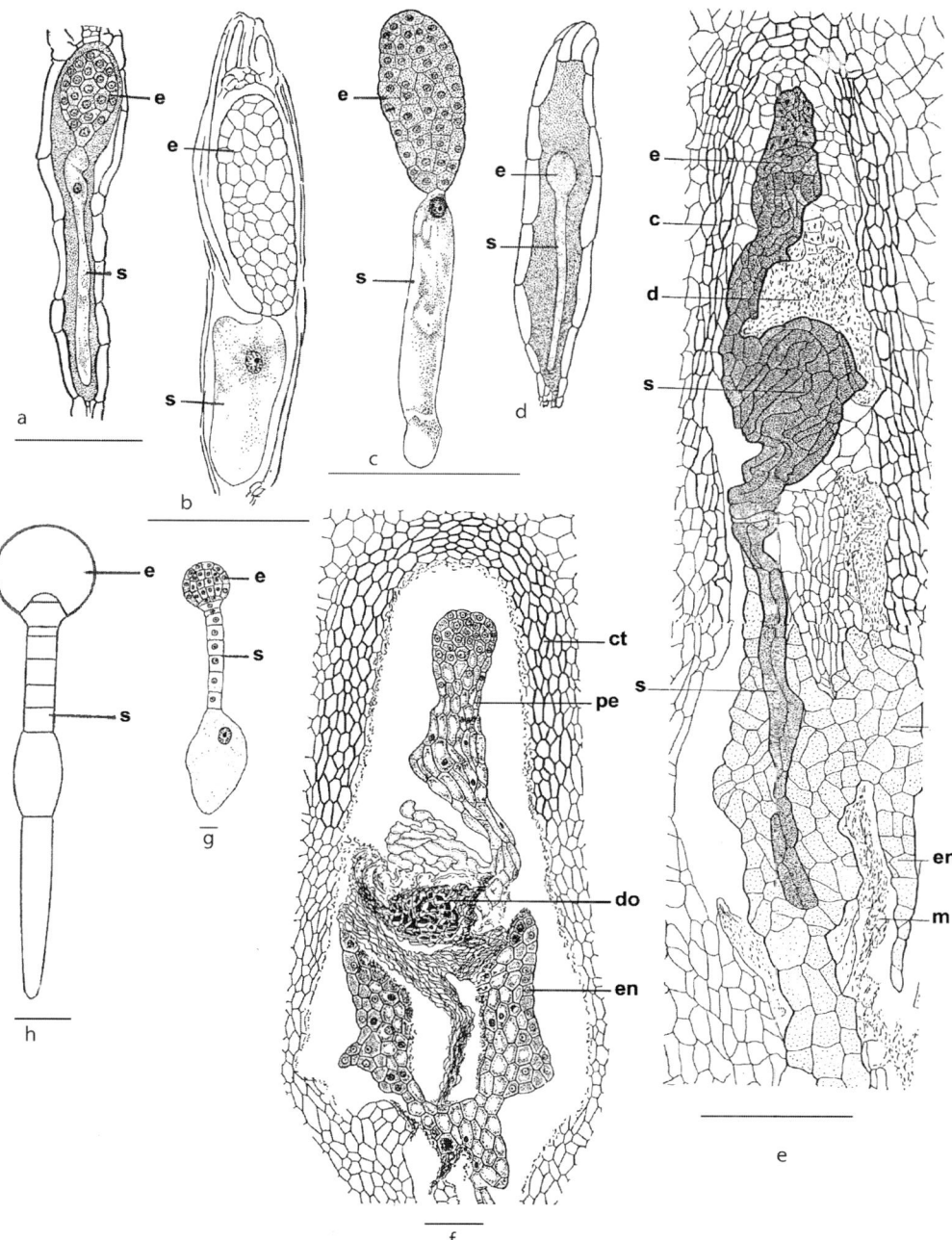

Fig. 4.1a–h Variations in suspensor morphology in representatives of different families. **a–d** Orchidaceae. **a** Unicellular, elongate suspensor attached to the embryo inside the seed of *Bulbophyllum mysorense*. **b** Unicellular, highly constricted suspensor attached to the embryo inside the seed of *Dendrobium barbatulum*. **c** Unicellular, slightly constricted suspensor attached to the embryo of *Spathoglottis plicata*. **d** Tubular suspensor inside the seed of *Peristeria elata*. (Reprinted from Swamy 1949). **e** Embryo-suspensor complex of *Macrosolen cochinsiensis* (Loranthaceae). (Reprinted from Maheshwari and Singh 1952). **f** Section of the ovule of *Peraxilla tetrapetala* (Loranthaceae) showing the embryo-suspensor complex. (Reprinted from Prakash 1960). **g** Filamentous suspensor and basal cell attached to the globular embryo of *Capsella bursa-pastoris* (Brassicaceae). (Reprinted from Schaffner 1906). **h** Filamentous suspensor attached to the globular embryo of *Diplotaxis erucoides* (Brassicaceae). (Modified from Simoncioli 1974). *c* Collenchyma, *ct* collenchymatous tube, *d* degenerated ovary cells enclosed in the collenchyma, *do* degenerated ovarian tissue, *e* embryo, *en* endosperm, *f* fruit wall, *m* degenerated cells of the mamelon (a structure which arises from the base of the ovary and terminates at the point of origin of the style), *pe* proembryo, *s* suspensor. *Bars* **a–c**, **e** 100 µm (bar in **a** applies also to **d**); **f** 7 µm; **g** 10 µm; **h** 40 µm

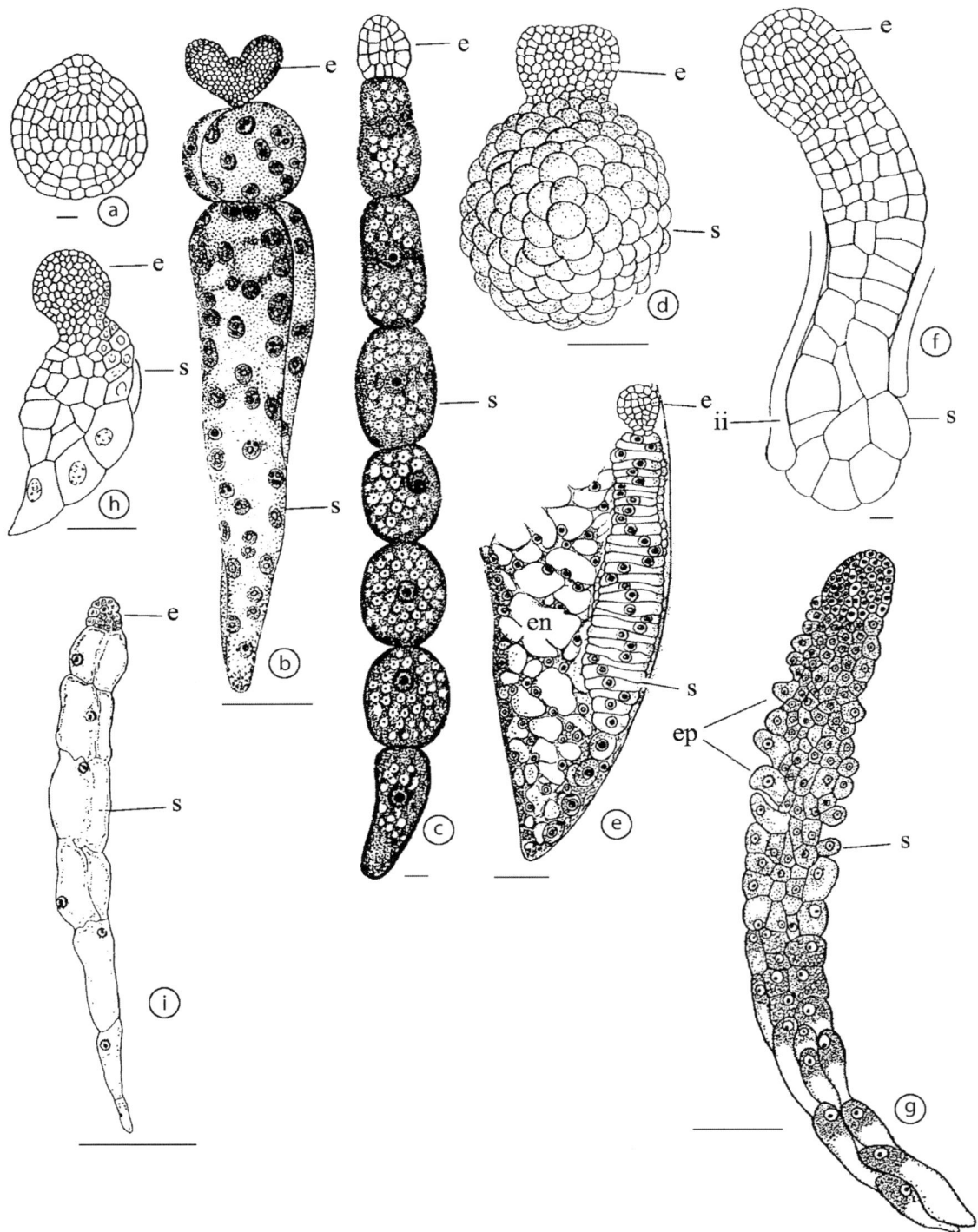

Fig. 4.2a–i Varations in suspensor morphology in the Fabaceae. **a** Suspensorless embryo of *Acacia retinodes*. **b** Enlarged, four-celled suspensor of *Lathyrus angustifolia*. **c** Highly inflated suspensor cells of *Ononis fruticosa*. **d** Suspensor of *Cytisus laburnum* consisting of a large number of inflated cells. **e** Suspensor of *Lupinus pilosus*, composed of short, broad cells. **f** Long, multiseriate suspensor of *Phaseolus multiflorus*. (Reprinted from Guignard 1881). **g** Suspensor of *Crotalaria verrucosa* with bulging epidermal cells. (Reprinted from Rau 1950). **h** Suspensor of *Sophora flavescens* with basal inflated cells. (Reprinted from Nagl 1962). **i** Biseriate filamentous suspensor of *Cicer soongaricum*. (Reprinted from Mercy et al. 1974). e Embryo, *en* endosperm, *ep* bulging epidermal cell, *ii* inner integument, *s* suspensor. Bars **a, c, f** 10 μm; **b, d, e, g, i** 100 μm, **h** 150 μm

Unquestionably, the family in which suspensors have attained their most varied and complex morphological forms is the Fabaceae (legumes). In recognition of these variations, some attention has been paid to separate suspensors of the genera assigned to this family into convenient groups and to identify trends in suspensor evolution (Lersten 1983). Members of two subfamilies, Mimosoideae and Caesalpinioideae, lack a suspensor altogether (Fig. 4.2a), and a few genera have only a rudimentary one. Among suspensor diversity described in the third subfamily, Faboideae, may be mentioned the four-celled suspensor, which enlarges to a great length and volume in *Lathyrus angustifolia*; the uniseriate, filamentous suspensor composed of one to eight highly inflated cells in *Ononis fruticosa*; the suspensor composed almost entirely of a large number of inflated cells in *Cytisus laburnum*; the long, biseriate suspensor composed of short, broad cells in *Lupinus pilosus*; the long, massive, multiseriate suspensor in *Phaseolus multiflorus* (Fig. 4.2b–f; Guignard 1881); the large suspensor with bulging epidermal cells in *Crotalaria verrucosa* (Fig. 4.2g; Rau 1950); the multicellular suspensor with inflated cells toward the base in *Sophora flavescens* (Fig. 4.2h; Nagl 1962); and the long, slender, biseriate, filamentous suspensor in *Cicer soongaricum* (Fig. 4.2i; Mercy et al. 1974). *Phaseolus* is a remarkable genus that shows a range of suspensor morphology embracing species such as *P. tenuiflorus*, which lacks a suspensor, and *P. coccineus*, *P. multiflorus*, and *P. vulgaris*, in which the suspensor attains large dimensions (Nagl 1974). Some correlations have been found between the increasingly complex morphology of the suspensor and the decreasing amount of endosperm in seeds of certain members of the Fabaceae, but the correlation does not hold true in many other members of the family.

Support for the presumed role of the suspensor in the absorption and translocation of nutrients to the growing embryo has come from the presence of haustorial outgrowths of suspensor cells that invade the endosperm and even maraud the extraembryonal tissues of the ovule. Members of Crassulaceae, Fumariaceae, Orchidaceae, Podostemaceae, Rubiaceae, Trapaceae, and Tropaeolaceae provide some of the best-studied examples of suspensor haustoria, but only a few selected examples will be described here. The suspensor of *Tropaeolum majus* (Tropaeolaceae) deserves mention, for its growth dwarfs that of the embryo and its massive haustoria invade virtually all parts of the seed (Fig. 4.3a). One end of the suspensor is attached to the embryo by a rosette of elongate cells, and at the other end there is a cellular mass from which two multicellular branches arise. The initial destination of one branch is the integument near the micropylar end and, from its vantage point, this branch grows around the ovule in the cells of the carpel to form a chalazal or carpel haustorium. The other branch traverses through the integument and funiculus into the vascular bundle of the placenta as the placental haustorium (Walker 1947; Nagl 1976b). In *Sedum ternatum* (Crassulaceae), the suspensor is a three-celled structure of which two cells are derived from the terminal cell of the first division of the zygote and the third cell is the basal cell formed from this division (Fig. 4.3b). A particularly intriguing behavior of the basal cell is that it becomes large and vesicular and, escaping through the micropyle, it grows as a tubular structure between the nucellus and the inner integument. That is not the end of the journey of the tube, as it finally ends up as a network of intracellular filaments in the raphe of the seed (Subramanyam 1963). In contrast to the spectacular growth of suspensor haustoria in the above two examples, in certain genera of Orchidaceae such as *Cymbidium* and *Geodorum*, the tubular haustoria that arise from the suspensor cells display a much shorter distance of activity (Fig. 4.3c) (Swamy 1942; Yeung et al. 1996). In *C. sinense*, changes in microtubule orientation from a longitudinal configuration parallel to the long axis to a transverse alignment accompany the elongation of suspensor cells (Huang et al. 1998).

4.1.1
Subcellular Morphology of the Suspensor

At first glance, the conclusion from the above observations that complex haustorial structures are involved in nutrient transfer appears to be justified. Considering that only a small part of the published work on suspensor morphology has been considered here, nowhere near as much attention has been devoted to the submicroscopic cytology morphology of the cells of the suspensor. This information is important to our understanding of the role of the suspensor and of its haustorial processes in the dynamics of nutrient transfer. Although this goal is far from being achieved, some aspects of the ultra-

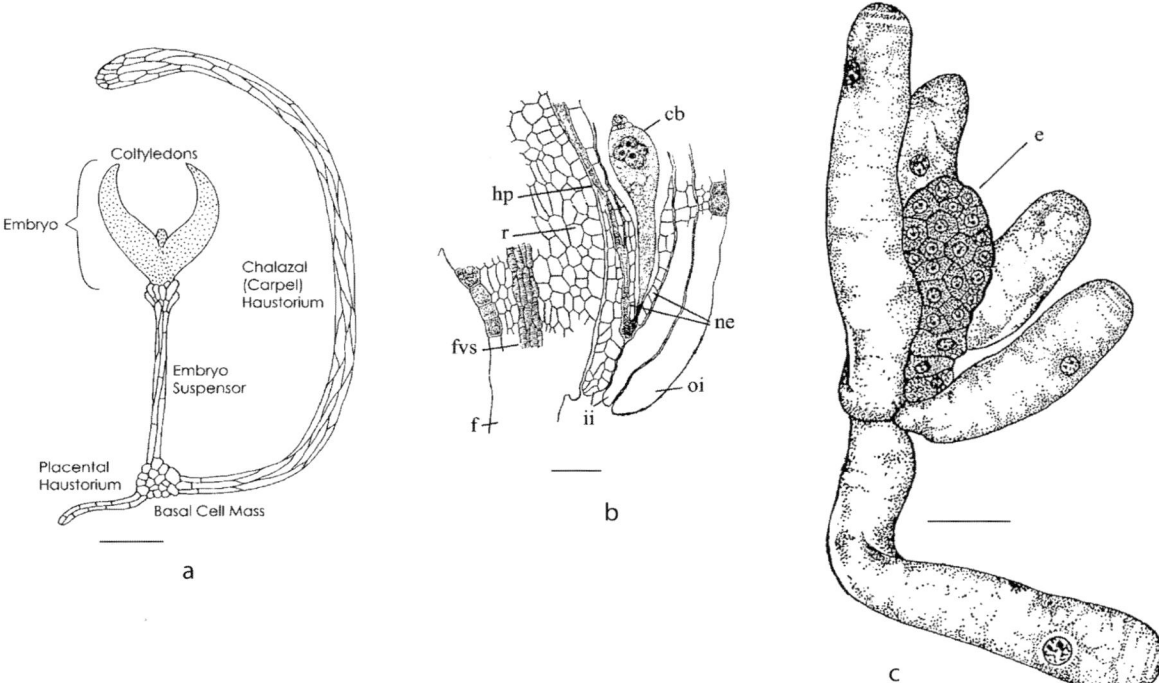

Fig. 4.3a–c Haustorial growth of suspensors. **a** Semi-diagrammatic representation of the embryo-suspensor complex of *Tropaeolum majus*. (Reprinted from Nagl and Kühner 1976). **b** Section of the ovule of *Sedum ternatum* showing the growth of the suspensor haustorium as a tubular structure from the basal cell and its ramifications into parts of the ovule. (Reprinted from Subramanyam 1963). **c** Embryo-suspensor complex of *Cymbidium bicolor*, showing the tube-like suspensor haustoria. (Reprinted from Swamy 1942). *cb* Basal cell, *e* embryo, *f* funicuus, *fvs* funicular vascular strand, *hp* tubular haustorium, *ii* inner integument, ne nucellar epidermis, *oi* outer integument, *r* raphe. Bars **a** 1 mm; **b**, **c** 100 µm

structure of suspensor cells pointing to their possible functional attributes have become clear and these are considered below.

The essential facts about the subcellular morphology of suspensor cells were established more than three decades ago by the work of Schulz and Jensen (1968, 1969) on *Capsella bursa-pastoris*, and the ideas on suspensor function generated by this work have stood the test of time well. Like cells of the embryo, cells of the suspensor have all the trappings of a typical plant cell. However, unlike cells of the embryo, suspensor cells are highly vacuolate; the latter also contain more ER and dictyosomes, but fewer ribosomes than the embryo cells. Ribosome density decreases further to vanishing point when cytoplasmic degeneration begins in cells of the suspensor of the heart-shaped embryo. The basal cell at the micropylar end of the suspensor has a huge vacuole, surrounded by a thin layer of cytoplasm. The chalazal end wall of the basal cell, as well as the end walls of other cells of the suspensor, have numerous plasmodesmata that maintain symplastic connection between the embryo, suspensor, and the basal cell, but there are no plasmodesmata in the walls separating the suspensor and the basal cell from the central cell. An especially fascinating feature of the suspensor cells and the basal cell is the presence of finger-like projections extruding from the cell walls (Fig. 4.4). These arise from the outer lateral walls of certain suspensor cells of the globular embryo and peak in number and complexity at the heart-shaped stage of the embryo. As for their more intimate details, the projections have a dense central core and a peripheral electron translucent matrix, very much like the filiform apparatus of the synergids. These projections, which abut the endosperm, are lined by a plasma membrane continuous with the ER of the endosperm and are closely associated with numerous and varied organelles, especially mitochondria and dictyosomes. An elaborate network of invaginations facing the cytoplasm decorates the inner wall of the micropylar and lateral parts of the basal

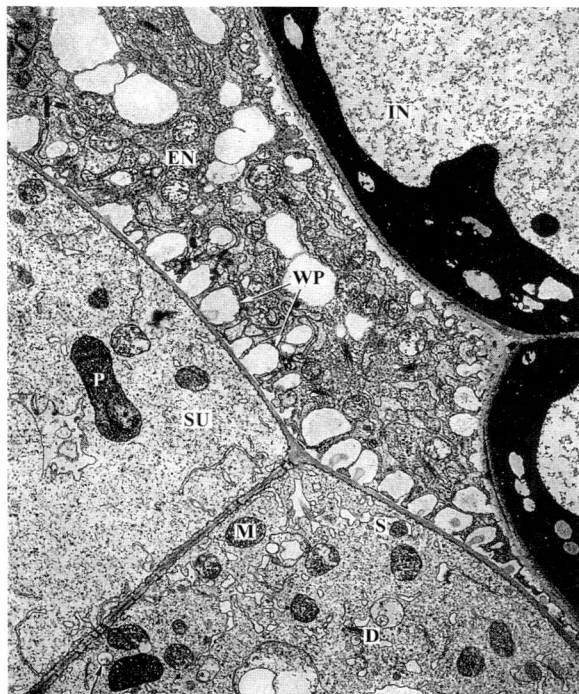

Fig. 4.4 An electron micrograph showing parts of cells of the suspensor, endosperm, and integument of *Capsella bursa-pastoris* at the heart-shaped embryo stage. Numerous plasmodesmata are seen on the end wall of the suspensor cell. Wall projections abutting into the endosperm are seen outside the lateral walls of the suspensor. The cytoplasm of the suspensor cell has begun to degenerate and contains many large membranous structures encircling vesicles and organelles. Dictyosomes, mitochondria, plastids, spherosomes, and multivesicular bodies (*black arrow, bottom left*) are present in the cytoplasm of the suspensor cell. *D* Dictyosome, *EN* endosperm, *IN* integument, *M* mitochondrion, *P* plastid, *S* spherosome, *SU* suspensor, *WP* wall projections. *Bar* 1 μm. (Reprinted from Schulz and Jensen 1969)

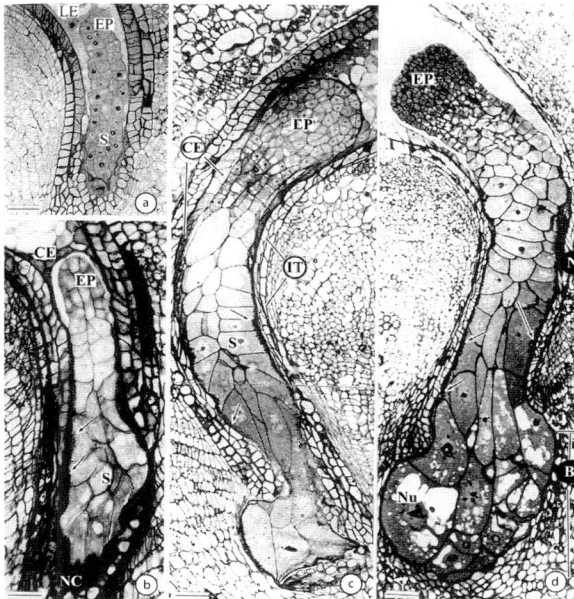

Fig. 4.5a–d Development of the embryo and suspensor in *Phaseolus coccineus*. **a** Early-stage proembryo showing very little structural differentiation between the embryo and suspensor. **b** Late-stage proembryo showing wall ingrowths in the suspensor (*arrows*). The basal region of the suspensor is pressed against the nucellar cap. **c** Early globular-stage embryo with suspensor. The suspensor is in an advanced stage of development and has ingrowths (*arrows*) along the wall adjacent to the integumentary tapetum and the cellular endosperm. **d** Early heart-shaped embryo with suspensor. The suspensor has well-developed wall ingrowths and there is a basal-terminal gradation in nuclear size in the suspensor cells, with the largest in the basal cells. *B* Basal region of the suspensor, *CE* cellular endosperm, *EP* embryo proper, *IT* integumentary tapetum, *LE* liquid endosperm, *N* neck of the suspensor, *NC* nucellar cap, *Nu* nucleus, *S* suspensor. *Bars* **a** 30 μm, **b** 16 μm, **c** 13 μm, **d** 22 μm. (Reprinted from Yeung and Clutter 1978)

cell; these wall labyrinths are lined by the plasma membrane of the basal cell. In fact, the beginnings of formation of invaginations from the wall of the basal cell can convincingly be traced to the wall at the micropylar end of the zygote. Ultrastructural investigation of the *Arabidopsis* suspensor is limited to the basal cell soon after its incarnation from the zygote, and the results are similar to those described in *C. bursa-pastoris*, at least insofar as the development of wall convolutions and the association of mitochondria and dictyosomes with the wall processes are concerned (Mansfield and Briarty 1991).

From other investigations on the subcellular morphology of suspensor cells published in the wake of the work on *C. bursa-pastoris*, two additional studies focused on the wall invaginations and their associated organelles are germane here. An investigation on *Phaseolus coccineus* refers to the interesting fact that there is little difference between cells of the embryo proper and of the suspensor in their general cytological features prior to the globular embryo stage. From this stage onwards, the development of plasma-membrane-lined wall invaginations from the suspensor cells in a polarized fashion proceeding from the basal cells to cells close to the embryo proper is a hallmark of suspensor ultrastructure (Fig. 4.5a–d). The wall projections first appear in the more basal suspensor cells of the globular embryo, and they attain their maximum concentration in the heart-shaped embryo. Beginning with the outer walls of suspensor cells adjacent to the integumentary tapetum, the projections are subsequently also formed in the inner walls of suspensor cells, with a sharp basal side restriction. Mitochondria, plastids, dictyosomes, polysomes, and smooth ER greatly increase in abundance in cells soon after the wall invaginations appear, but in the heart-shaped embryo, there is a decrease in their number and a change in the configuration of the ER. These changes suggest that the functional life of the suspensor is confined to a narrow window between the globular and heart-shaped embryos, although most of the suspensor cells persist through maturation of the embryo (Yeung and Clutter 1978, 1979).

Despite the fact that embryo development in *Stellaria media* (Caryophyllaceae) is of the Caryophyllad type in which suspensor function is attributed to derivatives of the terminal cell of the two-celled proembryo, it is the undivided basal cell of the proembryo that displays features in common with suspensor cells of *C. bursa-pastoris* and *P. coccineus*. The fact that the basal cell is several times larger than the two-celled embryo is of outstanding significance. First, it means that this cell functions in a mechanistic way in pushing the embryo deep into the central cell as it expands and extends beyond the micropyle into the nucellus. Second, the presence of large, highly differentiated plastids, microbodies, and wall projections at the micropylar end of the cell, and concentration of mitochondria near the wall projections, constitute clear evidence that the basal cell has a physiological role akin to suspensor cells. In the torpedo-shaped embryo, a suspensor composed of five to nine linearly arranged cells produced by division of the terminal cell of the two-celled proembryo is present; these cells are endowed with unique types of plastids, microbodies, and extensive profiles of dilated ER, which set them apart from the cells of the embryo (Newcomb and Fowke 1974). The array and concentration of organelles present in the basal cell and the suspensor cells, compared to those present in the embryo cells, suggests that the suspensor maintains a type of metabolism different from that of the embryo.

Two common subcellular features present in the suspensor cells of some plants that might lead to fundamental insights into their function are the presence of specialized plastids and the absence of deposits of cuticular material on the cell walls. First described in the suspensor cells of *Pisum sativum*, the plastids contain spherical bodies of intertwined proteinaceous tubules that are not converted into grana. The specialized plastids, which are absent from cells of the embryo, constitute the most abundant organelles of suspensor cells. The idea that these organelles may serve as sites for the accumulation of specific proteins that are subsequently used for the growth of the embryo is certainly appealing, but is in need of experimental evidence (Marinos 1970). Plastids described in suspensor cells of *Phaseolus vulgaris* (Schnepf and Nagl 1970), *Ipomoea purpurea* (Convolvulaceae; Ponzi and Pizzolongo 1973), *Stellaria media* (Newcomb and Fowke 1974), *Tropaeolum majus* (Nagl and Kühner 1976), *P. coccineous* (Yeung and Clutter 1979), *Medicago sativa* (Fabaceae) and *M. scutellata* (Sangduen et al. 1983), although differing in details, are somewhat analogous to those found in *P. sativum*. Among the changes found in the plastids of suspensor cells of *Vicia faba* (Fabaceae) is their transformation into chromoplasts containing crystals of carotenoids (Wredle et al. 2000). As established by fluorescent microscopy, the absence of cuticular substances on the walls of suspensor cells of certain plants contrasts with their presence on walls of the protoderm cells of the embryo, and is believed to facilitate nutrient uptake by the suspensor (Rodkiewicz et al. 1994; Lackie and Yeung 1996; Yeung et al. 1996).

4.1.2
Nuclear Cytology of the Suspensor

Nearly all ultrastructural studies have described the nucleus of suspensor cells as an undistinguished organelle, scarcely different in structure from the nucleus of other plant cells. The relatively simple structure of the nucleus, however, belies a surfeit of cytological changes to which this organelle is dedicated during the limited life span of the suspensor cells. Much of this work has been reviewed by D'Amato (1984) and Raghavan (1986) and only a few classical investigations are considered here.

It was mentioned earlier that embryo suspensors of members of the Fabaceae attain complex morphological forms. The variations in suspensor morphology that occur in members of this family are matched to some extent by the aberrant nuclear behavior of suspensor cells. A relatively simple nuclear phenomenon was uncovered by light microscopic observations in *Pisum sativum*, which produces just two long suspensor cells. Although nuclear divisions proceed normally in the newly cut off cells, cytokinesis is abruptly and permanently halted. The result is that by the time the suspensor cells attain their maximum length, the nucleus divides repeatedly to produce two cellular sacs of 64 free nuclei each. Disintegration of the nuclei precedes the eventual collapse of the suspensor in the mature embryo (Cooper 1938). Appearance of irregular numbers of free nuclei of different sizes in the two apical and the two basal cells of the suspensor in several species of *Lathyrus* has engendered the notion that, besides differences in the number of mitotic cycles, occurrence of restitution nuclei also accounts for nuclear abnormalities (Nagl 1962).

An unexpected finding that emerged from cytological investigations of the suspensor of *Phaseolus coccineus* is that after a multicellular suspensor is formed, the nuclei go through repeated DNA synthesis by endoreduplication, leading to the formation of giant chromosomes within a disproportionately large nucleus. These chromosomes also become polytenic due to the presence of the many parallel fibrils that result from repeated replication of chromatids that are huddled together in the chromosome without separating (Nagl 1962, 1967). Like polytene chromosomes in the salivary glands of certain insects, those of suspensor cells of *P. coccineus* and *P. vulgaris* also display, albeit in a much less striking way than in insects, puffs and loops that are simply chromosomal bands in which chromatin has undergone local decondensation, and hence represent sites of DNA synthesis and transcription (Nagl 1967, 1969; Avanzi et al. 1970). Based on measurements of nuclear volume, the maximum degree of endoreduplication in suspensor cells of *P. coccineus*, in terms of DNA content per nucleus, was estimated to be 4,096C, but later work using quantitative Feulgen microspectrophotometry showed that it could be as high as 8,192C. The geometric increase in the DNA content with progressive embryogenesis shows that an accurate mechanism to ensure the replication of all DNA in the nucleus at each round is operating in the cells. In individual suspensors, one can recognize a progressive increase in the level of endoreduplication beginning with low degree in cells between the junction of the embryo and suspensor, medium degree in cells in the neck region of the suspensor, and a very high degree of endoreduplication in the large cells of the basal part of the suspensor (Nagl 1962; Brady 1973). The level of endoreduplication is also reflected in the morphological appearance of nuclei: whereas nuclei of cells with a low degree of endoreduplication appear normal, those of cells showing a very high degree of endoreduplication have abnormal chromosomes condensed into a large mass of chromatin.

Comparative studies have disclosed the unsuspected fact that the degree of endoreduplication and polyteny in suspensor cells varies widely in different species of *Phaseolus*, ranging from 256C in *P. lunatus* and *P. tuberosus*, 1,024C in *P. acutifolius*, 4,096C in *P. hysterinus* and *P. multiflorus*, and 8,192C in *P. coccineus*. Even in the same species, some disturbingly low and high values have been scored, such as 16C and 512C in two cultivars of *P. mungo* and 8C, 128C, 512C, 1,024C, 2,048C, and 4,096C in several cultivars of *P. vulgaris* (Nagl 1974). Especially in these latter examples, one cannot exclude the possibility that a quantitative analysis of DNA content using plants grown under uniform and controlled conditions of growth will yield more convincing data than we have at present. A wide range of DNA values has also been documented in the endoreduplicated embryo suspensor cells of plants belonging to the Alismataceae [*Alisma lanceolatum* 128C (Hasitschka-Jenschke 1959); *A. plantago-aquatica* 1,024C (Bohdanowicz 1973)], Brassicaceae [*Eruca sativa* 75C (Corsi et al. 1973)], Caryophyllaceae (*Melandrium album*, *M. rubrum* 128C), Tunicaceae [*Tunica saxifraga* 32C (Nagl 1962)], Geraniaceae [*Geranium phaeum* 32C (Nagl 1962)], Trapaceae [*Trapa natans* 256C (Nagl 1962)], and Tropaeolaceae [*Tropaeolum majus* 2,048C (Nagl 1976a)], among others.

The function of the polytene chromosomes of suspensor cells is not yet entirely clear, although some molecular-cytological studies have implicated a role in gene amplification. In this context, the extrusion of micronucleoli from the nucleolus as well as from the condensed heterochromatic and the loose euchromatic regions and loops of the polytene chromosomes of suspensor cells of *P. coccineus* and *P. vulgaris* deserves to be mentioned. These micronucleoli presumably contain RNA and protein ensheathed by chromatin, and their final destination has been traced to the cytoplasm into which they are liberated by disintegration of the nuclear membrane (Avanzi et al. 1970; Nagl 1970a, 1973). In situ hybridization of ^{3}H-rRNA with DNA of the polytene chromosomes of *P. coccineus* has shown that the micronucleoli contain rDNA, encoding the 28S and 18S rRNA (Avanzi et al. 1972). This is considered as cytological evidence for the amplification of ribosomal cistrons during polytenization of chromosomes, although the possible amplification of nonribosomal RNA is not indubitably eliminated by this observation. The occurrence of selective amplification of DNA in polytene chromosomes has been deduced from the banding patterns of DNA isolated from roots, shoots, and suspensor cells of *P. coccineus* that appear following analytical ultracentrifugation. Whereas DNA from all three tissues appears as bands with buoyant densities of 1.700 g/ml and 1.692 g/ml, DNA from the suspensor contains a satellite band at 1.696 g/ml that is not detected in the other two tissues (Fig. 4.6). The appearance of the satellite DNA is believed to be in keeping with a selective amplification of genes other than rRNA that occurs in the suspensor cells (Lima-de-Faria et al. 1975). Although high transcriptional activity thus appears to be a hallmark of suspensor cells, the nature of the gene products produced, and their contribution to the function of the suspensor during its short life-span, require further study.

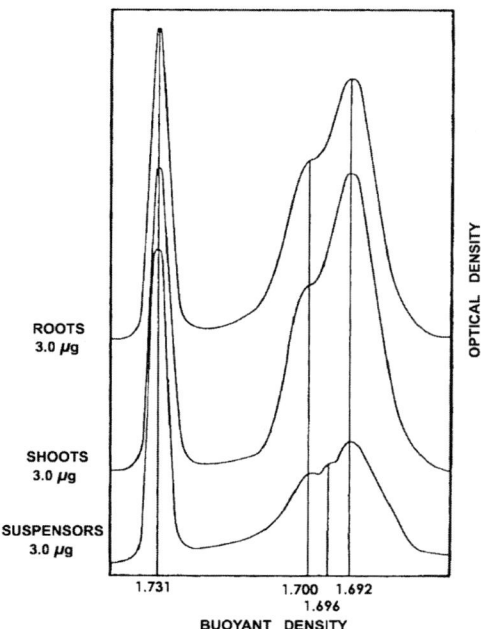

Fig. 4.6 Banding of DNA isolated from roots, shoots, and suspensors of *Phaseolus coccineus*. A satellite DNA with a buoyant density of 1.696 g/ml is present in the suspensors, but not in roots or shoots. (Reprinted from Lima-de-Faria et al. 1975)

4.1.3
Functional Physiology of the Suspensor

From the perspective of the possible physiological function of the suspensor, perhaps the most salient feature of its subcellular morphology is the presence of wall invaginations. The wall ingrowths assign the suspensor cells the task of the legendary transfer cells believed to facilitate the transport of solutes into and out of cells. The fact that the wall projections are lined with the plasma membrane is of significance in increasing the surface area of the cell for absorption, whereas the proximity to the wall projections of numerous organelles might be conjured up to support active transport of nutrients across the plasma membrane. The plasmodesmata connecting the suspensor cells to one another and to the embryo may serve to maintain an open channel of communication for the flow of solutes between the suspensor and the embryo. So, the view that has captured the imagination of plant embryologists regarding the role of suspensor is that it is a dynamic assemblage of cells that absorb metabolites from the endosperm, or from the surrounding diploid cells of the ovule, delivering them to the growing embryo. In reflecting upon the fact that the suspensor is a short-lived organ, the precise stage at which the embryo depends upon the suspensor for nutrition is a crucial element in any analysis of its function. In this context, the nature and importance of the metabolites released by degenerating suspensor cells for nurture of the embryo also cannot be ignored and requires study. Although the subcellular morphology of suspensor haustoria has not been investigated, there is little reasonable doubt that the haustoria interact with cells of the ovule, just as the suspensor cells do with cells of the endosperm.

Clearly, evidence for the potential function of the suspensor is mostly descriptive and comes from static electron micrographs; however, it is reassuring that limited physiological investigations have supported the role predicted by descriptive accounts. The importance of the suspensor may be easily demonstrated in experiments involving the culture of isolated embryos with and without an attached suspensor. Successful culture of globular embryos of *Eruca sativa* with or without the suspensor is difficult but early heart-shaped embryos grow and complete the embryogenic program in culture even in the absence of the suspensor, although the presence of the attached suspensor appreciably improves growth of the latter type of embryos. Interestingly, growth in culture of slightly older embryos is not promoted by the presence of the suspensor (Corsi 1972). Embryos of *Phaseolus*, which have a fairly robust suspensor, have offered promising material for testing the role of the suspensor in embryo growth and nutrition, and the fact that these embryos can be manipulated in culture has enhanced their usefulness. As in *E. sativa*, the requirement for an attached suspensor is stage-specific for growth in culture of embryos of *P. coccineus* and is acutely felt for heart-shaped embryos, but not for cotyledon-stage embryos; even a detached suspensor kept in contact with a cultured heart-shaped embryo is sufficient to potentiate respectable embryo growth (Yeung and Sussex 1979). Indicative of a functional relationship, the growth-promoting effect of the suspensor on heart-shaped embryos is found to coincide precisely with the attainment of maximum structural specialization by suspensor cells. Impressive support for the role of the suspensor in nutrient transport has come from studies on the uptake and incorporation of radioactively labeled metabolites such as sucrose and putrescine (a polyamine) into embryos of *P. coccineus* and *P. vulgaris*. When ^{14}C-sucrose is administered through pods, ovules, or isolated embryos of *P. vulgaris*, much of the radioactivity is found to accumulate in the suspensor or in the suspensor pole of late heart-shaped embryos. A similar pattern of uptake is observed when the label is introduced into the endosperm cavity of seeds of both species. It was also found that administration of ^{14}C-sucrose close to the embryo through the endosperm at the chalazal end of the ovule of *P. vulgaris* results in a higher uptake of the label in the suspensor than in the embryo. This would not happen if metabolites of the endosperm were absorbed directly by the growing embryo. A related observation that the uptake of ^{14}C-sucrose by the suspensor is sensitive to dinitrophenol is consistent with the view that the process is energy-dependent (Yeung 1980). Autoradiography of incorporation of ^{3}H- and ^{14}C-putrescine administered through the pod or directly to ovules of *P. coccineus* has also shown that the label is transported to the growing embryo through the suspensor (Nagl 1990).

Another possible factor involved in promotion of embryo growth by the suspensor is the array of its endogenous nutrients, especially hormones, that might be transmitted to the growing embryo. Analysis of hormone content by bioassays paid off initially with the identification of gibberellins or GA (Alpi et al. 1975), cytokinins (Lorenzi et al. 1978), and ABA (Perata et al. 1990) in the suspensor of *P. coccineus*, followed by the identification of IAA and/or GA in the suspensors of *Tropaeolum majus* and *Cytisus laburnum* (Przybyllok and Nagl 1977; Picciarelli et al. 1984). The best characterized hormone of the suspensor is GA. In later investigations, combined use of gas chromatography and mass spectrometry led to the demonstration of a diversity of gibberellins such as GA_1, GA_4, GA_5, GA_6, GA_8, and GA_{44} in the suspensor of *P. coccineus* (Alpi et al. 1979; Picciarelli and Alpi 1986; Piaggesi et al. 1989), GA_{63} in the suspensor of *T. majus* (Picciarelli and Alpi 1987), and GA_1 and an unknown GA in the suspensor of *C. laburnum* (Picciarelli et al. 1991). In *P. coccineus*, the GA content of the suspensor is nearly 30 times higher than that of the heart-shaped embryo, whereas at the cotyledonary stage there is a dramatic decrease in the hormone content of the suspensor and a significant increase in the organogenetic part of the embryo. This is considered to accord with the view that GA is transported from the suspensor to the embryo (Alpi et al. 1975). The ability of the suspensor to synthesize GA at the stage when it is presumed to be essential for the growth of the embryo is a crucial piece of evidence to support the view that the suspensor supplies the hormone to the embryo. That the suspensor functions as the site of GA synthesis was shown by the demonstration that cell-free extracts from suspensors can be primed to synthesize GA biosynthetic intermediates such as kaurene and ent-7α-hydroxykaurenoic acid and, finally, GA from appropriate exogenous precursors (Ceccarelli et al. 1979, 1981a, 1981b).

Culture of isolated embryos of *P. coccineus* has provided some clues as to the relationship of GA and cytokinins to suspensor function. Extirpation of the suspensor reduces the survival of heart-shaped embryos in a mineral salt medium; however, consistent with the high GA content of the suspensor described in the previous paragraph, addition of low concentrations of GA to the medium can substitute for the suspensor and rescue the embryos.

In contrast, the same levels of GA are inhibitory to the growth and survival of suspensor-deprived post-cotyledonary stage embryos as compared to intact embryos of the same age grown in a hormone-free medium. This observation is in harmony with the gibberellin autonomy attained by post-cotyledonary stage embryos (Cionini et al. 1976). It is known that polar cytokinins of low biological activity such as zeatin glucoside and zeatin riboside predominate in the heart-shaped embryos of *P. coccineus*, whereas cotyledonary stage embryos contain mostly the active cytokinins, zeatin and 2-isopentenyladenine. As a manifestation of the changing cytokinin requirements of embryos of different ages, growth of early-stage embryos is favored by high concentrations of zeatin and low concentrations of zeatin riboside in the medium, whereas late-stage embryos are mostly insensitive to the addition of any cytokinin to the medium (Lorenzi et al. 1978; Bennici and Cionini 1979).

The above observations raise the question of how hormones of suspensor origin promote the growth of the attached embryo. Preliminary data show that GA acts both at the transcriptional and translational levels in suspensor cells. Using cytological techniques, GA has been shown to induce the formation of micronucleoli and chromosomal puffs in the suspensor nuclei, and to promote the incorporation of ^{3}H-uridine into RNA of the suspensor cells of *P. coccineus* and *P. vulgaris* (Nagl 1970b; Forino et al. 1992). Translational activity has been invoked as an explanation for the stimulatory role of the suspensor in embryo growth, as well as for the effect of GA in inducing growth in culture of suspensorless embryos of *P. vulgaris*. This is based on observations such as a decrease in protein content of embryos cultured without the suspensor or cultured with the suspensor but not in organic connection, decreased amino acid incorporation into embryos cultured under the same conditions, and the ability of exogenous GA to restore the protein content and protein synthesizing activity of cultured suspensorless embryos (Brady and Walthall 1985; Walthall and Brady 1986). Now that we have begun to understand the spreading influence of GA on suspensor cells at the molecular and biochemical level, a more targeted approach to decipher the function of this and other hormones in the suspensor should receive attention.

4.1.4
Developmental Physiology and Programmed Death of Suspensor Cells

As part of its developmental program, the suspensor synthesizes a host of enzymes and macromolecules that keep the cells in prime metabolic condition during the early part of their life. Later, as the suspensor goes into decline, synthesis of another set of enzymes, accompanied by a series of distinct subcellular changes, collectively leads to the disintegration and death of the cells. These activities of the suspensor present insights as well as enigmas about its function.

Cytochemical detection of enzymes, especially those of oxidative metabolism, has revealed that suspensor cells of globular- to torpedo-shaped embryos of *Brassica campestris* are metabolically active, with peak enzyme titers corresponding to the termination of growth of the suspensor (Malik et al. 1976a, 1976b). Similarly, in *Tropaeolum majus*, enzyme activity in the pentose phosphate pathway is found to be higher in the suspensor and its haustoria than in the attached embryos at globular- to early cotyledonary-stages (Bhalla et al. 1981). By extrapolation of the results, the suspensor could be envisaged to use these enzymes in a variety of ways to facilitate the secretion of nutrients as well as the uptake and transport of metabolites from the endosperm to the growing embryo. In further pursuit of the developmental physiology of the suspensor, investigations on the comparative transcriptional and translational activities of the suspensor and embryo of *P. coccineus* have shown that, at most stages of embryo development analyzed, the suspensor cells contain more RNA and proteins, far in excess of their own needs, and synthesize them more efficiently than do cells of embryos of comparable age (Walbot et al. 1972; Sussex et al. 1973). Based on calculations made on the rate of RNA synthesis per unit of diploid gene copy, it seems that there is a gene dosage effect on transcription and that polyteny probably accounts for the synthesis of large quantities of RNA by suspensor cells of embryos at certain developmental stages (Clutter et al. 1974). As described in Chap. 2, indicative of a gene expression program different from that of the embryo, mRNAs that accumulate preferentially in the suspensor of *Arabidopsis* (Dornelas et al. 1999), maize (Ingram et al. 2000), and *P. coccineus* (Weterings et al. 2001), have been identified.

Sensitive immunohistological techniques have been employed to localize the storage proteins phaseolin, in cells of the fully developed suspensor of *P. coccineus* (Nagl et al. 1991), and 7S vicilin and 12S legumin in cells of the developing suspensor of *Vicia faba* (Panitz et al. 1995) and tobacco (Panitz et al. 1999). According to Panitz et al. (1995), seed storage protein genes coding for vicilin and legumin exhibit a biphasic expression pattern in suspensor cells of *V. faba*, as these immunologically detectable globulins appear transiently in suspensor cells of the globular embryo, and then disappear coincident with the synthesis of storage proteins in the endosperm. A legumin gene of *V. faba* transferred into tobacco is also expressed abundantly in cells of the fully-formed suspensor of transgenic embryos (Panitz et al. 1997). The high protein content of suspensor cells has raised the question as to whether they serve a transient storage function for reserve products. The appearance and disappearance of storage proteins at precisely defined stages in the development of the suspensor gives them a functional significance in embryo nutrition, as this occurs before the endosperm storage proteins become available.

Another point to consider about the developmental physiology of the suspensor relates to the inevitable failure of its cellular machinery leading to degeneration and death. Electron microscopy has been invaluable in uncovering changes in the organization of suspensor cells as they go into a decline. In *Capsella bursa-pastoris*, cytoplasmic degeneration is initiated when the suspensor attains its maximum number of ten cells, and a variety of cytoplasmic episodes, such as depletion of ribosomes, loss of nucleic acids and proteins, decrease in the size of nucleoli, replacement of long parallel strands of ER with a few short pieces, and appearance of blobs of cytoplasm engulfed by single membranes analogous to autophagic vacuoles, herald the changes by which the integrity of cells is compromised. Eventually, the cell wall is weakened and the suspensor is crushed by the growing embryo. The basal cell has a slightly longer life span than the suspensor cells, although it too succumbs to degenerative signals (Schulz and Jensen 1969). In *P. coccineus*, *P. vulgaris*, and *Tropaeolum majus*, autolytic processes begin in the basal suspensor cells. The sequestration of portions of the cytoplasm within swollen plastids by invagination of their double membrane and formation of multivesicular bodies bounded by the ER are considered as early hallmarks of suspensor dysfunction in *P. coccineus* and *P. vulgaris*. In the latter species, leucoplasts also undergo various changes to form multivesicular bodies. During the terminal phase of suspensor autolysis in *P. coccineus*, rupture of the double membrane of the modified plastids is followed by release of the disintegrating cytoplasm as autophagic vacuoles and separation of polytene chromosomes into individual units. Moreover, the lysis of suspensor cells in a polar fashion beginning with the large cells at the micropylar end is believed to ensure that materials of the lysed cells are utilized by the growing embryo (Nagl 1976c, 1977; Gärtner and Nagl 1980). In *T. majus*, swelling of mitochondria and their transformation into autophagic vacuoles, condensation of the cytoplasm, and release of organelles into the vacuole are viewed as early markers of suspensor cell lysis, which terminates with the rupture of the tonoplast, disorganization of the protoplast, and pycnosis of the nucleus (Gärtner and Nagl 1980). Evidence that hydrolases can play a causative role in the disintegration of suspensor cells has come from the localization of acid phosphatase activity in the degenerating suspensor cells of *P. vulgaris*. The high enzyme content of the modified plastids provides one explanation for the vulnerability of suspensor cells to signals for self-destruction (Gärtner and Nagl 1980). Besides acid phosphatase, other hydrolases such as acetylesterase and alkaline phosphatase are also involved in the autolysis of the suspensor and its haustoria in *T. majus*. The high enzyme activities observed in the suspensors of heart-shaped and cotyledonary stages of embryos coincide with a rapid decline in their protein contents, implying a role for hydrolases rather than proteases in the breakdown of proteins (Singh et al. 1980).

A well-established theme in developmental biology is that cell death is a regular accompaniment of ordered growth of multicellular organisms, most of which have also perfected genetic mechanisms to remove unwanted cells by pcd. The term apoptosis is often equated with pcd, although, strictly speaking, it refers to the stereotypical morphological and biochemical changes displayed by cells in which an intracellular death program is activated. Apoptosis in animal cells, where it has been studied intensely, incorporates a series of distinct subcellular changes, such as blebbing of the cell membranes, cell fragmentation, chromosome condensation, and orderly internucleosomal cleavage of DNA. The only suspensor system so far investigated that has lived up to expectations as a useful model for elucidating the ultrastructural and physiological pathways of pcd, including the demonstration of fragmentation of nuclear DNA by fluorescent end-labeling of the 3'-hydoxyl termini by TUNEL assay during death of suspensor cells, is *Vicia faba* (Wredle et al. 2001). However, critical data indicating internucleosomal fragmentation by the "laddering" of electrophoresed suspensor DNA combined with TUNEL assay for a definitive categorization of suspensor demise as apoptosis are lacking. Indeed, as will be seen in the following section, physiological and genetic studies in certain plants have unearthed a hidden developmental potential of the suspensor leading to a new round of cell divisions and morphogenesis replacing the programmed death of cells.

4.2
Genetic Control of Suspensor Form

Suspensors are always found attached to embryos, and this tight relationship with the embryo defines a suspensor. Whereas suspensorless embryos derived from both daughter cells of two-celled proembryos are the norm in a few plants, a suspensor alone, whether generated from the basal or terminal cell of the proembryo, has not been described as a functional unit in angiosperm embryogenesis. One key point highlighted by this observation is that the suspensor is not an isolated entity but owes its existence to the embryo, which apparently exercises regulatory control over the final form of the suspensor. This provocative link between suspensor and embryo has been amplified by genetic and molecular investigations, which have also yielded insights into the nature of the developmental interactions between the suspensor and the embryo.

In what appears to represent an important advance in our understanding of the control of suspensor fate, damage to proembryos caused by chronic irradiation of flowers, carpels, or ovules of *Nicotiana rustica* (Devreux and Scarascia Mugnozza 1962), *Capsella bursa-pastoris* (Devreux 1963), and *Arabidopsis* (Gerlach-Cruse 1969; Akhundova et al. 1978) by X- or γ-rays was found to potentiate the formation of additional cells in the suspensor. As seen in Fig. 4.7 (a–d), in contrast to the uniseriate file of about six cells in the suspensor of *Arabidopsis*, long and thick suspensors are formed by further transverse and longitudinal divisions of the original packet of cells when carpels were X-irradiated and subsequently pollinated by nonirradiated pollen grains (Gerlach-Cruse 1969). Hobbie et al. (2000) have described a variety of abnormalities in the suspensor of the auxin-insensitive *Arabidopsis* mutant, *axr6*. Although the mutation causes lesions in the orientation and timing of cell divisions in both the embryo and suspensor, the latter becomes especially vulnerable as it lacks a hypophysis and becomes two or more cells wide along most of its length. Aberrant divisions in the hypophysis combined with the formation of a long suspensor are the hallmarks of an embryo-lethal mutation caused by the loss of function of an auxin-binding protein in *Arabidopsis* (Chen et al. 2001). These observations suggest that auxin, either directly or indirectly through a receptor mediating auxin action, has a regulatory role in determining suspensor identity. Aberrant cell divisions seen on a modest scale in the suspensor cells of the *Arabidopsis* mutant *vacuoleless1* (*vcl1*), remarkable for the lack of vacuoles in suspensor cells, have been attributed to a derangement in the flow of nutrients resulting from the failure of vacuole biogenesis (Rojo et al. 2001).

Experiments using seeds of *Eranthis hiemalis* (Ranunculaceae), which contain an undifferentiated embryo and a fully differentiated suspensor, have revealed the morphogenetic potential of the latter to differentiate into a secondary embryo. This occurs when cells of the rudimentary embryo are killed by treatment of seeds with acidic buffer solutions at pH 4.0 (Haccius 1963). The conclusion from this experiment is that when the embryo is in an active mode of growth, suspensor growth remains repressed; removal of the source of inhibition unleashes to varying degrees the morphogenetic potential of the suspensor. This is an intriguing correlation that has also been supported by the identification of several embryo-lethal mutants of *Arabidopsis* that produce aborted seeds with abnormal suspensors. In collections of mutants obtained by mutagenesis of *Arabidopsis* seeds by

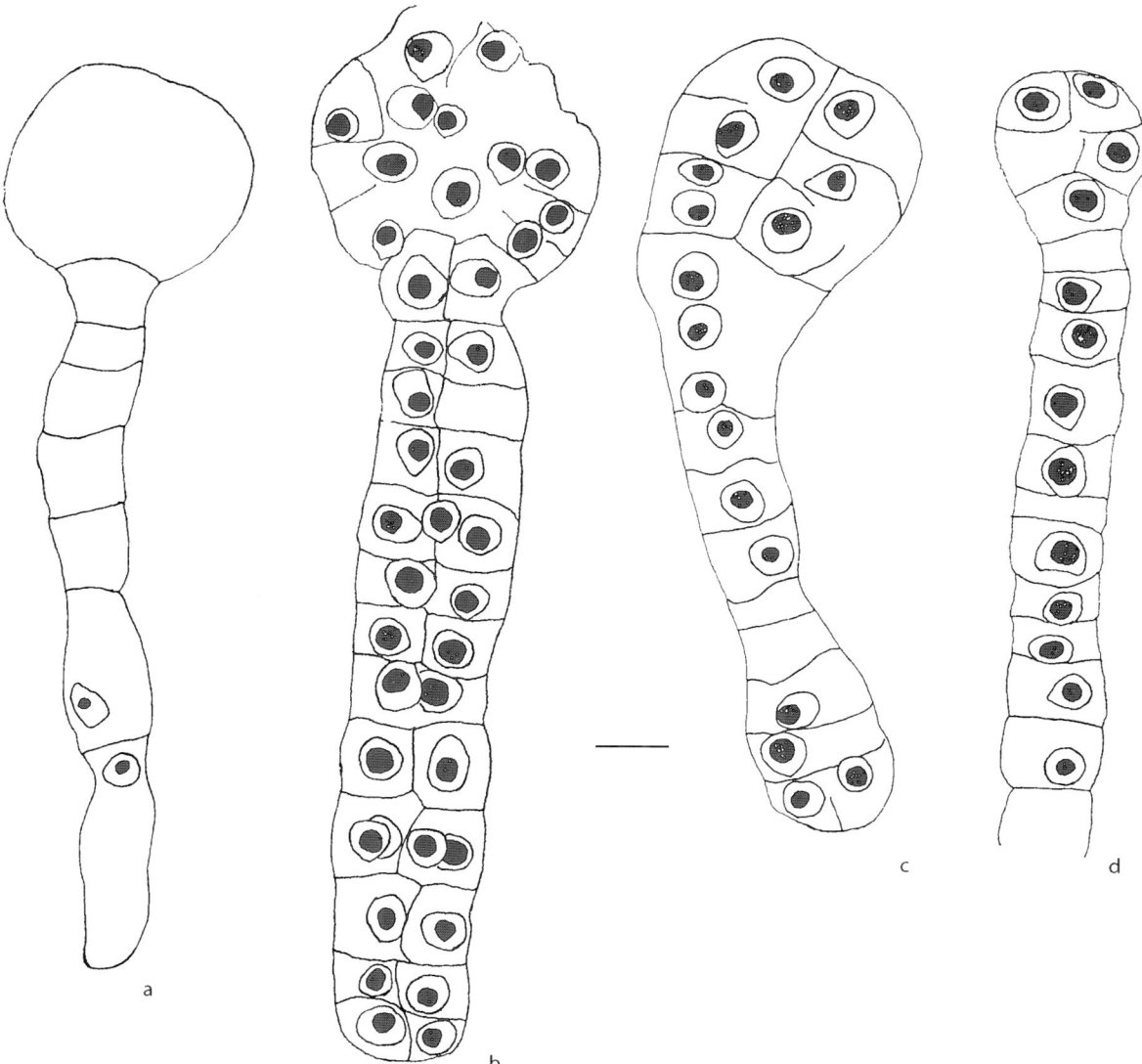

Fig. 4.7a–d Effect of X-irradiation of carpels of *Arabidopsis*, followed by pollination with nonirradiated pollen grains, on growth of the embryo and attached suspensor. **a** Globular embryo-suspensor complex of unirradiated control. **b** Degenerating globular embryo with a massive long suspensor from an irradiated carpel (12 krad), 8 days after pollination. **c** Globular embryo showing early signs of degeneration, with a suspensor showing division of cells at the micropylar end, from an irradiated carpel (4 krad), 4 days after pollination. **d** Early globular-stage embryo with a long suspensor from an irradiated carpel (2 krad), 6 days after pollination. *Bar* 10 µm. (Reprinted from Gerlach-Cruse 1969)

ethylmethane sulfonate (EMS) and by transferred-DNA (T-DNA) insertion following transformation with *Agrobacterium tumefaciens*, arrest of embryo growth accompanied by abnormal growth of the suspensor seems to be the invariant scenario in aborted seeds. From the geneticist's angle, this means that a missing gene product is required for the development of the embryo, but lack of this protein does not apparently impair the growth and further differentiation of the suspensor. In one class of EMS-generated mutants analyzed in detail, embryo growth is arrested around the globular stage, although the attached suspensor continues to proliferate to form a multitiered structure with as many as 160 cells. Growth of the suspensor is, however, terminated before the seed becomes filled, and signs of degeneration subsequently begin to appear in the suspensor cells (Marsden

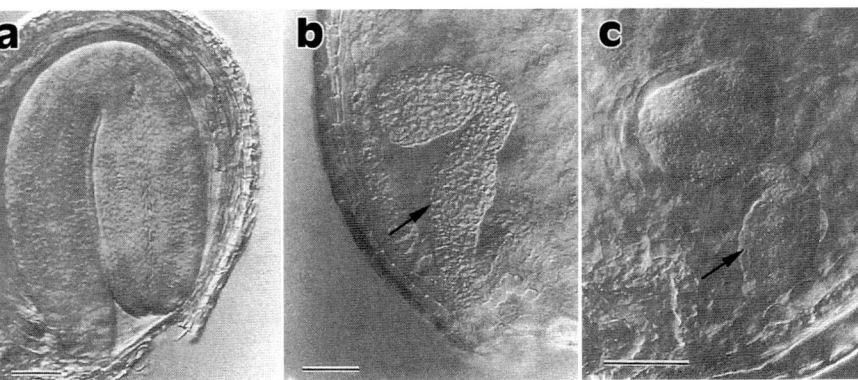

Fig. 4.8a–c Wild-type embryo and embryos of *sus* mutants of *Arabidopsis*. **a** Wild-type embryo at the cotyledon-stage. **b** Embryo of the *sus1-1* mutant. **c** Embryo of the *sus2-1* mutant. Mutant embryos correspond to the cotyledon-stage of the wild-type embryo. *Arrows-*Massive part of the suspensor. *Bars* 50 μm. (Reprinted from Schwartz et al. 1994)

and Meinke 1985). Multitiered suspensors are reported in embryos of the *lec1* mutant (Lotan et al. 1998), of transgenic plants in which the expression of the *L1L* (for *LEC1-LIKE*) gene is suppressed by RNA interference (Kwong et al. 2003), and of an embryo-lethal mutant designated as *embryo-defective development1* (*edd1*) identified by transposon mutagenesis (Uwer et al. 1998). Mutants such as *suspensor* (*sus*) and *raspberry* (*rsy*), obtained by chemical mutagenesis or by screening T-DNA mutagenized lines, also exhibit lesions in the globular stage embryo accompanied by the formation of massive, multiseriate suspensor. Intriguingly, cells of the abnormal suspensor acquire embryo-like characteristics such as storage products in *sus* mutants and transcripts of the *Arabidopsis* 12S storage protein gene in *rsy1* and *rsy2* mutants. The homology of the suspensor of *rsy1* and *rsy2* mutants to the embryo has been strengthened by the expression of the protoderm-specific gene *AtLTP1* in the suspensor cells in a pattern similar to that in the embryo. The suspensor proliferation observed in the *sus* mutant, shown in Fig. 4.8 (a–c), is similar to that seen in the *rsy* mutants (Yeung and Meinke 1993; Schwartz et al. 1994; Yadegari et al. 1994; Apuya et al. 2002). Based on the localization of the *RSY3* gene product in the chloroplasts, and the defective development of chloroplasts in the mutant, it has been predicted that the mutation disrupts a pathway for the production of signaling molecules in the chloroplasts for embryo development (Apuya et al. 2002). A central problem in analyzing the suspensor-embryo relationship in these mutants is whether the anarchic growth of the suspensor is due to disruption in the growth of the embryo. This appears to be the case, as morphogenetic defects in the embryos of *sus* mutants appear to precede visible changes in the suspensor (Schwartz et al. 1994). A partial developmental program of the embryo expressed in the modified suspensor cells of *sus* and *rsy* mutants is fully realized in *twin* (*twn*) mutants, in which the suspensor regenerates an additional one or two embryos in the ovule. In the *twn1* mutant, suspensor transformation occurs in the presence of an actively growing embryo, whereas in the *twn2* mutant, embryo growth is arrested following division of the zygote or of the two-celled embryo. Although the early division sequences of the suspensor cell in the embryogenic pathway were not followed, the final product appears to be similar to the embryo formed by the division of the zygote (Vernon and Meinke 1994; Zhang and Somerville 1997; Vernon et al. 2001). Two models based on a general theme uniting *sus* and *twn* mutants with altered suspensor morphology have at their core the idea that both activation of a suspensor-specific program and repression of an embryogenic program are involved in specifying the suspensor, and that products of specific genes such as *SUS* and *TWN* initiate signaling pathways required for normal embryo development and suspensor identity (Fig. 4.9a,b). The primary difference between the models is whether abnormality in the suspensor is due to defects in signal production in the embryo or in signal perception by the suspensor. According to one model, communication between the embryo and suspensor does not directly involve *SUS* gene products, which are necessary for embryo growth, but a signal produced in the embryo maintains suspensor cell identity. The second model envisages a direct role for the *SUS* gene in the production of signals for directing embryo development and suspensor morphology (Schwartz et al. 1994). A new idea as to how signals generated in the potential embryogenic cells are received and interpreted by suspensor cells has come from the analysis of embryogenesis in a *twn2* mutant isolated by T-DNA insertion at the 5'-untranslated region of a valyl-tRNA synthetase gene. When confronted with the mutation-induced failure of divisions in the apical cell of the two-celled embryo, suspensor cells are quick to proliferate to form secondary embryos. The inference is that a deficiency in an essential valyl-tRNA due to the mutation prevents the synthesis of adequate amounts of a signaling factor in the apical cells that suppresses the proliferation potential of the basal suspensor cells (Zhang and Somerville 1997). It appears that the isolation of mutants with defects in suspensor function is only a beginning toward a complete understanding of the dynamics of the suspensor-embryo relationship in *Arabidopsis* and other plants.

4.3 Concluding Comments

Although the organogenetic part of the embryo and the suspensor have their origin in the products of the first division of the zygote, the differences in their function are as striking as the differences in their final form. Our current understanding of the function of the suspensor owes much to its association with the growing embryo. Early in development, embryos of flowering plants are faced with the problem of their nutrition, a problem that has puzzled investigators for nearly a century. For part of this time, the debate centered on the role of the suspensor in channeling nutrients from the endosperm and ovular tissues to the growing embryo. Several ultrastructural studies of suspensors have spurred efforts to interpret the role of membrane-lined invaginations on their walls in the uptake of nutrients from the surrounding milieu and their transfer to the embryo. A breakthrough in understanding of the signaling between the suspensor and the embryo came with the discovery that when the growth of the embryo is stymied by a mutation, the suspensor begins to divide and even becomes embryo-like in structure. The nature of the signals produced in the embryo that maintain suspensor cell identity may prove to be the most intriguing part of this unfolding link and the next few years will probably witness a vigorous pursuit of this problem.

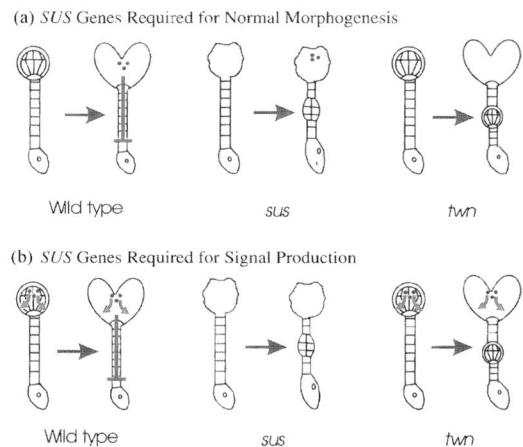

Fig. 4.9a,b Models of *SUS* gene action in embryogenesis of *Arabidopsis*. **a** In the model indicating a requirement for *SUS* genes in wild-type embryos, a signal (*dots*) is produced in the embryo and transported to the suspensor, resulting in maintenance of suspensor cell identity and inhibition of its further growth (*grey bar*). Disruption of embryo morphogenesis in the *sus* mutant results in failure of signal transduction to the suspensor, leading to additional cell divisions in the suspensor. In the *twn* mutant, the signal is either not produced or not received by the suspensor. **b** In the model indicating a requirement for *SUS* genes for signal production, the signal is required for early embryogenesis (*arrows*) and later to maintain suspensor cell identity (*grey bar*). Absence of signal in the *sus* mutant leads to disruption of embryo morphogenesis and initiation of additional divisions in the suspensor. The *twn* mutation is believed to disrupt signal transduction in the suspensor, but not in the embryo. (Reprinted from Schwartz et al. 1994)

REFERENCES

Akhundova GG, Grinikh LI, Shevchenko VV (1978) The embryogenesis in *Arabidopsis thaliana* following the γ-irradiation of the plants in the generative phase. Ontogenez 9:514–519

Alpi A, Tognoni F, D'Amato F (1975) Growth regulator levels in embryo and suspensor of *Phaseolus coccineus* at two stages of development. Planta 127:153–162

Alpi A, Lorenzi R, Cionini PG, Bennici A, D'Amato F (1979) Identification of gibberellin A_1 in the embryo suspensor of *Phaseolus coccineus*. Planta 147:225–228

Apuya NR, Yadegari R, Fischer RL, Harada JH, Goldberg RB (2002) *RASPBERRY3* gene encodes a novel protein important for embryo development. Plant Physiol 129:691–705

Avanzi S, Cionini PG, D'Amato F (1970) Cytochemical and autoradiographic analyses on the embryo suspensor cells of *Phaseolus coccineus*. Caryologia 23:605–638

Avanzi S, Durante M, Cionini PG, D'Amato F (1972) Cytological localization of ribosomal cistrons in polytene chromosomes of *Phaseolus coccineus*. Chromosoma 39:191–203

Bennici A, Cionini PG (1979) Cytokinins and in vitro development of *Phaseolus coccineus* embryos. Planta 147:27–29

Bhalla PL, Singh MB, Malik CP (1981) Characterization of pentose phosphate pathway in embryo suspensor of *Tropaeolum majus*. Biochem Physiol Pflanz 176:789–792

Bohdanowicz J (1973) Karyological anatomy of the suspensor in *Alisma* L. I. *Alisma plantago-aquatica* L. Acta Biol Cracov Ser Bot 16:235–246

Brady T (1973) Feulgen cytophotometric determination of the DNA content of the embryo proper and suspensor cells of *Phaseolus coccineus*. Cell Differ 2:65–75

Brady T, Walthall ED (1985) The effect of the suspensor and gibberellic acid on *Phaseolus vulgare* embryo protein content. Dev Biol 107:531–536

Ceccarelli N, Lorenzi R, Alpi A (1979) Kaurene and kaurenol biosynthesis in cell-free system of *Phaseolus coccineus* suspensor. Phytochemistry 18:1657–1658

Ceccarelli N, Lorenzi R, Alpi A (1981a) Gibberellin biosynthesis in *Phaseolus coccineus* suspensor. Z Pflanzenphysiol 102:37–44

Ceccarelli N, Lorenzi R, Alpi A (1981b) Kaurene metabolism in cell-free extracts of *Phaseolus coccineus* suspensors. Plant Sci Lett 21:325–332

Chen J-G, Ullah H, Young JC, Sussman MR, Jones AM (2001) ABP1 is required for organized cell elongation and division in *Arabidopsis* embryogenesis. Genes Dev 15:902–911

Cionini PG, Bennici A, Alpi A, D'Amato F (1976) Suspensor, gibberellin and in vitro development of *Phaseolus coccineus* embryos. Planta 131:115–117

Clutter ME, Brady T, Walbot V, Sussex I (1974) Macromolecular synthesis during plant embryogeny. Cellular rates of RNA synthesis in diploid and polytene cells in bean embryos. J Cell Biol 63:1097–1102

Cooper DC (1938) Embryology of *Pisum sativum*. Bot Gaz 100:123–132

Corsi G (1972) The suspensor of *Eruca sativa* Miller (Cruciferae) during embryogenesis in vitro. G Bot Ital 106:41–54

Corsi G, Renzoni GC, Viegi L (1973) A DNA cytophotometric investigation on the suspensor of *Eruca sativa* Miller. Caryologia 26:531–540

D'Amato F (1984) Role of polyploidy in reproductive organs and tissues. In: Johri BM (ed) Embryology of angiosperms. Springer, Berlin Heidelberg New York, pp 519–566

Devreux M (1963) Effets de l'irradiation gamma chronique sur l'embryogenèse de *Capsella bursa-pastoris* Moench. In: L'Energia Nucleare in Agricoltura, VI Congr Nucl (Roma). Comitato Nazionale Energia Nucleare, Rome, pp 199–217

Devreux M, Scarascia Mugnozza GT (1962) Action des rayons gamma sur les premiers stades de developpement de l'embryon de *Nicotiana rustica* L. Caryologia 15:279–291

Dornelas MC, Wittich P, von Recklinghausen I, van Lammeren A, Kreis M (1999) Characterization of three novel members of the *Arabidopsis SHAGGY*-related protein kinase (*ASK*) multigene family. Plant Mol Biol 39:137–147

Forino LMC, Tagliasacchi AM, Cavallini A, Cionini G, Giraldi E, Cionini PG (1992) RNA synthesis in the embryo suspensor of *Phaseolus coccineus* at two stages of embryogenesis, and the effect of supplied gibberellic acid. Protoplasma 167:152–158

Gärtner PJ, Nagl W (1980) Acid phosphatase activity in plastids (plastolysomes) of senescing embryo-suspensor cells. Planta 149:341–349

Gerlach-Cruse D (1969) Embryo- und Endospermentwicklung nach einer Röntgenbestrahlung der Fruchtknoten von *Arabidopsis thaliana* (L.) Heynh. Radiation Bot 9:433–442

Guignard L (1881) Recherches d'embryogénie végétale comparée. 1st Mémoire: Légumineuses. Ann Sci Nat Bot Ser 6, 12:5–166

Haccius B (1963) Restitution in acidity-damaged plant embryos – regeneration or regulation? Phytomorphology 13:107–115

Hanstein J (1870) Die Entwicklung des Keimes der Monokotylen und Dikotylen. Botanische Abhandlungen aus dem Gebiet der Morphologie und Physiologie, I. Marcus, Bonn

Hasitschka-Jenschke G (1959) Bemerkenswerte Kernstrukturen im Endosperm und im Suspensor zweier Helobiae. Österr Bot Z 106:301–314

Hobbie L, McGovern M, Hurwitz LR, Pierro A, Liu NY, Bandyopadhyay A, Estelle M (2000) The *axr6* mutants of *Arabidopsis thaliana* define a gene involved in auxin response and early development. Development 127:23–32

Huang B-Q, Ye X-L, Yeung EC, Zee SY (1998) Embryology of *Cymbidium sinense*: the microtubule organization of early embryos. Ann Bot 81:741–750

Ingram GC, Boisnard-Lorig C, Dumas C, Rogowsky PM (2000) Expression patterns of genes encoding HD-ZipIV homeodomain proteins define specific domains in maize embryos and meristems. Plant J 22:401–414

Jürgens G, Mayer U (1994) *Arabidopsis*. In: Bard J (ed) Embryos. Color atlas of development. Wolfe, London, pp 7–21

Kwong RW, Bui AQ, Lee H, Kwong LW, Fischer RL, Goldberg RB, Harada JJ (2003) LEAFY COTYLEDON1-LIKE defines a class of regulators essential for embryo development. Plant Cell 15:5–18

Lackie S, Yeung EC (1996) Zygotic embryo development in *Daucus carota*. Can J Bot 74:990–998

Lersten NR (1983) Suspensors in Leguminosae. Bot Rev 49:233–257

Lima-de-Faria A, Pero R, Avanzi S, Durante M, Ståhle U, D'Amato F, Granström H (1975) Relation between ribosomal RNA genes and the DNA satellites of *Phaseolus coccineus*. Hereditas 79:5–20

Lorenzi R, Bennici A, Cionini PG, Alpi A, D'Amato F (1978) Embryo-suspensor relations in *Phaseolus coccineus*: cytokinins during seed development. Planta 143:59–62

Lotan T, Ohto M, Yee KM, West MAL, Lo R, Kwong RW, Yamagishi K, Fischer RL, Goldberg RB, Harada JJ (1998) *Arabidopsis* LEAFY COTYLEDON1 is sufficient to induce embryo development in vegetative cells. Cell 93:1195–1205

Maheshwari P, Singh B (1952) Embryology of *Macrosolen cochinchinensis*. Bot Gaz 114:20–32

Malik CP, Singh M, Thapar N (1976a) Physiology of sexual reproduction: IV. Histochemical characteristics of suspensor of *Brassica campestris*. Phytomorphology 26:384–389

Malik CP, Vermani S, Bhatia DS (1976b) Physiology of sexual reproduction. III. Histochemical characteristics of suspensor during embryo development in *Brassica campestris* Linn. Var., Sarson. Acta Histochem 37:178–182

Mansfield SG, Briarty LG (1991) Early embryogenesis in *Arabidopsis thaliana*. II. The developing embryos. Can J Bot 69:461–476

Marinos NG (1970) Embryogenesis of the pea (*Pisum sativum*) II. An unusual type of plastid in the suspensor cells. Protoplasma 71:227–233

Marsden MPF, Meinke DW (1985) Abnormal development of the suspensor in an embryo-lethal mutant of *Arabidopsis thaliana*. Am J Bot 72:1801–1812

Mercy ST, Kakar SN, Varghese TM (1974) Embryology of *Cicer arietinum* and *C. soongaricum*. Bull Torrey Bot Club 101:26–30

Nagl W (1962) Über Endopolyploidie, Restitutionskernbildung und Kernstrukturen im Suspensor von Angiospermen und einer Gymnosperme. Österr Bot Z 109:431–494

Nagl W (1967) Die Riesenchromosomen von *Phaseolus coccineus* L.: Baueigentümlichkeiten, Strukturmodifikationen, zusätzliche Nukleolen und Vergleich mit den mitotischen Chromosomen. Österr Bot Z 114:171–182

Nagl W (1969) Puffing of polytene chromosomes in a plant (*Phaseolus vulgaris*). Naturwissenschaft 56:221–222

Nagl W (1970a) Temperature-dependent functional structures in the polytene chromosomes of *Phaseolus*, with special reference to the nucleolus organizers. J Cell Sci 6:87–107

Nagl W (1970b) Differentielle RNS-Synthese an pflanzlichen Riesenchromosomen. Ber Dtsch Bot Ges 83:301–309

Nagl W (1973) Origin and fate of the micronucleoli in the giant cells of the *Phaseolus* suspensor. Nucleus 16:100–109

Nagl W (1974) The *Phaseolus* suspensor and its polytene chromosomes. Z Pflanzenphysiol 73:1–44

Nagl W (1976a) Early embryogenesis in *Tropaeolum majus* L.: Evolution of DNA content and polyteny in the suspensor. Plant Sci Lett 7:1–6

Nagl W (1976b) Early embryogenesis in *Tropaeolum majus* L.: ultrastructure of the embryo-suspensor. Biochem Physiol Pflanz 170:253–260

Nagl W (1976c) Ultrastructural and developmental aspects of autolysis in embryo-suspensors. Ber Dtsch Bot Ges 89:301–311

Nagl W (1977) "Plastolysomes" – plastids involved in the autolysis of the embryo-suspensor in *Phaseolus*. Z Pflanzenphysiol 85:45–51

Nagl W (1990) Translocation of putrescine in the ovule, suspensor and embryo of *Phaseolus coccineus*. J Plant Physiol 136:587–591

Nagl W, Kühner S (1976) Early embryogenesis in *Tropaeolum majus* L.: diversification of plastids. Planta 133:15–19

Nagl W, Seeliger T, Hartmann A (1991) Immunohistochemical localization of the storage protein phaseolin in the embryo-suspensor of *Phaseolus coccineus*. Eur J Basic Appl Histochem 35:175–183

Newcomb W, Fowke LC (1974) *Stellaria media* embryogenesis: the development and ultrastructure of the suspensor. Can J Bot 52:607–614

Panitz R, Borisjuk L, Manteuffel R, Wobus U (1995) Transient expression of storage-protein genes during early embryogenesis of *Vicia faba*: synthesis and metabolization of vicilin and legumin in the embryo, suspensor and endosperm. Planta 196:765–774

Panitz R, Manteuffel R, Bäumlein H, Wobus U (1997) Biphasic expression of a *Vicia faba* legumin B gene in developing seeds of transgenic tobacco. J Plant Physiol 150:115–126

Panitz R, Manteuffel R, Wobus U (1999) Tobacco embryogenesis: storage-protein-accumulating cells of embryo, suspensor, and endosperm are able to undergo cytokinesis. Protoplasma 207:31–42

Perata P, Picciarelli P, Alpi A (1990) Pattern of variations in abscisic acid content in suspensors, embryos, and integuments of developing *Phaseolus coccineus* seeds. Plant Physiol 94:1776–1780

Piaggesi A, Picciarelli, P, Lorenzi R, Alpi A (1989) Gibberellins in embryo-suspensor of *Phaseolus coccineus* seeds at the heart stage of embryo development. Plant Physiol 91:362–366

Picciarelli P, Alpi A (1986) Gibberellins in suspensors of *Phaseolus coccineus* L. seeds. Plant Physiol 82:298–300

Picciarelli P, Alpi A (1987) Embryo-suspensor of *Tropaeolum majus*: identification of gibberellin A_{63}. Phytochemistry 26:329–330

Picciarelli P, Alpi A, Pistelli L, Scalet M (1984) Gibberellin-like activity in suspensors of *Tropaeolum majus* L. and *Cytisus laburnum* L. Planta 162:566–568

Picciarelli P, Piaggesi A, Alpi A (1991) Gibberellins in suspensor, embryo and endosperm of developing seeds of *Cytisus laburnum*. Phytochemistry 30:1789–1792

Poddubnaya-Arnoldi VA (1967) Comparative embryology of the Orchidaceae. Phytomorphology 17:312–320

Ponzi R, Pizzolongo P (1973) Ultrastructure of plastids in the suspensor cells of *Ipomoea purpurea* Roth. J Submicrosc Cytol 5:257–263

Prakash S (1960) Morphological and embryological studies in the family Loranthaceae-VI. *Peraxilla tetrapetala* (Linn. F.) van Tiegh. Phytomorphology 10:224–234

Przybyllok T, Nagl W (1977) Auxin concentration in the embryo and suspensors of *Tropaeolum majus*, as determined by mass fragmentation (single ion detection). Z Pflanzenphysiol 84:463–465

Raghavan V (1986) Embryogenesis in angiosperms. A developmental and experimental study. Cambridge University Press, New York

Raghavan V (2001) Life and times of the suspensor of angiosperm embryos. In: Rangaswamy NS (ed) Phytomorphology Golden Jubilee issue 2001: Trends in plant sciences. International Society of Plant Morphologists, Delhi, pp 251–276

Rau MA (1950) The suspensor haustoria of some species of *Crotalaria* Linn. Ann Bot 14:557–562

Rodkiewicz B, Fyk B, Szczuka E (1994) Chlorophyll and cutin in early embryogenesis in *Capsella*, *Arabidopsis*, and *Stellaria* investigated by fluorescent microscopy. Sex Plant Reprod 7:287–289

Rojo E, Gillmor CS, Kovaleva V, Somerville CR, Raikhel NV (2001) VACUOLELESS1 is an essential gene required for vacuole formation and morphogenesis in *Arabidopsis*. Dev Cell 1:303–310

Sangduen N, Kreitner GL, Sorensen EL (1983) Light and electron microscopy of embryo development in perennial and annual *Medicago* species. Can J Bot 61:837–849

Schaffner M (1906) The embryology of the shepherd's purse. Ohio Nat 7:1–8

Schnepf E, Nagl W (1970) Über einige Strukturbesonderheiten der Suspensorzellen von *Phaseolus vulgaris*. Protoplasma 69:133–143

Schulz P, Jensen WA (1969) *Capsella* embryogenesis: the suspensor and the basal cell. Protoplasma 67:139–163

Schulz R, Jensen WA (1968) *Capsella* embryogenesis: the egg, zygote, and young embryo. Am J Bot 55:807–819

Schwartz BW, Yeung EC, Meinke DW (1994) Disruption of morphogenesis and transformation of the suspensor in abnormal *suspensor* mutants of *Arabidopsis*. Development 120:3235–3245

Schwartz BW, Vernon DM, Meinke DW (1997) Development of the suspensor: differentiation, communication, and programmed cell death during plant embryogenesis. In: Larkins BA, Vasil IK (eds) Cellular and molecular biology of plant seed development. Kluwer, Dordrecht, pp 53–72

Simoncioli C (1974) Ultrastructural characteristics of "*Diplotaxis erucoides* (L.) DC" suspensor. G Bot Ital 108:175–189

Singh MB, Bhalla PL, Malik CP (1980) Activity of some hydrolytic enzymes in autolysis of the embryo suspensor in *Tropaeolum majus* L. Ann Bot 45:523–527

Subramanyam K (1963) Embryology of *Sedum ternatum* Michx. J Indian Bot Soc (Maheshwari comm vol.) 52A:259–275

Sussex I, Clutter M, Walbot V, Brady T (1973) Biosynthetic activity of the suspensor of *Phaseolus coccineus*. Caryologia 25[Suppl]:261–272

Swamy BGL (1942) Female gametophyte and embryogeny in *Cymbidium bicolor* Lindl. Proc Indian Acad Sci 15B:194–201

Swamy BGL (1949) Embryological studies in the Orchidaceae. II. Embryogeny. Am Midl Nat 41:202–232

Uwer U, Willmitzer L, Altmann T (1998) Inactivation of a glycyl-tRNA synthetase leads to an arrest in plant embryo development. Plant Cell 10:1277–1294

Vernon DM, Meinke DW (1994) Embryogenic transformation of the suspensor in *twin*, a polyembryonic mutant of *Arabidopsis*. Dev Biol 165:566–573

Vernon DM, Hannon MJ, Le M, Forsthoefel NR (2001) An expanded role for the *TWN1* gene in embryogenesis: defects in cotyledon pattern and morphology in the *twn1* mutant of *Arabidopsis* (Brassicaceae). Am J Bot 88:570–582

Walbot V, Brady T, Clutter M, Sussex I (1972) Macromolecular synthesis during plant embryogeny: rates of RNA synthesis in *Phaseolus coccineus* embryos and suspensors. Dev Biol 29:104–111

Walker RI (1947) Megasporogenesis and embryo development in *Tropaeolum majus* L. Bull Torrey Bot Club 74:240–249

Walthall ED, Brady T (1986) The effect of the suspensor and gibberellic acid on *Phaseolus vulgaris* embryo protein synthesis. Cell Differ 18:37–44

Wardlaw CW (1955) Embryogenesis in plants. Methuen, London

Weterings K, Apuya NR, Bi Y, Fischer RL, Harada JJ, Goldberg RB (2001) Regional localization of suspensor mRNAs during early embryo development. Plant Cell 13:2409–2425

Wredle U, Walles B, Hakman I (2000) Chromoplasts are formed in *Vicia faba* suspensor cells. Int J Plant Sci 161:713–719

Wredle U, Walles B, Hakman I (2001) DNA fragmentation and nuclear degradation during programmed cell death in the suspensor and endosperm of *Vicia faba*. Int J Plant Sci 162:1053–1063

Yadegari R, de Paiva GR, Laux T, Koltunow AM, Apuya N, Zimmerman JL, Fischer RL, Harada JJ, Goldberg RB (1994) Cell differentiation and morphogenesis are uncoupled in *Arabidopsis raspberry* embryos. Plant Cell 6:1713–1729

Ye XL, Zee SY, Yeung EC (1997) Suspensor development in the nun orchid, *Phaius tankervilliae*. Int J Plant Sci 158:704–712

Yeung EC (1980) Embryogeny of *Phaseolus coccineus*: the role of the suspensor. Z Pflanzenphysiol 96:17–28

Yeung EC, Clutter ME (1978) Embryogeny of *Phaseolus coccineus*: growth and microanatomy. Protoplasma 94:19–40

Yeung EC, Clutter ME (1979) Embryogeny of *Phaseolus coccineus*: the ultrastructure and development of the suspensor. Can J Bot 57:120–136

Yeung EC, Meinke DW (1993) Embryogenesis in angiosperms: development of the suspensor. Plant Cell 5:1371–1381

Yeung EC, Sussex IM (1979) Embryogeny of *Phaseolus coccineus*: the suspensor and the growth of the embryo-proper in vitro. Z Pflanzenphysiol 91:423–433

Yeung EC, Zee SY, Ye XL (1996) Embryology of *Cymbidium sinense*: embryo development. Ann Bot 78:105–110

Zhang JZ, Somerville CR (1997) Suspensor-derived polyembryony caused by altered expression of valyl-tRNA synthetase in the *twn2* mutant of *Arabidopsis*. Proc Natl Acad Sci USA 94:7349–7355

5 Genetic and Molecular Control of Embryogenesis – Role of Nonzygotic and Zygotic Genes

The greatest difficulty in the study of genic actions in controlling development is met with when we try to visualize how they are interlocked in order to produce their effects locally, that is, in the three dimensions of space and in the fourth dimension of time, at a given moment. To understand normal development in terms of genic action we must integrate all the individual sources of information on the action of mutant loci and infer the action of the normal genic material. In doing so it should not make any difference what views we hold on the nature of the genic material – the classic gene concept or the modern pattern idea. In any case the information we have is derived almost exclusively from the results of interference of mutant loci with the normal course of development (and the analysis of the factors of development by experimental embryology). Therefore, whatever the constitution of the normal genic material, it must act in the way revealed by mutant interference with its action.

R.B. Goldschmidt 1955

5.1 Asymmetry in Parental Genome Contributions 102
5.1.1 Evidence for Maternal-effect Genes 103
5.1.2 Silencing of Paternal Genes 105
5.2 Gene Activity during Progressive Embryogenesis 106
5.2.1 Gene Expression during Early Embryogenesis 108
5.2.2 Gene Expression during Late Embryogenesis and Transition to Germination 112
5.3 Embryo Gene Expression Program Studied by Mutant Screening 115
5.3.1 Embryo-lethal Mutants of *Arabidopsis* 115
5.3.2 The World of Cytokinesis-Defective Mutants 117
5.3.3 Embryo-defective Mutants of *Arabidopsis* 120
5.3.4 Embryo-Defective Mutants of Maize and Rice 121
5.4 Concluding Comments 123
References 123

In contemplating how an angiosperm embryo with its well-defined shoot and root apical meristems has evolved from a single-celled zygote, it is hard to avoid postulating a role for gene action at successive stages of embryogenesis. The genes activated come either from parts of the parental genomes or the zygotic genome or from both. Indeed, it has long been clear that genetic factors are intrinsically responsible for establishing the polarity and body plan of the early embryo, and are involved in programming the morphogenetic and tissue differentiation processes, general house-keeping chores, and seed protein accumulation at appropriate embryogenic stages. More recently, gene action has been implicated in the lapse into dormancy of embryos during their final stage of development. Analyses of embryo ontogeny in interspecific and intergeneric crosses and in spontaneously occurring mutants generated the first line of evidence showing that genes are in fact providing critical cues during embryogenesis, and a few investigations along these lines have become classics in the plant embryology literature. It was shown in Chap. 3 that the use of modern tools of genetics and molecular biology has led to the identification of genes regulating the development of the shoot and root apical meristems, and those involved in establishment of the apicobasal and radial patterning elements in embryos. The major insights into regulatory programs critical for continued growth and morphogenesis of the embryo from its single-celled beginning are considered in this chapter. Accounts illuminating the role of genetic and molecular mechanisms involved in the maturation of embryos and in the induction of dormancy will be presented in the next chapter.

This chapter begins with the current understanding of the role of non-zygotic, parental genes during early embryogenesis, a topic that has remained latent for a long time, but has suddenly been reawoken and is yielding insightful investigations by geneticists and molecular biologists.

5.1
Asymmetry in Parental Genome Contributions

An established concept in animal embryology is that the unfertilized egg cytoplasm is blessed with templates of stored mRNAs to code for the first proteins necessary to guide the initial development of the embryo. After fertilization, the influence of the zygotic genome over further development generally begins with the blastula stage of the embryo. Initial support for the presence of stored mRNAs in animal eggs came from investigations showing nearly normal development in parthenogenetically activated enucleated eggs of sea urchins, and partial development of fertilized eggs of insects, sea urchins, and amphibians treated with inhibitors of mRNA synthesis such as actinomycin-D and α-amanitin. Following confirmation of the presence of stored templates by in vitro translation of mRNAs extracted from unfertilized eggs of model organisms, further biochemical and molecular studies of these systems have provided a detailed chronology of the changes in messenger abundance in the egg, zygote, and early stage embryos, disappearance of stored templates, and assumption of transcriptional control by the embryo genome. The decisive experiments in this context are reviewed by Davidson (1986). As embryogenesis in flowering plants occurs within the privileged confines of the embryo sac, which itself is embedded in the sporophytic tissues of the ovule, the inaccessibility of egg cells, zygotes, and early-stage embryos has hindered biochemical and molecular investigations in search of stored mRNAs and their disposition during development. However, there is considerable indirect evidence both in favor of, and against, the involvement of the maternal genome in angiosperm embryogenesis. Claims for an independent role for maternal transcripts in directing early embryo development in the absence of fertilization include the stimulation of division of the egg cell by a pollen tube from which sperm have been inactivated by physical and chemical agents (Lacadena 1974), and the purported origin presumably of haploid embryos from egg cells of cultured, unpollinated ovules or ovaries (Yang and Zhou 1982). In a broad sense, a role for maternal programming of embryo development independent of fertilization can be invoked to explain diplosporous and aposporous types of apomixis – where an embryo arises parthenogenically from an unreduced egg cell or from a somatic cell of the ovule – with the caveat that apomicts might be considered as special cases where the sexual processes have been changed over evolutionary time to the apomictic type (Ramachandran and Raghavan 1992; Koltunow 1993; Koltunow and Grossniklaus 2003). A substantial contribution of the maternal genome in promoting embryogenesis and plant formation is seen in interspecific hybrids of *Hordeum vulgare* × *Hordeum bulbosum*. The appearance of *H. vulgare*-like haploid progeny in the cross, coupled with cytological analysis of embryos, led to the conclusion that, following fertilization, the chromosomes of *H. bulbosum* are lost, leaving the zygote at the mercy of the maternal genome to complete its development (Kasha and Kao 1970). Grimanelli et al. (2005) have shown by microarray analyses that the early divisions of the zygote in sexual maize plants, and in apomictic hybrids between maize and its wild relative *Tripsacum*, occur before the onset of changes in the transcript population present in the unfertilized ovules. These findings have bolstered the view that the unfertilized angiosperm egg cell has the potential to initiate the developmental program of the embryo using maternal transcripts. On the other hand, the convergence of discoveries of the production of embryo-like structures in the absence of the maternal environment, namely, somatic embryogenesis (embryogenic pathway followed by somatic cells in tissue culture) and pollen embryogenesis (the transformation of pollen grains of cultured anthers into embryo-like structures), topics both covered in Chap. 9, has led to the assertion that the maternal genome might not be strictly necessary for early divisions in the embryogenic pathway (Russinova and de Vries 2000). That both parental alleles of a large number of genes strewn throughout the embryo genome control embryo development is, however, suggested by the isolation of many recessive embryo-defective and embryo-lethal mutations from *Arabidopsis* and maize, to be described later in this chapter.

5.1.1
Evidence for Maternal-effect Genes

An appreciation of the classical view that formation of viable seeds in flowering plants depends upon the coordinated development of the embryo and endosperm within the haploid female gametophyte, and of the diploid sporophytic ovular tissues surrounding the female gametophyte, has considerably strengthened the evidence for maternal programming of embryogenesis in angiosperms by so-called "maternal-effect" genes, but not without a few twists (Garcia et al. 2005; Grimanelli et al. 2005). The importance of cellular interactions between the embryo, endosperm, and the maternal gametophyte and sporophyte has led to the interpretation that seed development entails possible nonzygotic influences in the form of gene action from sporophytic and gametophytic parts of the ovule and from the endosperm. This means that any mutations in seed development affecting the signature of maternal genes can be caused by either gametophytic or sporophytic genes of maternal origin. The involvement of a maternal gene controlling embryo development first emerged from the genetic analysis of embryogenesis in the recessive *sin1* mutant of *Arabidopsis*. The mutant ovules, which also have abnormal integuments, fail to form a normal embryo sac due to aberrant megasporocyte meiosis, and are thus female-steriles (Robinson-Beers et al. 1992; Schauer et al. 2002). Combining speed and reliability, a series of crosses between flowers either homozygous or heterozygous for the wild-type *SIN1* allele as the female parent (+/+, +/*SIN1*) and either wild-type or *sin1* mutant plants as pollen donors showed that embryos of homozygous mutants are normal in every respect when they develop within the embryo sac of a heterozygous mutant maternal sporophyte; however, when the maternal sporophyte is a homozygous mutant (*sin1/sin1*), defects confined mainly to the cotyledons are observed in the surviving embryos. From these results it has been reasoned that the *sin1* mutation displaying a maternal effect on embryogenesis is sporophytic in nature, and that the *SIN1* gene product might influence embryo development by the production of a signaling molecule from tissues of the ovule lining the embryo sac (Ray et al. 1996). Another study has shown that the down-regulated expression of two MADS-box genes, *FLORAL BINDING PROTEIN7* (*FBP7*) and *FBP11*, in transgenic *Petunia hybrida* leads to the production of shrunken ovules with partially or totally disintegrated endosperm and slow-growing, occasionally arrested embryos. Here also, genetic analysis has established that the shrunken ovule phenotype is a maternal sporophytic effect that indirectly causes a major lesion in endosperm development (Colombo et al. 1997).

In Chap. 1, reference was made to the isolation of *fis* class mutants of *Arabidopsis* with fertilization defects. According to Grossniklaus et al. (1998), one of the wild-type genes of this class (*MEA*) is expressed in the female gametophyte of *Arabidopsis* before fertilization, and is required for normal post-fertilization development of the embryo and endosperm. Genetic analysis of the effect of the mutant gene on seed formation has given tantalizing glimpses of its activity: the constellation of developmental defects observed, such as delayed morphogenesis, excessive cell proliferation in the embryo, and reduced free nuclear divisions in the endosperm caused by the mutation, has been attributed to disruption of gene action transmitted through the female gametophyte. The basis for this conclusion is the observation by a traditional genetic approach that when the mutant heterozygote (*mea/+*) is self-fertilized, nearly 50% of the seeds house defective embryos. Since half of the haploid female gametes generated in the cross carry the mutant allele, maternal gametophytic control of embryogenesis is apparent here. Normal seed set occurs when wild-type females are pollinated with *mea/+* pollen, but nearly 50% of seeds derived from mutant eggs in the reciprocal cross collapse late in ontogeny by suffering significant embryo and endosperm developmental defects (see Plate 8, Fig. a–c). As the oversized embryos derived from mutant eggs succumb irrespective of the nature and dosage of the paternal contribution, completion of embryogenesis and formation of viable seeds appear to depend upon the presence of a wild-type *MEA* allele in the female gametophyte – specifically to reduce cell proliferation in the embryo and to promote the same in the endosperm. A possible endosperm effect in causing embryo lethality in the *mea* mutant was eliminated by showing that two paternal copies of the *MEA* gene in the endosperm, generated by crossing *mea/+* females with pollen from a tetraploid line, could not overcome the 50% abortion rate in seeds. By genetic analysis of the inheritance pattern of two mutant *mea* alleles, Kiyosue et al. (1999) have independently confirmed a requirement for only the maternal, gametophyte-specific wild-type *MEA* allele, and the dispensability of the paternal allele for normal embryo and endosperm development in *Arabidopsis*. Like the *mea* mutant, phenotypes of *fie* and *fis1* mutants cause partial endosperm development and inflict embryo lethality only when the mutant alleles are inherited from the female parent, and thus resemble maternal effect defects. In addition to the seed abortion phenotype, the three mutants display at a low frequency precocious endosperm proliferation before fertilization (Ohad et al. 1996; Chaudhury et al. 1997; Kiyosue et al. 1999). The expression of transcripts of *MEA* and *FIE* genes in the central cell of the unfertilized embryo sac and subsequently in the developing embryo and endosperm is also compatible with their function in repressing proliferation of the polar fusion nucleus and controlling embryo and endosperm development (Viella-Calzada et al. 1999; Spillane et al. 2000). Other later characterized mutants that display a gametophytic maternal effect on embryo and endosperm development in *Arabidopsis* are *demeter* (*dme*) (Choi et al. 2002), *msi1* [for multicopy suppressor of IRA (inhibitory regulator of Harvey sarcoma virus oncogene RAS-cAMP pathway)] (Köhler et al. 2003a), *msi1-2*, and *borgia* (*bga*) (Guitton et al. 2004). It has also been shown that, in addition to the maternal sporophytic effect described earlier, the pattern of inheritance of post-zygotic expression of the *SIN1* gene in *Arabidopsis* is suggestive of a maternal gametophytic effect (Golden et al. 2002).

The product of the *MEA* gene is a member of a subgroup of the polycomb group of proteins of *Drosophila melanogaster*; the hallmark of proteins of this subgroup is a 130-amino acid motif known as the SET domain – a family of regulatory proteins encoded by genes such as *SUPPRESSOR OF VARIEGATION, ENHANCER OF ZESTE*, and *TRITHORAX*. Polycomb proteins are a structurally disparate set of proteins that have the intriguing ability to function as gene silencers by controlling the normal one-way traffic of transcription factors to DNA. The protein products of the *MEA*, *FIE*, and *FIS1* genes have close affinities, thus reinforcing the view that they are part of a complex that determines the expression of regulatory genes during seed development. The *FIE* gene product shares strong similarities with a second subgroup of polycomb proteins whose defining characteristic is the presence of a WD-domain – a 40- to 60-amino acid repeat unit that usually ends with a tryptophan-aspartic acid pair. Proteins in this subgroup are encoded by *Drosophila EXTRA SEX COMBS* (*ESC*) and mouse and human *EMBRYONIC ECTODERM DEVELOPMENT* (*EED*) genes (Ohad et al. 1999). The polycomb protein with the SET-domain has resurfaced in *FIS1* (considered as allelic to *MEA*), *F644* (a *FIE*-like gene), and *EMB173*, a previously reported gene that causes defects in embryo development, reassigned as an *MEA* gene allele (Castle et al. 1993; Kiyosue et al. 1999; Luo et al. 1999). The only holdout from the polycomb net seems to be the *FIS2* gene, whose protein is predicted to contain a zinc finger motif and three nuclear localization signals, suggesting that it is linked to the transcriptional machinery (Luo et al. 1999).

In general terms, the polycomb proteins may be thought to regulate, by hitherto unknown mechanisms, target genes involved in cell proliferation in the embryo and endosperm of developing seeds of *Arabidopsis*. Köhler et al. (2003b) have assigned the MADS-box gene *PHERES1* (*PHE1*) the key role of a downstream target for transcriptional repression by FIS-class protein products, based on the transient expression of *PHE1* gene in seeds of wild-type plants containing preglobular stage embryos and the considerably high level of *PHE1* expression in the seeds of *fis*-class mutants. With focus on the *mea* mutant, this work also showed that high levels of expression of the *PHE1* gene in the mutant is causally linked with seed abortion, and that it is possible to rescue the seed abortion phenotype in the mutant by reducing the *PHE1* expression level. These results fit well with the proposed role of polycomb proteins in *Arabidopsis* embryogenesis. Another investigation has shown that the *PHE1* gene displays a preferential paternal expression, with the maternal allele remaining repressed by the effect of the *MEA* gene (Köhler et al. 2005).

5.1.2 Silencing of Paternal Genes

In the work on the *mea* mutant described earlier, in situ hybridization showed the presence of *MEA* mRNA in the synergids, egg, and central cell before fertilization, indicative of maternal transcription of the *MEA* gene in the female gametophyte of *Arabidopsis*. There was a persistent presence of transcripts in the cells of the embryo and endosperm after fertilization, probably due to zygotic transcription. An improved in situ hybridization procedure that allowed quantitation and detection of nuclear dots associated with nascent gene transcripts also revealed that nuclear dots present in the triploid primary endosperm nucleus are of the maternally inherited *MEA* allele, and not of the paternal allele. This raises the question as to when the paternal genome becomes active during the post-fertilization interlude. That paternally inherited *MEA* alleles are silenced during development of the embryo and endosperm was shown by examining *MEA* gene expression by reverse-transcription polymerase chain reaction (RT-PCR) analysis of RNA prepared from embryo-bearing siliques of reciprocal crosses between wild-type and a *mea* mutant allele (Vielle-Calzada et al. 1999). One question raised by the differential functioning of the maternal and paternal genes is how the development of the endosperm and embryo can proceed by inheritance of a wild-type allele from the female, but not from the male gametophyte. Additional experiments have provided clear evidence to show that the expression of paternal alleles is frequently delayed during embryogenesis and seed development, and that the silencing occurs at the transcriptional level by genomic imprinting. This process, which is almost universally relevant in animal systems, but much less so in plants, represents a situation where the two parental alleles of a gene may show differential activity during development of the zygote, leading in extreme cases to some genes being expressed predominantly from one of the parental chromosomes only; the genome of the other parent is kept transcriptionally inert by the silencing mechanism, which presumably blocks the normal flow of transcription factors. This results in genes being expressed or silenced according to their parental origin (Grossniklaus et al. 2001). Obviously, genomic imprinting contravenes the expectation of equal participation of the genome inherited from both parents in development and, in the absence of paternally inherited alleles, early divisions of the zygote are presumed to be programmed by their maternal copies (maternal imprinting). One critical piece of evidence for genomic imprinting has come from the analysis of parental chromosome-specific expression of a cluster of 20 genes during embryogenesis and seed development in reciprocal crosses between wild-type and transposants of *Arabidopsis* that harbor a reporter gene construct to monitor gene expression. It was found that, following initial transcriptional inactivity, paternal copies of the genes become active only after seed development has progressed for more than 3 or 4 days after fertilization when the embryo has produced 32–64 cells. Making a case for a global paternal gene silencing during embryogenesis, the protein products of some of the genes showing delayed paternal expression have been associated with important cellular functions such as cell cycle regulation, transcription, and the assembly of protein secondary structure. These observations make a compelling case that the molecular effect on the embryo of inheriting maternal alleles of *MEA* and other genes is the silencing of paternally inherited alleles by genomic imprinting (Vielle-Calzada et al. 2000). Another study showed that, in the progeny of crosses between two *Arabidopsis* ecotypes, only the maternal *MEA* allele is detected in the endosperm of seeds harboring torpedo-shaped and older embryos, but both parental alleles are expressed in embryos of similar age. The implication of these results is that, in seed development, genomic imprinting directly affects the endosperm, but not the embryo, whose abortion in the *mea* mutant is probably engineered by some defective endosperm function (Kinoshita et al. 1999). Using a reporter gene construct to monitor gene expression, in addition to *MEA*, *FIS2* and *FIE* (*FIS3*) genes have also been shown to be imprinted in the embryo and endosperm nurtured in the same embryo sac (Luo et al. 2000). Since DNA methylation is known to play a key role in gene silencing, experiments demonstrating embryo rescue in *mea* mutants by a recessive mutation in the *DECREASE IN DNA METHYLATION1* (*DDM1*) gene – which affects chromatin conformation and reduces methylation levels in the genome – are of interest. The crucial observation is that when *mea/MEA* heterozygous plants are pollinated with homozygous *ddm1/ddm1* pollen, the suppression of *mea* seed abortion by *ddm1* mutation allows many embryos to complete morphogenesis, and even surpass growth of wild-type embryos (Vielle-Calzada et al. 1999; Yadegari et al. 2000). It is not known whether activating silenced genes by the *ddm1* mutation can overcome defective endosperm development in *fis* mutants; as described in Chap. 8, other crosses leading to methylation changes have been effective in restoring normal endosperm growth in some *fis* mutants. Transcriptional delay of paternal alleles of 16 embryo-/endosperm-expressed genes during grain development in maize has recently been reported (Grimanelli et al. 2005).

Despite evidence from the studies cited above, the precise mechanism by which an imprint is conferred on the maternal genes to the exclusion of paternal alleles remains elusive. Moreover, the existence of a differential genome-wide parental effect on early development of the embryo and endosperm as a global or as an all-or-none phenomenon has been questioned, and evidence has been presented for an early, but low paternal effect in embryos of *Arabidopsis* (Baroux et al. 2001; Vielle-Calzada et al. 2001; Weijers et al. 2001a) and in maize zygotes obtained by in vitro fertilization (Scholten et al. 2002). The *PROLIFERA* (*PRL*) gene of *Arabidopsis* encodes a protein that regulates DNA replication in dividing cells, and on the basis of genetic evidence appears to be preferentially transcribed from the maternally contributed genome; expression of the gene from both paternally and maternally supplied alleles in the developing embryo and endosperm demonstrated by in situ hybridization using a reporter gene has nevertheless ruled out imprinting of this gene (Springer et al. 2000). The *capulet* (*cap1* and *cap2*) mutants of *Arabidopsis* have been found to be female gametophytic, displaying maternal effects on embryo and endosperm development. Unambiguous evidence is, however, found wanting to support imprinting of the *CAP* genes; rather, several genetic and molecular criteria have led to the suggestion that these genes represent true female gametophyte genes required to initiate divisions in the products of double fertilization (Grini et al. 2002).

In summary, evidence is piling up to show that, far from being under the control of both paternal and maternal genomes, the latter exercises an overriding control over the first few rounds of divisions of the zygote and the endosperm nucleus during seed formation (Reyes and Grossniklaus 2003). In most cases the absence of prefertilization expression of maternal alleles has not been demonstrated convincingly enough to conclude whether gene activity observed in the zygote or in the primary endosperm nucleus is caused by newly transcribed maternal mRNAs or by transcripts present in the cells of the prefertilization embryo sac. Undoubtedly, understanding the role of maternal transcripts in initiating development in flowering plants has become more complex than in animal systems, in part because of the occurrence of double fertilization and the genetically intractable nature of the plant life cycle, alternating between dominant sporophytic and relatively inconspicuous gametophytic generations.

5.2
Gene Activity during Progressive Embryogenesis

The molecular mechanisms involved in the regulation of gene expression during transformation of the fertilized egg into the embryo in a precise temporal framework have long been a major focus of interest of experimental plant embryologists. By employing molecular and genetic approaches to bear on this problem, only lately has significant progress been made toward identifying the genetic hierarchies critically involved in embryo development in flowering plants. Investigations by Goldberg et al. (1981a, 1981b) and Galau and Dure (1981), comparing the mRNA complexity of embryos of soybean and cotton, respectively, by DNA-RNA hybridizations with different embryo RNA populations, have been particularly influential in providing a quantitative angle to embryo gene expression programs. Both plants also provide ideal model systems to monitor changes in the population of mRNAs during comparable periods characterizing major morphological and physiological landmarks of embryogenesis. A general theme that emerged from those studies was that, as in vegetative plant organs, about 15,000–20,000 diverse genes are expressed at the mRNA level in embryos of both plants. Since many of these genes may be involved in performing the same functions in a developmental framework, such duplicates need to be eliminated through mutations to arrive at a reasonably accurate estimate of the number of genes with essential functions during embryogenesis. However, the presence of approximately the same number of mRNAs in embryos spanning early and late stages of development has supported the notion that tissue and organ formation associated with embryogenesis proceeds with minimal changes in structural gene information. It is even more remarkable to learn from these studies that substantial amounts of the messages that pattern embryo morphogenesis also survive in the lineages that are present in dormant seed embryos and are detected post-germinationally in the seedling plant. However, what gives the embryo its distinct flavor is the presence of a small number of embryo-specific mRNAs that become prevalent at different stages of embryogenesis. The changes in these and other abundant mRNA sets during embryogenesis and germination of cotton embryos are shown in Fig. 5.1; superimposed on this diagram are the abundancy changes in storage protein messages in soybean embryos (group 4, from top). Modulations in the abundance of mRNAs of cotton embryos were followed in two-dimensional gels by identifying stainable extant proteins whose mRNAs are no longer present, and radioactively labeling of in vivo synthesized proteins and their corresponding mRNAs as detected by in vitro protein synthesis. The presence of a protein in the in vivo synthesis catalog does not prove on its own that the protein emanates from abundant mRNAs. The additional linking of its presence to the in vitro synthesis catalog has been used as evidence to confirm this assumption. Of the cotton embryo mRNAs thus identified, those of group 1, encoding actin, tubulin, and calmodulin, have strong survival instincts, as they persist throughout embryogenesis and germination. Because what is required for early embryogenesis is likely to be different from what is required for late embryogenesis, no functions have been assigned for proteins derived from mRNAs of groups 2, 3, and 6. In contrast, the function of transcripts encoding storage proteins, trypsin inhibitors, and lectin, abundantly found in soybean embryos, is fine-tuned to be optimal during embryo maturation when food reserves accumulate. One of the most remarkable mRNAs is included in the subset that encodes the LATE EMBRYOGENESIS ABUNDANT (LEA) proteins (group 5), found in greater abundance in mature embryos than in young or germinated embryos. Since the functions of LEA proteins are discussed in a later chapter, suffice it here to state that these proteins, first characterized in cotton embryos, have also been identified in embryos of other plants in roles designed to overcome desiccation-related stress (Dure et al. 1981; Galau et al. 1986; Goldberg et al. 1989).

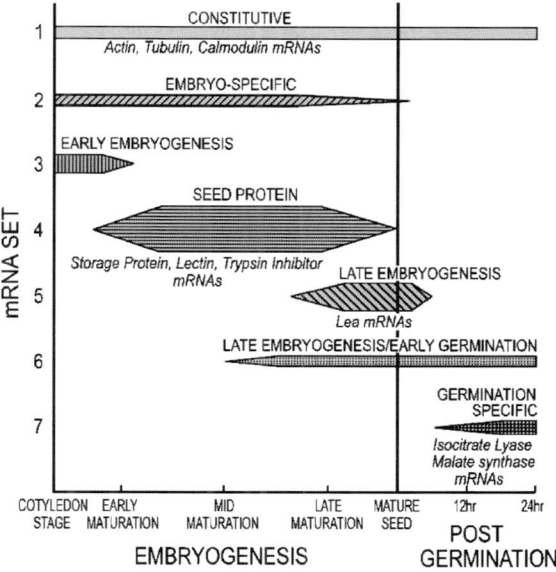

Fig. 5.1 Regulation of prevalent mRNA sequence sets during embryogenic and post-embryogenic development in cotton and soybean. The thickness of each bar represents the prevalence of each mRNA set. Tapering of each bar shows periods of accumulation or decay of the mRNA set. (Reprinted from Goldberg et al. 1989)

Early efforts at understanding the role of genes during embryogenesis focused on the spatial and temporal patterns of accumulation of distinct sets of mRNAs and proteins in developing embryos of various plants, in the hope that such information will yield important clues about the function of these macromolecules in embryo morphogenesis. Since the expression patterns of marker mRNAs during embryogenesis represent the combination of a few temporal components, the modularity observed may be considered to reflect the existence of sequential gene expression programs, each activated by a distinct regulatory factor. In latter years, a flurry of activity using cloned genes has strengthened this generalization, as more and more genes encoding house-keeping proteins and transcription factors involved in embryo development and embryo-expressed genes encoding proteins involved in maintaining the general architecture of the adult plant, have come to light. These investigations have been summarized in an earlier publication (Raghavan 1997). A few of the genes included in these groups that have later been characterized functionally or at the expression level in *Arabidopsis* are the *KANADI* (*KAN*), *PHABULOSA* (*PHB*), *REVOLUTA* (*REV*), *TCP* (for Teosinte branched 1, Cycloidea, and PCF1), and *ARGONAUTE1* (*AGO1*) genes implicated in the perception of radial and adaxial-abaxial positional information in the leaf primordium expressed in most stages of developing embryos (Kerstetter et al. 2001; McConnell et al. 2001; Emery et al. 2003; Palatnik et al. 2003; Kidner and Martienssen 2004); *Arabidopsis thaliana HOMEOBOX8* (*Athb-8*), a homeobox gene expressed as an early marker of vascular tissues in the torpedo-shaped embryo (Baima et al. 1995); *BREVIPEDICELLUS* (*BP*), which controls stem and leaf architecture and is turned on initially in the presumptive hypocotyl tissue of late globular stage embryo and persists in the hypocotyl of later stage embryos (Douglas et al. 2002); the MADS-box gene *AGAMOUS-Like21* (*AGL21*) expressed in globular to torpedo-shaped embryos (Burgeff et al. 2002); *PLS*, modulating root growth expressed in heart-shaped and older embryos (Casson et al. 2002); *JAGGED* (*JAG*), necessary for proper lateral organ shape, expressed in cells of the cotyledon primordia of transition-shaped and older embryos (Ohno et al. 2004); and *BLADE-ON-PETIOLE1* (*BOP1*), which regulates the proximal-distal differentiation of the leaf primordium, and is expressed at the base of developing cotyledons close to the shoot apical meristem of torpedo-shaped and older embryos (Ha et al. 2004). Interestingly, the function of the *TCP* and *AGO1* genes that control organ polarity beginning in the developing embryos is believed to be regulated by newly discovered small RNA molecules known as microRNAs (Palatnik et al. 2003; Kidner and Martienssen 2004).

5.2.1
Gene Expression during Early Embryogenesis

The molecular mechanisms that regulate pattern-specific differentiation during embryogenesis were reviewed in Chap. 3; here, we consider studies that have shaped our views on gene expression during region-specific differentiation of embryos, and on the gene products essential for this process. This might appear to be a formidable undertaking as high density filter array hybridizations have revealed the existence of nearly 300 transcription factor genes alone controlling *Arabidopsis* embryogenesis (de Folter et al. 2004) and a smaller number of diverse candidate genes controlling rice embryogenesis (Lan et al. 2004). In the past, gene expression studies in the egg cell and the nascent zygote have been hampered by the difficulty of extracting these cells from the embryo sac. The development of techniques to isolate gametes principally from maize and fuse them in vitro to form viable zygotes represents a major advance in the study of gene expression programs during the transformation of egg into zygote. Using this system, Sauter et al. (1998) followed the expression of A1 (*Zeama*; *CycA1*;1), B1 (*Zeama*; *CycB1*;2), and B2 (*Zeama*; *CycB2*;1) groups of mitotic *CYCLIN* (*CYC*) genes in isolated unfertilized egg cells and in in vitro fertilized eggs of maize; this work showed that, of the three cyclins, only the A1 *CYC* gene is expressed in the egg cell. After a transient decrease immediately after fertilization, A1 *CYC* gene transcripts reappear later in the zygote preparatory to its first division. Transcripts of the B2 gene accumulate subsequent to those of the A1 gene during zygote development, whereas expression of transcripts of the B1 gene is highly restricted to three short intervals before the first zygotic division. It is likely that the presence of *CYC* genes in the egg after fertilization reflects the role of zygotic mRNAs, rather than maternally – derived messages, in driving the divisions of the zygote and early embryo. Use of an RT/PCR method to generate a cDNA library from limited amounts of material has introduced a hopeful note into the arduous task of isolating and identifying the genes involved in events immediately before and after fertilization in angiosperms. Although a cDNA library of maize egg cells was constructed using this method, expression analysis of the isolated clones in the zygote and early embryos of maize unearthed no important regulatory genes controlling zygote growth and division (Dresselhaus et al. 1994; Richert et al. 1996; Cordts et al. 2001). However, this approach has led to the isolation of two new MADS-box genes that are expressed in the egg cell as well as in zygotes formed normally in the embryo sac and after in vitro fertilization (Heuer et al. 2001). As is well-known, MADS-box genes play critical roles in floral meristem specification and floral organ identity, and encode proteins that function as transcription factors. From a cDNA library constructed using in vitro fertilized zygotes of maize, Dresselhaus et al. (1996) isolated a full-length clone of calreticulin, a major Ca^{2+} storage protein of the ER. A role for calreticulin in the synthesis of new cell wall materials and proteins for the impending division of the zygote is supported by the observation that transcripts of this gene are more strongly expressed in the zygote than in the egg cell. Expression studies of ribosomal genes obtained by screening cDNA libraries made from egg cells and from in vitro fertilized zygotes have also been informative in showing that different modes of regulation of these genes operate during zygotic and somatic cell cycles of maize (Dresselhaus et al. 1999). Construction of cell-type-specific cDNA libraries using isolated egg cells and defined stages of zygotes and proembryos has recently been extended to wheat to identify several fertilization-induced transcripts (Sprunck et al. 2005).

A new *MADS*-box gene from rapeseed (*Brassica napus*), designated *AGL15*, has entered the stage to advance the analysis of gene action during embryogenesis. The suggestion that the *AGL15* gene is centrally important in embryo specification has been consolidated by the finding that transcripts of the gene accumulate in a highly preferential manner in all embryogenic tissues of *Arabidopsis*, *B. napus*, and maize, beginning as early as the globular stage. A complementary insight provided by AGL15-specific antibodies and immunohistochemistry showing localization of the AGL15 protein in the nuclei of embryos has also been instrumental in supporting its role as a transcription factor. This approach has further revealed the presence of relatively high levels of AGL15 protein in the nuclei of cells of developing embryos beginning with the globular through the torpedo-shaped stages (see Plate 8, Fig. d–g). A correlation was also observed between the accumulation of AGL15 protein and development of cells into asexual embryos by apomixis, somatic embryogenesis, and pollen embryogenesis in diverse plants. The target that responds to regulation by AGL15 protein appears to be a gene regulating GA metabolism in the embryo. Based on these results, the AGL15 protein is appealingly portrayed to participate in the regulation of the program from precursor cells active during early embryo developmental stages. The observed expression of gene transcripts and/or the gene product in all cells of developing embryos, however, weighs against *AGL15* gene function in the specification of particular embryo tissues or organs (Heck et al. 1995; Rounsley et al. 1995; Perry et al. 1996, 1999; Wang et al. 2004). Loss of the AGL15 protein from embryos of *Arabidopsis* homozygous for the *lec1-2* mutation that transforms cotyledons into leaves has cast new light on the role of the protein in the maintenance of the cotyledon pathway in developing embryos. The decline in AGL15 protein levels precedes the first detection of mutational defects, resulting in the inactivation of several embryo-specific programs in the heart-shaped/torpedo-shaped transition stage embryos (Meinke et al. 1994; Perry et al. 1996). Thus, we have little idea of the exact function of the *AGL15* gene in embryogenesis or how defects in its function affect the process.

A gene that plays a key role as a molecular marker for the globular to heart-shaped transition of developing embryos of *Arabidopsis* is *PEI1*. An intriguing observation is that the gene encodes a Cys_3His zinc finger domain-containing protein, generally associated with certain animal and fungal transcription factors. This discovery, along with the finding that transcripts of the gene are expressed throughout embryogenesis, with maximum expression from heart-shaped through cotyledon stages, has proved to be an important requisite for assigning *PEI1* the status of an embryo-specific gene. A stringent characterization of a cloned gene requires the use of antisense or cosuppression (perturbing the expression of a gene situated in one place in the genome by inactivation of another copy of the same gene situated elsewhere) strategies to affect its function in developing embryos. In keeping with this, transgenic plants expressing an antisense construct of the *PEI1* gene were found to produce ovules in which embryo development was stymied at the heart-shaped stage, with the caveat that the effects of the antisense gene were complex, rather than subtle (see Plate 9, Fig. a,b). From a detailed analysis of abnormal embryo phenotypes, the function of the gene has been couched in terms of the transcription factor acting in the apical domain of the embryo leading to the failure of cotyledon development, while root growth is unhindered (Li and Thomas 1998).

Some other recruits to the expanding suite of genes implicated in embryo development in *Arabidopsis* identified by their expression patterns include a new D-type of cyclin (*CYCD*), *YABBY* (*YAB*), *ASKdzeta* (*ASKζ*), and *SMT*. Besides embryos, each of these genes functions in a range of adult plant organs, activating the same physiological responses, perhaps through the same signaling pathway. D-Type cyclins have an important role in the G1-to-S transition during the cell cycle across the board from growth in cell suspension cultures to histogenesis and organogenesis in the whole plant. Consequently, it comes as no surprise that in situ hybridization revealed clusters of cells throughout the heart-shaped and torpedo-shaped embryos of *Arabidopsis* strongly expressing transcripts of the *CYCD4-1* gene, but expression was very weak in mature embryos (de Veylder et al. 1999). Members of the *YAB* gene family appear to be central components in the specification of abaxial cell fate in lateral organs in the mature plant such as leaves, floral meristems, and floral organs. Included in this gene family are *FILAMENTOUS FLOWER* (*FIL*), *YAB2*, and *YAB3*, which carry encoded within their structure the recipe for transcription factors. Transcripts of both *FIL* and *YAB3* and, to a lesser extent, those of *YAB2* show similar patterns of expression, targeted solely to the abaxial domain of the cotyledons of embryos at all developmental stages except the mature stage, and can be considered to specify abaxial cell fate (Siegfried et al. 1999). The action of these genes in specifying the abaxial fate of lateral plant organs is complemented by other genes such as *KAN*, *REV*, and *PHB*, which are also expressed in developing embryos (Kerstetter et al. 2001; McConnell et al. 2001; Emery et al. 2003). The protein kinase-encoding *ASKζ* gene is expressed in the whole embryo throughout its development and contrasts with the suspensor-restricted expression (Chap. 2) of the related *ASKη* gene (Dornelas et al. 1999). Three *SMT* genes (*SMT1*, *SMT2*, and *SMT3*) isolated from *Arabidopsis* have been shown to encode an S-adenosylmethionine-dependent C-24 sterol methyltransferase that catalyzes transmethylation. Distinct patterns of expression of transcripts of these genes are seen in developing *Arabidopsis* embryos: strong *SMT1* expression is confined to the region below the embryonic cotyledons, *SMT2* is expressed throughout the developing embryo, and *SMT3* is excluded from the central vascular tissues, but is expressed strongly in the apical part of the cotyledons, in the hypocotyl, and root primordium. Loss of *SMT* gene function by a mutation perturbs embryogenesis significantly, leading to the production of a formless mass of cells. It is believed that SMT proteins act during embryogenesis as precursors for membrane lipid components and steroid hormones (Diener et al. 2000).

A novel role attributed to the *Arabidopsis* floral organ identity gene *APETALA2* (*AP2*) is in controlling seed size and weight by acting through the maternal sporophyte. It has been shown that a loss-of-function mutation in this gene causes an increase in seed weight, in part through its effects on the endosperm, and on embryo cell number, size, and storage protein accumulation (Jofuku et al. 2005; Ohto et al. 2005). At the physiological level, the *ap2* mutation apparently causes changes in the ratio of hexose to sucrose, resulting in high hexose/sucrose ratios during seed development. The change in sugar metabolism might in turn provoke increased cell division activity in the embryo (Ohto et al. 2005).

Consideration of the unique morphology of the grass embryo has provided a context in which to understand the role of specific genes and their encoded proteins in the formation of embryonic organs. A paradigm linking gene expression with organ formation during embryogenesis in members of the Poaceae comes from studies on maize, rice, and wheat. It was mentioned in Chap. 3 that homeobox genes such as *KN1* and the related *OSH1* (*Oskn1*), which contain the signature motif encoding transcriptional regulators, are involved in the maintenance of shoot apical meristem identity in developing embryos of maize and rice, respectively. Genes such as *ZmHox* isolated from maize embryos are additional members of the homeobox family that probably control developmental decisions affecting organ and cell fate in the embryo. Transcripts of these genes begin to appear as early as the proembryo stage in maize, and persist in dif-

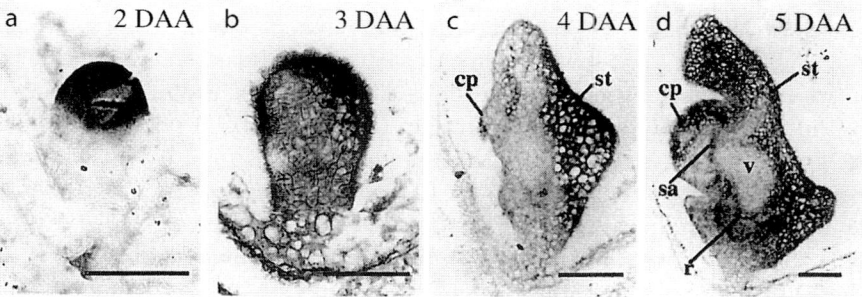

Fig. 5.2
In situ localization of transcripts of the *RINO1* gene in developing rice embryos 2 days after anthesis (DAA) (**a**), 3 DAA (**b**), 4 DAA (**c**) and 5 DAA (**d**). *cp* Coleoptile, *r* radicle, *sa* shoot apical meristem, *st* scutellum, *v* procambium. *Bars* 50 µm. (Reprinted from Yoshida et al. 1999)

ferentiating organs such as the shoot and root apical meristems and leaf primordia, and meristematic tissues such as the procambial strands and protoderm, but are absent in the scutellum and coleoptile (Klinge and Werr 1995). Based on the cloning and sequencing of calmodulin cDNAs from transition-stage maize embryos, progressive embryogenesis is thought to reflect high calmodulin mRNA levels in the early stages and a decrease upon embryo maturation (Breton et al. 1995). The omnipresence of calmodulin in Ca^{2+} signaling processes suggests that Ca^{2+}-mediated signaling is part of the regulatory decision-making process during embryogenesis. The expression pattern of transcripts of the *Oskn1* gene at the boundaries of the emerging epiblast and around the root apex in rice embryos has suggested an additional role for this gene in implementing positional information for the development of these organs. Interestingly, the genes *Oskn2* and *Oskn3* have seemingly insignificant roles in specifying the embryonic shoot apical meristem (Chap. 3). Initial clues to their involvement in the genesis of other embryonic organs have been provided by the expression of transcripts of *Oskn2* in the epiblast and scutellum and of *Oskn3* at the base of the coleoptile (Postma-Haarsma et al. 1999). The fact that expression domains are established before overt morphological differentiation suggests that these genes might be responsible for specifying the identity of organs. Another early molecular marker of the scutellum in rice embryo is the *myo*-inositol-1-phosphate synthase gene, *RINO1*, whose transcripts accumulate in the cells of the globular embryo destined to become the scutellum (Fig. 5.2). Accumulation of gene transcripts corresponds with the appearance of phytin-containing particles or globoids in the scutellum, suggesting that *myo*-inositol-1-phosphate synthase, catalyzing formation of the first intermediate in inositol metabolism, is the principal arbiter of the phytin-biosynthesis pathway in the embryo (Yoshida et al. 1999).

Immunolocalization of wheat embryo lectin, known as wheat germ agglutinin, as a molecular marker has yielded two important insights into the role of this protein in the regulation of organ differentiation in the wheat embryo. Wheat germ agglutinin does not accumulate uniformly throughout the developing embryo, but is detected first in the radicle and coleorhiza and subsequently in the epiblast and coleoptile during the period of rapid embryo growth. Interestingly, there is also a cell-specific pattern of expression of wheat germ agglutinin confined to one or two layers of epidermal cells of the root meristem, root cap, and coleorhiza, the outermost layer of cells of the coleoptile, and virtually all cells of the epiblast. Other evidence pointing to a role for this lectin in cell differentiation includes localization of transcripts of a cloned lectin gene in differentiating embryonic organs. However, given the fact that the major sites of persistent accumulation of transcripts are the epidermal layers of the radicle and coleorhiza, it is possible that wheat germ agglutinin detected in the coleoptile and epiblast is a product of another lectin gene (Raikhel and Quatrano 1986; Raikhel et al. 1988). The combined results of the studies reviewed above suggest that specification of organs of the grass embryo is likely to be complex and to involve redundant and/or parallel pathways.

ARABIDOPSIS EMBRYOGENESIS, ACCORDING TO *LEC* GENES

A set of genes that includes *LEC1*, *LEC2*, *FUSCA3* (*FUS3*), and *L1L*, collectively known as *LEC* genes, appear to be critical regulators of embryogenesis and dormancy in *Arabidopsis*. On the basis of mutational effects, the *LEC1* gene was first described for its role in the specification of cotyledon identity and in the completion of embryo maturation (Meinke 1992). Subsequent isolation of *lec2*, *fus3*, and *l1l* mutants, followed by the cloning of the *LEC1*, *LEC2* and *L1L* genes has provided access to the powerful molecular tools of ectopic expression to infer that these genes might be involved in embryogenesis by regulating the transcription of other genes (Meinke et al. 1994; Lotan et al. 1998; Stone et al. 2001; Kwong et al. 2003). The protein products of *LEC1* and *L1L* are related to the heme-activated protein 3 (HAP3) subunit of the CCAAT-box-binding transcription factor (CBF), and have as their signature feature a highly conserved central B domain of the HAP3 subunit (Lotan et al. 1998; Kwong et al. 2003). The *LEC2* and *FUS3* genes both encode B3 domain transcription factors functional in seed maturation (Luerßen et al. 1998; Stone et al. 2001). A close examination of the B domain of *LEC* genes has served to refine its function in embryogenesis. It appears that *Arabidopsis* HAP3 regulators (AHAP3) belong to two classes – the LEC1 and L1L subunits constituting the LEC1-type subunit and the remaining subunits making up the non-LEC1-type – and that genes encoding the LEC1-type proteins are critical in embryogenesis (Kwong et al. 2003). Based on the ability of different combinations of LEC1-type and non-LEC1-type HAP3 domains to overcome the effects of the *lec1* mutation, Lee et al. (2003) have identified a single amino acid, Asp-55, specific to the LEC1 B domain that is required to confer partial *LEC1* gene activity.

The most detailed information on the role of *LEC* genes in embryo development in *Arabidopsis* has come from the study of the *lec1* mutant, which causes defects in two seemingly unrelated processes of embryogenesis. The distinguishing feature of this mutant is the reversion of cotyledons to a leaf-like state by the development of trichomes and stomata on their adaxial side. The mutant seeds also fail to display features of embryo maturation, such as acquisition of desiccation tolerance, loss of chlorophyll, accumulation of storage proteins, and entry into dormancy (Meinke 1992). The wider role of *LEC* genes in embryogenesis, especially in inducing embryogenic competence, came to be realized by in situ hybridization studies showing that transcripts of *LEC1*, *FUS3*, and *L1L* genes are expressed in both the embryo proper and the suspensor of early-stage embryos, with a preferential localization in the protoderm of the heart-shaped embryo (Lotan et al. 1998; Kwong et al. 2003; Tsuchiya et al. 2004). Ectopic expression of *LEC1* and *LEC2* genes in *Arabidopsis* leads to the production of morphologically abnormal plants with embryo-like structures or somatic embryos on them (Lotan et al. 1998; Stone et al. 2001). Other mutational defects impinge on the accumulation of storage reserves, such as reduced accumulation of lipid and protein reserves and enhanced accumulation of starch, as well as the expressional activity of genes encoding these functions (Keith et al. 1994; Meinke et al. 1994), and underexpression of genes encoding 12 S and 2 S storage proteins and oleosins (proteins associated with oil-body membrane) (Bäumlein et al. 1994; Kirik et al. 1996; Parcy et al. 1997) and LEA proteins (Vicient et al. 2000) in embryos of *lec1* and *fus3* mutants. A new function attributed to the *FUS3* gene is to negatively regulate the *TTG1* gene (required for trichome abundance on leaves), with the result that introduction of the loss-of-function *ttg1* mutation into a *fus3* mutant partially rescues the *fus3* phenotypic trait of ectopic trichome production on cotyledons (Tsuchiya et al. 2004).

The role of *LEC* genes in the regulation of dormancy of *Arabidopsis* seeds is described in Chap. 6. A theme that has emerged from a cumulative analysis of these investigations is that, individually and collectively, the action of *LEC* genes has great consequences for orderly development of the embryo. The protein products of these genes probably act as central regulators of almost the entire range of embryogenic processes, beginning with the single-celled zygote, pausing with the dormant embryo, and extending into somatic cells to trigger a latent embryogenic program. No other genes considered thus far come anywhere close to *LEC* genes in terms of the multifaceted roles they play in embryogenesis.

5.2.2
Gene Expression during Late Embryogenesis and Transition to Germination

Although embryogenesis and seed germination are traditionally considered separately, unraveling the regulatory circuits that operate during germination has inexplicably provided important links to genes expressed during late embryogenesis. Particularly interesting are studies on the molecular mechanisms that regulate the synthesis of the glyoxylate cycle enzymes, isocitrate lyase (ICL) and malate synthase (MS), during transition from late embryogenesis to germination in *Brassica napus*. The main import of these studies is that, although ICL and MS and their mRNAs are very abundantly expressed in seedlings, the enzymes and their transcripts are also prevalent in late-stage embryos. Using isolated nuclei in assays to monitor transcription rates, it was found that genes for the enzymes remain transcriptionally competent in both embryos and seedlings, but are repressed to some extent in late-stage embryos; these surprising results suggest that genes encoding ICL and MS in late-stage embryos are regulated at the transcriptional level, with the caveat that post-transcriptional processes also affect their mRNA levels (Comai et al. 1989). Translational or post-translational processes are also believed to be critically involved in maintaining steady-state levels of ICL and MS in embryos, as the relative accumulation of enzymes and changes in their activities do not correspond to the mRNA levels (Ettinger and Harada 1990).

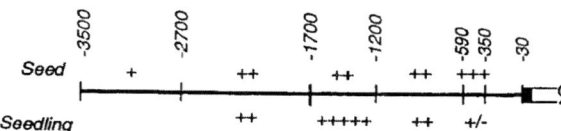

Fig. 5.3 A summary diagram showing the effectiveness of DNA sequences in the 5′ flanking region of the isocitrate lyase (ICL) gene of *Brassica napus* (*horizontal line*) in regulating the expression of a reporter gene in embryos (seeds) and seedlings of transgenic *Arabidopsis*. *Plus signs* indicate the relative influence of the fragments on promoter activity. (Reprinted from Zhang et al. 1996)

The pervasiveness of germination-related genes in embryos has raised the question of whether the modulations in the levels of ICL and MS and their transcripts that occur during embryogenesis and germination are stringently regulated solely by changes in transcription and translation rates. What causes the repression of genes encoding these enzymes in late-stage embryos and their activation during germination? Although there are several ICL and MS genes in the *B. napus* genome, at least one gene of each enzyme is active in both embryos and seedlings. The notion that the MS gene is regulated differently in embryos and seedlings comes from the observation that transcripts of the gene present in the procambium of mature embryos are lost as it differentiates into vascular tissues in seedlings (Comai et al. 1992; Zhang et al. 1993). Evidence showing a similar differential regulation of the ICL gene in embryos and seedlings has emerged through identification of segments of the 5′ flanking regions of the gene that activate the promoter in reporter gene constructs in transient assays in *B. napus* and in transgenic *Arabidopsis*. A summary of the analyses in transgenic *Arabidopsis* is given in Fig. 5.3. Whereas, in global terms, the results show that the same DNA sequences (between positions −1700 and −1200) are involved in regulating the ICL gene in both embryos and seedlings, discrete domains that are highly expressed in embryos (between positions −590 and −350) but affect promoter activity at a low level in seedlings are also present (Zhang et al. 1996). Thus, a general propensity to coordinate activities of the ICL gene at two different phases in the sporophytic life of the plant – embryo and seedling – appears to be achieved by different unidentified physiological signals with their own distinct *cis*-acting elements.

AX92 is another gene expressed discontinuously in embryos and seedlings of *B. napus*, but the mechanism of its action has been couched in terms of a common regulatory signal functioning in both stages of the sporophyte. It turns out that when a chimeric gene consisting of the 5′ and 3′ untranslated and flanking regions of *AX92* fused to a reporter gene is expressed in transgenic *B. napus*, DNA sequences located 3′ of the protein coding region are found responsible for gene activation in the root cortex of both the embryo and seedling. Conservation of sequences with similar functions at different stages of the sporophytic life cycle is taken as evidence to indicate that identical mechanisms underlie the expression of this gene during embryogenesis and germination (Dietrich et al. 1992). The many as yet unidentified genes that are expressed during embryogenesis and transition to germination and their impact on the sporophyte make it likely that activation and spatial control of these genes might employ mechanisms similar to those seen with genes encoding ICL and MS or with *AX92*.

It was stated in Chap. 2 that when immature embryos are cultured, they skip the latter part of embryogenesis and germinate into rudimentary, weak seedlings by precocious germination. This abnormal development provides immature embryos with challenges that are often met by specific metabolic changes. In several plants, the identification of a narrow window spanning the period before the embryo has played out its maturation program and when it becomes competent to respond to germination cues, has made it possible to address the question of how a common set of genes control the choice between continued embryo development and germination. Highly informative is the model of temporally separated transcription and translation patterns of two enzymes involved in the degradation of stored proteins and lipids in normal and precociously germinated embryos of cotton. Remarkably, the proteolytic enzyme carboxypeptidase is not present in developing embryos; however, appearing after 24 h of normal germination, it continues to increase for another 72 h, after which there is a decrease. The suggestion that this enzyme is synthesized on templates of mRNAs stored in the cotyledons during embryogenesis has been validated by the insensitivity of enzyme synthesis to actinomycin-D, as well as being supported by the pattern of enzyme synthesis in precociously germinating, progressively younger, embryos in a medium containing actinomycin-D. It was found that enzyme appearance and precocious germination are inhibited by the drug only at a point before the embryo attains about two-thirds of its final size; after this developmental stage is reached, cells of the embryo begin to synthesize mRNAs for the enzyme. This led to the view that, although mRNAs for carboxypeptidase are transcribed during maturation of the embryo (the period beyond two-thirds of the final size), in some manner they maintain an inactive translational status during this period. It was also possible to implicate ABA in the synthesis of this enzyme by showing that translation of mRNAs encoding carboxypeptidase in precociously germinating embryos is inhibited by this hormone. Diffusion of ABA into the embryo from the ovular tissues thus precluding translation of mRNAs, and precocious germination of the embryo in ovulo in the fruit might explain these results in the natural setting of the plant. The regulation of ICL has highlighted its similarity to that of carboxypeptidase in normal and precociously germinated embryos (Ihle and Dure 1969, 1972).

Monitoring the fate of storage proteins has proved to be a significant improvement over enzyme markers as a means to establishing the end of the embryo-specific program and the beginning of a germination episode. Although embryogenesis in *B. napus* is developmentally separated from precocious germination, the morphological changes and the regulation of synthesis of the 12 S storage protein cruciferin and 2 S napin have provided evidence of an overlap between embryogenesis-related and germination programs. During embryogenesis, the accumulation of cruciferin continues until embryo maturity, whereas that of napin stops before maturity. Embryos of mature seeds germinate to produce the first pair of leaves at the same time as they completely degrade the stored cruciferin. Mid-cotyledon-stage embryos germinating precociously by radicle extension and cotyledon expansion, however, remain resolutely embryogenic by the formation of secondary cotyledons instead of leaves, and by the capacity to synthesize and accumulate cruciferin (Crouch and Sussex 1981; Finkelstein and Crouch 1984). There is also a strong correlation between the production of extra cotyledons in cultured embryos and the expression of transcripts of cruciferin and napin in the cells of primordial structures that assume embryogenic identity (Fernandez 1997). These findings are consistent with the operation of molecular switches at the cellular level to regulate the embryogenic and germination modes of growth in immature embryos. Other studies have implicated ABA in maintaining embryogenic development and in stimulating the accumulation of cruciferin and napin in cultured embryos prone to germinate precociously. However, ABA has a greater effect on the accumulation of storage proteins than on their respective transcripts in cultured embryos; the reasons for this effect and its molecular relevance are not clear. Although there are unpredictable swings in the ABA content of developing embryos, as in cotton, endogenous ABA seems to suppress the germination instincts of *B. napus* embryos during their normal development, coincident with storage protein accumulation (Finkelstein et al. 1985).

It must be remembered that different morphological and physiological changes occur during embryogenesis and seed germination, and that the involvement of different sets of genes and proteins becomes necessary. Notwithstanding this reservation, the above account shows that there are certain biochemical pathways that, in their basic tenets, are common to embryogenesis and germination, yet are regulated differently.

5.3 Embryo Gene Expression Program Studied by Mutant Screening

During the past two decades, mutational approaches have provided much information about the molecular underpinnings of embryo development in flowering plants, although, unwittingly, they had also the effect of focusing on the depths of our ignorance in this area. Beginning with the simple principle that perturbing embryogenesis by mutations makes it possible to identify sets of genes required for specific embryo developmental episodes, mutations have been used for the genetic dissection of the morphological processes of embryogenesis, and finally to comprehend the developmental program of progressive embryogenesis. How developmental pattern is initiated and maintained in *Arabidopsis* embryos by pattern-forming genes, uncovered by screening of seedling mutants following saturation mutagenesis, was described in Chap. 3. Before pattern-forming genes came to the fore, this model plant had served as the main workhorse for characterizing genes that regulate morphogenetic changes and housekeeping functions in the embryo; analyses of maize and rice mutants have also identified numerous genetic loci that, when mutated, yield developmentally abnormal embryos in stages ranging from proembryos to mature embryos. Even a cursory consideration of the genes implicated in embryogenesis by mutant screening makes one realize that the molecular and genetic basis for the maintenance of the mature plant body is laid down during embryogenesis.

5.3.1 Embryo-lethal Mutants of *Arabidopsis*

Embryonic mutants isolated from *Arabidopsis* following X-irradiation, chemical seed mutagenesis, transposon tagging, enhancer trapping, and T-DNA transfer by *Agrobacterium*-mediated seed transformation, are globally defective in seed maturation and specifically exhibit a wide range of developmental and metabolic defects during embryogenesis (McElver et al. 2001). These mutants are now designated as *emb* and the current view tends to classify them as either embryo-lethals, embryo-defective, or pigment mutants. The embryo-lethals are reminiscent of similar mutants in animals and produce embryos that are blocked in early development with abnormal phenotypes, are doomed for extinction, are desiccation-intolerant, and do not grow in culture into normal seedlings. Defects that surface throughout embryogenesis predominate in the phenotypes of the embryo-defective class of mutants. However, mutant embryos remain viable without differentiating normal organs and tissues of the wild-type embryo, have survival instincts, and are amenable to culture or transplantation in the soil to give rise to abnormal seedlings. Both embryo-lethal and embryo-defective mutants may also exhibit altered pigmentation, resulting in embryos lacking chlorophyll or accumulating anthocyanins; these are included in the class of pigment mutants (Errampalli et al. 1991; Meinke 1991a, 1994). Of the three mutant classes, by far the largest number and most interesting phenotypes have been identified among embryo-lethals and embryo-defectives, and include seeds containing green blimps, twin and bloated embryos, embryos with fused or single cotyledons, enlarged shoot apices, split hypocotyls, reduced hypocotyls, altered patterns, abnormally large suspensors, distorted epidermal layers, and embryos that prematurely germinate; because of the limited degree of variability displayed by some mutant types, they are hard to tell apart (Meinke 1991b, 1994). Based on the linkage between the mutation and the phenotype, some estimates suggest that, in *Arabidopsis*, 500–1,000 genes may be essential for the completion of embryogenesis and seed development (McElver et al. 2001).

The main problem in analyzing lethal mutations in embryos is that their effects are complex and they impinge on other aspects of plant growth. One of the first closely investigated embryo-lethal phenotypes in *Arabidopsis* turned out to be a biotin auxotroph, implying that embryo lethality is due to a disruption in the biosynthetic pathway of this essential vitamin. This seemed to be true, although two *bio* (*bio1* and *bio2*) mutants analyzed showed seemingly sharp lesions in different steps in the biotin pathway. Embryos of the *bio1* mutant appeared pale throughout development, with a lethal phase between the globular and cotyledonary stages of embryogenesis; although embryos contained reduced levels of biotin, they could be rescued in culture by exogenous biotin, dethiobiotin or 7,8-diaminopelargonic acid, the latter two being intermediates in the biotin biosynthetic pathway (Schneider et al. 1989; Shellhammer and Meinke 1990). A simple model based on this observation envisages that the biotin auxotroph is defective in the conversion of 7-keto-8-aminopelargonic acid to 7,8-diaminopelargonic acid in the biotin pathway. Consistent with this model, it was possible to render the auxotroph a prototroph by introducing as a transgene a functional copy of the bacterial gene *BIOA*, which encodes the enzyme 7,8-diaminopelargonic acid transferase (Patton et al. 1996). Following a variety of defects in cell division patterns, embryos of the *bio2* mutant succumb as early as the globular stage. From this point on, the story that unfolded, with some surprises, was that, unlike *bio1* embryos, arrested *bio2* embryos are rescued not by dethiobiotin, but by biotin. This was aptly attributed to the inability of the mutant embryos to convert dethiotin to biotin, a step catalyzed by the enzyme biotin synthase. The molecular characterization of the biotin synthase (*BIO2*) gene in wild-type and mutant plants has resulted in a key finding that the mutation is due to a deletion of the entire genomic coding region for this critical enzyme (Patton et al. 1998). Analysis of additional embryo-lethal biotin auxotrophs, when combined with a detailed knowledge of the vegetative and reproductive development of *Arabidopsis*, offers tremendous potential to complete a genetic dissection of the entire biotin biosynthetic pathway in plants.

A paradigm linking an enzyme in trehalose metabolism in embryogenesis comes from the work on the embryo-lethal *trehalose phosphate synthase1* (*tps1*) mutant disrupted in a gene encoding trehalose-6-phosphate synthase. The phenotypic consequence of this mutation is the presence of wrinkled seeds containing embryos arrested at the torpedo-shaped stage, on the verge of storage protein accumulation. Taking a traditional developmental approach, it was possible to rescue the mutant embryos by growing them in a medium containing sucrose at a suboptimal level normally required for the growth of wild-type embryos. The parallel between rescuing yeast *tps* mutants and the *Arabidopsis* embryo mutant by restricting the influx of sucrose argues that sustaining a sugar balance underlies the action of the *TPS* gene product (Eastmond et al. 2002).

Mutation in the *GLOBULAR ARREST1* (*GLA1*) gene has been shown to cause embryo lethality beginning at the globular stage due to a defect in folate biosynthesis. The defect is however overcome in mutant embryos of transgenic lines that overproduce *GLA1* transcripts, suggesting that the GLA1 protein is routed to the embryo from the maternal tissues (Ishikawa et al. 2003). According to Collinge et al. (2004), an embryo-lethal mutation in the *ORIGIN RECOGNITION COMPLEX* (*ORC*) gene connected with DNA replication and maintenance of chromosome structure during cell division, causes early seed abortion. The aborted seeds harbor embryos consisting, at best, of eight cells born out of irregular cell divisions of the zygote. A study of another embryo-lethal mutant has highlighted the importance of ubiquitin metabolism in embryogenesis, as functional disruption of a gene that recycles multiple ubiquitin chains back to ubiquitin monomers does not allow embryo development to proceed beyond the globular stage. Because of its wide spectrum of talents, ubiquitin is involved in a number of cellular processes ranging from cell division to cell death. Although the substrate for gene action was not identified in the mutant embryo, that the gene is essential for embryogenesis was confirmed by the rescue of the mutant phenotype by complementation with the wild-type gene in a transgenic setting (Doelling et al. 2001).

A few other embryo-lethal mutants of *Arabidopsis* characterized at the molecular level have lesions in cellular activities such as ribosome and chloroplast functions and peroxisome biogenesis. A gene causing disruption in ribosome functions in the embryo identified by insertion of a transposon is designated as *SMALL SUBUNIT RIBOSOMAL PROTEIN S16* (*SSR16*). Mutant embryos appear to abort at the globular to heart-shaped transition stage. The *SSR16* gene encodes a ribosomal protein with homology to the ribosomal protein S16 (RPS16) of certain bacteria, mitochondria of the fungus *Neurospora crassa*, and plastids of higher plants, and is known to be a key player in the assembly and stability of ribosomes in bacteria (Tsugeki et al. 1996). It is not determined whether embryo lethality is due to reduced transcription of the *SSR16* gene. Weijers et al. (2001b) have described a semi-dominant mutation named *Arabidopsis Minute-like1* (*aml1*) in another ribosomal protein gene of *Arabidopsis*; the mutated gene is designated *AtRPS5*. The observation that embryo development shows a semi-dominant arrest at an early stage in the homozygous mutant is remarkable, but not surprising, given the need for efficient protein translation for progressive embryogenesis. A mutation in the *DOMINO1* (*DOM1*) gene slows down embryo growth, and the globular embryos that are formed come with unusually large nucleoli. The nucleolar defect is seen as early as the zygote stage and has been attributed to a defect in ribosome biogenesis (Lahmy et al. 2004). Indicative of the importance of the functional integrity of chloroplasts for embryo development, arrest of embryo growth between the globular and heart-shaped stages accompanied by the failure to accumulate chlorophyll in the plastids is frequent in the transposon-induced mutant, *edd1*. Although the culture of mutant embryos produces small non-green plantlets, lethality of the mutation becomes obvious when the regenerated plantlets fail to become fertile, and turn into calluses. The *EDD1* gene encodes a novel plastidic form of glycyl-tRNA synthetase (GlyRS) that shows significant homology to a structurally diverse class of enzymes that catalyze the ligation of amino acids to specific mRNA molecules during protein synthesis. Since the EDD1 protein is able to direct a marker protein into isolated pea chloroplasts, it has been proposed that the *edd1* mutation might have detrimental effects on chloroplast biogenesis and metabolism during embryogenesis by interfering with the translational machinery of the chloroplast or with the exchange of signals between the plastid and the nucleus (Uwer et al. 1998). Like *edd1*, the *schlepperless* (*slp*) mutation generated by T-DNA insertion inflicts defects in embryogenesis accompanied by a lesion in chloroplast development in cells. Perhaps the most devastating features of the mutant phenotype are retardation of embryo development before the heart-shaped stage, formation of short cotyledons, appearance of white, stunted seedlings in culture and failure to form mature plants. By using genetic and physical mapping, the mutated gene has been tracked down to *CHAPERONIN-60α*, encoding the plastid chaperonin-60 subunit protein. Since chaperonin is presumably involved in the folding and assembly of proteins in a manner finely tuned to attain their proper conformations, it has been postulated that embryo lethality results from the disruption of chloroplast development due to the lack of a functionally appropriate protein such as chaperonin-60α (Apuya et al. 2001). Failure to produce normal chloroplasts is the cause of embryo lesion in the *rsy3* mutant; RSY3 protein is believed to be a novel protein localized within the chloroplast and necessary for its differentiation (Apuya et al. 2002). A pathway mediated by the *TARGET OF RAPAMYCIN* (*TOR*) gene is known to be involved in the generation of form in animal embryos by relaying the perception of nutrients to the growing tissues. In line with this, premature arrest of embryo development is frequent in *Arabidopsis* mutants with a disrupted *TOR* gene (Menand et al. 2002). Reflecting on the significance of the diverse repertoire of peroxisomes in plant cells, mutational interference of peroxisome biogenesis has been reported to evoke dramatic changes such as abnormalities in the ER, lipid bodies, and protein bodies in the subcellular cytology of mutant embryos. These effects can be equally well interpreted in terms of secondary responses of processes disrupted by the mutation, rather than as the direct result of peroxisome dysfunction (Schumann et al. 2003; Sparkes et al. 2003). A requirement for small GTPases for maintaining embryo viability has been unexpectedly demonstrated in *Antirrhinum majus* (Scrophulariaceae), a model plant used for the isolation of floral organ identity genes. Ingram et al. (1998) showed that mutation in the *ERA-RELATED GTPases* (*ERG*) gene of *A. majus* causes embryo lethality soon after fertilization; since the ERG protein contains mitochondrial localization signals, embryo lethality is thought to be caused by failure of mitochondrial divisions. Although investigations on the mutants described in this paragraph have provided insights into embryo lethality, the severity of these mutations makes it likely that mutants have underlying defects in other, as yet unidentified, cellular functions.

5.3.2
The World of Cytokinesis-Defective Mutants

A group of embryo-lethal mutants of *Arabidopsis*, designated as cytokinesis-defective, has been used to address a clear set of cell-biological problems involving regulation of cell cycle progression during embryogenesis. The mutants included in this group have cell wall stubs and gapped cell walls as a result of incomplete cytokinesis during embryogenic divisions, and multiple nuclei in dividing cells (Nacry et al. 2000; Söllner et al. 2002). Investigations into cytokinesis-defective mutants were influenced in part by the well-established paradigms associated with partitioning of the cytoplasm (cytokinesis) resulting in the formation of a cell plate during cell division in higher plants. At late anaphase or early telophase of mitosis, a dynamic array of microtubules called the phragmoplast is formed between the two daughter nuclei. Concomitantly, secretory vesicles derived from the Golgi-complex carrying cell wall materials are recruited by the phragmoplast to the equator of the dividing cell where their fusion gives rise to the incipient cell plate. The cell plate grows laterally by continuous vesicle fusion to its margins until it joins with the walls of the mother cell. The expanding cell plate also undergoes a complex process of maturation involving the formation of the middle lamella, flanking plasma membranes, and primary walls. This scenario implies that division of plant cells requires a cadre of genes to encode proteins implicated in cytoskeletal dynamics, vesicle delivery, their docking and fusion, and membrane and cell wall biogenesis.

Important insights into the genes involved in cytokinesis during embryogenic divisions have come from analysis of the *titan* and *pilz* groups of *Arabidopsis* mutants. At least nine genetic loci have been identified in the *TITAN* (*TTN*) gene that, when mutated, causes dramatic enlargement and polyploidy of the endosperm nuclei and varying degrees of embryo lethality (Tzafrir et al. 2002). Phenotypic characterization of *ttn1* and *ttn5* showed that the mutations lead to the early demise of the embryo, which produces, at most, four grossly enlarged, sometimes multinucleate cells with giant nuclei. Although *ttn2*, *ttn7*, *ttn8*, and *ttn9* embryos are also arrested in early development, lethality is not accompanied by cell enlargement, whereas the *ttn3* embryo is relatively normal with viable cells that survive seed desiccation. Embryo lethality is the norm in *ttn4* and *ttn6* mutants; in the former the embryo generates a few additional cells late in development before it succumbs, whereas in the latter the embryo attains a globular stage constituted of abnormal cells before lethality sets in (Liu and Meinke 1998; McElver et al. 2000; Tzafrir et al. 2002). These phenotypic classes of *ttn* mutants are shown diagrammatically in Fig. 5.4. Cloning and characterization of protein products of a host of *TTN* genes highlights their importance in fundamental aspects of microtubule assembly, and has made it possible to thread together an explanation for the mutant phenotypes. The protein product of the *TTN5* gene has been identified as a relative of the ARF family of GTP-binding proteins known as ARL2 that probably regulates a pathway in membrane transport for microtubule assembly (McElver et al. 2000). The *TTN1* gene encodes a regulatory protein related to the tubulin-folding cofactor D, involved in tubulin dynamics (Tzafrir et al. 2002), whereas the *TTN3*, *TTN7*, and *TTN8* genes encode the proteins – condensins and cohensins – required for chromosome functions such as their structural maintenance, condensation, chromatid separation, and dosage compensation during mitosis (Liu et al. 2002). Thus, it seems likely that *ttn* phenotypes result from the disruption of some aspects of normal chromosome function during mitosis or microtubule organization during cytokinesis in developing embryos.

Mutations in four genes, namely, *CHAMPIGNON* (*CHO*), *HALLIMASCH* (*HAL*), *PFIFFERLING* (*PFI*), and *PORCINO* (*POR*), collectively dubbed the *PILZ* group, induce phenotypic changes resulting in mushroom-shaped embryos constituted of a few enlarged cells containing one to several nuclei surrounded by noncellular endosperm with grossly enlarged nuclei, very much like *ttn* mutants (Fig. 5.5a–d). Phenotypic identity and map locations have hinted that *cho* corresponds to *ttn1* and *hal* is allelic to *ttn5* (Mayer et al. 1999; McElver et al. 2000). Although an overall slowdown of the cell cycle is implied in the appearance of mutant embryos, normal, wild-type expression of subunits of cell cycle genes such as *CYCLIN-DEPENDENT KINASE* (*CDK*) and *CYC* showed that progression of the cell cycle is not arrested in mutant embryos, whose cells thus apparently remain division-competent. Based on the expression of *KN* mRNA and its encoded protein, syntaxin, in the cells of mutant embryos in a pattern similar to that in wild-type embryos, it was concluded that mutant cells respond to the signal for cell division but fail to organize cell plates. In this vein, a general defect in microtubule assembly, such as absence of the typical microtubule arrays in interphase and mitotic stages, and the presence of stubs of microtubules in the latter was observed in the cells of mutant embryos, suggesting that proteins disrupted by mutation in the *PILZ* group of genes are deployed for microtubule organization during mitosis and for cytokinesis, but not for progression of the cell cycle (Mayer et al. 1999). The four *PILZ* group genes encode orthologs of mammalian tubulin-folding cofactors (TFC) C, D, and E, and an ARL2, which collectively mediate the synthesis of a dimer of α and β tubulin subunits constituting the basic microtubule building block. Another gene with a related embryo-lethal phenotype, *KIESEL* (*KIS*) has been found to encode a TFC ortholog A. These findings have contributed additional evidence for a role for microtubules in cell division and vesicle trafficking during cytokinesis in the developing embryo (Steinborn et al. 2002).

However, it is difficult to determine whether the affected genes have any overlapping functions during cell division in wild-type embryos. The cytokinesis-defective phenotypes of *TTN* and *PILZ* group mutants are also shared by the *hinkel* (*hik*) and *Arabidopsis Dynamin-like Proteins1* (*adl1*) mutants. The *HIK* gene encodes a plant-specific kinesin-related protein that is believed to function in the reorganization of the phragmoplast microtubules during cell plate formation (Strompen et al. 2002). Embryo-lethal defects in homozygous *adl1A/adl1E* double mutants are manifest in the presence of multinucleate cells with incomplete cell walls. Since dynamin and dynamin-related proteins are believed to play a central role in the secretory pathway, particularly in the release of vesicles from the plasma membrane

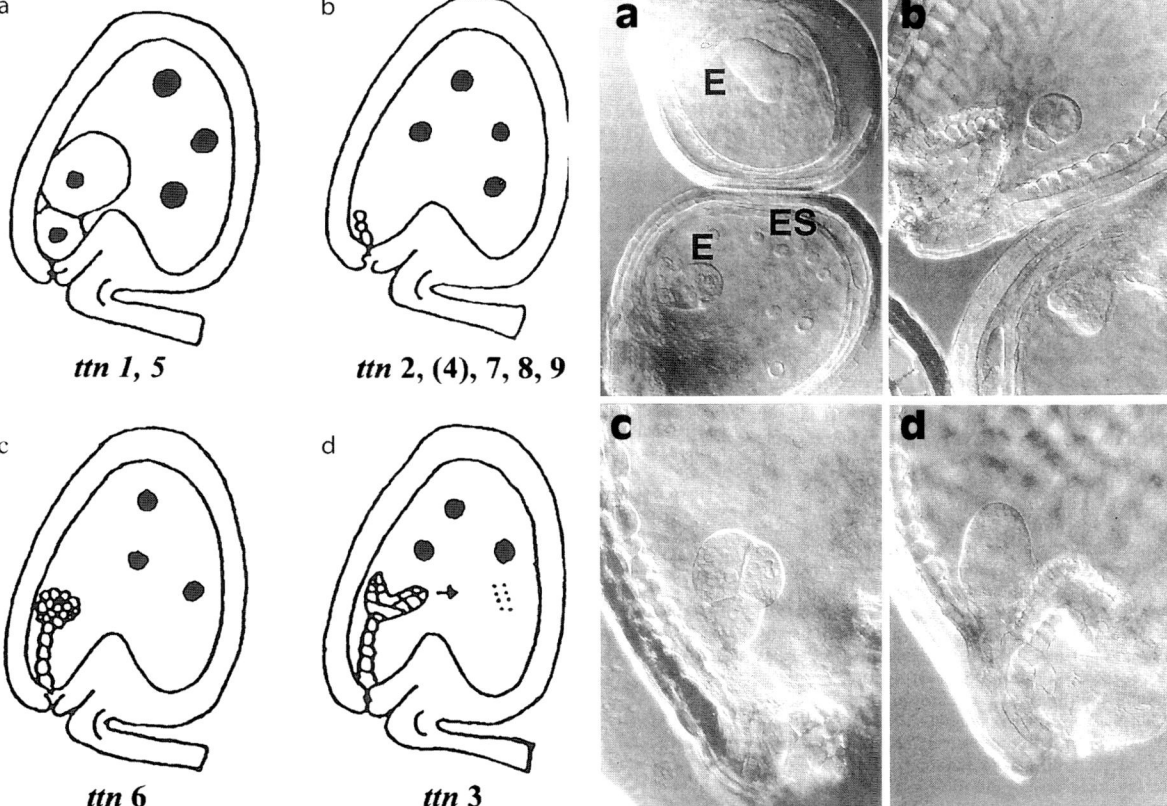

Fig. 5.4 Phentotypic classes of *ttn* mutants of *Arabidopsis* as seen in diagrams of ovules. *Large black dots* Enlarged nucleoli, *small dots* (*ttn3* endosperm) condensed mitotic chromosomes, *arrow* continued embryo development in *ttn3* seeds. An intermediate embryo phenotype is observed in *tt4* seed late in development. (Reprinted from Tzafrir et al. 2002)

Fig. 5.5a–d Nomarski optics of whole mounts of wild-type and *pilz* mutant embryos and endosperm of *Arabidopsis*. **a** *pfi* mutant embryo (*E, bottom*) and wild-type torpedo-shaped embryo (*E, top*) from the same silique; *ES* large endosperm nuclei. **b** *hal* mutant embryo (*top*) and wild-type heart-shaped embryo (*bottom*) from the same silique. **c** *cho* mutant embryo, consisting of four to six large cells. **d** *por* mutant embryo consisting of one large cell. (Reprinted from Mayer et al. 1999)

and their sorting to various subcellular destinations, the defects in the mutant embryos might demonstrate a requirement for these proteins in vesicle formation and fusion, membrane dynamics, and the intricate regulation of these processes necessary for cell plate assembly (Kang et al. 2003).

Cell division abnormalities during embryogenesis have also been described in mutations in *TON, FS* (Chap. 2), *GN, KN, KEU* (Chap. 3), and several other previously uncharacterized genes of *Arabidopsis*, with the difference that these mutations do not result in embryo lethality (Nacry et al. 2000; Söllner et al. 2002). In the cytokinesis-defective mutant *prl*, the zygote-to-embryo transition is occasionally marked by repeated failure of cytokinesis, leading to the formation of a large multinucleate zygote (Holding and Springer 2002), whereas in the *Arabidopsis thaliana cullin1* (*Atcul1*) mutant generated by T-DNA insertions in the *AtCUL1* gene, even the first division of the zygote is blocked (Shen et al. 2002). Analysis of embryogenesis in the *Arabidopsis thaliana Skp-like1* (*ask1*) and *ask2* double mutant also points to the involvement of defective cell divisions during the globular to heart-shaped transition stage (Liu et al. 2004). The fact that mutations in genes that disrupt mitosis and cytokinesis have conspicuous phenotypic effects on embryos, ranging from abnormalities in the zygote, embryo, and seedling shape to embryo lethality, implies that these cellular processes have an overriding effect in virtually all phases of embryogenesis, but the mechanism by which defects in the same cellular processes cause a range of abnormal syndromes remains puzzling.

Disruption of cytokinesis accompanied by patchy cell wall formation during embryogenesis are the defining features of a T-DNA-generated mutation designated as *cytokinesis-defective1* (*cyt1*). The mutant embryo becomes a disorganized mass of enlarged cells resulting from incomplete cytokinesis and failure to synthesize a normal cell wall (Nickle and Meinke 1998). The *CYT1* gene encodes mannose-1-phosphate guanylyltransferase, an enzyme catalyzing the synthesis of GDP-mannose; it is believed that a decrease in GDP-mannose leads to secondary defects such as changes in cell wall composition and deficiency in N-glycosylation. In concert with this fact, mutant embryos are found deficient in N-glycosylation and have an altered composition of cell wall polysaccharides, resulting in a considerable decrease in the level of cellulose (Lukowitz et al. 2001). Embryo developmental arrest and defects in cytokinesis have also been reported in another *Arabidopsis* mutant, *glucosidase1* (*gcs1*) in which N-glycosylation is affected somewhat indirectly (Boisson et al. 2001). The formation of abnormally thick cell walls along with the accumulation of amorphous aggregates and fibrillar material in the walls of the *Arabidopsis* embryo-lethal mutant *vcl1* is, at first sight, difficult to reconcile with the fact that the primary lesion in the mutant embryo cells is the absence of normal vacuoles. The mutational lesions are seen in a clearer light when it is realized that the aberrant cell wall might be due to missecretion of normally vacuole-localized proteins to the extracellular matrix (Rojo et al. 2001). Hall and Cannon (2002) have shown that a mutation in the gene *ROOT-SHOOT-HYPOCOTYL-DEFECTIVE* (*RSH*) can provoke random orientations of planes of divisions of embryos beginning with the zygote stage. The result is the formation of morphologically defective embryos constituted of cells of irregular sizes and shapes. The *RSH* gene encodes a hydroxyproline-rich glycoprotein-type cell wall protein that strengthens the cell wall. The increasing list of cytokinesis-defective embryo mutants will continue to drive intense interest in further characterizing novel genes implicated in cell plate formation and cell wall biosynthesis in plants.

5.3.3
Embryo-defective Mutants of *Arabidopsis*

It is estimated that more than 2,000 *Arabidopsis* mutants with a wide range of defects in embryo development (embryo-defectives) and pigment accumulation have been isolated by different groups of investigators (McElver et al. 2001). In attempts to map these *EMB* genes relative to visible and molecular markers, about 110 mutant genes have been assigned places on the genetic map of *A. thaliana* (Patton et al. 1991; Franzmann et al. 1995). Of the different embryo-defective mutants already described are those with altered suspensor morphology (Chap. 4) and abnormalities in the shoot and root apical meristems, and in the apicobasal and radial patterning of embryos (Chap. 3). As will be seen in the next chapter, several *emb* mutants are disrupted in the maturation of embryos and the associated accumulation of anthocyanins. Early work on *emb* mutants fueled speculation that genes that function critically during embryogenesis are zygotically transcribed. Based on the nonrandom distribution of seeds harboring mutant embryos along the length of heterozygous siliques, it was subsequently concluded that many genes that control early stages of embryogenesis are expressed prior to fertilization and are of gametophytic origin (Meinke1982, 1985). It was mentioned in Chap. 3 that establishment of the apicobasal axis is defective in embryos of the *gn/emb30* mutant. Since an examination of the allele-specific pattern of expression of *GN/EMB30* gene during embryogenesis did not reveal transcripts from the paternal allele, it is easy to imagine an exclusive role for the maternal allele of this gene in the initial development of the embryo (Vielle-Calzada et al. 2000). It will be interesting to see whether this type of asymmetry in the expression of paternal and maternal genes will emerge as a common theme in the control of gene expression during embryogenesis.

Focused studies using tissue culture techniques have been important in advancing the analysis of the *emb* mutants and in casting them in a new light. The majority of *emb* mutants have been rescued by culture of seeds, ovules, or isolated embryos in a nutrient medium, where they show a variety of morphogenetic changes such as callus growth, shoot and root formation from callus, and flowering of regenerated plants. Although embryos arrested at the globular stage do not resume growth when seeds are cultured, seeds enclosing embryos defective prior to the heart-shaped stage produce some callus from embryos. Other embryos arrested at the globular to heart-shaped stages produce abnormal shoots or callus with branched trichomes. Formation of extensive callus, followed in some cases by the regeneration of roots and shoots is frequently observed upon culture of seeds harboring embryos defective from the torpedo-shaped to mature cotyledon stages. Various classes of rescued mutant plants also produce normal or abnormal flowers in culture. A general conclusion derived from these observations is that many *emb* mutants have incurred lesions in genes required for both embryogenesis and vegetative growth (Baus et al. 1986; Franzmann et al. 1989). Protein bodies are useful developmental markers as they appear only toward the final stages of embryogenesis. Ultrastructural studies have revealed the impact of mutations on the cellular machinery of embryos: some mutant embryos contain normal protein bodies similar to those of the wild-type, whereas others lack mature protein bodies. This picture of protein body dimorphism is also correlated with the failure of mutant embryos to accumulate the 12S and 2S storage proteins found in wild-type embryos. Evidently, the mutations not only disrupt morphogenesis, but also compromise cellular differentiation in embryos (Heath et al. 1986; Patton and Meinke 1990). Culture of seedlings of some *keu*-like cytokinesis-defective mutants produced abnormal plants, indicating that *KEU*-like loci are required not only for cytokinesis during embryogenic divisions, but also during vegetative development (Söllner et al. 2002). While the large number of *emb* mutants isolated offers a reliable approach to identify genes that play an essential role in embryogenesis, apart from the characterization of several cytokinesis-defective and pattern-forming genes and their protein products (Chap. 3), very little is known about other genes or their encoded proteins that cause defects in embryo development. Albert et al. (1999) have described an embryo-defective mutation (*emb506*) that appears to affect the post-globular development of embryos but still allows cell divisions that result in giant globular embryos in the mature seed. The gene has been cloned and shown to encode an ankyrin-repeat-containing protein. Ankyrin repeats represent conserved domains of 33 amino acids thought to mediate protein-protein interactions. The role of this protein has been studied by using the promoter of the *ABA-INSENSITIVE* (*ABI3*) gene to direct expression of the *EMB506* gene during embryogenesis, but not during vegetative growth, of transgenic *Arabidopsis*. By providing the wild type protein only during embryogenesis, it was possible to show in this work that although the partially complemented plants are fertile, they display lack of pigmentation in leaves and inflorescences due to defective chloroplast biogenesis (Despres et al. 2001). This observation has profound implications for future studies on the function of *EMB* genes, because a protein that is essential for embryogenesis also appears to be required for another vital function in the plant.

5.3.4
Embryo-Defective Mutants of Maize and Rice

The genetic tractability of maize, combined with the presence of several transposable element systems, has made this plant an attractive model for mutational analysis of embryogenesis. However, unlike *Arabidopsis*, where molecular cloning of genes has been a sequel to mutant isolation, study of maize embryo mutants has not progressed appreciably beyond descriptive accounts. Most embryo-defective maize mutants analyzed have been isolated by pollinating ears with chemically mutagenized pollen grains or by screening stocks that exhibit *Mutator* (*Mu*) transposon activity as a result of gene disruption by direct insertional mutagenesis. The latter approach, which has a long and successful tradition in maize genetics, has yielded a set of mutants for use in molecular cloning of genes. An abundant class of mutants known as *defective kernel* (*dek*) isolated by these methods shows lesions in both endosperm and embryo development. Typically, embryo growth in *dek* mutants is arrested over a wide range of stages, before, during, and after the establishment of the embryo axis. Embryos of the majority of mutants are of the nutritional type as they grow in culture, although there are differences in the capacity of defective embryos to resume normal growth in a mineral salt medium with or without organic additives. Other *dek* mutants are permanently blocked at some stage of embryo development and are hence developmental mutants (Sheridan and Neuffer 1980, 1982; Scanlon et al. 1994; Becraft et al. 2002). Retarded embryo growth beginning at the coleoptilar stage noted in the developmental mutant *empty pericarp2* (*emp2*) is correlated with increased expression of heat-shock protein genes. Since the *EMP2* gene encodes a heat-shock binding protein that apparently functions as a negative regulator of the heat shock response, a feedback mechanism resulting in the overaccumulation of nonfunctional transcripts of the mutant gene has been proposed to account for the mutant phenotype (Fu et al. 2002). Based on the analysis of a *dek* mutant, *lachrima*, a gene encoding a novel transmembrane protein has been isolated. Whereas this gene is expressed mainly in the meristems of developing wild-type embryos, no expression is detected in any part of mutant embryos. Since mutant embryos are blocked at the mid-transition stage of embryogenesis, a suggested requirement for a transmembrane protein involved in auxin transport at this critical stage of embryogenesis is worthy of further investigation (Stiefel et al. 1999). Another *DEK* gene (*DEK1*) encodes a transmembrane protein that shares high homology with calpains, a family of animal cytosolic enzymes involved in signal transduction (Lid et al. 2002). Although characterization of the *DEK1* gene product as a signaling molecule at this level hardly addresses its mechanism of action, the broader biological significance of the gene is clearly in focus.

A second class of mutants includes exclusively *Mu*-induced mutations, designated as *embryo-specific* (*emb*), whose morphogenetic effects are specific to the embryo while the endosperm is normal in its development. A detailed phenotypic characterization of 51 *emb* mutants showed that the majority are blocked during differentiation of the embryo axis and scutellum and initiation of the first leaf primordium, while others are blocked either during formation of the organogenetic part of the embryo and suspensor or during elaboration of embryonic structures preparatory to dormancy (Clark and Sheridan 1991; Sheridan and Clark 1993). Some *emb* mutants have been further characterized to reveal novel genetic and phenotypic differences between mutants, which may prove useful for the isolation and molecular analysis of the genes affected. Confocal laser scanning microscopic images of propidium-iodide-stained embryos of two loss-of-function *emb* mutants are shown in Plate 9, Fig. c–j. The current view is that different genes are probably affected in each of the *emb* mutations, and that early stages of maize embryogenesis, like those of *Arabidopsis*, are controlled by a battery of genes (Heckel et al. 1999; Elster et al. 2000; Consonni et al. 2003; Magnard et al. 2004). Analysis of a group of maize mutants showing abnormal seedling morphology, designated as *defective seedling* (*des*), isolated from active *Mu* stocks, has suggested the possibility that disruption of seedling morphology might be a consequence of mutational defects during embryogenesis, and that these mutants might serve as a springboard for further studies to identify genes active during embryogenesis (Gavazzi et al. 1993; Dolfini et al. 1999; Landoni et al. 2000). In the only successful cloning of a gene whose function is disrupted by an *emb* mutation, Magnard et al. (2004) have shown that the protein product of the affected gene (*Zea mays PLASTID RIBOSOMAL PROTEIN L35*; *ZmPRPL35-1*) has similarity to protein L35 of the large subunit of plastid ribosomes. Conceptually, this information suggests that the *ZmPRPL35-1* gene product might function as a structural protein for the plastid translational machinery of the embryo, and that the presence of functional plastids is more important in embryogenesis than in endosperm development in maize. Emerging from a protracted phase of isolating and cataloging embryo-lethal mutants, the genetics of maize embryogenesis appears now to be entering a mature phase of increasingly active investigations.

A broad spectrum of mutants with lesions at different stages of embryogenesis ranging from the absence of embryonic organs, to abnormal or abnormally placed organs, to embryos lacking either shoot or root, has been obtained by chemical mutagenesis of rice grains. Mutant embryos rarely differentiated a normal shoot without a root, indicating that the same genetic loci probably control shoot and root differentiation episodes (Nagato et al. 1989; Kitano et al. 1993; Hong et al. 1995b, 1996; Scarpella et al. 2003). A gene for shoot development has also been identified, as a mutation in this gene produces embryos lacking a coleoptile, but with an underdeveloped epiblast, a flat shoot apex, and abnormal first three leaves; the mutation also affects plastochron, phyllotaxis, and leaf structure. As the abnormalities gradually disappear, leading to the establishment of a normal shoot, the mutated gene is thought to function at an early stage of shoot development (Tamura et al. 1992). A particularly interesting case of interaction between the embryo and the endosperm uncovered in the developing rice grain involves a temperature-sensitive mutant, *embryoless1* (*eml1*). The most characteristic feature of temperature sensitivity is that rearing plants at a day/night temperature of 30/25°C following pollination causes the formation of grains with no embryos, or of grains with malformed embryos but with a large quantity of endosperm. Grains with large embryos and poorly developed endosperm result when plants are reared at a constant temperature of 18°C or 20°C (Hong et al. 1995a). The subtle effects of the mutant gene on the embryo and endosperm regulated by temperature changes suggest a role for the *EML1* gene product in the continued development of these parts of the grain. Since temperature has no selective role in the development of the embryo or endosperm, identification of the gene product and the temperature-induced changes in the conformation of the protein should help identify the potential mechanism of the temperature effect in the mutant. A challenge facing researchers with maize and rice embryo-defective mutants is to tackle, by comparative gene expression studies, the morphology of certain embryonic organs of members of the Poaceae that do not have any counterparts in other flowering plants.

5.4 Concluding Comments

As the survey in this chapter shows, there are plenty of interesting genes that play a role in the fabrication of the flowering plant embryo from its single-celled origin, and which orchestrate embryo growth. Identification of these genes by direct isolation has been less fruitful than their identification by mutant screening, yet significant advances have been made in our understanding of the central players in the control of the genetic programs of embryogenesis. Much current thinking, based on mutational analysis of embryogenesis in *Arabidopsis*, indicates that early divisions of the zygote are under maternal control by genomic imprinting. In the continued growth of embryos, an important role has been ascribed to genes that encode transcription factors and proteins involved in cellular signaling. Since mutational approaches have also drawn our attention to the role of a number of essential genes in embryogenesis lacking any assigned function, identifying their precise role is central to a fuller understanding of the genetic and molecular control of embryogenesis. The ease of manipulation of *Arabidopsis* for this type of research, now combined with the complete sequencing of its entire genome, ensures that this plant will remain at the forefront of deciphering progressive embryogenesis in flowering plants in terms of coherent gene expression programs.

REFERENCES

Albert S, Després B, Guilleminot J, Bechtold N, Pelletier G, Delseny M, Devic M (1999) The *EMB 506* gene encodes a novel ankyrin repeat containing protein that is essential for the normal development of *Arabidopsis* embryos. Plant J 17:169–179

Apuya NR, Yadegari R, Fischer RL, Harada JJ, Zimmerman JL, Goldberg RB (2001) The *Arabidopsis* embryo mutant *schlepperless* has a defect in the *Chaperonin-60α* gene. Plant Physiol 126:717–730

Apuya NR, Yadegari R, Fischer RL, Harada JH, Goldberg RB (2002) *RASPBERRY3* gene encodes a novel protein important for embryo development. Plant Physiol 129:691–705

Baima S, Nobili F, Sessa G, Lucchetti S, Ruberti I, Morelli G (1995) The expression of the *Atbh-8* homeobox gene is restricted to provascular cells in *Arabidopsis thaliana*. Development 121:4171–4182

Baroux C, Blanvillain R, Gallois P (2001) Paternally inherited transgenes are down-regulated but retain low activity during early embryogenesis in *Arabidopsis*. FEBS Lett 509:11–16

Bäumlein H, Miséra S, Luerßen H, Kölle K, Horstmann C, Wobus U, Müller A (1994) The *FUS3* gene of *Arabidopsis thaliana* is a regulator of gene expression during late embryogenesis. Plant J 6:379–387

Baus AD, Franzmann L, Meinke DW (1986) Growth in vitro of arrested embryos from lethal mutants of *Arabidopsis thaliana*. Theor Appl Genet 72:577–586

Becraft PW, Li K, Dey N, Asuncion-Crabb Y (2002) The maize *dek1* gene functions in embryonic pattern formation and cell fate specification. Development 129:5217–5225

Boisson M, Gomord V, Audran C, Berger N, Dubreucq B, Granier F, Lerouge P, Faye L, Caboche M, Lepiniec L (2001) *Arabidopsis glucosidase 1* mutants reveal a critical role of N-glycan trimming in seed development. EMBO J 20:1010–1019

Breton C, Chaboud A, Matthys-Rochon E, Bates EEM, Cock JM, Fromm H, Dumas C (1995) PCR-generated cDNA library of transition-stage maize embryos: cloning and expression of calmodulin genes during early embryogenesis. Plant Mol Biol 27:105–113

Burgeff C, Liljegren SJ, Tapia-López R, Yanofsky MF, Alvarez-Buylla ER (2002) MADS-box gene expression in lateral primordial, meristems and differentiated tissues of *Arabidopsis thaliana* roots. Planta 214:365–372

Casson SA, Chilley PM, Topping JF, Evans IM, Souter MA, Lindsey K (2002) The *POLARIS* gene of *Arabidopsis* encodes a predicted peptide required for correct root growth and leaf vascular patterning. Plant Cell 14:1705–1721

Castle LA, Errampalli D, Atherton TL, Franzmann LH, Yoon ES, Meinke DW (1993) Genetic and molecular characterization of embryonic mutants identified following seed transformation in *Arabidopsis*. Mol Gen Genet 241:504–514

Chaudhury AM, Ming L, Miller C, Craig S, Dennis ES, Peacock WJ (1997) Fertilization-independent seed development in *Arabidopsis thaliana*. Proc Natl Acad Sci USA 94:4223–4228

Choi Y, Gehring M, Johnson L, Hannon M, Harada JJ, Goldberg RB, Jacobsen SE, Fischer RL (2002) *DEMETER*, a DNA glycosylase domain protein, is required for endosperm gene imprinting and seed viability in *Arabidopsis*. Cell 110:33–42

Clark JK, Sheridan WF (1991) Isolation and characterization of 51 *embryo-specific* mutations of maize. Plant Cell 3:935–951

Collinge MA, Spillane C, Köhler C, Gheyselinck J, Grossniklaus U (2004) Genetic interaction of an origin recognition complex subunit and the polycomb group gene *MEDEA* during seed development. Plant Cell 16:1035–1046

Colombo L, Franken J, van der Krol AR, Wittich PE, Dons HJM, Angenent GC (1997) Downregulation of ovule-specific MADS box genes from *Petunia* results in maternally controlled defects in seed development. Plant Cell 9:703–715

Comai L, Dietrich RA, Maslyar DJ, Baden CS, Harada JJ (1989) Coordinated expression of transcriptionally regulated isocitrate lyase and malate synthase genes in *Brassica napus* L. Plant Cell 1:293–300

Comai L, Matsudaira KL, Heupel RC, Dietrich RA, Harada JJ (1992) Expression of a *Brassica napus* malate synthase gene in transgenic tomato plants during the transition from late embryogeny to germination. Plant Physiol 98:53–61

Consonni G, Aspesi C, Barbante A, Dolfini S, Giuliani C, Giuliani A, Hansen S, Brettschneider R, Pilu R, Gavazzi G (2003) Analysis of four maize mutants arrested in early embryogenesis reveals an irregular pattern of cell division. Sex Plant Reprod 15:281–290

Cordts S, Bantin J, Wittich PE, Kranz E, Lörz H, Dresselhaus T (2001) *ZmES* genes encode peptides with structural homology to defensins and are specifically expressed in the female gametophyte of maize. Plant J 25:103–114

Crouch ML, Sussex IM (1981) Development and storage-protein synthesis in *Brassica napus* L. embryos in vivo and in vitro. Planta 153:64–74

Davidson EH (1986) Gene activity in early development, 3rd edn. Academic Press, Orlando, FL

de Folter S, Busscher J, Colombo L, Losa A, Angenent GC (2004) Transcript profiling of transcription factor genes during silique development in *Arabidopsis*. Plant Mol Biol 56:351–366

Despres B, Delseny M, Devic M (2001) Partial complementation of *embryo defective* mutations: a general strategy to elucidate gene functions. Plant J 27:149–159

de Veylder L, de Almeida Engler J, Burssens S, Manevski A, Lescure B, van Montagu M, Engler G, Inzé D (1999) A new D-type cyclin of *Arabidopsis thaliana* expressed during lateral root primordia formation. Planta 208:453–462

Diener AC, Li H, Zhou W-X, Whoriskey WJ, Nes WD, Fink GR (2000) *STEROL METHYLTRANSFERASE 1* controls the level of cholesterol in plants. Plant Cell 12:853–870

Dietrich RA, Radke SE, Harada JJ (1992) Downstream DNA sequences are required to activate a gene expressed in the root cortex of embryos and seedlings. Plant Cell 4:1371–1382

Doelling JH, Yan N, Kurepa J, Walker J, Vierstra RD (2001) The ubiquitin- specific protease UBP14 is essential for early embryo development in *Arabidopsis thaliana*. Plant J 27:393–405

Dolfini S, Landoni M, Consonni G, Rascio N, Dalla Vecchia F, Gavazzi G (1999) The maize *lilliputian* mutation is responsible for disrupted morphogenesis and minute stature. Plant J 17:11–17

Dornelas MC, Wittich P, von Recklinghausen I, van Lammeren A, Kreis M (1999) Characterization of three novel members of the *Arabidopsis SHAGGY*-related protein kinase (*ASK*) multigene family. Plant Mol Biol 39:137–147

Douglas SJ, Chuck G, Dengler RE, Pelecanda L, Riggs CD (2002) *KNAT1* and *ERECTA* regulate inflorescence architecture in *Arabidopsis*. Plant Cell 14:547–558

Dresselhaus T, Lörz H, Kranz E (1994) Representative cDNA libraries from few plant cells. Plant J 5:605–610

Dresselhaus T, Hagel C, Lörz H, Kranz E (1996) Isolation of a full-length cDNA encoding calreticulin from a PCR library of in vitro zygotes of maize. Plant Mol Biol 31:23–34

Dresselhaus T, Cordts S, Heuer S, Sauter M, Lörz H, Kranz E (1999) Novel ribosomal genes from maize are differentially expressed in the zygotic and somatic cell cycles. Mol Gen Genet 261:416–427

Dure L III, Greenway SC, Galau GA (1981) Developmental biochemistry of cottonseed embryogenesis and germination: changing messenger ribonucleic acid populations as shown by in vitro and in vivo protein synthesis. Biochemistry 20:4162–4168

Eastmond PJ, van Dijken AJH, Spielman M, Kerr A, Tissier AF, Dickinson HG, Jones JDG, Smeekens SC, Graham IA (2002) Trehalose-6-phosphate synthase 1, which catalyses the first step in trehalose synthesis, is essential for *Arabidopsis* embryo maturation. Plant J 29:225–235

Elster R, Bommert P, Sheridan WF, Werr W (2000) Analysis of four *embryo-specific* mutants in *Zea mays* reveals that incomplete radial organization of the proembryo interferes with subsequent development. Dev Genes Evol 210:300–310

Emery JF, Floyd SK, Alvarez J, Eshed Y, Hawker NP, Izhaki A, Baum SF, Bowman JL (2003) Radial patterning of *Arabidopsis* shoots by class III HD-ZIP and KANADI genes. Curr Biol 13:1768–1774

Errampalli D, Patton D, Castle L, Mickelson L, Hansen K, Schnall J, Feldmann K, Meinke D (1991) Embryonic lethals and T-DNA insertional mutagenesis in *Arabidopsis*. Plant Cell 3:149–157

Ettinger WF, Harada JJ (1990) Translational or post-translational processes affect differentially the accumulation of isocitrate lyase and malate synthase proteins and enzyme activities in embryos and seedlings of *Brassica napus*. Arch Biochem Biophys 281:139–143

Fernandez DF (1997) Developmental basis of homeosis in precociously germinating *Brassica napus* embryos: phase change at the shoot apex. Development 124:1149–1157

Finkelstein RR, Crouch ML (1984) Precociously germinating rapeseed embryos retain characteristics of embryogeny. Planta 162:125–131

Finkelstein RR, Tenbarge KM, Shumway JE, Crouch ML (1985) Role of ABA in maturation of rapeseed embryos. Plant Physiol 78:630–636

Franzmann L, Patton DA, Meinke DW (1989) In vitro morphogenesis of arrested embryos from lethal mutants of *Arabidopsis thaliana*. Theor Appl Genet 77:609–616

Franzmann LH, Yoon ES, Meinke DW (1995) Saturating the genetic map of *Arabidopsis thaliana* with embryonic mutations. Plant J 7:341–350

Fu S, Meeley R, Scanlon MJ (2002) *empty pericarp2* encodes a negative regulator of the heat shock protein response and is required for maize embryogenesis. Plant Cell 14:3119–3132

Galau GA, Dure L III (1981) Developmental biochemistry of cottonseed embryogenesis and germination: changing messenger ribonucleic acid populations as shown by reciprocal heterologous complementary deoxyribonucleic acid-messenger ribonucleic acid hybridization. Biochemistry 20:4169–4178

Galau GA, Hughes DW, Dure L III (1986) Abscisic acid induction of cloned cotton late embryogenesis-abundant (*Lea*) mRNAs. Plant Mol Biol 7:155–170

Garcia D, Fitz Gerald JN, Berger F (2005) Maternal control of integument cell elongation and zygotic control of endosperm growth are coordinated to determine seed size in *Arabidopsis*. Plant Cell 17:52–60

Gavazzi G, Dolfini S, Galbiati M, Helentjaris T, Landoni M, Pelucchi N, Todesco G (1993) Mutants affecting germination and early seedling development in maize. Maydica 38:265–274

Goldberg RB, Hoschek G, Ditta GS, Breidenbach RW (1981a) Developmental regulation of cloned superabundant embryo mRNAs in soybean. Dev Biol 83:218–231

Goldberg RB, Hoschek G, Tam SH, Ditta GS, Breidenbach RW (1981b) Abundance, diversity, and regulation of mRNA sequence sets in soybean embryogenesis. Dev Biol 83:201–217

Goldberg RB, Barker SJ, Perez-Grau L (1989) Regulation of gene expression during plant embryogenesis. Cell 56:149–160

Golden TA, Schauer SE, Lang JD, Pien S, Mushegian AR, Grossniklaus U, Meinke DW, Ray A (2002) *SHORT INTEGUMENTS1/SUSPENSOR1/CARPEL FACTORY*, a dicer homolog, is a maternal effect gene required for embryo development in *Arabidopsis*. Plant Physiol 130:808–822

Goldschmidt RB (1955) Theoretical genetics. University of California Press, Berkeley, CA

Grimanelli D, Perotti E, Ramirez J, Leblanc O (2005) Timing of the maternal-to-zygotic transition during early seed development in maize. Plant Cell 17:1061–1072

Grini PE, Jürgens G, Hülskamp M (2002) Embryo and endosperm development is disrupted in the female gametophytic *capulet* mutants of *Arabidopsis*. Genetics 162:1911–1925

Grossniklaus U, Vielle-Calzada J-P, Hoeppner MA, Gagliano WB (1998) Maternal control of embryogenesis by *MEDEA*, a *polycomb* group gene in *Arabidopsis*. Science 280:446–450

Grossniklaus U, Spillane C, Page DR, Köhler C (2001) Genomic imprinting and seed development: endosperm formation with and without sex. Curr Opin Plant Biol 4:21–27

Guitton A-E, Page DR, Chambrier P, Lionnet C, Faure J-E, Grossniklaus U, Berger F (2004) Identification of new members of fertilisation independent seed polycomb group pathway involved in the control of seed development in *Arabidopsis thaliana*. Development 131:2971–2981

Ha CM, Jun JH, Nam HG, Fletcher JC (2004) BLADE-ON-PETIOLE1 encodes a BTB/POZ domain protein required for leaf morphogenesis in *Arabidopsis thaliana*. Plant Cell Physiol 45:1361–1370

Hall Q, Cannon MC (2002) The cell wall hydroxyproline-rich glycoprotein RSH is essential for normal embryo development in *Arabidopsis*. Plant Cell 14:1161–1172

Heath JD, Weldon R, Monnot C, Meinke DW (1986) Analysis of storage proteins in normal and aborted seeds from embryo-lethal mutants of *Arabidopsis thaliana*. Planta 169:304–312

Heck GR, Perry SE, Nichols KW, Fernandez DE (1995) AGL15, a MADS domain protein expressed in developing embryos. Plant Cell 7:1271–1282

Heckel T, Werner K, Sheridan WF, Dumas C, Rogowsky PM (1999) Novel phenotypes and developmental arrest in early *embryo specific* mutants of maize. Planta 210:1–8

Heuer S, Hansen S, Bantin J, Brettschneider R, Kranz E, Lörz H, Dresselhaus T (2001) The maize MADS box genes *ZmMADS3* affects node number and spikelet development and is co-expressed with *ZmMADS1* during flower development, in egg cells, and early embryogenesis. Plant Physiol 127:33–45

Holding DR, Springer PS (2002) The *Arabidopsis* gene *PROLIFERA* is required for proper cytokinesis during seed development. Planta 214:373–382

Hong SK, Aoki T, Kitano H, Satoh H, Nagato Y (1995a) Temperature-sensitive mutation, *embryoless 1*, affects both embryo and endosperm development in rice. Plant Sci 108:165–172

Hong SK, Aoki T, Kitano H, Satoh H, Nagato Y (1995b) Phenotypic diversity of 188 rice embryo mutants. Dev Genet 16:298–310

Hong SK, Kitano H, Satoh H, Nagato Y (1996) How is embryo size genetically regulated in rice? Development 122:2051–2058

Ihle JN, Dure L III (1969) Synthesis of a protease in germinating cotton cotyledons catalyzed by mRNA synthesized during embryogenesis. Biochem Biophys Res Commun 36:705–710

Ihle JN, Dure LS III (1972) The developmental biochemistry of cottonseed embryogenesis and germination. III. Regulation of the biosynthesis of enzymes utilized in germination. J Biol Chem 247:5048–5055

Ingram GC, Simon R, Carpenter R, Coen ES (1998) The Antirrhinum ERG gene encodes a protein related to bacterial small GTPases and is required for embryonic viability. Curr Biol 8:1079–1082

Ishikawa T, Machida C, Yoshioka Y, Kitano H, Machida Y (2003) The *GLOBULAR ARREST1* gene, which is involved in the biosynthesis of folates, is essential for embryogenesis in *Arabidopsis thaliana*. Plant J 33:235–244

Jofuku KD, Omidyar PK, Gee Z, Okamuro JK (2005) Control of seed mass and seed yield by the floral homeotic gene *APETALA2*. Proc Natl Acad Sci USA 102:3117–3122

Kang B-H, Busse JS, Bednarek SY (2003) Members of the *Arabidopsis* dynamin-like gene family, ADL1, are essential for plant cytokinesis and polarized cell growth. Plant Cell 15:899–913

Kasha KJ, Kao KN (1970) High frequency haploid production in barley (*Hordeum vulgare* L.). Nature 225:874–876

Keith K, Kraml M, Dengler NG, McCourt P (1994) *fusca3*: a heterochronic mutation affecting late embryo development in *Arabidopsis*. Plant Cell 6:589–600

Kerstetter RA, Bollman K, Taylor RA, Bomblies K, Poethig RS (2001) *KANADI* regulates organ polarity in *Arabidopsis*. Nature 411:706–709

Kidner CA, Martienssen RA (2004) Spatially restricted microRNA directs leaf polarity through ARGONAUTE1. Nature 428:81–84

Kinoshita T, Yadegari R, Harada JJ, Goldberg RJ, Fischer RL (1999) Imprinting of the *MEDEA* polycomb gene in the *Arabidopsis* endosperm. Plant Cell 11:1945–1952

Kirik V, Kölle K, Balzer H-J, Bäumlein H (1996) Two new oleosin isoforms with altered expression patterns in seeds of the *Arabidopsis* mutant *fus3*. Plant Mol Biol 31:413–417

Kitano H, Tamura Y, Satoh, H, Nagato Y (1993) Hierarchical regulation of organ differentiation during embryogenesis in rice. Plant J 3:607–610

Kiyosue T, Ohad N, Yadegari R, Hannon M, Dinneny J, Wells D, Katz A, Margossian L, Harada JJ, Goldberg RB, Fischer RL (1999) Control of fertilization-independent endosperm development by the *MEDEA* polycomb gene in *Arabidopsis*. Proc Natl Acad Sci USA 96:4186–4191

Klinge B, Werr W (1995) Transcription of the *Zea mays* homeobox (*ZmHox*) genes is activated early in embryogenesis and restricted to meristems of the maize plant. Dev Genet 16:349–357

Köhler C, Hennig L, Bouveret R, Gheyselinck J, Grossniklaus U, Gruissem W (2003a) *Arabidopsis* MSI1 is a component of the MEA/FIE polycomb group complex and required for seed development. EMBO J 22:4804–4814

Köhler C, Hennig L, Spillane C, Pien S, Gruissem W, Grossniklaus U (2003b) The *polycomb*-group protein MEDEA regulates seed development by controlling expression of the MADS-box gene *PHERES1*. Genes Dev 17:1540–1553

Köhler C, Page DR, Gagliardini V, Grossniklaus U (2005) The *Arabidopsis thaliana* MEDEA polycomb group protein controls expression of PHERES1 by parental imprinting. Nat Genet 37:28–30

Koltunow AM (1993) Apomixis: embryo sacs and embryos formed without meiosis or fertilization in ovules. Plant Cell 5:1425–1437

Koltunow AM, Grossniklaus U (2003) Apomixis: a developmental perspective. Annu Rev Plant Biol 54:547–574

Kwong RW, Bui AQ, Lee H, Kwong LW, Fischer RL, Goldberg RB, Harada JJ (2003) *LEAFY COTYLEDON1-LIKE* defines a class of regulators essential for embryo development. Plant Cell 15:5–18

Lacadena J-R (1974) Spontaneous and induced parthenogenesis and androgenesis. In: Kasha KJ (ed) Haploids in higher plants. Advances and potential. University of Guelph, Ontario, pp 13–32

Lahmy S, Guilleminot J, Cheng C-M, Bechtold N, Albert S, Pelletier G, Delseny M, Devic M (2004) *DOMINO1*, a member of a small plant-specific gene family, encodes a protein essential for nuclear and nucleolar functions. Plant J 39:809–820

Lan L, Chen W, Lai Y, Suo J, Kong Z, Li C, Lu Y, Zhang Y, Zhao X, Zhang X, Zhang Y, Han B, Cheng J, Zue Y (2004) Monitoring of gene expression profiles and isolation of candidate genes involved in pollination and fertilization in rice (*Oryza sativa* L.) with a 10K cDNA microarray. Plant Mol Biol 54:471–487

Landoni M, Gavazzi G, Rascio N, Dalla Vecchia F, Consonni G, Dolfini S (2000) A maize mutant with an altered vascular pattern. Ann Bot 85:143–150

Lee H, Fischer RL, Goldberg RB, Harada JJ (2003) *Arabidopsis* LEAFY COTYLEDON1 represents a functionally specialized subunit of the CCAAT binding transcription factor. Proc Natl Acad Sci USA 100:2152–2156

Li Z, Thomas TL (1998) *PEI1*, an embryo-specific zinc finger protein gene required for heart-stage embryo formation in *Arabidopsis*. Plant Cell 10:383–398

Lid SE, Gruis D, Jung R, Lorentzen JA, Ananiev E, Chamberlin M, Niu X, Meeley R, Nichols S, Olsen O-A (2002) The *defective-kernel 1* (*dek1*) gene required for aleurone cell development in the endosperm of maize grains encodes a membrane protein of the calpain gene superfamily. Proc Natl Acad Sci USA 99:5460–5465

Liu C, Meinke DW (1998) The *titan* mutants of *Arabidopsis* are disrupted in mitosis and cell cycle control during seed development. Plant J 16:21–31

Liu C, McElver J, Tzafrir I, Joosen R, Wittich P, Patton D, van Lammeren AAM, Meinke D (2002) Condensin and cohesin knockouts in *Arabidopsis* exhibit a *titan* phenotype. Plant J 29:405–415

Liu F, Ni W, Griffith ME, Huang Z, Chang C, Peng W, Ma H, Xie D (2004) The *ASK1* and *ASK2* genes are essential for *Arabidopsis* early development. Plant Cell 16:5–20

Lotan T, Ohto M, Yee KM, West MAL, Lo R, Kwong RW, Yamagishi K, Fischer RL, Goldberg RB, Harada JJ (1998) *Arabidopsis* LEAFY COTYLEDON1 is sufficient to induce embryo development in vegetative cells. Cell 93:1195–1205

Luerßen H, Kirik V, Herrmann P, Miséra S (1998) *FUSCA3* encodes a protein with a conserved VP1/ABI3-like B3 domain which is of functional importance for the regulation of seed maturation in *Arabidopsis thaliana*. Plant J 15:755–764

Lukowitz W, Nickle TC, Meinke DW, Last RL, Conklin PL, Somerville CR (2001) *Arabidopsis cyt1* mutants are deficient in a mannose-1-phosphate guanylyltransferase and point to a requirement of N-linked glycosylation for cellulose biosynthesis. Proc Natl Acad Sci USA 98:2262–2267

Luo M, Bilodeau P, Koltunow A, Dennis ES, Peacock WJ, Chaudhury AM (1999) Genes controlling fertilization-independent seed development in *Arabidopsis thaliana*. Proc Natl Acad Sci USA 96:296–301

Luo M, Bilodeau P, Dennis ES, Peacock WJ, Chaudhury A (2000) Expression of parent-of-origin effects for *FIS2*, *MEA*, and *FIE* in the endosperm and embryo of developing *Arabidopsis* seeds. Proc Natl Acad Sci USA 97:10637–10642

Magnard J-L, Heckel T, Massonneau A, Wisniewski J-P, Cordelier S, Lassagne H, Perez P, Dumas C, Rogowsky PM (2004) Morphogenesis of maize embryos requires *ZmPRPL35-1* encoding a plastid ribosomal protein. Plant Physiol 134:649–663

Mayer U, Herzog U, Berger F, Inzé D, Jürgens G (1999) Mutations in the *PILZ* group genes disrupt the microtubule cytoskeleton and uncouple cell cycle progression from cell division in *Arabidopsis* embryo and endosperm. Eur J Cell Biol 78:100–108

McConnell JR, Emery J, Eshed Y, Bao N, Bowman J, Barton MK (2001) Role of *PHABULOSA* and *PHAVOLUTA* in determining radial patterning in shoots. Nature 411:709–713

McElver J, Patton D, Rumbaugh M, Liu C, Meinke D (2000) The *TITAN5* gene of *Arabidopsis* encodes a protein related to the ADP ribosylation factor family of GTP binding proteins. Plant Cell 12:1379–1392

McElver J, Tzafrir I, Aux G, Rogers R, Ashby C, Smith K, Thomas C, Schetter A, Zhou Q, Cushman MA, Tossberg J, Nickle T, Levin JZ, Meinke D, Patton D (2001) Insertional mutagenesis of genes required for seed development in *Arabidopsis thaliana*. Genetics 159:1751–1763

Meinke DW (1982) Embryo-lethal mutants of *Arabidopsis thaliana*: evidence for gametophytic expression of the mutant genes. Theor Appl Genet 63:381–386

Meinke DW (1985) Embryo-lethal mutants of *Arabidopsis thaliana*: analysis of mutants with a wide range of lethal phases. Theor Appl Genet 69:543–552

Meinke DW (1991a) Embryonic mutants of *Arabidopsis thaliana*. Dev Genet 12:382–392

Meinke DW (1991b) Perspectives on genetic analysis of plant embryogenesis. Plant Cell 3:857–866

Meinke DW (1992) A homeotic mutant of *Arabidopsis thaliana* with leafy cotyledons. Science 258:1647–1650

Meinke DW (1994) Seed development in *Arabidopsis thaliana*. In: Meyerowitz EM, Somerville CR (eds) *Arabidopsis*. Cold Spring Harbor Laboratory Press, Cold Spring Harbor, New York, pp 253–295

Meinke DW, Franzmann LH, Nickle TC, Yeung EC (1994) *Leafy cotyledon* mutants of *Arabidopsis*. Plant Cell 6:1049–1064

Menand B, Desnos T, Nussaume L, Berger F, Bouchez D, Meyer C, Robaglia C (2002) Expression and disruption of the *Arabidopsis TOR* (target of rapamycin) gene. Proc Natl Acad Sci USA 99:6422–6427

Nacry P, Mayer U, Jürgens G (2000) Genetic dissection of cytokinesis. Plant Mol Biol 43:719–733

Nagato Y, Kitano H, Kamijima O, Kikuchi S, Satoh H (1989) Developmental mutants showing abnormal organ differentiation in rice embryos. Theor Appl Genet 78:11–15

Nickle TC, Meinke DW (1998) A cytokinesis-defective mutant of *Arabidopsis* (*cyt1*) characterized by embryonic lethality, incomplete cell walls and excessive callose accumulation. Plant J 15:321–332

Ohad N, Margossian L, Hsu Y, Williams C, Repetti P, Fischer RL (1996) A mutation that allows endosperm development without fertilization. Proc Natl Acad Sci USA 93:5319–5324

Ohad N, Yadegari R, Margossian L, Hannon M, Michaeli D, Harada JJ, Goldberg RB, Fischer RL (1999) Mutations in *FIE*, a WD polycomb group gene, allow endosperm development without fertilization. Plant Cell 11:407–415

Ohno CK, Reddy GV, Heisler MGB, Meyerowitz EM (2004) The *Arabidopsis JAGGED* gene encodes a zinc finger protein that promotes leaf tissue development. Development 131:1111–1122

Ohto M, Fischer RL, Goldberg RB, Nakamura K, Harada JJ (2005) Control of seed mass by *APETALA2*. Proc Natl Acad Sci USA 102:3123–3128

Palatnik JF, Allen E, Wu X, Schommer C, Schwab R, Carrington JC, Weigel D (2003) Control of leaf morphogenesis by microRNAs. Nature 425:257–263

Parcy F, Valon C, Kohara A, Miséra S, Giraudat J (1997) The *ABSCISIC ACID-INSENSITIVE3*, *FUSCA3*, and *LEAFY COTYLEDON1* loci act in concert to control multiple aspects of *Arabidopsis* seed development. Plant Cell 9:1265–1277

Patton DA, Meinke DW (1990) Ultrastructure of arrested embryos from lethal mutants of *Arabidopsis thaliana*. Am J Bot 77:653–661

Patton DA, Franzmann LH, Meinke DW (1991) Mapping genes essential for embryo development in *Arabidopsis thaliana*. Mol Gen Genet 227:337–347

Patton DA, Volrath S, Ward ER (1996) Complementation of an *Arabidopsis thaliana* biotin auxotroph with an *Escherichia coli* biotin biosynthetic gene. Mol Gen Genet 251:261–266

Patton DA, Schetter AL, Franzmann LH, Nelson K, Ward ER, Meinke DW (1998) An embryo-defective mutant of *Arabidopsis* disrupted in the final step of biotin synthesis. Plant Physiol 116:935–946

Perry SE, Nichols KW, Fernandez DE (1996) The MADS domain protein AGL15 localizes to the nucleus during early stages of seed development. Plant Cell 8:1977–1989

Perry SE, Lehti MD, Fernandez DE (1999) The MADS-domain protein AGAMOUS-like 15 accumulates in embryonic tissues with diverse origins. Plant Physiol 120:121–129

Postma-Haarsma AD, Verwoert IIGS, Stronk OP, Koster J, Lamers GEM, Hoge JHC, Meijer AH (1999) Characterization of the KNOX class homeobox genes *Oskn2* and *Oskn3* identified in a collection of cDNA libraries covering the early stages of rice embryogenesis. Plant Mol Biol 39:257–271

Raghavan V (1997) Molecular embryology of flowering plants. Cambridge University Press, New York

Raikhel NV, Quatrano RS (1986) Localization of wheat-germ agglutinin in developing wheat embryos and those cultured in abscisic acid. Planta 168:433–440

Raikhel NV, Bednarek SY, Wilkins TA (1988) Cell-type-specific expression of a wheat-germ agglutinin gene in embryos and young seedlings of *Triticum aestivum*. Planta 176:406–414

Ramachandran C, Raghavan V (1992) Apomixis in distant hybridization. In: Kalloo G, Chowdhury JB (eds) Distant hybridization in crop plants. Springer, Berlin Heidelberg New York, pp 106–121

Ray S, Golden T, Ray A (1996) Maternal effects of the *short integument* mutation on embryo development in *Arabidopsis*. Dev Biol 180:365–369

Reyes JC, Grossniklaus U (2003) Diverse functions of *polycomb* group proteins during plant development. Sem Cell Dev Biol 14:77–84

Richert J, Kranz E, Lörz H, Dresselhaus T (1996) A reverse transcriptase-polymerase chain reaction assay for gene expression studies at the single cell level. Plant Sci 114:93–99

Robinson-Beers K, Pruitt RE, Gasser CS (1992) Ovule development in wild-type *Arabidopsis* and two female-sterile mutants. Plant Cell 4:1237–1249

Rojo E, Gillmor CS, Kovaleva V, Somerville CR, Raikhel NV (2001) *VACUOLELESS1* is an essential gene required for vacuole formation and morphogenesis in *Arabidopsis*. Dev Cell 1:303–310

Rounsley SD, Ditta GS, Yanofsky MF (1995) Diverse roles for MADS box genes in *Arabidopsis* development. Plant Cell 7:1259–1269

Russinova E, de Vries SC (2000) Parental contribution to plant embryos. Plant Cell 12:461–463

Sauter M, von Wiegen P, Lörz H, Kranz E (1998) Cell cycle regulatory genes from maize are differentially controlled during fertilization and first embryonic cell division. Sex Plant Reprod 11:41–48

Scanlon MJ, Stinard PS, James MG, Myers AM, Robertson DS (1994) Genetic analysis of 63 mutations affecting maize kernel development isolated from *mutator* stocks. Genetics 136:281–294

Scarpella E, Rueb S, Meijer AH (2003) The *RADICLELESS1* gene is required for vascular pattern formation in rice. Development 130:645–658

Schauer SE, Jacobsen SE, Meinke DW, Ray A (2002) *DICER-LIKE1*: blind men and elephants in *Arabidopsis* development. Trends Plant Sci 7:487–491

Schneider T, Dinkins R, Robinson K, Shellhammer J, Meinke DW (1989) An embryo-lethal mutant of *Arabidopsis thaliana* is a biotin autotroph. Dev Biol 131:161–167

Scholten S, Lörz H, Kranz E (2002) Paternal mRNA and protein synthesis coincides with male chromatin decondensation in maize zygotes. Plant J 32:221–231

Schumann U, Wanner G, Veenhuis M, Schmid M, Gietl C (2003) AthPEX10, a nuclear gene essential for peroxisome and storage organelle formation during *Arabidopsis* embryogenesis. Proc Natl Acad Sci USA 100:9626–9630

Shellhammer J, Meinke D (1990) Arrested embryos from the *bio1* auxotroph of *Arabidopsis thaliana* contain reduced levels of biotin. Plant Physiol 93:1162–1167

Shen W-H, Parmentier Y, Hellmann H, Lechner E, Dong A, Masson J, Granier F, Lepiniec L, Estelle M, Genschik P (2002) Null mutation of AtCUL1 causes arrest in early embryogenesis in *Arabidopsis*. Mol Biol Cell 13:1916–1928

Sheridan WF, Clark JK (1993) Mutational analysis of morphogenesis of the maize embryo. Plant J 3:347–358

Sheridan WF, Neuffer MG (1980) Defective kernel mutants of maize. II. Morphological and embryo culture studies. Genetics 95:945–960

Sheridan WF, Neuffer MG (1982) Maize developmental mutants. Embryos unable to form leaf primordia. J Hered 73:318–329

Siegfried KR, Eshed Y, Baum SF, Otsuga D, Drews GN, Bowman JL (1999) Members of the *YABBY* gene family specify abaxial cell fate in *Arabidopsis*. Development 126:4117–4128

Söllner R, Glässer G, Wanner G, Somerville CR, Jürgens G, Assaad FF (2002) Cytokinesis-defective mutants of *Arabidopsis*. Plant Physiol 129:678–690

Sparkes IA, Brandizzi F, Slocombe SP, El-Shami M, Hawes C, Baker A (2003) An *Arabidopsis pex10* null mutant is embryo lethal, implicating peroxsomes in an essential role during plant embryogenesis. Plant Physiol 133:1809–1819

Spillane C, MacDougall C, Stock C, Köhler C, Vielle-Calzada J-P, Nunes S, Grossniklaus U, Goodrich J (2000) Interaction of the *Arabidopsis* polycomb group proteins FIE and MEA mediates their common phenotypes. Curr Biol 10:1535–1538

Springer PS, Holding DR, Groover A, Yordan C, Martienssen RA (2000) The essential Mcm7 protein PROLIFERA is localized to the nucleus of dividing cells through the G_1 phase and is required maternally for early *Arabidopsis* development. Development 127:1815–1822

Sprunck S, Baumann U, Edwards K, Langridge P, Dresselhaus T (2005) The transcript composition of egg cells changes significantly following fertilization in wheat (*Triticum aestivum* L.). Plant J 41:660–672

Steinborn K, Maulbetsch C, Priester B, Trautmann S, Pacher T, Geiges B, Küttner F, Lepiniec L, Stierhof Y-D, Schwarz H, Jürgens G, Mayer U (2002) The *Arabidopsis PILZ* group genes encode tubulin-folding cofactor orthologs required for cell division but not cell growth. Genes Dev 16:959–971

Stiefel V, López Becerra E, Roca R, Bastidas M, Jharmann T, Graziano E, Puigdomènech P (1999) *TM20*, a gene coding for a new class of transmembrane proteins expressed in the meristematic tissues of maize. J Biol Chem 274:27734–27739

Stone SL, Kwong LW, Yee KM, Pelletier J, Lepiniec L, Fischer RL, Goldberg RB, Harada JJ (2001) *LEAFY COTYLEDON2* encodes a B3 domain transcription factor that induces embryo development. Proc Natl Acad Sci USA 98:11806–11811

Strompen G, El Kasmi F, Richter S, Lukowitz W, Assaad FF, Jürgens G, Mayer U (2002) The *Arabidopsis HINKEL* gene encodes a kinesin-related protein involved in cytokinesis and is expressed in a cell cycle-dependent manner. Curr Biol 12:153–158

Tamura Y, Kitano H, Satoh H, Nagato Y (1992) A gene profoundly affecting shoot organization in the early phase of rice development. Plant Sci 82:91–99

Tsuchiya Y, Nambara E, Naito S, McCourt P (2004) The *FUS3* transcription factor functions through the epidermal regulator *TTG1* during embryogenesis in *Arabidopsis*. Plant J 37:73–81

Tsugeki R, Kochieva EZ, Fedoroff NV (1996) A transposon insertion in the *Arabidopsis SSR16* gene causes an embryo-defective lethal mutation. Plant J 10:479–489

Tzafrir I, McElver JA, Liu C, Yang LJ, Wu JQ, Martinez A, Patton DA, Meinke DW (2002) Diversity of TITAN functions in *Arabidopsis* seed development. Plant Physiol 128:38–51

Uwer U, Willmitzer L, Altmann T (1998) Inactivation of a glycyl-tRNA synthetase leads to an arrest in plant embryo development. Plant Cell 10:1277–1294

Vicient CM, Bies-Etheve N, Delseny M (2000) Changes in gene expression in the *leafy cotyledon1* (*lec1*) and *fusca3* (*fus3*) mutants of *Arabidopsis thaliana* L. J Exp Bot 51:995–1003

Vielle-Calzada J-P, Thomas J, Spillane C, Coluccio A, Hoeppner MA, Grossniklaus U (1999) Maintenance of genomic imprinting at the *Arabidopsis medea* locus requires zygotic *DM1* activity. Genes Dev 13:2971–2982

Vielle-Calzada J-P, Baskar R, Grossniklaus U (2000) Delayed activation of the paternal genome during seed development. Nature 404:91–94

Vielle-Calzada J-P, Baskar R, Grossniklaus U (2001) Early paternal gene activity in *Arabidopsis* – reply. Nature 414:710

Wang H, Caruso LV, Downie AB, Perry SE (2004) The embryo MADS domain protein AGAMOUS-Like 15 directly regulates expression of a gene encoding an enzyme involved in gibberellin metabolism. Plant Cell 16:1206–1219

Weijers D, Geldner N, Offringa R, Jürgens G (2001a) Early paternal gene activity in *Arabidopsis*. Nature 414:709–710

Weijers D, Franke-van Dijk M, Vencken R-J, Quint A, Hooykaas P, Offringa R (2001b) An *Arabidopsis* minute-like phenotype caused by a semi-dominant mutation in a *RIBOSOMAL PROTEIN S5* gene. Development 128:4289–4299

Yadegari R, Kinoshita T, Lotan O, Cohen G, Katz A, Choi Y, Katz A, Nakashima K, Harada JJ, Goldberg RB, Fischer RL, Ohad N (2000) Mutations in the *FIE* and *MEA* genes that encode interacting polycomb proteins cause parent-of-origin effects on seed development by distinct mechanisms. Plant Cell 12:2367–2381

Yang HY, Zhou C (1982) In vitro induction of haploid plants from unpollinated ovaries and ovules. Theor Appl Genet 63:97–104

Yoshida KT, Wada T, Koyama H, Mizobuchi-Fukuoka R, Saito S (1999) Temporal and spatial patterns of accumulation of the transcript of *myo*-inositol-1-phosphate synthase and phytin-containing particles during seed development in rice. Plant Physiol 119:65–72

Zhang JZ, Gomez-Pedrozo M, Baden CS, Harada JJ (1993) Two classes of isocitrate lyase genes are expressed during late embryogeny and postgermination in *Brassica napus* L. Mol Gen Genet 238:177–184

Zhang JZ, Santes CM, Engel ML, Gasser CS, Harada JJ (1996) DNA sequences that activate isocitrate lyase gene expression during late embryogenesis and during postgerminative growth. Plant Physiol 110:1069–1079

6 Maturation and Dormancy – Survival Strategies of the Embryo

The plant embryo is capable of germination during its late development but is constrained from doing so while attached to the mother plant. Although a variety of evidence suggests that something actively maintains embryogenesis until a termination switch is thrown that permits subsequent germination, there is little agreement about the identity of these putative maintenance factors and switches or when they act. Part of the disagreement reflects apparent species-specific and developmental stage-specific differences in embryo competence as well as different emphasis placed on the results of the descriptive, experimental and genetic approaches to the problem. Some of this debate arises from the incomplete description of embryo development and the lack of standard criteria and nomenclature. Nonetheless, recent results suggest that we should tentatively reject several ideas about the regulation of late embryo development and consider alternative interpretations of the well-established observations that have dominated current thought.

G.A. Galau et al. 1991

6.1 Embryo Maturation   132
6.1.1 Synthesis of Maturation Proteins   132
6.1.2 Is Embryo Maturation in the Seed Developmentally Regulated by ABA?   134
6.1.3 Genetic Regulation of Embryo Maturation by ABA   136
6.2 Embryo Dormancy   138
6.2.1 Carbohydrates in Desiccation Tolerance   141
6.2.2 Proteins in Desiccation Tolerance   143
6.3 Concluding Comments   145
References   146

The termination of cell divisions in the developing embryo signifies completion of morphogenesis and establishment of the embryo body plan. This sets the stage for the embryo to enter the maturation phase and eventually lapse into quiescence or dormancy. Although maturation, quiescence, and dormancy of embryos insulated within the protective layers of the seed coats are enduring themes in plant physiology, the causes and consequences of these phenomena often vex plant biologists. In general terms, embryo maturation (alternatively called mid-embryogenesis phase) is equated with seed maturation, and includes such tell-tale signs as the arrest of further embryo morphogenesis and the accumulation of an acervate complex of storage reserves, including proteins, carbohydrates, and lipids. This is followed by the late-embryogenesis phase, characterized by cessation of metabolic activity in the embryo, suppression of premature germination, and acquisition of desiccation tolerance (the ability to germinate after drying), enabling the embryo and seed to resist adverse environmental conditions and yet remain viable. The late-embryogenesis phase, which generally follows ovule abscission, is considered as the quiescent or dormant stage of the embryo, although quiescence and dormancy are traditionally used to describe the state of the whole seed rather than that of the embryo enclosed within. Thus, quiescence is considered as a state of arrested development of the embryo in a nondormant seed that is easily overcome by providing suitable environmental prerequisites for initiating germination growth and development, such as water, a favorable temperature, and the normal composition of the atmosphere; in contrast, dormancy is a temporary failure of the embryo of a viable seed to germinate under conditions that favor germination of embryos of quiescent seeds. Both quiescence and dormancy of seeds have survival values for the plant because they force the embryo to wait passively for improved conditions for germination to maximize chances of seedling survival. For this reason, much of the physiological, genetic, and molecular research on the entry into and release from quiescence and dormancy of embryos has been undertaken with whole seeds rather than with isolated embryos, and a discussion based on such studies on seeds does not necessarily exclude the effects of the maternally-derived seed coats and the interplay of genome dosages of the diploid embryo and the triploid endosperm on the developmental state of the embryo favoring germination. Because of the element of uncertainty arising from the use of seeds in reaching conclusions on the state of the enclosed embryo, the term dormancy is used in this chapter to denote the arrested developmental state of the embryo following a period of maturation, irrespective of whether the seeds are quiescent or dormant.

These introductory comments are admittedly simplified, and are intended to capture the essence of classical views on maturation, quiescence, dormancy, and survival strategy of embryos. Although there is an endless diversity of seeds, investigations involving a healthy dose of molecular biology on selected model systems have now established that a set of common gene products are responsible for the physiological processes that characterize embryo maturation, and that complex gene expression programs control both maturation and preparation for dormancy of the embryo enclosed in the seed. This chapter will focus primarily on the role of growth hormones, proteins, and carbohydrates in the regulation of maturation of embryos, and in the processes that temper embryos to withstand desiccation and lapse into dormancy. For practical purposes, there is no sharp line separating maturation and dormancy of embryos, as these phases proceed in a partially overlapping fashion following completion of embryo morphogenesis. Indeed, an interesting implication of the relationship between the two phases is that the same hormones, genes, and mechanisms seem to control them in a temporal way.

A large body of research with molecular overtones conducted on seed maturation and dormancy is reviewed by Harada (1997); perspectives on the genetic programs and control signals of seed maturation are highlighted by Wobus and Weber (1999); studies on the genetics of seed dormancy and germination in *Arabidopsis* are summarized by Bentsink and Koornneef (2002).

6.1
Embryo Maturation

The transition of embryo from the morphogenetic to the maturation phase is characterized and quantified by an increase in dry weight due to the accumulation of storage reserves. Not easily quantified, but nonetheless important, are the execution of a large number of physiological and molecular changes that initiate and maintain the regulatory programs of the maturation phase. Experimental analysis has focused on the plant hormone ABA, in particular, changes in its concentration and on the protein products of a few unique *Arabidopsis* genes involved in ABA perception, as controlling factors in embryo maturation. As described later in this section, studies of mutant phenotypes of *Arabidopsis* are beginning to reshape our thinking as to how a cluster of genes impact on the network of processes involved in embryo maturation and seed dormancy.

6.1.1
Synthesis of Maturation Proteins

The maturation phase of the embryo, which follows the proliferation phase, involves a long period of growth by cell expansion, during which rapid synthesis and accumulation of storage proteins and other reserves occurs. As mentioned in Chap. 2, the major embryo storage proteins include the 7S and 12S globulins and the 2S albumin groups. There are a number of globulins named after the plants or families in which they were first described, such as vicilin, convicilin, and legumin from *Vicia faba* and *Phaseolus vulgaris*; conglycinin and glycinin from *Glycine max*; phaseolin from *P. vulgaris*; and cruciferin (12S) from *Brassica napus*. Included in the albumins is napin (2S) from *B. napus*. A variety of prevalent proteins such as lectins and trypsin inhibitors, whose functions are not clearly understood, also accumulate during the maturation phase. The stored reserves are hydrolyzed during germination of the seed to serve as a source of nitrogen for the developing seedling (Goldberg et al. 1981).

In legumes and other plants, storage proteins accumulate mostly in the parenchymatous cells of the cotyledons and embryo axis in the form of large protein bodies consisting of an amorphous protein matrix bounded by two electron-dense phospholipid layers of a unit membrane. These proteins are encoded by several multigene families, and expression of these genes is regulated temporally during embryogenesis (Goldberg et al. 1989). One intriguing aspect of the temporal regulation of storage protein genes is that their transcripts accumulate to high levels only during particular stages of embryogenesis, and are generally absent at other stages. They are also differentially expressed within specific embryo cells and tissues, irrespective of the mechanisms that establish seed protein mRNA accumulation patterns. For example, in mature embryos of *Arabidopsis*, transcripts of a 12S storage protein gene are present mostly in the parenchyma cells of cotyledons and hypocotyl, and much less so in the procambial cells of these organs (Pang et al. 1988). In soybean, transcripts of the *KTi1/2* gene are localized primarily in the periphery of cotyledons of midmaturation stage embryos, and are not detectable in the cells of the embryo axis; this contrasts with the high abundance of transcripts of the *KTi3* gene in the embryo axis as well as in cotyledons. Moreover, the onset of accumulation of transcripts of the *KTi* and β-conglycinin genes in soybean embryos moves in a wave-like pattern from the outer to the inner face of the cotyledon (Perez-Grau and Goldberg 1989). A wave-like pattern in the accumulation of legumin and vicilin and their mRNAs, directly correlated with mitotic gradients, is also observed in the developing cotyledons of *Vicia faba* (Borisjuk et al. 1995). In *Brassica napus*, cruciferin mRNAs accumulate later in embryo development than napin transcripts, which appear as early as the late heart-shaped stage; however, at later stages of embryogenesis, transcripts of cruciferin and napin are both localized in different regions of the embryo, such as accumulation beginning in the cortical cells of the embryo axis of late heart-shaped embryos, in the abaxial face of cotyledons in the torpedo-shaped embryos, and in the adaxial face of cotyledons at later stages of development (Fernandez et al. 1991). These cell commitment patterns for the accumulation of storage protein transcripts might well be regulated by precisely-timed embryo-specific signaling events. Correlative analyses on the role of metabolites such as sugars and amino acids transported from the mother plant into developing seeds, and of cellular calcium as a signaling molecule during embryogenesis, are beginning to address this issue (Wobus and Weber 1999).

A study in cotton of the accumulation kinetics of five cloned storage protein gene transcripts, and changes in their concentrations during embryogenesis and germination, showed that they are expressed in unique, but overlapping patterns, involving a period of increase in abundance during maturation, followed by a precipitous decrease before the desiccation phase, with further decline occurring during early germination. A second set of four transcripts is also high in abundance during embryo maturation followed by a rapid decline, but showing a recovery in abundance during the desiccation phase. Such transcripts encoding proteins expressed in high abundance invariably mark the maturation phase of embryogenesis and are designated as *MATURATION* (*MAT*) mRNAs (Galau et al. 1987; Hughes and Galau 1989). A global view of the *MAT* class of genes is that they represent the most abundantly transcribed genes whose abundance is restricted to a narrow window between the morphogenesis and desiccation phases of the embryo. Based on the analysis of developing siliques of *Arabidopsis*, transcripts of *Arabidopsis thaliana 2S ALBUMIN* (*At2S3*), *CRUCIFERIN C* (*CRC*), *PAP85*, and *OLEOSIN1* (*OLEO1*) genes encoding a napin-like protein, a cruciferin, a vicilin-like protein, and a homolog of *B. napus* oleosin, respectively, are assumed to be the typical mRNA markers of the *MAT* class in the enclosed embryos. As shown in Fig. 6.1, transcripts of these genes are expressed almost simultaneously in developing siliques, and become the most abundant mRNAs during later stages of silique development; interestingly, *PAP85* transcripts continue to accumulate beyond the maturation phase (Parcy et al. 1994). Transcripts of two additional genes, designated *ARABIDOPSIS THALIANA SEED* (*ATS1* and *ATS3*) are not expressed in embryos at the same high level as storage protein gene transcripts. The difference between the two *ATS* genes is, however, reflected in their distinctive expression patterns: whereas *ATS1* transcripts accumulate in the mature embryo, particularly in the protoderm and vascular tissues and in the shoot and root apical meristems, *ATS3* transcripts are excluded from the vascular tissues and meristems. Although protein products of the two genes accumulate in embryos in a spatial pattern corresponding to their respective transcripts, the identity of the products remains unknown. Like the storage protein genes, the overall function of the *ATS* genes could be to confer a maturation-related function on the embryo (Nuccio and Thomas 1999). The genome organization of storage protein genes and regulation of their expression in *Arabidopsis* have been reviewed by Fujiwara et al. (2002).

The past two decades have uncovered much information about storage proteins in embryos of various economically important eudicots by cloning and sequencing the genes and characterizing their genetic organization. The spatial and temporal expression of genes in transgenic plants has been one of the most intellectually rewarding and avidly pursued areas in the molecular biology of storage proteins. This work is fueled by the growing appreciation of the potential practical benefits to be derived from reshuffling genes in novel combinations to encode useful proteins in seeds of plants of otherwise limited economic use (Raghavan 1997).

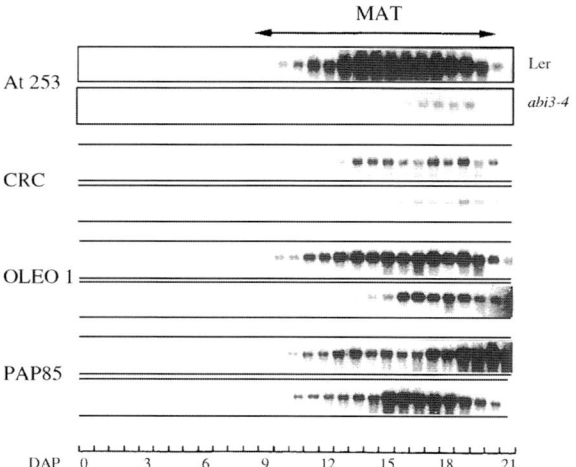

Fig. 6.1 Gel blot analysis of the expression pattern of *MATURATION* (*MAT*) mRNAs during silique development in wild-type and *abi3-4* mutant *Arabidopsis*. Silique development lasted 21 days from anthesis (0 DAP) to dry seed stage (21 DAP). Each probe was hybridized to 1.5 µg total RNA. (Reprinted from Parcy et al. 1994)

6.1.2
Is Embryo Maturation in the Seed Developmentally Regulated by ABA?

Until recently, the predominant view was that ABA is an important factor in the control of embryo maturation in the seed, implying that the maturation period in embryogenesis is marked by changes in the endogenous levels of this hormone. One set of experiments supporting this view showed that, in embryos of *Phaseolus vulgaris* (Prevost and Le Page-Degivry 1985), maize (Neill et al. 1987), barley (Robertson et al. 1989), and soybean (Chang and Walling 1991), ABA levels are high at the time of maximum fresh or dry weight accumulation, and that a decrease in hormone concentration occurs coincident with the final stages of maturation. In embryos of rapeseed (Finkelstein et al. 1985), cotton (Galau et al. 1987), and pea (Wang et al. 1987), two peaks of ABA maxima are seen, one of which appears to be correlated with the high water potential of seeds or the time at which their water potential is on the decline. A different pattern of changes in the concentration of the hormone, which attains a maximum midway through seed development, has been documented in siliques of *Arabidopsis*. The need to trace the origin of ABA in seeds led to reciprocal crosses between ABA-deficient mutants and wild-type *Arabidopsis* followed by germination tests on seeds. The results supported a dual origin of ABA in developing siliques, one fraction, maternal in origin, regulated by the genome of the mother plant, accounting for the peak in ABA content halfway through seed development, and a second fraction, regulated by the embryonic genome, which controls embryo maturation and induction of seed dormancy (Karssen et al. 1983). However, it has not been established whether the hormone is synthesized by the embryo itself or is imported from the maternal tissues of the ovule, or whether hormone accumulation is a consequence or a cause of maturation-related events.

As mentioned in Chaps. 2 and 5, precocious germination is a fundamental process displayed by immature embryos of many plants, which skip the maturation and dormancy programs of embryogenesis upon culture and, instead, leap ahead into germination mode. Whereas the focus of early investigations into precocious germination was on the reinstatement of the maturation phase in cultured embryos by manipulation of the medium composition, impetus for later investigations arose from a desire to decipher the physiological and molecular basis of this phenomenon. A satisfying outcome of these latter studies has been the discovery that immature embryos of cotton (Ihle and Dure 1970; Choinski et al. 1981), maize (Robichaud et al. 1980), rapeseed (Crouch and Sussex 1981), wheat (Triplett and Quatrano 1982), and soybean (Eisenberg and Mascarenhas 1985), among others, cultured in medium containing ABA, not only fail to germinate precociously but also undergo certain aspects of the physiological and molecular differentiation characteristic of the maturation program (Fig. 6.2). An enticing model based on these results is that ABA is a natural factor that suppresses precocious germination during the normal course of embryogenesis *in planta*. In support of this model, it was shown that precocious germination and the synthesis of a germination-specific proteolytic enzyme, carboxypeptidase, involved in the mobilization of stored food reserves of cotyledons, triggered by the culture of immature embryos of cotton, are reversed by the addition of either an aqueous extract of the ovule or ABA to the culture medium (Ihle and Dure 1970). Also crucial to the action of ABA in maintaining the maturation program in cultured cotton embryos is the inhibition of synthesis of the glyoxylate-cycle enzyme ICL, and the promotion of synthesis of a subset of proteins, including a battery of enzymes, appearing late in embryogenesis (Ihle and Dure 1972; Choinski et al. 1981). As referred to in Chap. 5, the dynamics of accumulation of the storage proteins cruciferin and napin, and the increased transcriptional activities of the genes encoding them in immature embryos of rapeseed cultured in the presence of ABA in the medium, emphasize the role of this hormone in reinstating embryo maturation processes. Storage proteins and/or their transcripts have also figured as biochemical markers to reinforce the role of ABA in maintaining the maturation program in cultured immature embryos of wheat (Triplett and Quatrano 1982) and soybean (Eisenberg and Mascarenhas 1985). Studies on the regulation of gene expression in developing maize embryos have shown that a few polypeptides in the 23–

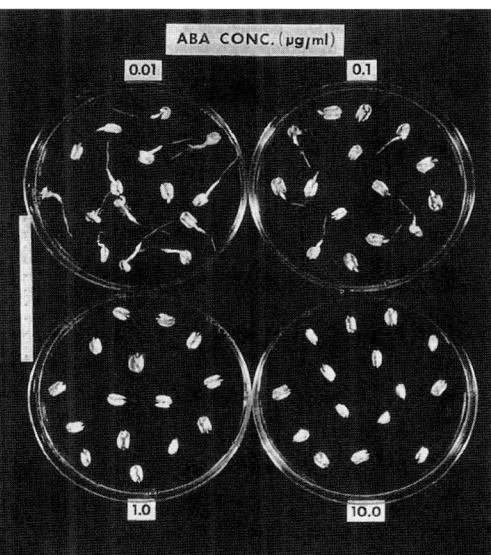

Fig. 6.2 Photographs of cotton embryos cultured for 4 days in media containing various concentrations of abscisic acid (ABA). *Bar* (left) 10 cm. (Reprinted from Choinski et al. 1981)

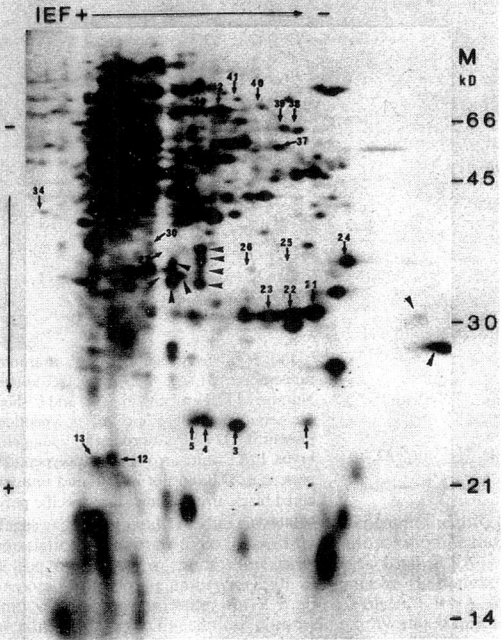

Fig. 6.3 Two-dimensional electrophoretic separation of proteins synthesized by immature maize embryos in medium containing ABA. Embryos were excised from grains 20 days after pollination, incubated for 21 h in medium containing 1 μM ABA, and subsequently labeled with ^{35}S-methionine for 2 h. *Numbered black arrows* Polypeptides that are not synthesized at this stage but are induced by ABA treatment, *unnumbered black arrowheads* polypeptides present in immature embryos after induction with ABA and not identified in embryos during maturation. (Reprinted from Sánchez-Martínez et al. 1986)

25 kDa range that appear in mature embryos and disappear during germination, are rapidly induced in immature embryos by ABA treatment (Fig. 6.3). Transcripts of an ABA-inducible gene isolated from immature maize embryos are found to increase progressively during embryogenesis, reaching a peak at the onset of desiccation, and are also induced precociously in immature embryos treated with the hormone. A glycine-rich protein encoded by this gene is considered to be an RNA-binding protein with putative structural and protective functions (Sánchez-Martínez et al. 1986; Gómez et al. 1988; Ludevid et al. 1992).

The above results have left some uncertainties leading to the interpretation that ABA does not play a major role in initiating or maintaining the embryo maturation program. Harada (1997) has collated evidence indicating that, despite the enhanced expression of several storage protein genes in immature embryos of a variety of plants cultured in the presence of exogenous ABA, many of the gene products are detected at low levels in embryos nurtured in hormone-free basal medium. This suggests that ABA cannot be solely responsible for the activation of storage protein genes that accompanies embryo maturation. Seed protein gene expression profiles in cultured embryos of cotton have raised an additional question. The accumulation kinetics of cloned mRNAs showed that cultured immature embryos do not initiate the maturation program, but simultaneously traverse the post-abscission and germination programs. This occurs in the absence of ABA in the medium, but the post-abscission and germination programs are enhanced by exogenous ABA. A possible scenario suggested for sustaining the maturation program of the embryo *in planta* is the provision of a maternal maturation factor other than ABA that is lost upon culture. In the context of this model, endogenous ABA has been relegated to function in the lowering of the water potential of the embryo to inhibit germination (Galau et al. 1991; Hughes and Galau 1991). It seems likely that there are many roles for ABA during embryogenesis, and perhaps this diversity defies the formulation of a single unifying hypothesis. In the long run, these experiments might challenge not so much the current hypothesis on the role of ABA in embryo maturation as the logic of alternative interpretations.

6.1.3
Genetic Regulation of Embryo Maturation by ABA

The genetic control of embryo maturation has been examined through the isolation and characterization of mutants of *Arabidopsis* with a reduced capacity for ABA synthesis, or which display insensitivity to the hormone during germination. Seeds of the mutant line designated as *ABA-deficient* (*aba*) germinate precociously to the full extent under supportive conditions in which wild-type seeds do not germinate in appreciable numbers. A good correlation is also observed between the absence of ABA in mutant seeds during all stages of their development and their proclivity for premature germination (Karssen et al. 1983). However, the impact of the suggestion from these results, i.e., that ABA-related processes may be involved in the transition of the embryo from the morphogenetic to the maturation stage, was dwarfed by the later isolation and characterization of genes from two diverse gene families – *ABI* and *LEC* – and the identification of their encoded proteins as transcriptional regulators with a link to ABA signaling. Analysis of the mutant phenotypes has shown that they are remarkable, not only because of the pleiotropic effects on seed development, but also because of their role as informative genetic models for the study of gene action during embryo maturation.

Five classes of *abi* mutants (*abi1*–*abi5*) of *Arabidopsis* were selected on the basis of the ability of seeds to germinate in the presence of moderate-to-high concentrations of ABA that are inhibitory to seeds of the wild-type. In addition, seeds of *abi1*, *abi2*, and *abi3* mutants have normal or increased levels of endogenous ABA, and all mutant seeds except *abi4* and *abi5* exhibit reduced dormancy (Koornneef et al. 1984; Finkelstein 1994). Besides displaying seed-specific developmental disruptions, the phenotypes of these mutants show defects in one or more aspects of vegetative and reproductive growth, including response to water stress, root branching, plastid differentiation, meristem quiescence, and floral initiation (Koornneef et al. 1984; Brocard et al. 2002). Several alleles of the *abi3* mutant that affect, to varying degrees, embryo maturation and the acquisition of dormancy and desiccation tolerance by seeds, have been described.

Representative seeds and dissected embryos of the wild-type and a mutant allele are shown in Plate 10, Fig. a–h. After seemingly normal development up to about 8 days after flowering, mutant seeds and their enclosed embryos diverge from the wild-type in their inability to degreen and dehydrate. Indicative of a defect in their response to ABA, seeds of *abi3* mutant alleles readily germinate in the presence of a concentration of exogenous ABA that is inhibitory to germination of their wild-type counterparts; seeds of some mutant alleles are several orders of magnitude less sensitive to ABA than those of the wild-type (Nambara et al. 1992; Ooms et al. 1993; Nambara et al. 1994, 1995, 2000, 2002). In a severe mutant allele, besides the usual symptoms associated with desiccation intolerance of seeds, embryos display precocious activation of the shoot apical meristem, and premature differentiation of the vascular tissues typical of seedlings. Seeds of this mutant allele also show a high level of expression of the *cab* gene encoding chlorophyll *a/b* binding protein; this gene is normally expressed at a low level in seeds and at a high level in seedlings of wild-type *Arabidopsis*. A particularly important observation is that the expressional abundance of genes for the 12S and 2S storage proteins found during embryo maturation in wild-type plants is severely down-regulated in this mutant allele (Nambara et al. 1992, 1995). Parcy et al. (1994) found that mutation in a strong *ABI3* allele inhibits the expression of maturation-specific *At2S3* and *CRC* mRNAs in siliques, with a less pronounced effect on the expression of *OLEO1* and *PAP85* gene transcripts (see Fig. 6.1). Other studies have shown that double mutants generated by crossing the parental *abi3* and *aba* mutants produce underdeveloped, green, nondormant, desiccation-intolerant seeds, dramatically altered in their maturation properties. Despite the absence of any detectable signs of germination, double mutant seeds also synthesize germination-related proteins, showing a trend to initiate a premature germination program (Koornneef et al. 1989; Meurs et al. 1992). A prediction resulting from these several lines of investigation is that the *abi3* mutation might cause a defect in embryo maturation at the same time as premature germination is induced (Nambara et al. 1995). In this sense, the *ABI3* locus might be considered to repress germination characteristics during embryo dormancy.

Cloning of *ABI* genes and biochemical characterization of their protein products highlight a recurrent theme in hormone recognition in plants involving a complex regulatory network that coordinates perception of the signal to cellular responses. In this respect, hormones such as ABA might elicit a cascade of events by interacting with a specific receptor site(s) in the plant. Apart from its effect on seed maturation, dormancy, and germination, ABA plays a major role in various other aspects of plant growth and development, ranging from alteration of ion fluxes in stomata to tolerance to salt, cold, and drought stresses. Consequently, analysis of this expanded panel of *ABI* genes might help to further elucidate the different layers of regulatory information controlling ABA action in a wide range of plant growth processes. The *ABI3* gene was isolated by map-based positional walking, and clues to the molecular basis of *ABI3* function have come from the analysis of its predicted protein sequence, which shows features of a putative transcriptional activator. Based on shared regions of sequence similarity between ABI3 protein and the protein encoded by the maize *VIVIPAROUS1* (*VP1*) gene, ABA actions during a subset of the seed developmental program impinging on the embryo are thought to be governed by a modulating role of ABI3 protein in the signaling pathway (Giraudat et al. 1992). The cloning and characterization of the *ABI1* gene revealed that it encodes components of a protein composed of a novel N-terminal segment and a C-terminal domain with membership in the 2C class of protein serine/threonine phosphatases (PP2Cs), suggesting that it is a Ca^{2+}-modulated phosphatase involved in ABA signal transduction (Leung et al. 1994; Meyer et al. 1994). Since *abi1* and *abi2* mutants share several common phenotypes, it is not surprising to find that the structure of the *ABI2* gene is closely related to that of the *ABI1* gene, with their protein products belonging to distinct branches of the ABA-signaling network (Leung et al. 1997). That the *ABI4* gene is a new element of the signal transduction pathway of the ABI class of genes was evident from the homology of its protein product to the family of transcription regulators characterized by a conserved DNA-binding domain known as the AP2 domain (Finkelstein et al. 1998). The predicted gene product of the *ABI5* gene shows structural similarities to the basic leucine zipper (bZIP) class of transcriptional regulators (Finkelstein and Lynch 2000; Lopez-Molina and Chua 2000). Overall, these results highlight the multiplicity of signaling processes that impinge on seed maturation mediated by ABA action. Several genetic and molecular experiments on *Arabidopsis* involving *ABI* genes have provided further insights into the complexities of pathway interactions during seed maturation, although the question of how cross-talk between the different signaling circuits takes place remains to be elucidated. Indeed, there are also some observations that argue against an exclusive role for ABA mediated by *ABA* and *ABI* genes in embryo maturation (Koornneef and Karssen 1994).

As mentioned in Chap. 5, mutations in the *LEC* genes cause seedling development to be advanced, bypassing embryo maturation and seed dormancy. Although most *fus* mutants are defective in light response pathways, seeds of two *fus3* alleles, along with those of the *lec1* mutant, fail to exhibit several features of maturation, such as desiccation tolerance, lapse into dormancy, loss of chlorophyll, and accumulation of storage products in the embryo (Meinke 1992; Bäumlein et al. 1994; Keith et al. 1994). Unlike seeds of *lec1* and *fus3* mutants, seeds of the *lec2* mutant survive desiccation, although they lose viability with storage (Meinke et al. 1994). Embryos of the three mutants express post-germination characteristics such as premature activation of the shoot apical meristem and the presence of differentiated vascular tissues in the cotyledons. In contrast to wild-type seedlings, in which leaves are programmed to produce trichomes, a surprising sideline to the maturation defects in the *lec* mutants is the production of trichomes by seedling cotyledons; cotyledons of mutant embryos also accumulate anthocyanins, causing seeds to appear highly pigmented (Meinke et al. 1994; Bäumlein et al. 1994; Keith et al. 1994). Phenotypic observations of seeds of double and parental single *lec* and *abi3* mutants combined with expression studies in embryos of genes characteristic of both embryonic and post-embryonic development indicate that *ABI3* and *LEC* genes play a fundamental role in regulating the maturation of *Arabidopsis* seeds by a complex choreography involving distinct and broadly overlapping pathways activated in the embryo (Meinke et al. 1994; West et al. 1994). The protein products of *LEC* genes are logical candidates to oversee the regulatory mechanisms that coordinate the developmental events during embryo maturation as they all encode transcription factors (Chap. 5). The presence of a conserved B3 domain transcription factor in FUS3 and LEC2 proteins reinforces the functional similarity of the *LEC* genes to the *VP1*- and *ABI3*-encoded transcription factors (Luerßen et al. 1998; Stone et al. 2001).

Some additional observations have added weight to the existence of links between *ABI3* and *LEC* genes in the control of embryo maturation in *Arabidopsis*. Based on quantitative analyses of the developmental responses of *abi3*, *lec1*, and *fus3* single mutants and double mutants combining weak or severe *abi3* mutation with either *lec1* or *fus3* mutations, it appears that the effects of the *ABI3* gene on the accumulation of chlorophyll and anthocyanins, expression of members of the cruciferin storage protein gene family in embryos, and on the germination sensitivity of seeds to ABA, are controlled by both the *LEC1* and *FUS3* loci. These findings have raised the possibility that, rather than acting in independent regulatory pathways, these genes have a broad biological responsibility in controlling seed maturation processes in *Arabidopsis* by functioning synergistically through cross-connected signal transduction networks (Parcy et al. 1997).

6.2
Embryo Dormancy

Genes that regulate embryo maturation in *Arabidopsis* are simultaneously involved in two other distinct but related functions, namely, activating embryo dormancy and repressing premature germination. The overlapping effects of these genes on pre- and post-germination embryos make it a challenging task to identify regulators of embryo dormancy and, consequently, reproducible associations of specific genes with dormancy induction and maintenance in mature embryos have proved elusive. As has become evident from the previous section, the *ABI3*, *LEC1*, *LEC2*, and *FUS3* genes have received much attention due to their prominent involvement in multiple processes of late embryogenesis including accumulation of storage proteins as well as acquisition of dormancy and desiccation tolerance. Dormancy is overcome, and the embryo initiates the germination program, when the positive action of

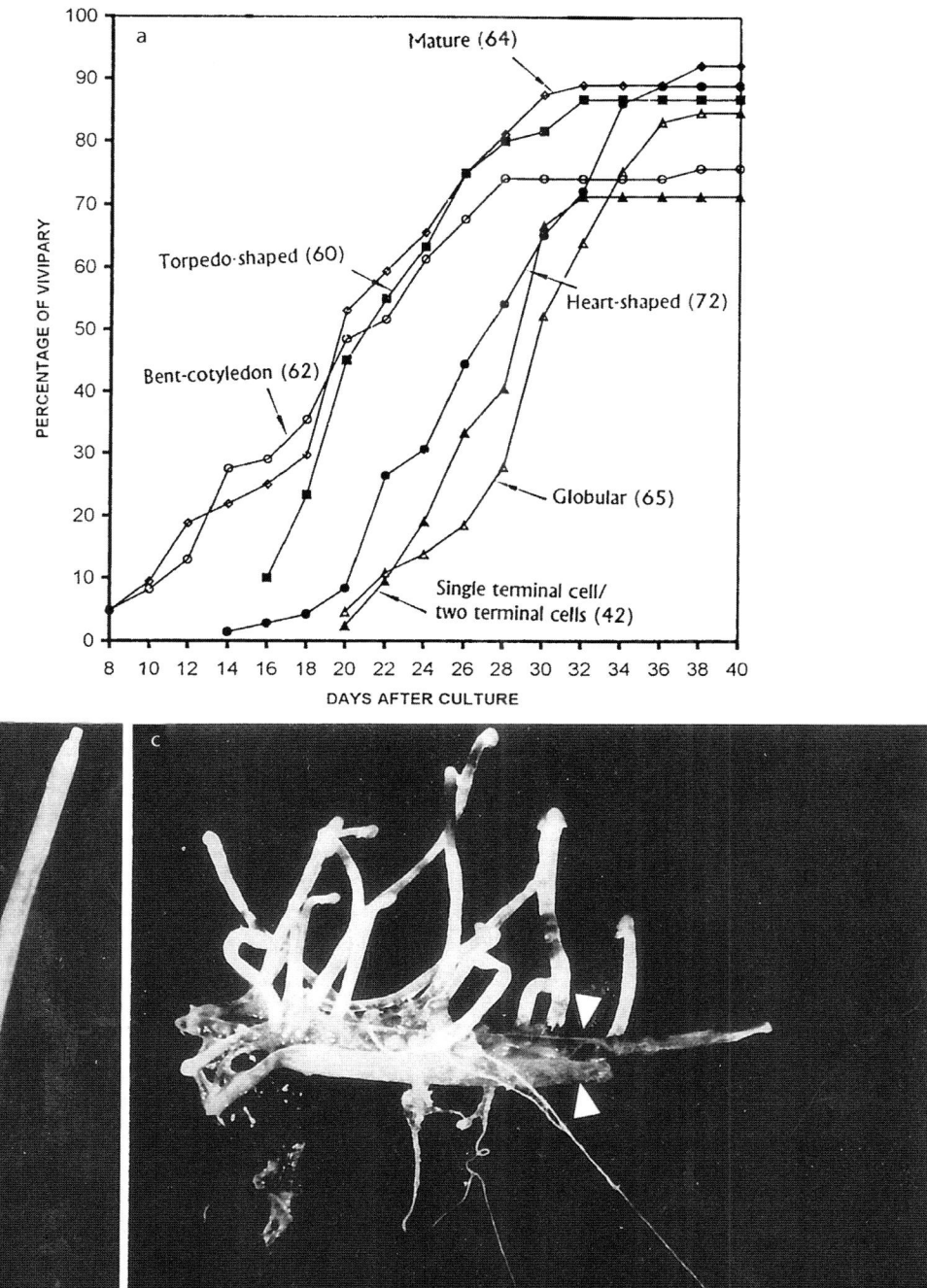

Fig. 6.4a–c Vivipary in cultured siliques of *Arabidopsis*. **a** Time course of vivipary in dark-cultured siliques enclosing embryos of various developmental stages. Mature embryos are green and are enclosed in green ovules. Data-points indicate the number of siliques (expressed as additive percentages of the number of successful cultures) showing the first signs of vivipary on the days indicated. Figures in parenthesis at each embryo developmental stage indicate the total number of cultures counted. **b** A silique containing torpedo-shaped embryos cultured for 18 days in the dark showing the outgrowth of the plumule (*arrow*) as the first sign of vivipary. **c** A silique containing torpedo-shaped embryos cultured for 25 days in the dark and in light for 1 day, showing the emergence of numerous seedlings. *Arrowheads* point to the two separate halves of the silique. *Bars* 1 mm. (Reprinted from Raghavan 2002)

these genes is reversed under natural conditions or by mutations. Seeds of some *Arabidopsis* mutants, designated *reduced-dormancy*, are characterized by a reduced dormancy trait reflected in the high germination percentages soon after harvest, but display otherwise wild-type behavior (Léon-Kloosterziel et al. 1996; Peeters et al. 2002). At the opposite end of the developmental spectrum are seeds of the *comatose* (*cts*) mutant, which do not germinate and exhibit almost permanent dormancy. Since mutant embryos do not show any disruption of cell or tissue organization, it has been suggested that the mutation reduces the capacity of seeds to respond to the dormancy-breaking stimulus (Russell et al. 2000). This view has gained firm support from the demonstration that the *CTS* gene encodes a homolog of the human X-linked adrenoleukodystrophy protein (ALDP), a peroxisomal protein of the ATP-binding cassette (ABC) associated with the transport of long-chain fatty acids into the peroxisome. Because of the striking inability of cotyledons of the mutant embryo to break down lipid bodies, a product of lipid metabolism might be considered to accumulate in cells, causing derangement of the signal transduction pathway, or of a crucial metabolic step, leading to continued maintenance of the dormant state (Footitt et al. 2002).

A study of embryo growth in relation to the developmental timing in some of the *Arabidopsis* mutants described above has provided a molecular foundation for the delineation of two distinct developmental processes affecting the maintenance of, and exit from, dormancy of embryos. One striking finding is that torpedo-shaped and older embryos of certain alleles of *lec1*, *lec2*, and *fus3* mutants grow into full-fledged seedlings in a culture medium that does not support growth of wild-type embryos of the same age; hence, these mutants are also characterized as embryo growth arrest mutants. Moreover, replicative DNA synthesis continues in the mutant embryos after it is turned off in their wild-type counterparts. As mutations in *LEC1*, *LEC2*, and *FUS3* genes result in reduced embryo growth arrest, the results point to a genetic control of cell division frequency leading to inhibition of embryo growth in wild-type plants. In contrast, the behavior in culture of embryos of *aba1* and *abi3* mutants, as well as their DNA synthesis profile, are similar to wild-type embryos, indicating that these mutants are not defective in embryo growth arrest. However, the arrest of cell division in *aba1* and *abi3* mutant embryos is found to coincide with their ability to germinate prematurely in siliques cut open and placed on a water-agar medium; this signifies the absence of dormancy. These results have led to a model (see Plate 11, Fig. a) for the control of seed maturation, dormancy, and germination of *Arabidopsis* seeds in which, *LEC1*, *LEC2*, and *FUS3* gene products impose arrest of embryo growth, and *ABA1* and *ABI3* gene products signal the onset of dormancy (Raz et al. 2001). The involvement of *LEC2* and *FUS3* genes in the control of GA biosynthesis during embryogenesis indicated in the model has been substantiated by the work of Curaba et al. (2004), which shows that, relative to wild-type levels, there is an increase in the level of active gibberellins in developing seeds of *lec2* and *fus3* mutants.

The phenomena of dormancy and germination of seeds have a close affinity in that they both concern the fate of the enclosed embryo. From this point of view, culture of isolated siliques has provided a physiological perspective of how the embryo responds to the dormancy-enforced growth arrest and resumption of growth caused by vivipary (premature embryo germination). Culture of excised siliques of different ages has shown that, whereas early stage and immature embryos enclosed in ovules complete their full development and germinate viviparously, vivipary is not observed in cultured siliques enclosing brown ovules with dormant, mature yellowish embryos (Fig. 6.4a–c). To the extent that culture of the silique makes the developing embryo nondormant, silique culture seems to have the same effect as mutation in genes that induce dormancy (Raghavan 2002). Since DNA replication is arrested in embryos developing in cultured siliques at the same time as in normally developing embryos, the culture environment may be said to eliminate a requirement for dormancy following embryo growth arrest.

Mutations that interfere with dormancy of embryos of maize induce vivipary, typically observed when the embryo begins to germinate and form a seedling, even though the grain is still attached to the ear on the mother plant. Mutations in the *VP1* locus inflict pleiotropic effects on grain maturation, including a reduced sensitivity of embryos to exogenous ABA, indicating that, along with ABA, the VP1 protein is required for maize embryo maturation. Consequently, mutant embryos are easily coaxed to grow in culture in the presence of ABA concentrations inhibitory to embryos of the wild-type grain (Robichaud and Sussex 1986; McCarty and Carson 1991). Details of the molecular mechanism underlying vivipary in maize have come from cloning of the *VP1* gene and functional analysis of its protein product, which has been identified as a transcriptional activator. Transient assays using protoplasts derived from suspension-cultured maize cells have shown that the VP1 protein is required for the ABA-induced activation of expression of the *EARLY METHIONINE LABELED* (*Em*) gene of wheat (a *LEA* gene) and *C1* gene of the maize anthocyanin pathway (McCarty et al. 1991; Hattori et al. 1992). Additionally, VP1 protein has been shown to have a specific role in inhibiting the precocious induction of genes for α-amylase necessary for the hydrolysis of endosperm starch into sugars during germination of the grains (Hoecker et al. 1995). A synergistic effect of the VP1 protein is seen on the induction by ABA of the rice homolog of the *Em* gene in a transient expression system using rice cell protoplasts (Hattori et al. 1995). It is also known that the VP1 transcription factor is homologous to the product of the *ABI3* gene, indicating that VP1/ABI3 class proteins probably control embryo dormancy by acting as transcriptional regulators. In support of this view, it has been found that expression of the *VP1* gene driven by a CaMV 35S promoter in an *abi3* mutant allele of *Arabidopsis* can partially overcome mutational lesions, yielding seeds morphologically and physiologically similar to wild-type seeds (Suzuki et al. 2001). The strong functional conservation between the *VP1* gene of maize and the *ABI3* gene of *Arabidopsis* suggested by these results has been strengthened by a microarray analysis of *VP1*-regulated gene expression in transgenic *Arabidopsis* carrying 35S:*VP1* in an *abi3* null mutant background; this study has shown that the *VP1* gene is sufficient to confer ABA induction of a broad range of seed protein genes (Suzuki et al. 2003). Reinforced by the additional information that the transcriptionally active amino acid residue region of the VP1 protein shares sequence identity with the B3 domain of the LEC proteins (Lotan et al. 1998; Luerßen et al. 1998; Stone et al. 2001), these data are generally consistent with the existence of molecular links between *VP1*, *ABI3*, and *LEC* genes in the control of embryo maturity, dormancy, and germination in *Arabidopsis* and maize.

6.2.1 Carbohydrates in Desiccation Tolerance

As a prelude to quiescence or dormancy, seed desiccation occurs during the late embryogenesis phase and, in extreme cases, the moisture content of the seed decreases dramatically to less than 10%. Desiccation of the embryo is thus a normal programmed event in the final phase of seed development, but in other phases of the plant life cycle, desiccation is akin to the proverbial "kiss-of-death". A fully desiccated seed invokes the notions of space, time, and gene activity as it positions the embryo to germinate upon rehydration. Even a cursory consideration of desiccation of the seed, which allows the embryo to survive an extremely low cellular water content, forces the realization that multiple protective physiological processes must be in operation in developing desiccation tolerance. In particular, work undertaken during the past two decades has seen the characterization of groups of carbohydrates and proteins that function to forestall some of the damage incurred by desiccation.

The most unexpected molecules implicated in desiccation tolerance of seeds are various carbohydrates, especially soluble sugars. A breakdown of starch, or its transient accumulation followed by its depletion, in cells of embryos has been identified as a regular feature associated with the acquisition of desiccation tolerance in seeds of soybean (Rosenberg and Rinne 1987), *Sinapis alba* (Brassicaceae; Fischer et al. 1988), and *Brassica campestris* (Leprince et al. 1990). Starch depletion is closely coupled with an increase in soluble sugars such as sucrose in soybean, glucose in *S. alba*, and stachyose and sucrose in *B. campestris*. Precocious maturation involving controlled dehydration of developing seeds has served as a useful experimental approach to unravel the changes associated with desiccation tolerance in naturally maturing embryos. For example, during slow drying of soybean embryos, stachyose and sucrose levels attain values nearly three times higher than those reported for naturally matured embryos (Blackman et al. 1992). Studies on embryos of wild-type and *vp* mutants of maize have also brought raffinose accumulation into the equation for acquisition of desiccation tolerance. Developing maize embryos accumulate mostly sucrose and raffinose as their soluble non-reducing sugars. In a field-grown hybrid maize, desiccation tolerance is gradually acquired only after the sucrose to raffinose mass ratio in whole grains, as well as in isolated embryos, drops to less than 20:1, whereas a sucrose to raffinose ratio of 10:1 is favored for complete desiccation tolerance. This relationship is maintained even in whole grains and isolated embryos induced to acquire desiccation tolerance by slow or fast drying (Brenac et al. 1997a).

Nondormant, desiccation-intolerant embryos of maize *vp* mutants do not express the *VP1* gene. Comparative studies on the acquisition of desiccation tolerance in relation to the accumulation of sucrose and raffinose in the wild-type *Vp1-R* and its mutant allele *vp1-R* showed that, whereas embryos of the former become tolerant to precocious drying in association with a sucrose:raffinose mass ratio of 10:1 or lower, the sucrose:raffinose mass ratio in embryos of the mutant, which do not acquire desiccation tolerance to drying, is nowhere near approaching 10:1 or even 20:1. Not surprisingly, in contrast to wild-type embryos, mutant embryos accumulate only trace amounts of raffinose. Differences are also seen between wild-type and mutant embryos in their patterns of accumulation of sucrose and total soluble carbohydrates (Fig. 6.5a–j). The favored interpretation of these results is that the wild-type gene regulates raffinose biosynthesis in the embryo preparatory to desiccation. However, studies on germination of grains of certain nonviviparous mutant alleles of the *VP1* gene in relation to raffinose accumulation has generated contradictory results indicating that, despite the association of desiccation tolerance with raffinose accumulation, raffinose biosynthesis might occur even in the absence of a fully functional VP1 protein (Brenac et al. 1997b).

The availability of desiccation-sensitive genotypes of *Arabidopsis* has allowed functional predictions to be made regarding accumulation of carbohydrates in relation to desiccation tolerance of seeds. According to Ooms et al. (1993), dormancy of seeds of the desiccation-tolerant wild-type and a weak allele of the *abi3* mutant (*abi3-1*) of *Arabidopsis* is foreshadowed by the accumulation of raffinose and stachyose. On the other hand, seeds of the desiccation-sensitive *aba1-1/abi3-1* double mutant and a strong allele of the *abi3* monogenic mutant (*abi3-5*) contain abundant sucrose instead of raffinose and stachyose, which are present in minimal amounts. It is however questionable whether a high soluble carbohydrate content in cells contributes to desiccation tolerance, since seeds of mutants that are desiccation-sensitive accumulate three to five times more soluble sugars than seeds of the highly desiccation-tolerant wild-type. It was also found that when desiccation tolerance is induced in seeds of the *aba1-1/abi3-1* double mutant by incubating them in a medium containing ABA and sucrose or trehalose, seeds that recover desiccation tolerance contain high concentrations of raffinose without any change in their monosaccharide content. These results have led to the suggestion that a low ratio of mono- to oligo-saccharides, rather than the total carbohydrate content, might control acquisition of desiccation tolerance in *Arabidopsis* seeds (Meurs et al. 1992; Ooms et al. 1994).

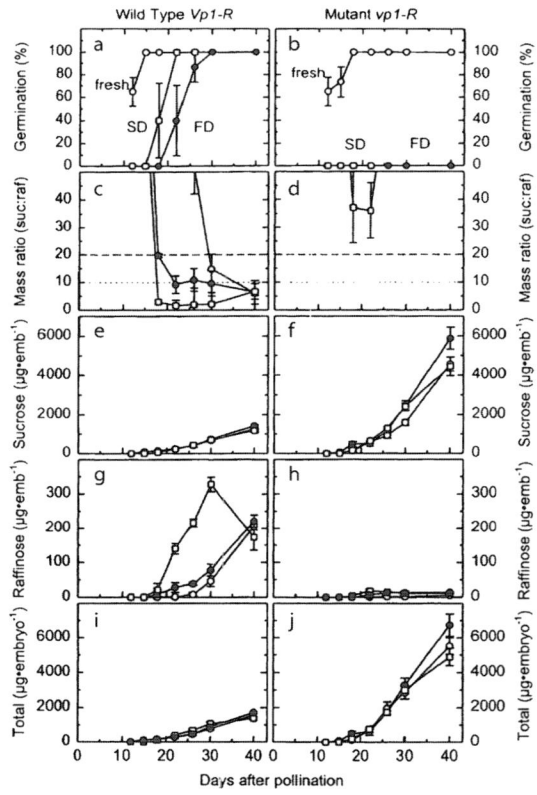

Fig. 6.5a–j Changes during germination of wild-type (*Vp1-R*) and mutant (*vp1-R*) grains of maize. Germination percentages (**a**, **b**), mass ratio of sucrose to raffinose (**c**, **d**), and sucrose (**e**, **f**), raffinose (**g**, **h**), and total soluble carbohydrate (**i**, **j**) contents of embryos before drying (fresh, ○), after fast drying (FD, ●), and after slow drying (SD, □) are shown. None of the grains subjected to fast and slow drying were desiccation-tolerant. Values are mean ± standard error. (Reprinted from Brenac et al. 1997)

Explaining how soluble sugars may help to ensure survival of cells of the embryo during desiccation represents a great challenge to plant physiologists. One view that has gained some ground is that sucrose and raffinose function as natural components to promote a vitrified or glassy state that limits solute crystallization in the cytoplasm and its total dehydration (Williams and Leopold 1989). In support of this view, comparison of the cytoplasmic viscosities of cells of desiccation-tolerant and desiccation-intolerant embryo axes of soybean has indeed shown glass-transitions in cells of the desiccation-tolerant embryo, but not in susceptible cells of the desiccation-intolerant embryo (Bruni and Leopold 1991). The stable glass formation that occurs at room temperature in the presence of raffinose in a sucrose milieu under in vitro conditions, similar to that found in desiccation-tolerant embryo axes in vivo, is also considered as supporting evidence (Koster 1991). A great deal more remains to be learned about the role of soluble sugars as adaptive agents in conferring desiccation tolerance in embryos, and it appears that sucrose and raffinose will continue to lead the way in future investigations.

6.2.2
Proteins in Desiccation Tolerance

Historically, proteins have played a leading role as candidate molecules for conferring desiccation tolerance in seeds, as synthesis of new proteins is pervasive in embryos of maturing seeds. The initial clues linking the acquisition of desiccation tolerance to specific gene products came from a study of the patterns of in vivo and in vitro proteins synthesized by desiccation-intolerant and desiccation-tolerant embryos of barley. The result identified a set of 25–30 proteins and mRNAs that are newly synthesized, or whose synthesis is enhanced, during the stage of embryogenesis leading to desiccation tolerance (Bartels et al. 1988). Embryos excised from mature soybean seeds (approximately 70 days after flowering) synthesize a plethora of nonstorage proteins, designated as "maturation polypeptides" and their corresponding mRNAs; two of these polypeptides, 128 kDa and 31 kDa in mass, are also found to accumulate in embryos with seed maturity. The same protein and mRNA profiles are seen in embryos when seeds are precociously matured through controlled dehydration. However, neither the maturation polypeptides nor their transcripts accumulate in embryos of immature seeds harvested in mid-stage development (35 days after flowering). So, one way to think about seed maturation is to imagine that it represents a metabolically active phase of the nongrowing embryo (Rosenberg and Rinne 1988).

Given the extent of protein accumulation in embryos of mature seeds, one might suspect that these proteins are causally related to desiccation tolerance of seeds. Suggestive of a role for stabilization of protein structure in desiccation tolerance of seeds, Wolkers et al. (1998b) have shown that slow drying of immature maize embryos, which confers desiccation-tolerance, causes changes in the cytoplasmic profile of proteins, such as the formation of secondary structures, very similar to those found in embryos of desiccation-tolerant mature grains. In another study by these authors (Wolkers et al. 1998a), a high protein stability in wild-type seeds of *Arabidopsis* and a decreasingly lower stability in progressively maturation-defective mutant seeds was also demonstrated. As mentioned earlier, the cellular cytoplasm of embryos of desiccation-prone seeds forms a glassy matrix; hydrogen bonding involving sugars and proteins has a major impact on glass formation. Arising out of these considerations, it appears that the protection provided to cytoplasmic proteins and cell membranes is central to the desiccation tolerance of embryos, which presumably depends to a large extent on their raffinose, sucrose, and possibly LEA protein contents (Walters et al. 1997; Wolkers et al. 1999).

The beginnings of the concept of LEA proteins, which have become the glamour proteins of embryo desiccation in flowering plants, can be dated rather precisely. A two-dimensional gel electrophoretic analysis of the changing mRNA populations during embryogenesis in cotton, based on comparisons of the extant, in vivo and in vitro synthesized proteins of the cotyledons, provided the first indication of the presence of a set of 14 polypeptides associated with the onset of embryo maturation and dormancy (Dure et al. 1981). Another fundamental observation made in this work was that these protein sequences disappear during germination of mature embryos and precociously accumulate in young embryos cultured in a medium containing ABA. The dynamic complex of mRNAs and their corresponding protein sequences, which are significantly more abundant in mature embryos than in young embryos, was named LEA (mnemonic for "late embryogenesis abundant") by Galau et al. (1986). The accumulation kinetics of 18 cloned cotton *LEA* mRNAs showed that their concentrations increase at least 10- to 1,700-fold during embryogenesis coincident with the time of ovule abscission, and decline 15- to 220-fold during the first day of germination. Using a complementary approach, transcript abundance of the cotton *LEA* genes is found to be highest during an 8-day interval between embryo maturation and desiccation (Galau et al. 1987; Hughes and Galau 1989). In another study, using several heterologous cotton cDNA probes, *LEA* mRNAs were detected in desiccation-prone embryos and/or seeds of rape seed, soybean, and tobacco, and their expression shown to be enhanced by ABA in cultured immature embryos, to provide a highly influential model of regulation of *LEA* mRNA expression in dicots (Jakobsen et al. 1994). Based on their expression patterns, *LEA* mRNAs of *Arabidopsis* are subdivided into two classes (Fig. 6.6a,b): those belonging to the *LEA* class, which begin to accumulate about 18 days after pollination and decline concomitantly with *MAT* gene mRNAs (see Fig. 6.1), and those belonging to the *LEA-A* class, which begin to accumulate about 5 days earlier than the LEA class (Parcy et al. 1994). The observation that *LEA* transcripts and their polypeptide products are most abundant in embryos just prior to desiccation prompted the prophetic statement by Galau et al. (1987) that "some of the LEA polypeptides may function as desiccation protectants, binding or replacing water during the drying process in a fashion which allows rapid recovery during germination". This prediction was amply justified by the physical characteristics of LEA proteins, such as their extreme hydrophilicity and resistance to denaturation, and by the later discovery that various LEA proteins accumulate in other plant organs under conditions of water deficit, cold, salt and osmotic stress, and exposure to ABA (Skriver and Mundy 1990). The physical and structural properties of a purified soybean LEA protein have attested to their potential role in preventing freezing, desiccation, or osmotic stress damage (Soulages et al. 2002).

A classification of LEA proteins based on their commonly shared amino acid sequence domains, and expression patterns of the relevant genes, has recognized three groups (Dure et al. 1989; Dure 1993). The first, and best characterized, LEA gene from monocots is the *Em* gene from wheat embryos, assigned to group 1. Expression of transcripts of this gene at low levels during early embryogenesis and at high levels in late-stage embryos was the key feature that led to its identification as a *LEA* gene (Williamson et al. 1985). A rice embryo *LEA* gene designated as *RESPONSIVE TO ABA21* (*RAB21*) included in group 2, which progressively accumulates in developing embryos, is also expressed in roots, leaves, and suspension-cultured cells under stress (Mundy and Chua 1988). Although group 3 originally included a protein each from cotton, barley, and rape, and two from carrot (Dure et al. 1989; Franz et al. 1989), the group has been strengthened by later additions of two proteins from wheat (Curry et al. 1991; Curry and Walker-Simmons 1993) and five from soybean (Hsing et al. 1995). As discussed elsewhere (Raghavan 1997), a number of sequences that fit the bill for *LEA* gene transcripts and proteins have been described from other plants, but are not included in the list; a later addition to the list is endosperm of castor bean (*Ricinus communis;* Euphorbiaceae) – a seed tissue that contains stable *LEA* gene transcripts (Han et al. 1997). Two *LEA* genes, designated as *AtEm1* and *AtEm6* (for *Arabidopsis thaliana Em*), encoding two different proteins homologous to the EM protein of wheat, were first isolated and characterized from *Arabidopsis* (Finkelstein 1993; Gaubier et al. 1993); later investigations using a cDNA probe from radish (*Raphanus sativus;* Brassicaceae) led to the isolation of two additional *LEA* genes (Raynal et al. 1999). Altogether, the number of *LEA* genes isolated thus far from *Arabidopsis* is obviously low, as a cDNA library prepared from mature seeds has provided evidence for the presence of many *LEA* genes (Delseny et al. 2001).

The pervasiveness of LEA proteins in embryos of many plants, and their accumulation in other plant organs exposed to various kinds of stresses, suggest that these proteins developed during angiosperm evolution to regulate desiccation tolerance. Despite the fact that several studies have shown that application of ABA can induce synthesis of LEA proteins in the absence of environmental stress, conclusions regarding the role of ABA in the modulation of LEA protein synthesis remain contradictory. Two reviews have critically evaluated some of the results arguing for and against a role for ABA as an endogenous regulator of LEA protein synthesis based on embryo culture investigations and expression patterns of marker mRNAs (Galau et al. 1991; Hughes and Galau 1991); these articles make it clear that it no longer makes any sense to consider ABA as a sole developmental regulator of the synthesis of desiccation-related proteins in embryos. A different perspective on the role of ABA in LEA protein synthesis is provided by analysis of *AtEm* gene expression in ABA-deficient and ABA-insensitive mutants of *Arabidopsis*. Compared to *aba*, *abi1*, and *abi2* mutants, expression of the *AtEm6* gene is considerably reduced in seeds of a weak allele of the *abi3* mutant (Finkelstein 1993); in a strong allele, expression of both *AtEm1* and *AtEm6* is drastically impaired (Parcy et al. 1994). In an intermediate allele, expression of the *AtEm6* gene is found to be more severely affected than that of the *AtEm1* gene (Bies-Etheve et al. 1999). Expression of the *AtEm6* gene is also almost completely disrupted in the *abi5* mutant (Finkelstein 1994; Carles et al. 2002). Collectively, these results indicate a dependence of the accumulation of *AtEm* gene transcripts on the transcription factor encoded by *ABI* genes leading to ABA synthesis, even though in some cases the extent of reduction of *AtEm* gene expression is not correlated with the reduction in ABA content. Disruption of both *MAT* and *LEA* genes in *abi3* mutant alleles makes the role of the *ABI3* transcription factor unique as a global regulator of seed maturation in *Arabidopsis* (Bensmihen et al. 2002).

To understand how the ABI transcription factors regulate *LEA* genes, several homologs of the *ABI5* gene that encode bZIP proteins were identified. Analysis of mutation in one of these genes named *ENHANCED Em LEVEL* (*EEL*) showed that, compared to wild-type seeds, expression of *AtEm1* and *AtEm6* gene transcripts is enhanced in mutant seeds. In a detailed study of the expression of *AtEm1* RNA under various experimental conditions, it was established that both ABI5 and EEL proteins compete for the same binding sites within the *AtEm1* promoter. This observation suggests that homologous transcription factors can be envisaged to play antagonistic roles to tightly control expression of *LEA* genes during embryogenesis (Bensmihen et al. 2002). Such fine-tuning using two transcription factors targeted at the same DNA-binding module might be necessary to ensure that the *AtEm1* RNA

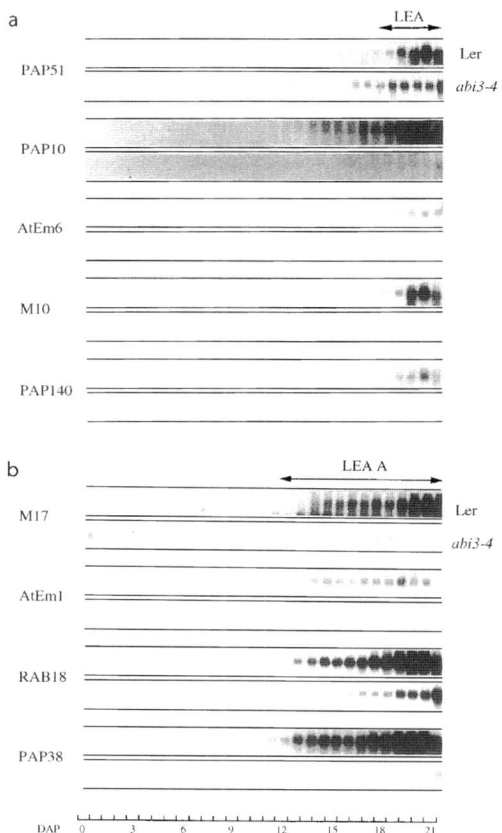

Fig. 6.6a,b Gel blot analysis of the expression pattern of the two classes of *LEA* gene transcripts during silique development in wild-type and *abi3-4* mutant *Arabidopsis*. Silique development lasted 21 days from anthesis (0 DAP) to dry seed stage (21 DAP). **a** *LEA* mRNAs. **b** *LEA-A* mRNAs. Each probe was hybridized with 1.5 µg total RNA. (Reprinted from Parcy et al. 1996)

level reaches its peak at the onset of embryo desiccation.

6.3
Concluding Comments

Early evidence for the maintenance of regulatory circuits during embryogenesis leading to embryo maturation and in the lapse of seeds into a quiescent or dormant state came from physiological investigations. An important principle emerging from the analysis of embryo maturation and dormancy during the latter half of the past century is that the plant hormone ABA is involved in regulating both the onset and the maintenance of the dormant state. A promising fruitful approach to close the gap between physiological observations and the molecular basis of maturation and dormancy of the embryo has been the isolation of *Arabidopsis* mutants impaired in their dormancy and sensitivity to ABA, which has facilitated identification of genes involved in metabolic and regulatory pathways of embryo maturation and dormancy. It is difficult to discuss embryo dormancy without evoking the next phase in the life of the embryo, namely, germination. The ecological importance of germination of seeds highlights the potential impact that a deeper genetic and molecular understanding of the phenomenon of embryo dormancy will have on species survival. Where the ongoing research on these topics is headed next is far from clear, but the strategy is likely to lead to identification of putative factors that maintain seeds of agronomically important plants in the dormant state and trigger their germination under propitious conditions.

REFERENCES

Bartels D, Singh M, Salamini F (1988) Onset of desiccation tolerance during development of the barley embryo. Planta 175:485–492

Bäumlein H, Miséra S, Luerßen H, Kölle K, Horstmann C, Wobus U, Müller AJ (1994) The *FUS3* gene of *Arabidopsis thaliana* is a regulator of gene expression during late embryogenesis. Plant J 6:379–387

Bensmihen S, Rippa S, Lambert G, Jublot D, Pautot V, Granier F, Giraudat J, Parcy F (2002) The homologous ABI5 and EEL transcription factors function antagonistically to fine-tune gene expression during late embryogenesis. Plant Cell 14:1391–1403

Bentsink L, Koornneef M (2002) Seed dormancy and germination. In: Somerville CR, Meyerowitz EM (eds) The *Arabidopsis* Book. American Society of Plant Biologists, Rockville, MD, doi/10.1199/tab.0050, http://www.aspb.org/publications/arabidopsis/

Bies-Etheve N, da Silva Conceicao A, Giraudat J, Koornneef M, Léon-Kloosterziel K, Valon C, Delseny M (1999) Importance of the B2 domain of the *Arabidopsis* ABI3 protein for Em and 2S albumin gene regulation. Plant Mol Biol 40:1045–1054

Blackman SA, Obendorf RL, Leopold AC (1992) Maturation proteins and sugars in desiccation tolerance of developing soybean seeds. Plant Physiol 100:225–230

Borisjuk L, Weber H, Panitz R, Manteuffel R, Wobus U (1995) Embryogenesis of *Vicia faba* L.: Histodifferentiation in relation to starch and storage protein synthesis. J Plant Physiol 147:203–218

Brenac P, Horbowicz M, Downer SM, Dockerman AM, Smith ME, Obendorf RL (1997a) Raffinose accumulation related to desiccation tolerance during maize (*Zea mays* L.) seed development and maturation. J Plant Physiol 150:481–488

Brenac P, Smith ME, Obendorf RL (1997b) Raffinose accumulation in maize embryos in the absence of a fully functional *Vp1* gene product. Planta 203:222–228

Brocard IM, Lynch TJ, Finkelstein RR (2002) Regulation and role of the *Arabidopsis abscisic acid-insensitive 5* gene in abscisic acid, sugar, and stress response. Plant Physiol 129:1533–1543

Bruni F, Leopold AC (1991) Glass transitions in soybean seed. Relevance to anhydrous biology. Plant Physiol 96:660–663

Carles C, Bies-Etheve N, Aspart L, Léon-Kloosterziel KM, Koornneef M, Echeverria M, Delseny M (2002) Regulation of *Arabidopsis thaliana Em* genes: role of ABI5. Plant J 30:373–383

Chang YC, Walling LL (1991) Abscisic acid negatively regulates expression of chlorophyll *a/b* binding protein genes during soybean embryogeny. Plant Physiol 97:1260–1264

Choinski JS Jr, Trelease RN, Doman DC (1981) Control of enzyme activities in cotton cotyledons during maturation and germination. III. In-vitro embryo development in the presence of abscisic acid. Planta 152:428–435

Crouch ML, Sussex IM (1981) Development and storage-protein synthesis in *Brassica napus* L. embryos in vivo and in vitro. Planta 153:64–74

Curaba J, Moritz T, Blervaque R, Parcy F, Raz V, Herzog M, Vachon G (2004) *AtGA2ox2*, a key gene responsible for bioactive gibberellin biosynthesis, is regulated during embryogenesis by *LEAFY COTYLEDON2* and *FUSCA3* in *Arabidopsis*. Plant Physiol 136:3660–3669

Curry J, Walker-Simmons MK (1993) Unusual sequence of group 3 LEA (II) mRNA inducible by dehydration stress in wheat. Plant Mol Biol 21:907–912

Curry J, Morris CF, Walker-Simmons MK (1991) Sequence analysis of a cDNA encoding a group 3 LEA mRNA inducible by ABA or dehydration stress in wheat. Plant Mol Biol 16:1073–1076

Delseny M, Bies-Etheve N, Carles C, Hull G, Vicient C, Raynal M, Grellet F, Aspart L (2001) Late embryogenesis abundant (LEA) protein gene regulation during *Arabidopsis* seed maturation. J Plant Physiol 158:419–427

Dure L III (1993) Structural motifs in Lea proteins. In: Close TJ, Bray EA (eds) Plant responses to cellular dehydration during environmental stress. American Society of Plant Physiologists, Rockville, MD, pp 91–103

Dure L III, Greenway SC, Galau GA (1981) Developmental biochemistry of cottonseed embryogenesis and germination: changing messenger ribonucleic acid populations as shown by in vitro and in vivo protein synthesis. Biochemistry 20:4162–4168

Dure L III, Crouch M, Harada J, Ho T-HD, Mundy J, Quatrano R, Thomas T, Sung ZR (1989) Common amino acid sequence domains among the LEA proteins of higher plants. Plant Mol Biol 12:475–486

Eisenberg AJ, Mascarenhas JP (1985) Abscisic acid and the regulation of synthesis of specific seed proteins and their messenger RNAs during culture of soybean embryos. Planta 166:504–514

Fernandez DE, Turner FR, Crouch ML (1991) In situ localization of storage protein mRNAs in developing meristems of *Brassica napus* embryos. Development 111:299–313

Finkelstein RR (1993) Abscisic acid-insensitive mutations provide evidence for stage-specific signal pathways regulating expression of an *Arabidopsis* late embryogenesis-abundant (*lea*) gene. Mol Gen Genet 238:401–408

Finkelstein RR (1994) Mutations at two new *Arabidopsis* ABA response loci are similar to the *abi3* mutations. Plant J 5:765–771

Finkelstein RR, Lynch TJ (2000) The *Arabidopsis* abscisic acid response gene *ABI5* encodes a basic leucine zipper transcription factor. Plant Cell 12:599–609

Finkelstein RR, Tenbarge KM, Shumway JE, Crouch ML (1985) Role of ABA in maturation of rapeseed embryos. Plant Physiol 78:630–636

Finkelstein RR, Wang ML, Lynch TJ, Rao S, Goodman HM (1998) The *Arabidopsis* abscisic acid response locus *ABI4* encodes an APETALA2 domain protein. Plant Cell 10:1043–1054

Fischer W, Bergfeld R, Plachy C, Schäfer R, Schopfer P (1988) Accumulation of storage materials, precocious germination and development of desiccation tolerance during seed maturation in mustard (*Sinapis alba* L.). Bot Acta 101:344–354

Footitt S, Slocombe SP, Larner V, Kurup S, Wu Y, Larson T, Graham I, Baker A, Holdsworth M (2002) Control of germination and lipid mobilization by *COMATOSE*, the *Arabidopsis* homologue of human ALDP. EMBO J 21:2912–2922

Franz G, Hatzopoulos P, Jones TJ, Krauss M, Sung ZR (1989) Molecular and genetic analysis of an embryonic gene, DC8, from *Daucus carota* L. Mol Gen Genet 218:143–151

Fujiwara T, Nambara E, Yamagishi,K, Goto DB, Naito S (2002) Storage proteins. In: Somerville CR, Meyerowitz EM (eds) The *Arabidopsis* Book. American Society of Plant Biologists, Rockville, MD, doi/10.1199/tab.00202, http://www.aspb.org/publications/arabidopsis/

Galau GA, Hughes DW, Dure L III (1986) Abscisic acid induction of cloned cotton late embryogenesis-abundant (*Lea*) mRNAs. Plant Mol Biol 7:155–170

Galau GA, Bijaisoradat N, Hughes DW (1987) Accumulation kinetics of cotton late embryogenesis-abundant mRNAs and storage protein mRNAs: coordinate regulation during embryogenesis and the role of abscisic acid. Dev Biol 123:198–212

Galau GA, Jakobsen KS, Hughes DW (1991) The control of late dicot embryogenesis and early germination. Physiol Plant 81:280–288

Gaubier P, Raynal M, Hull G, Huestis GM, Grellet F, Arenas C, Pagès M, Delseny M (1993) Two different *Em*-like genes are expressed in *Arabidopsis thaliana* seeds during maturation. Mol Gen Genet 238:409–418

Giraudat J, Hauge BM, Valon C, Smalle J, Parcy F, Goodman HM (1992) Isolation of the *Arabidopsis ABI3* gene by positional cloning. Plant Cell 4:1251–1261

Goldberg RB, Hoschek G, Ditta GS, Breidenbach RW (1981) Developmental regulation of cloned superabundant embryo mRNAs in soybean. Dev Biol 83:218–231

Goldberg RB, Barker SJ, Perez-Grau L (1989) Regulation of gene expression during plant embryogenesis. Cell 56:149–160

Gómez J, Sánchez-Martínez D, Stiefel V, Rigau J, Puigdomènech P, Pagès M (1988) A gene induced by the plant hormone abscisic acid in response to water stress encodes a glycine-rich protein. Nature 334:262–264

Han B, Hughes DW, Galau GA, Bewley JD, Kermode AR (1997) Changes in late-embryogenesis-abundant (LEA) messenger RNAs and dehydrins during maturation and premature drying of *Ricinus communis* L. seeds. Planta 201:27–35

Harada JJ (1997) Seed maturation and control of germination. In: Larkins BA, Vasil IK (eds) Cellular and molecular biology of plant seed development. Kluwer, Dordrecht, pp 545–592

Hattori T, Vasil V, Rosenkrans L, Hannah LC, McCarty DR, Vasil IK (1992) The *Viviparous-1* gene and abscisic acid activate the *C1* regulatory gene for anthocyanin biosynthesis during seed maturation in maize. Genes Dev 6:609–618

Hattori T, Terada T, Hamasuna S (1995) Regulation of the *Osem* gene by abscisic acid and the transcriptional activator VP1: analysis of *cis*-acting promoter elements required by regulation by abscisic acid and VP1. Plant J 7:913–925

Hoecker U, Vasil IK, McCarty DR (1995) Integrated control of seed maturation and germination programs by activator and repressor functions of Viviparous-1 of maize. Genes Dev 9:2459–2469

Hsing Y-I, Chen Z-Y, Shih M-D, Hsieh J-S, Chow T-Y (1995) Unusual sequences of group 3 LEA mRNA inducible by maturation or drying in soybean seeds. Plant Mol Biol 29:863–868

Hughes DW, Galau GA (1989) Temporally modular gene expression during cotyledon development. Genes Dev 3:358–369

Hughes DW, Galau GA (1991) Developmental and environmental induction of *Lea* and *LeaA* mRNAs and the postabscission program during embryo culture. Plant Cell 3:605–618

Ihle JN, Dure L III (1970) Hormonal regulation of translation inhibition requiring RNA synthesis. Biochem Biophys Res Commun 38:995–1001

Ihle JN, Dure LS III (1972) The developmental biochemistry of cottonseed embryogenesis and germination. III. Regulation of the biosynthesis of enzymes utilized in germination. J Biol Chem 247:5048–5055

Jakobsen KS, Hughes DW, Galau GA (1994) Simultaneous induction of postabscission and germination mRNAs in cultured dicotyledonous embryos. Planta 192:384–394

Karssen CM, Brinkhorst-van der Swan DLC, Breekland AE, Koornneef M (1983) Induction of dormancy during seed development by endogenous abscisic acid: studies on abscisic acid deficient genotypes of *Arabidopsis thaliana* (l.) Heynh. Planta 157:158–165

Keith K, Kraml M, Dengler NG, McCourt P (1994) *fusca3*: a heterochronic mutation affecting late embryo development in *Arabidopsis*. Plant Cell 6:589–600

Koornneef M, Karssen CM (1994) Seed dormancy and germination. In: Meyerowitz EM, Somerville CR (eds) Arabidopsis. Cold Spring Harbor Laboratory Press, Cold Spring Harbor, New York, pp 313–334

Koornneef M, Reuling G, Karssen CM (1984) The isolation and characterization of abscisic acid-insensitive mutants of *Arabidopsis thaliana*. Physiol Plant 61:377–383

Koornneef M, Hanhart CJ, Hilhorst HWM, Karssen CM (1989) In vivo inhibition of seed development and reserve protein accumulation in recombinants of abscisic acid biosynthesis and responsiveness mutants in *Arabidopsis thaliana*. Plant Physiol 90:463–469

Koster KL (1991) Glass formation and desiccation tolerance in seeds. Plant Physiol 96:302–304

Léon-Kloosterziel KM, van de Bunt GA, Zeevaart JAD, Koornneef M (1996) *Arabidopsis* mutants with a reduced seed dormancy. Plant Physiol 110:233–240

Leprince O, Bronchart R, Deltour R (1990) Changes in starch and soluble sugars in relation to the acquisition of desiccation tolerance during maturation of *Brassica campestris* seed. Plant Cell Environ 13:539–546

Leung J, Bouvier-Durand M, Morris P-C, Guerrier D, Chefdor F, Giraudat J (1994) *Arabidopsis* ABA responsegene *ABI1*: features of a calcium-modulated protein phosphatase. Science 264:1448–1452

Leung J, Merlot S, Giraudat J (1997) The *Arabidopsis AB-SCISIC ACID-INSENSITIVE2* (*ABI2*) and *ABI1* genes encode homologous protein phosphatases 2C involved in abscisic acid signal transduction. Plant Cell 9:759–771

Lopez-Molina L, Chua N-H (2000) A null mutation in a bZIP factor confers ABA-insensitivity in *Arabidopsis thaliana*. Plant Cell Physiol 41:541–547

Lotan T, Ohto M, Yee KM, West MAL, Lo R, Kwong RW, Yamagishi K, Fischer RL, Goldberg RB, Harada JJ (1998) *Arabidopsis* LEAFY COTYLEDON1 is sufficient to induce embryo development in vegetative cells. Cell 93:1195–1205

Ludevid MD, Freire MA, Gómez J, Burd CG, Albericio F, Giralt E, Dreyfuss G, Pagès M (1992) RNA binding characteristics of a 16 kDa glycine-rich protein from maize. Plant J 2:999–1003

Luerßen H, Kirik V, Herrmann P, Miséra S (1998) *FUSCA3* encodes a protein with a conserved VP1/ABI3-like B3 domain which is of functional importance for the regulation of seed maturation in *Arabidopsis thaliana*. Plant J 15:755–764

McCarty DR, Carson CB (1991) The molecular genetics of seed maturation in maize. Physiol Plant 81:267–272

McCarty DR, Hattori T, Carson,CB, Vasil V, Lazar M, Vasil IK (1991) The *Viviparous-1* developmental gene of maize encodes a novel transcriptional activator. Cell 66:895–905

Meinke DW (1992) A homeotic mutant of *Arabidopsis thaliana* with leafy cotyledons. Science 258:1647–1650

Meinke DW, Franzmann LH, Nickle TC, Yeung EC (1994) *Leafy Cotyledon* mutants of *Arabidopsis*. Plant Cell 6:1049–1064

Meurs C, Basra AS, Karssen CM, van Loon LC (1992) Role of abscisic acid in the induction of desiccation tolerance in developing seeds of *Arabidopsis thaliana*. Plant Physiol 98:1484–1493

Meyer K, Leube MP, Grill E (1994) A protein phosphatase 2C involved in ABA signal transduction in *Arabidopsis thaliana*. Science 264:1452–1455

Mundy J, Chua N-H (1988) Abscisic acid and water-stress induce the expression of a novel rice gene. EMBO J 7:2279–2286

Nambara E, Naito S, McCourt P (1992) A mutant of *Arabidopsis* which is defective in seed development and storage protein accumulation is a new *abi3* allele. Plant J 2:435–441

Nambara E, Keith K, McCourt P, Naito S (1994) Isolation of an internal deletion mutant of the *Arabidopsis thaliana ABI3* gene. Plant Cell Physiol 35:509–513

Nambara E, Keith K, McCourt P, Naito S (1995) A regulatory role for the *ABI3* gene in the establishment of embryo maturation in *Arabidopsis thaliana*. Development 121:629–636

Nambara E, Hayama R, Tsuchiya Y, Nishimura M, Kawaide H, Kamiya Y, Naito S (2000) The role of *ABI3* and *FUS3* loci in *Arabidopsis thaliana* on phase transition from late embryo development to germination. Dev Biol 220:412–423

Nambara E, Suzuki M, Abrams S, McCarty DR, Kamiya Y, McCourt P (2002) A screen for genes that function in abscisic acid signaling in *Arabidopsis thaliana*. Genetics 161:1247–1255

Neill SJ, Horgan R, Rees AF (1987) Seed development and vivipary in *Zea mays* L. Planta 171:358–364

Nuccio ML, Thomas TL (1999) *ATS1* and *ATS3*: two novel embryo-specific genes in *Arabidopsis thaliana*. Plant Mol Biol 39:1153–1163

Ooms JJJ, Léon-Kloosterziel KM, Bartels D, Koornneef M, Karssen CM (1993) Acquisition of desiccation tolerance and longevity in seeds of *Arabidopsis thaliana*. Plant Physiol 102:1185–1191

Ooms JJJ, Wilmer JA, Karssen CM (1994) Carbohydrates are not the sole factor determining desiccation tolerance in seeds of *Arabidopsis thaliana*. Physiol Plant 90:431–436

Pang PP, Pruitt RE, Meyerowitz EM (1988) Molecular cloning, genomic organization, expression and evolution of 12S seed storage protein genes of *Arabidopsis thaliana*. Plant Mol Biol 11:805–820

Parcy F, Valon C, Raynal M, Gaubler-Comella P, Delseny M, Giraudat J (1994) Regulation of gene expression programs during *Arabidopsis* seed development: roles of the *ABI3* locus and of endogenous abscisic acid. Plant Cell 6:1567–1582

Parcy F, Valon C, Kohara A, Miséra S, Giraudat J (1997) The *ABSCISIC ACID-INSENSITIVE3*, *FUSCA3*, and *LEAFY COTYLEDON1* loci in concert to control multiple aspects of *Arabidopsis* seed development. Plant Cell 9:1255–1277

Peeters AJM, Blankestijn-de Vries H, Hanhart CJ, Léon-Kloosterziel KM, Zeevaart JAD, Koornneef M (2002) Characterization of mutants with reduced seed dormancy at two novel *rdo* loci and a further characterization of *rdo1* and *rdo2* in *Arabidcpsis*. Physiol Plant 115:604–612

Perez-Grau L, Goldberg RB (1989) Soybean seed protein genes are regulated spatially during embryogenesis. Plant Cell 1:1095–1109

Prevost L, Le Page-Degivry MT (1985) Changes in abscisic acid content in axis and cotyledons of developing *Phaseolus vulgaris* embryos and their physiological consequences. J Exp Bot 36:1900–1905

Raghavan V (1997) Molecular embryology of flowering plants. Cambridge University Press, New York

Raghavan V (2002) Induction of vivipary in *Arabidopsis* by silique culture: implications for seed dormancy and germination. Am J Bot 89:766–776

Raynal M, Guilleminot J, Gueguen C, Cooke R, Delseny M, Gruber V (1999) Structure, organization and expression of two closely related novel *Lea* (late-embryogenesis-abundant) genes in *Arabidopsis thaliana*. Plant Mol Biol 40:153–165

Raz V, Bergervoet JHW, Koornneef M (2001) Sequential steps for developmental arrest in *Arabidopsis* seeds. Development 128:243–252

Robertson M, Walker-Simmons M, Munro D, Hill RD (1989) Induction of α-amylase inhibitor synthesis in barley embryos and young seedlings by abscisic acid and dehydration stress. Plant Physiol 91:415–420

Robichaud C, Sussex IM (1986) The response of viviparous-1 and wild type embryos of *Zea mays* to culture in the presence of abscisic acid. J Plant Physiol 126:235–242

Robichaud CS, Wong J, Sussex IM (1980) Control of in vitro growth of viviparous embryo mutants of maize by abscisic acid. Dev Genet 1:325–330

Rosenberg LA, Rinne RW (1987) Changes in seed constituents during germination and seedling growth of precociously matured soybean seeds (*Glycine max*). Ann Bot 60:705–712

Rosenberg LA, Rinne RW (1988) Protein synthesis during natural and precocious soybean seed (*Glycine max* [L.] Merr.) maturation. Plant Physiol 87:474–478

Russell L, Larner V, Kurup S, Bougourd S, Holdsworth M (2000) The *Arabidopsis COMATOSE* locus regulates germination potential. Development 127:3759–3767

Sánchez-Martínez D, Puigdomènech P, Pagès M (1986) Regulation of gene expression in developing *Zea mays* embryos. Protein synthesis during embryogenesis and early germination of maize. Plant Physiol 82:543–549

Skriver K, Mundy J (1990) Gene expression in response to abscisic acid and osmotic stress. Plant Cell 2:503–512

Soulages JL, Kim K, Walters C, Cushman JC (2002) Temperature-induced extended helix/random coil transitions in a group 1 late embryogenesis-abundant protein from soybean. Plant Physiol 128:822–832

Stone SL, Kwong LW, Yee KM, Pelletier J, Lepiniec L, Fischer RL, Goldberg RB, Harada JJ (2001) *LEAFY COTYLEDON2* encodes a B3 domain transcription factor that induces embryo development. Proc Natl Acad Sci USA 98:11806–11811

Suzuki M, Kao C-Y, Cocciolone S, McCarty DR (2001) Maize VP1 complements *Arabidopsis abi3* and confers novel ABA/auxin interaction in roots. Plant J 28:409–418

Suzuki M, Ketterling MG, Li Q-B, McCarty DR (2003) Viviparous1 alters global gene expression patterns through regulation of abscisic acid signaling. Plant Physiol 132:1664–1677

Triplett BA, Quatrano RS (1982) Timing, localization, and control of wheat germ agglutinin synthesis in developing wheat embryos. Dev Biol 91:491–496

Walters C, Reid JL, Walker-Simmons MK (1997) Heat-soluble proteins extracted from wheat embryos have tightly bound sugars and unusual hydration properties. Seed Sci Res 7:125–134

Wang TL, Cook SK, Francis RJ, Ambrose MJ, Hedley CL (1987) An analysis of seed development in *Pisum sativum*. VI. Abscisic acid accumulation. J Exp Bot 38:1921–1932

West MAL, Yee KM, Danao J, Zimmerman JL, Fischer RL, Goldberg RB, Harada JJ (1994) *LEAFY COTYLEDON1* is an essential regulator of late embryogenesis and cotyledon identity in *Arabidopsis*. Plant Cell 6:1731–1745

Williams RJ, Leopold AC (1989) The glassy state in corn embryos. Plant Physiol 89:977–981

Williamson JD, Quatrano RS, Cuming AC (1985) E_m polypeptide and its messenger RNA levels are modulated by abscisic acid during embryogenesis in wheat. Eur J Biochem 152:501–507

Wobus U, Weber H (1999) Seed maturation: genetic programmes and control signals. Curr Opin Plant Biol 2:33–38

Wolkers WF, Alberda M, Koornneef M, Léon-Kloosterziel KM, Hoekstra FA (1998a) Properties of proteins and the glassy matrix in maturation-defective mutant seeds of *Arabidopsis thaliana*. Plant J 16:133–143

Wolkers WF, Bochicchio A, Selvaggi G, Hoekstra FA (1998b) Fourier transform infrared microspectroscopy detects changes in protein secondary structure associated with desiccation tolerance in developing maize embryos. Plant Physiol 116:1169–1177

Wolkers WF, Tetteroo FAA, Alberda M, Hoekstra FA (1999) Changed properties of the cytoplasmic matrix associated with desiccation tolerance of dried carrot somatic embryos. An in situ Fourier transform infrared spectroscopic study. Plant Physiol 120:153–163

7 Developmental and Functional Biology of the Endosperm – A Medley of Cellular Interactions

The endosperm was early interpreted as a second, but abortive, embryo – at first, because the union of the polar nuclei was considered fertilization and, later, when union of the second male nucleus with the polar nuclei was discovered. Still later, the endosperm was considered a delayed, complex type of nutritive gametophytic tissue, not an abortive structure resulting from a fertilization. The discovery that the endosperm, in early stages, exists as markedly different types (cellular or nuclear) and varies in nature and number of constituent cells has greatly complicated interpretation of its nature. It can probably best be termed – as it has been several times – a "new structure", one of complex morphological nature, characteristic of the angiosperms only.

A.J. Eames 1961

7.1 Cellular Organization of the Endosperm 152
7.1.1 The Odyssey of Free Nuclei to a Cellular Tissue 154
7.1.2 Development of the Endosperm in *Arabidopsis* 156
7.2 Biochemical Organization of the Endosperm 158
7.2.1 DNA Amplification 158
7.2.2 Accumulation of Storage Products 160
7.2.3 Programmed Cell Death of the Endosperm 161
7.3 Role of the Endosperm in Embryo Nutrition 162
7.3.1 Structural Modifications of the Endosperm 162
7.3.2 Physiological Considerations 164
7.3.3 Genetic Considerations 165
7.4 Concluding Comments 167
References 168

Following double fertilization, the endosperm develops from the product of fusion of the two polar nuclei with one male gamete. The fusion product, termed the primary endosperm nucleus, becomes committed to a program of differentiation to form an amorphous, nutrient-rich tissue of the endosperm with its own triploid genetic make-up, separate from the diploid embryo or the surrounding tissues. Numerous studies on endosperm ontogeny have shown that, despite the almost simultaneous nature of the two fusion events during double fertilization, the timing of the initial divisions of the fusion products is variable, with the primary endosperm nucleus commonly beginning to divide first. It is frequently observed that seeds of some plants, such as beans and peas, that have an endosperm at early stages of development, do not possess the tissue at maturity, whereas grains of wheat, corn, and other cereals, and castor bean seeds have copious amounts of endosperm, which serves for the nurture of the embryo and seedling during germination. This dilemma has traditionally been resolved by the explanation that in the former seed types, the endosperm is depleted and used up for nourishment of the embryo and that reserve substances for nutrition of the seedling are stored in the cotyledons. As a repository of reserve food materials used in human and animal nutrition, endosperms of cereal grains continue to provide much of the inspiration for current research aimed at improving the quality and quantity of this tissue in economically important plants. For obvious reasons, this is a project that will never really be finished, but it is a goal that is gradually approaching.

After more than 100 years of latency, the fundamental question associated with the evolutionary origin of the endosperm has been revitalized. The two early debated hypotheses, one proposing that the endosperm is a modified second embryo and the other considering the endosperm as being evolutionarily homologous to the female gametophyte, were entwined with the shifting views on angiosperm phylogeny and, consequently, there was little way to distinguish empirically between these two hypotheses (Friedman 1998). New data described in Chap. 1 and reviewed by Friedman and Williams (2004) have shown that, in contrast to the triploid endosperm found in the overwhelming majority of flowering plants, a diploid endosperm, originating from a four-celled, four-nucleate female gametophyte with a haploid central cell, predominates in the limited number of their ancient lineages so far investigated. Based on this observation, it has been inferred that, over evolutionary time, addition of a male nucleus by a second fertilization event in a seven-celled, eight-nucleate female gametophyte with a diploid central cell would have provided the specific genetic and developmental event required to transform a diploid biparental endosperm into a triploid one. This is a seductively simple idea but, considering the past vicissitudes of the current hypotheses, there are likely to be surprises ahead in this field before a conclusion is reached.

Several features of the endosperm make it a useful model for cell biological, genetic, and molecular studies as a snapshot of events in a single tissue. First, unlike the embryo, the endosperm in most eudicots consists of only one or two uniform cell types that are programmed mainly for the accumulation of starch and protein storage reserves. In many species studied, cytokinesis is uncoupled from the nuclear division cycle as the endosperm goes through a stage of a multinucleate mass of protoplasm, or syncytium. Although eventual wall formation takes place to generate a cellular tissue, in several respects, the mechanism of placement and growth of walls in the syncytium has turned out to be unusual. Isolation of genes involved in the development of the endosperm is beginning to provide insights into the molecular mechanisms of cell fate specification and cell differentiation in this tissue.

Since the fate of cells in the developing endosperm is not lineage-dependent, all sorts of mutant and wild type cells can be analyzed in the clones generated in appropriate experimental systems. Many recently characterized endosperm mutants offer interesting experimental systems with which to study the mechanisms involved in genomic imprinting. Last, but not least, insight into the mechanism by which genes for the synthesis of storage proteins are regulated in the developing endosperm affords great potential to the genetic engineering of cereal grains with improved nutritional qualities.

All of the above-noted studies are beginning to blossom through a combination of choice of experimental systems and creativity in the use of high resolution structural and molecular techniques as well as classical genetic screens. Given the expected surge of further research in the coming years, the goal of this and the next chapter is to provide a framework describing recent achievements in these areas of endosperm development against a background of earlier studies. The critical events that constitute stepwise processes in the different aspects of endosperm ontogeny, and the role of the endosperm in embryo nutrition will be the focus of this chapter; Chap 8 will deal principally with the genetics and molecular biology of the endosperm in model systems such as *Arabidopsis* and cereal grains. Earlier studies on the comparative morphology and cytology, and developmental biology of the endosperm have been admirably handled in reviews by Brink and Cooper (1947), Bhatnagar and Sawhney (1981), Vijayaraghavan and Prabhakar (1984), Lopes and Larkins (1993), and DeMason (1997), whereas the reviews by Brown et al. (2002) and Olsen (2001, 2004) provide excellent syntheses of more recent studies.

7.1
Cellular Organization of the Endosperm

The use of conventional histological methods to monitor the fate of the primary endosperm nucleus in fertilized ovules of diverse species facilitated identification of the pathways involved in the final configuration of the endosperm in seeds, and led to the recognition of nuclear, cellular, and helobial types of endosperm development in flowering plants (Maheshwari 1950). In the nuclear mode, the primary endosperm nucleus undergoes several cycles of divisions without cytokinesis. The newly formed daughter nuclei remain embedded in the peripheral cytoplasm surrounding the vacuole of the large central cell for a variable period of time before wall formation occurs to give rise to a cellular tissue. The nuclear type pathway is by far the most common in endosperm development, and occurs in model eudicots such as *Arabidopsis* (Schneitz et al. 1995; Herr 1999; Brown et al. 1999), *Capsella bursa-pastoris* (Schulz and Jensen 1974), and cotton (Schulz and Jensen 1977), cereals such as barley (Bosnes et al. 1992), rice (Brown et al. 1996b), wheat (Mares et al. 1975; Fineran et al. 1982), and maize (Randolph 1936; Olsen 2001), and legumes such as *Phaseolus vulgaris* (Yeung and Cavey 1988) and soybean (Dute and Peterson 1992). During free nuclear divisions, the migration of nuclei within the central cell is not random, but occurs in relatively predictable directions. By direct observations of the division planes of the primary endosperm nucleus, and by clonal analysis, McClintock (1978, for review) showed that it is possible to create a fate map for the endosperm, enabling one to trace even as little as a one-eighth sector of the mature tissue to a lineage of the primary endosperm nucleus. A crucial parameter in the endosperm structure of cereal grains is the precise pattern of cellularization that leads to the differentiation of two layers of endosperm initials. In this scenario, cells of the outermost layer of the endosperm give rise to the protein-rich aleurone, and the inner cells, after repeated divisions, become filled with starch to form the starchy endosperm. Grains of wheat and other cereals contain a single layer of aleurone cells but in barley the aleurone is at least three layers thick.

In the cellular type of endosperm, mitosis and cytokinesis are coupled, with the result that the initial and subsequent divisions of the primary endosperm nucleus are followed throughout the entire course of development by cell plate formation. No consolidated list of families exhibiting the cellular type of endosperm development has been published, although the type is known to occur in some advanced families such as Acanthaceae, Lobeliaceae, Scrophulariaceae, Gesneriaceae, and Loranthaceae (Bhatnagar and Sawhney 1981; Vijayaraghavan and Prabhakar 1984); it has also been described in isolated members of basal angiosperms included in the families Amborellaceae, Nymphaeaceae, and Illiciaceae (Floyd and Friedman 2000, 2001). The helobial type of endosperm was slow to be recognized as a distinct type and is generally accorded an intermediate position between the nuclear and cellular types. However, a phylogenetic analysis of variations in endosperm development described in flowering plants has disputed the idea that the helobial endosperm is an evolutionary intermediate between the other two ontogenetic types (Bharathan 1999). In the current understanding of helobial endosperm ontogeny, the division of the primary endosperm nucleus separates the central cell into a large micropylar cell and a small chalazal cell. Free nuclear divisions occur in the micropylar cell before cellularization sets in, whereas the nucleus of the chalazal cell either remains undivided or divides only occasionally. The helobial type of endosperm is found only in monocotyledons and, as the name implies, is prevalent in the order Helobiae; endosperm of *Haemanthus katherinae* (Amaryllidaceae), from which the first complete motion-pictures of mitosis and cytokinesis in living cells were produced, boasts of having the helobial type (Swamy and Parameswaran 1963; Bajer 1965; Newcomb 1978).

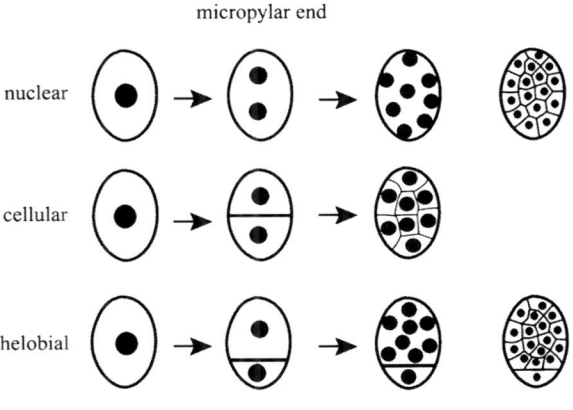

Fig. 7.1 Diagrammatic representations of the nuclear, cellular, and helobial types of endosperm development. Endosperm development is initiated by the division of the primary endosperm nucleus shown in the first diagram on the left for each type. The diagrams on the far right indicate that wall formation occurs around the free nuclei in both nuclear and helobial types, which eventually become cellular

It appears from the above that no clear line exists between the three types of endosperm development and, regardless of their ontogenetic mode, all types ultimately become, for the most part, cellular (Fig. 7.1). No serious attempt has been made to explain the evolutionary rationale behind the origin of the three modes of endosperm ontogeny; a notion, based on endosperm development in basal angiosperms and analysis of endosperm developmental patterns across flowering plants in a phylogenetic context, is that the cellular endosperm is the ancestral type from which the nuclear type evolved, probably by the disruption of some aspects of the normal cytokinetic process (Floyd et al. 1999; Floyd and Friedman 2000, 2001; Bharathan 1999; Geeta 2003). Although this view on the evolution of endosperm ontogeny does not fit comfortably into the present-day framework of molecular development of the tissue, it attests to the generality of a phenomenon based on three variable states of a single character.

As described in Chap. 9, autonomous development of the endosperm frequently occurs in certain apomicts, although the genetic constitution of the tissue produced differs from that of sexual plants (Koltunow 1993; Koltunow and Grossniklaus 2003). By mechanisms that are poorly understood, endosperm development without fertilization occurs in certain plants following pollination by irradiated pollen (Musial and Przywara 1998), and when unfertilized ovules and ovaries are cultured in vitro (Mól et al. 1995; Wijowska et al. 1999). Of course, the discovery of endosperm development independent of fertilization in the *fis*-class of mutants of *Arabidopsis* (Chap. 8), and establishment of the role of genomic imprinting in endosperm development in these mutants, have become classics in the recent plant developmental biology literature.

To aid research on the underlying genetic and molecular program for cell fate specification, in addition to the aleurone cells and starchy cells alluded to earlier, two other major cell types, namely, embryo-surrounding region and transfer layer cells, have been identified in the endosperm. The embryo surrounding region consists of cells confined to a restricted area within the starchy endosperm in the immediate vicinity of the embryo. Although a cellular or noncellular embryo surrounding region can probably be identified in endosperms of most flowering plants, only lately has attention been drawn to its occurrence as a distinct cellular region. Schel et al. (1984) found that a few cells interconnected by plasmodesmata surrounding a small part of the globular embryo of maize have a dense cytoplasm enriched with rough ER but, as the embryo grows out of this stage, these cells just wrap around the suspensor, while the region surrounding the embryo proper remains free of them. A distinctive gene expression pattern has also been found to be characteristic of these cells (Opsahl-Ferstad et al. 1997). A well-defined group of cells (transfer cell layer) characterized by heavy wall ingrowths typical of transfer cells has been described under various names in the endosperms of many plants, especially those of cereal grains (Thompson et al. 2001); the favored name for these cells is basal endosperm transfer layer. The location of these cells generally close to the vascular bundle of the maternal tissues is consistent with the view that they aid in the uptake of nutrients. Aberrant endosperm development, including failure of formation of the basal endosperm transfer layer, associated with embryo abortion and reduced grain filling observed in interploidy crosses in maize attests to the importance of these cell layers for nutrient transfer to the endosperm (Charlton et al. 1995).

7.1.1
The Odyssey of Free Nuclei to a Cellular Tissue

The free-nuclear stage in the nuclear and helobial types of endosperm described above should reflect expression of a remarkably efficient genetic system because once the primary endosperm nucleus in the nuclear type, or the nucleus of the micropylar chamber in the helobial type, is triggered to divide, divisions continue until a mass of free nuclei is produced. The free nuclei can thus be envisioned to have been generated by the activation of a genetic program for cell cycle arrest in the dividing nuclei. In the nuclear type of endosperm in barley, it has been shown that the rate of RNA synthesis increases six-fold during the syncytium stage, indicating that this stage is driven by transcripts synthesized by the free nuclei (Bosnes and Olsen 1992). At present, little is known about the genes that are activated in the endosperm syncytium; in the only published report, a cDNA clone isolated from the free-nuclear stage of barley endosperm by differential screening was found to encode an unknown protein (Doan et al. 1996).

The cytological basis for the delayed formation of walls between isolated nuclei dispersed in an amorphous cytoplasm as seen in the nuclear mode of endosperm development is also an important yet poorly understood problem. Although most descriptions of wall placement and deposition during cellularization of the nuclear endosperm based on light and electron microscopic observations have in some way implicated phragmoplasts and cell plates in the process, the work of Brown et al. (1994) on barley endosperm using immunolocalization techniques combined with three-dimensional imaging provided the first clear account of the involvement of microtubule arrays in preparing the cytoplasm for cellularization. Figure 7.2 (a–d) presents a summary diagram of wall patterning and its relationship to microtubular cytoskeleton during endosperm development, beginning with the wall-less syncytium and ending with the formation of the aleurone layer (Olsen et al. 1995). An important observation is that, in preparation for cellularization during the late syncytium stage, radial arrays of microtubules that proliferate from the nuclear surface organize the syncytium into units defined as nuclear-cytoplasmic domains. Shortly thereafter, wall materials in the form of phragmoplast configurations are deposited at the interstices of the nuclear-cytoplasmic enclaves, which thus serve to establish the initial pattern of cellularization in the form of 'free growing' anticlinal walls (see Plate 11, Fig. b–d). Eventually, the newly crafted walls grow into the vacuole of the central cell from the peripheral syncytium to subdivide the cytoplasm into open-ended compartments or alveoli. During the deposition of cell walls, the nuclear-cytoplasmic domains become polarized along the plane perpendicular to the embryo sac wall. This is accompanied by a dramatic rearrangement of the nuclear-based radial microtubules, which now appear to arise from both ends of the nuclei. Almost simultaneously, adventious phragmoplasts are formed at the interfaces of these opposing microtubule systems (Brown et al. 1996a; Olsen 2004). This stage is followed by continued centripetal growth of walls, and sealing of the open ends of alveoli mediated by adventitious phragmoplasts. Next, the alveoli are partitioned into two layers of cells by a round of divisions with spindles oriented at right angles to the embryo sac wall. Cell plates directed by phragmoplasts of the type that arise between daughter nuclei (interzonal phragmoplasts) produce periclinal walls (parallel with the embryo sac wall) that join with the anticlinal walls to form the first complete layer of cells cut off at the periphery of the embryo sac. This layer of cells develops into initials of the aleurone layer and the inner cells form the starchy endosperm. The aleurone initials divide anticlinally with the aid of a full panoply of the typical cytokinetic apparatus, namely, hoop-like, well-ordered cortical arrays of microtubules during interphase, a preprophase band of microtubules, and interzonal phragmoplasts. In rice, an unexpected difference between wall properties of the endosperm cells and cells of the surrounding maternal tissues has been revealed by the observation that the former have a uniform distribution of callose (a polymer of 1→3-β-glucans) in contrast to the preponderance of (1→3, 1→4)-β-glucans in the cells of the latter (Brown et al. 1997).

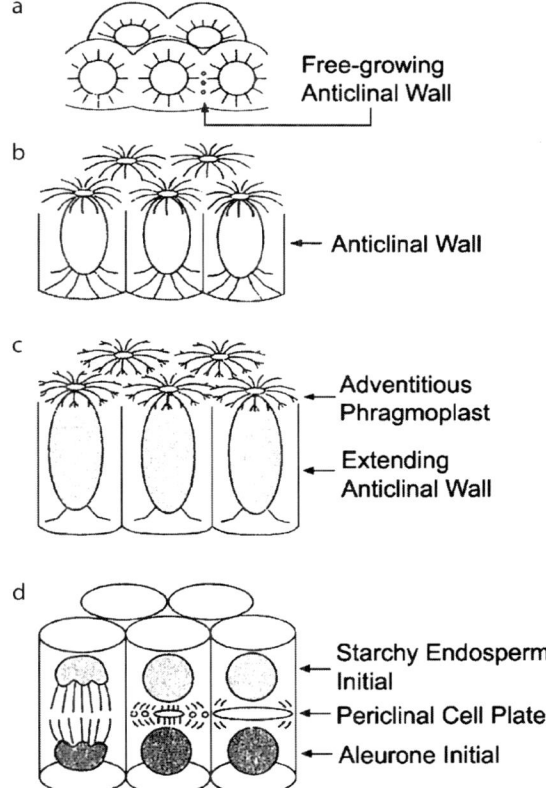

Fig. 7.2a–d Diagrams showing cellularization of nuclear-type endosperm. **a** Syncytial stage, when mitosis occurs without accompanying cytokinesis. The free nuclei organize into nuclear cytoplasmic domains. **b** Rearrangement of microtubules and elongation of anticlinal walls. **c** Continuing anticlinal wall growth associated with adventitious phragmoplasts formed at interfaces of opposing microtubule systems. **d** Mitosis followed by typical cytokinesis (*left* to *right*), giving rise to the peripheral aleurone layer of cells (*solid nuclei*) and inner starchy cells (*stippled nuclei*). (Reprinted from Olsen et al. 1995)

Investigations using fluorescence immunohistochemistry of endosperm development in rice and wheat (Brown et al. 1996a, 1996b, 1997; Tian et al. 1998) among cereals, and in *Brassica napus* (van Lammeren et al. 1996), *Arabidopsis* (Brown et al. 1999), and *Coronopus didymus* (Brassicaceae; Nguyen et al. 2001, 2002) among eudicots, have confirmed that, with minor variations in detail, the overall pattern of endosperm development, including the absence of normal interzonal phragmoplasts during inception of the anticlinal walls, is similar to that described in barley. In *C. didymus*, both microtubules and F-actin are coaligned in preparation for cellularization, but the precise role of actin in the process is unclear (Nguyen et al. 2001, 2002).

As part of the seed, the mature endosperm may vary in appearance from a formless mass of vacuolate cells to a compact tissue filled with diverse storage materials. In *Arabidopsis* and some other members of the Brassicaceae, specialization of the mature endosperm results in the formation of a cyst in the chalazal region. Although the cyst is initiated early in development of the endosperm, it remains essentially distinct from the cells of the endosperm that begin to accumulate storage products. A comparative study of cysts in developing seeds of members of the Brassicaceae has revealed that cysts perform the remarkable feat of compressing the nuclear-cytoplasmic domains of several chalazal endosperm cells into a stratified mass comprising an apical zone with a concentration of plastids, nuclei, and mitochondria, a middle zone enriched in endomembranes, and a basal zone constituted of labyrinthine wall projections with associated mitochondria (Brown et al. 2004).

There are numerous reports of the progression of cellular changes during endosperm development in maize (Kowles and Phillips 1988). Following the success obtained in fusing isolated eggs and sperm of maize in vitro, Kranz et al. (1998) have monitored the development of the endosperm produced in vitro by fusing a sperm nucleus with the polar fusion nucleus of the central cell of maize. The surprising finding from this study was that the in vitro crafted primary endosperm nucleus had no surprises of its own as it developed in a predictable way, passing from a coenocytic to a cellular endosperm in a manner very similar to that of the in vivo formed tissue. Nonetheless, use of this system in future studies will provide important insights into the early cellular and molecular events of sperm-polar fusion nucleus interaction.

7.1.2
Development of the Endosperm in *Arabidopsis*

The endosperm of *Arabidopsis* has been at the forefront of new observations providing basic information on the cytoskeletal dynamics around free nuclei in preparation for cellularization. The early process of cellularization has been resolved in great detail using high-pressure freezing fixation and electron microscopy to show that, after the nuclear-cytoplasmic domains are established, small groups of overlapping microtubules that radiate from neighboring nuclei initially assemble into mini-phragmoplasts. The mini-phragmoplasts are put together in a patchwork way to generate a novel kind of cell plate – the syncytial-type cell plate (Otegui and Staehelin 2000a, 2000b). A reconstruction of high-pressure frozen/freeze-substituted sample of the developing endosperm aided by high voltage electron tomography (an image technology for obtaining three-dimensional information by electron microscopy) has indicated the participation of Golgi-derived vesicles transported along the phragmoplast microtubules, probably mediated by kinesin-like motor proteins, in the formation of the syncytial-type cell plate (Otegui et al. 2001). By examination of whole mounts of ovules to gain a global view of progressive development of living endosperm, Boisnard-Lorig et al. (2001) showed that the syncytial endosperm up to the stage of cellularization can be divided into nine substages, each defined by the total number of nuclei produced. According to Brown et al. (2003), a theme underlying the division of endosperm nuclei in *Arabidopsis* is that the precise patterns of cytoskeletal behavior define the early development of the syncytial endosperm. Whereas for the first two divisions of the primary endosperm nucleus the mitotic spindles are oriented parallel to the long axis of the central cell, there is a change in the orientation of spindles to being mostly oblique at the four-nucleate stage. At the fourth division, another change in the pattern of division from synchronous to successive occurs, with the mitotic wave progressing from the micropylar to the chalazal region. It is also now established that, in *Arabidopsis*, far from being a homogeneous mass of free nuclei, differences in the frequency of divisions, nuclear shape, cytoskeletal arrays, and cytoplasmic features mark the separation of the syncytium into micropylar, central/peripheral, and chalazal developmental domains (Brown et al. 1999, 2003; Boisard-Lorig et al. 2001). Of fundamental importance to the differentiation of these domains of the endosperm is the presence of a system of parallel microtubules around nuclei in the micropylar chamber, radial microtubules around nuclei in the central chamber, and a reticulum of microtubules and actin filaments around nuclei in the chalazal chamber (Brown et al. 2003). Development of the chalazal domain is also characterized by a diverse array of structural and cytological changes, such as nuclear fusions resulting in giant nuclei, formation of polyploid nuclei and stacks of ER, accumulation of Golgi bodies, plastids, mitochondria and vesicles, and the formation of a callose wall (Boisard-Lorig et al. 2001; Otegui et al. 2002; Baroux et al. 2004). Other striking features of the chalazal domain are the oriented migration of nuclei of the syncytium, organization of individual nodules formed by fusion of nuclear-cytoplasmic domains, formation of a large multinucleate cyst that incorporates the fused nuclear-cytoplasmic domains, and finally, the differentiation of the cyst into a dome-shaped apical region and a basal haustorium, the branches of which penetrate into the chalazal region of the ovule (Nguyen et al. 2000; Sørensen et al. 2001; Guitton et al. 2004). Although most of the endosperm is used up by the growing embryo, the single peripheral layer that persists in the mature seed has been considered equivalent in location and ontogeny to the aleurone layer of cereal grains (Brown et al. 1999). As will be discussed later in this chapter, the structural versatility of the chalazal endosperm chamber translates into a potential for functional versatility for embryo nutrition, acting as a conduit for channeling maternal nutrients into the developing seed. This account of endosperm development in *Arabidopsis* is probably an oversimplification of what must be a complex network of developmental controls acting on the products of the first division of the primary endosperm nucleus, and which continue beyond into the cellularization stage.

Two approaches involving molecular markers for cell types have been used to identify stages of endosperm development as well as endosperm compartments in developing ovules of *Arabidopsis*. One is in situ hybridization, which has shown that, of the several MADS-box *AGL* genes tested, only the *AGL18* gene is specifically expressed in the endosperm. Correlating with the structural differences between the micropylar and chalazal chambers of the endosperm, gene transcripts are found associated with the nodules of the chalazal endosperm (Alvarez-Buylla et al. 2000). In a second approach, β-glucuronidase (GUS)-marker and GFP-marker lines have been generated by artificially tethering the reporter gene to different gene promoters, or by screening promoter trap lines displaying GUS or GFP expression. Genes such as *MEA* (*FIS1*), *FIS2*, and *FIE* (*FIS3*), which repress endosperm development in the absence of pollination, were the first whose promoters were coupled to the *GUS* gene to follow gene activity patterns in the developing endosperm (see Plate 12, Fig. a–f). By their association with the polar nuclei before fertlization, and with the primary endosperm nucleus and free endosperm nuclei after fertilization, these genes are considered to represent specific markers of early nuclear endosperm development in *Arabidopsis*. Following cellularization, gene activity is restricted to the chalazal chamber (Luo et al. 2000). Fluctuations in GFP expression in the endosperm of a transgenic line created by random insertion of a T-DNA::mGFP5 construct has revealed that, after uniform fluorescence in all parts of the endosperm early in development, fluorescence gradually diminished in the micropylar and central domains, persisting strongly only in the cyst of the chalazal domain (Sørensen et al. 2001). Included in a new set of 16 GUS-expressing promoter trap lines are markers for the chalazal and micropylar endosperm compartments, although gene expression using these markers is also detected in the diploid tissues of the integuments and embryo (Stangeland et al. 2003).

Altogether, these findings provide compelling evidence to show that much remains to be done to obtain a complete cytological picture of endosperm development. With the techniques currently available, the prognosis looks good for laying a sound foundation to study the mechanisms that control the establishment and maintenance of polarity in the central cell for the programmed migration of the endosperm nuclei, the prevention of phragmoplast formation that unleashes free nuclear divisions of the primary endosperm nucleus, and the initiation of periclinal mitotic divisions in the alveoli.

7.2 Biochemical Organization of the Endosperm

Two intriguing aspects of the biochemical organization of the endosperm as a distinct tissue are the ability of the cell nuclei to undergo DNA amplification, and the propensity of the cytoplasm to accumulate an acervate complex of storage products. Amplification of DNA is functionally diverse as revealed by its occurrence not only in the cells of the endosperm proper, but also in the haustoria that appear from its chalazal or micropylar ends (D'Amato 1984). Reserve carbohydrates and storage proteins, which constitute by far the major storage products of the endosperm, accumulate for the sole purpose of providing carbon and nitrogen sources, respectively, for the embryo during seed germination. Although storage proteins typically have a high amide content and are occasionally rich in sulfur-containing amino acids, many other proteins, such as protease inhibitors, α-amylase inhibitors, lectins, thionins, ribosome-inactivating proteins, and certain enzymes, also pile up in the endosperm cells of diverse plants (Lopes and Larkins 1993).

7.2.1 DNA Amplification

The obvious connection between ploidy level of the primary endosperm nucleus, ranging from diploid in the Oenothera type of embryo sac, triploid in the Polygonum, Allium, Drusa, and Adoxa types, pentaploid in the Fritillaria, Penaea, Plumbago, and Plumbagella types, and $9n$ to $15n$ in the Peperomia type, has engendered the notion that, with the exception of the small number of plants with the Oenothera type of embryo sac, in the majority of flowering plants, the presence of nuclei with more than the diploid number of chromosomes is a way of life for the endosperm generated following double fertilization (Maheshwari 1950; D'Amato 1984). It is now known that, besides this natural diversity in chromosome number, endosperm nuclei exhibit a capacity to increase in size due to endoreduplication or polyteny (Fig. 7.3a,b). The contemporary view tends to be that endoreduplication allows for amplification of nuclear DNA and occurs through endonuclear chromosome duplication without accompanying mitosis. Although this process, in which repeated rounds of DNA synthesis occurring in an intact nucleus, leads to the production of chromatids, the chromosome number itself remains unchanged. Polytene chromosomes presumably arise by endoreduplication, which results in the presence of many parallel fibrils in the chromatids (D'Amato 1984; Brachet 1985).

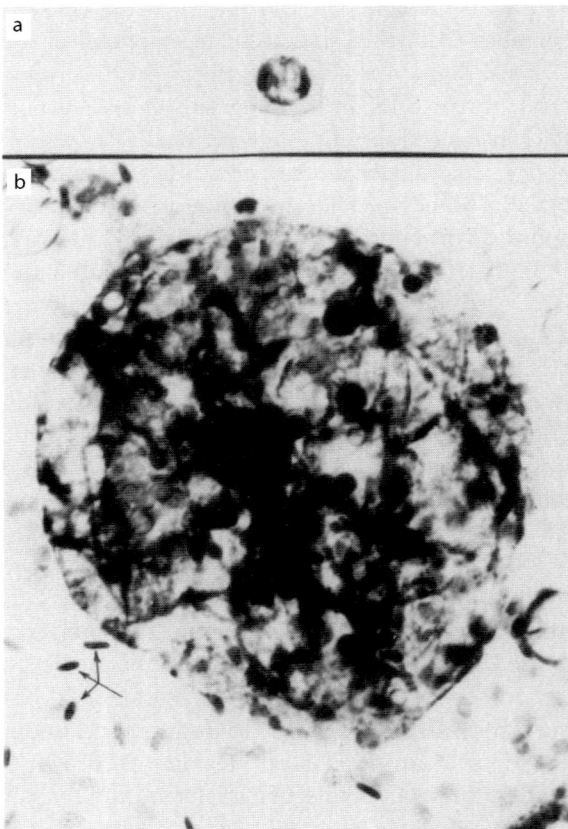

Fig. 7.3a,b Increase in size of the endosperm nucleus of maize due to endoreduplication. A maize root tip cell in **a** is compared with an endosperm nucleus in **b** (22 days after pollination), against a background of chicken erythrocyte nuclei (*arrows*). (Reprinted from Kowles and Phillips 1988)

Given the prolonged cellular phase of cereal endosperms when progressive changes in DNA content occur, it is perhaps not surprising that endoreduplication has been quantified by monitoring changes in DNA levels of cereal endosperm nuclei during development. Although these studies show a typical pattern of a dramatic surge in nuclear DNA content following cessation of mitotic activity, there is also considerable heterogeneity in the maximum DNA content attained, ranging from 10C in barley (Giese 1992), 13C in maize (Kowles et al. 1990), 24C in *Triticale*, rye (*Secale cereale*; Poaceae), and wheat (*Triticum aestivum* and *T. durum*; Herz and Brunori 1985; Chojecki et al. 1986; Brunori et al. 1989), and 42C in rice (Ramachandran and Raghavan 1989). Research on the mechanism of endoreduplication of the endosperm genome has been dominated by the work done on maize, in which DNA contents of cells of some genotypes reach maximum levels of 384C and 690C (Kowles and Phillips 1988; Cavallini et al. 1995; Larkins et al. 2001). One of the key questions concerning endoreduplication is how events leading to the induction of DNA synthesis are coordinated with those resulting in the arrest of mitosis. Undoubtedly, an answer to this question will be very much linked to regulation of the cell cycle, nudging the dividing cells from one phase to another by the activity of ubiquitous protein kinases. As elucidated in *Drosophila* and in single-celled eukaryotes such as yeast, a catalytic subunit called CDK and a regulatory subunit, cyclin, are components of protein kinases. An additional level of complexity in the functioning of CDKs is attributed to the presence of a network of several distinct cyclins, each binding with a CDK to form an active protein kinase. The temporal activity of CDKs has received much attention as it is driven by the type of associated cyclin, as well as by multiple reactions that affect CDK phosphorylation and their association with inhibitors (Morgan 1997; Grafi 1998). The direct involvement of a CDK in the regulation of endoreduplication was demonstrated by the reduction of endoreduplication in endosperm cells of transgenic maize following ecotpic expression of a gene encoding a dominant negative form of the CDK. This latter work showed that, whereas overexpression of the wild-type CDK did not affect endoreduplication, the defective enzyme lowered kinase activity and significantly reduced the DNA content of endosperm nuclei (Leiva-Neto et al. 2004).

In an earlier study, Grafi and Larkins (1995) demonstrated that two closely related protein kinases, similar in function to protein kinases controlling mammalian cell cycle regulation, are involved in the orderly progression of maize endosperm cells through repeated cycles of DNA synthesis without accompanying mitosis. This work showed that DNA synthesis during early stages of endosperm development is maintained by an increase in the amount and activity of S-phase-related protein kinases, and that the endoreduplicated cells contain a factor that suppresses the activity of the M-phase promoting factor. Endoreduplication is thus thought to proceed via induction of S-phase-related protein kinases and concomitant inhibition of the M-phase promoting factor. The fact that the level and phosphorylation state of a retinoblastoma-related protein named pocket protein isolated from maize endosperm cells changes in association with the onset of endoreduplication has provided strong evidence for a role for the pocket family of proteins in this process (Grafi et al. 1996). A new mitotic cyclin belonging to the subgroup *Zeama*, *CycB1*, designated as *CycZme1*, has been identified in maize endosperm cells; a down-regulation of this cyclin accompanying the onset of endoreduplication has suggested its lack of involvement, or only peripheral involvement, perhaps in association with an inhibitor of CDK activity, in endoreduplication (Sun et al. 1999b). A protein kinase that has emerged as a controlling factor in the endoreduplication of maize endosperm cells is Wee1. This kinase was originally identified in *Schizosaccharomyces pombe* by virtue of its effects in delaying mitosis by phosphorylating the M-phase-promoting factor. In keeping with its function in other systems, a maize Wee1 homolog (ZmWee1), which accumulates during endoreduplication, has been invoked to inhibit CDK by phosphorylation (Sun et al. 1999a). Thus, there are likely to be multiple regulatory pathways involving different protein kinases and cyclins that enable endosperm cells to bypass mitosis and continue unabated DNA synthesis.

Several additional observations have provided some remarkable insights into the functional and biological significance of DNA amplification in the endosperm. Different approaches used to estimate the levels of endoreduplication in endosperm nuclei of inbred lines of maize and their reciprocal crosses have proved effective in drawing attention to the existence of maternal genetic control of endoreduplication (Cavallini et al. 1995; Kowles et al. 1997;

Dilkes et al. 2002). In another approach, Leblanc et al. (2002) have shown that in interploidy crosses in maize, tipping the parental genome ratio in the endosperm toward maternal excess stimulates mitotic arrest and endoreduplication, whereas paternal excess delays the DNA amplification process. The local hormonal environment in the endosperm seems to impact endoreduplication, as the increase in nuclear DNA content coincides with the IAA content of the endosperm; exogenous application of 2,4-D also increases the DNA content of the endosperm (Lur and Setter 1993). Zhao and Grafi (2000) have shown that the shift of maize endosperm cells from mitotic mode to endoreduplication mode is associated with a reduction in the concentration of the consummate transcription-repressing histone H1 protein, and a concomitant increase in the transcription-activating high mobility group (HMG) protein. This observation presumes that by increasing the number of DNA templates, endoreduplication increases the transcriptional activity of cells. This was tested by determining the efficiency of the purified maize endosperm HMG protein to bind to the promoter sequence of a gene encoding the maize storage protein zein, and, as predicted, the binding activity became high at stages corresponding to intense endoreduplication. It is well-known that the increase in kernel volume and mass that occurs in cereal grains due to the synthesis and accumulation of starch and storage proteins in the endosperm is temporally correlated with endoreduplication, implying that endoreduplication might assist in the rapid synthesis of these storage products during grain development.

7.2.2
Accumulation of Storage Products

The economic importance of the endosperm, especially that of cereal grains, which provide a major source of starch and proteins for the human population and for domesticated animals, is paramount. As reliable transformation systems are now available for the 'big three' cereals, namely wheat (Vasil and Vasil 1999), rice (Datta 1999), and corn (Gordon-Kamm et al. 1999), an understanding of the nature of the storage products in the endosperm of these grains, and the cytological mechanism by which their synthesis and accumulation are regulated, will be of great value in the genetic engineering of crops with improved nutritional qualities. In considering gene action in the synthesis of starch, most analyses have been performed with spontaneously occurring mutants, such as *shrunken* (*sh*), *waxy* (*wx*), *brittle* (*bt*), and *sugary* (*su*), that affect the quality and quantity of this carbohydrate in maize endosperm. The legitimacy of the names given to these common mutants is underscored by the phenotypic or chemical nature of the endosperm, which is shrunken and opaque due to a high sugar and reduced starch content in *sh*, is opaque in *wx* and consists exclusively of amylopectin instead of a mixture of amylose and amylopectin as in the wild-type endosperm, contains extremely reduced amounts of starch and high sugar in *bt*, and is wrinkled and glassy with a high sugar content in *su* (Creech 1965). As in maize, *wx* endosperm of rice lacks amylose and contains primarily amylopectin (Sano 1984). The biochemical lesions caused by these mutations have been considered elsewhere (Raghavan 1997).

In addition to albumins and globulins, which also constitute the storage proteins of the embryo, the main storage proteins of the endosperm are prolamins (soluble in alcohol) and glutelins (soluble in dilute acid or alkali). The three major cereals are identified by the type of storage proteins that accumulate in their endosperm: prolamins in maize, prolamins and glutelins in wheat, and glutelins, prolamins, and globulins in rice. In most flowering plants, synthesis of storage proteins and their accumulation in protein bodies are the most crucial steps associated with the cellular phase of endosperm development. Electron microscopy has revealed that storage proteins of cereal grains are deposited in different cellular compartments. The prolamin type of maize endosperm storage proteins known as zeins are currently classified into four structurally distinct types, namely, α-zein (19 and 22 kDa), β-zein (15 kDa), γ-zein (28 kDa), and δ-zein (10 kDa). Mobilization of zeins in the endosperm commences when cells are in the division mode and continues until grain maturity. The protein bodies, which contain aggregates of zein, become evident as membrane-enclosed spherical deposits, and accumulate in small vesicles produced by localized dilations of the ER (Khoo and Wolf 1970). In support of the role of ER in zein synthesis, it has been found that the rough ER directs the in vitro synthesis of proteins corresponding in molecular mass to in vivo synthesized zeins (Larkins and Hurkman 1978).

Questions have been raised as to whether the spatial organization of zeins within the protein body is due to specific interactions between individual zeins, or to preferential targeting of mRNAs of some zeins to specific regions of the rough ER. From an analysis of the distribution of mRNAs encoding the 22-kDa α-zein and the 28 kDa γ-zein on cisternal and rough ER membranes, Kim et al. (2002) have concluded that transcripts of both zeins are distributed more or less randomly on both regions of the ER. This finding implies that interactions between different zeins, rather than sorting of their respective mRNAs, might account for protein body assembly in the maize endosperm. Rice endosperm contains two types of storage proteins, prolamins known as oryzins and globulin-like glutelins, each of which is synthesized on a different configuration of the rough ER. Oryzins are synthesized and retained in the lumen of the ER, where they form protein bodies, whereas glutelins, synthesized in the cisternal ER, are transported to the vacuoles through the Golgi complex to form protein bodies. The targeting of these proteins into distinct protein bodies is also associated with an asymmetric distribution of their respective mRNAs in specific ER membranes (Li et al. 1993). The current view of the nature of the signal responsible for sorting of the oryzin transcripts to a specific ER subdomain is that it resides in the 3'-noncoding sequences of the mRNA (Choi et al. 2000).

Unlike in maize and rice, the prolamins of wheat are deposited in protein bodies inside vacuoles. Although the proteins are synthesized on the rough ER, the exact mode of their transport to the vacuole has been a topic of controversy. Apparently, transport of proteins to vacuoles via the Golgi does not hold true for wheat prolamins; one view is that the protein body becomes engulfed in a vacuole formed by the fusion of small vesicles attached to its surface by a process essentially analogous to autophagy (Levanony et al. 1992). On the other hand, based on immunoelectron microscopy and cell fractionation, details of the appearance of 7S lectins and 2S albumins, two of the major storage proteins of castor bean endosperm, in the storage vacuoles seem to fit with the concept of trafficking of precursors of these proteins through the Golgi apparatus to reach their site of storage (Jolliffe et al. 2004).

The structure of the endosperm storage protein genes of several cereals is progressively being unraveled, and some have now been successfully expressed in transgenic systems (Raghavan 1997).

7.2.3
Programmed Cell Death of the Endosperm

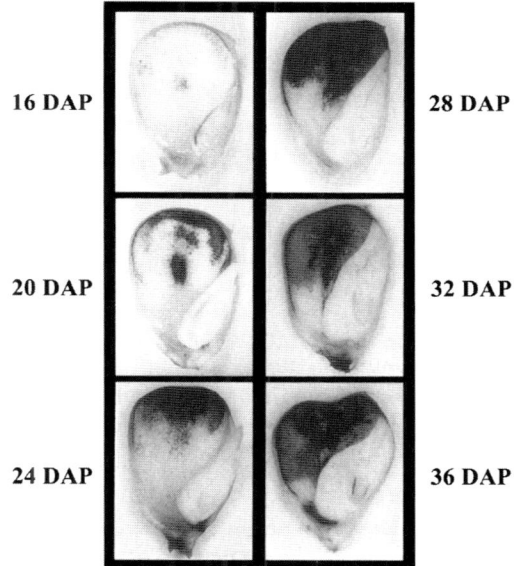

Fig. 7.4 Progression of programmed cell death (pcd) in the endosperm of maize as indicated by Evans blue stain exclusion. Dark staining areas indicate dead cells. *DAP* Days after pollination. (Reprinted from Young and Gallie 2000)

The starchy endosperm of cereal grains has a limited life-span as it is fated to die prior to maturation of the grain. In contrast, in mature grains the aleurone layer is considered to constitute the living cells of the endosperm, remaining alive through grain maturity (Esau 1965). During their ontogeny, cells of a plant tissue undergo either pcd, involving molecular components of an active, gene-dependent process characterized by changes in nuclear morphology, internucleosomal cleavage of nuclear DNA, and activation of nucleases and proteases, or localized cell death (necrosis), generally resulting from injury, and characterized by irregular clumping of chromatin, random decay of DNA, mitochondrial swelling, dissolution of ribosomes, and membrane rupture (Kerr et al. 1995). Starchy endosperm cells of cereal grains generally undergo pcd. Viability staining of developing maize kernels by a dye-exclusion method showed that pcd is initiated by cells within the central endosperm early during grain development, followed soon after by a basipetal wave beginning at the top of the kernel that engulfs the entire starchy endosperm cells late in development (Fig. 7.4). However, initiation of pcd in wheat endosperm is not associated with a specific locus but is a random process that gradually consumes all the starch-containing cells. The suicidal tendencies of the endosperm cells of maize and wheat have key similarities, such as the orderly internucleosomal degradation of the genome, the presence of high levels of nuclease activities, and induction by ethylene (Young et al. 1997; Young and Gallie 1999, 2000). Interestingly, even after pcd is initiated, endosperm cells continue to accumulate storage products, which are hydrolyzed during germination. In eudicots such as *Vicia faba*, pcd is initiated in the endosperm cells almost simultaneously with the same process in the suspensor cells (Wredle et al. 2001).

The degradation of endosperm during the germination of seeds and grains also exhibits hallmarks of pcd. Data with a high degree of resolution have been obtained showing that germination of barley grains is accompanied by DNA fragmentation and other cellular changes in the aleurone cells that mimic apoptotic cell death in animals (Wang et al. 1996; Bethke et al. 1999). DNA fragmentation of cells of castor bean endosperm is coincident with the development of small organelles known as ricinosomes. By its accumulation of cysteine endoprotease, the ricinosome has been recognized as a key player in pcd of the endosperm, as the mature cysteine endopeptidase necessary for the degradation of cytoplasmic components is released from this organelle during cellular disintegration of the endosperm (Schmid et al. 1999; Than et al. 2004). To allow for the possibility that pcd in endosperm cells might be explained in terms of a gene-regulated process, identification of genes promoting this process is necessary.

7.3
Role of the Endosperm in Embryo Nutrition

Historically, the endosperm has been assigned the twin functions of nurturing the embryo during its heterotrophic phase of growth, and accumulating combustible sources of energy for sustaining embryo-to-seedling growth during seed germination. Much of the current rudimentary understanding of the role of the endosperm as a nurse tissue for the growing embryo is based on a variety of structural, physiological, and genetic observations, which are considered below under three rubrics.

7.3.1
Structural Modifications of the Endosperm

As mentioned earlier, following double fertilization the primary endosperm nucleus gets a head-start over the zygote in initiating mitotic divisions. Labyrinthine wall projections that arise from the inner wall of the embryo sac and grow into the endosperm, as described in pea (Marinos 1970), *Helianthus annuus* (Newcomb and Steeves 1971), *Stellaria media* (Newcomb and Fowke 1973), cotton (Schulz and Jensen 1977), *Haemanthus katherinae* (Newcomb 1978), *Medicago sativa* (Sangduen et al. 1983), *Vigna sinensis* (Fabaceae; Hu et al. 1983), and *Vicia faba* (Johansson and Walles 1994), probably suggest an indirect role for the endosperm in facilitating cell-to-cell transfer of metabolites from the ovular tissues for embryo nutrition. Structurally, the wall projections from embryo sacs of different species look much alike by electron microscopy; the amplified plasma membrane lining them raises the profile of wall projections as it confers transfer cell functions on them involving active transport. In the same vein, the presence of wall ingrowths in maize endosperm, specifically, close to the basal cells of early-stage embryos, in the placentochalazal region of late-stage embryos (Schel et al. 1984; Davis et al. 1990; Charlton et al. 1995), and in the outer periclinal walls of castor bean endosperm intruding into the crushed nucellar cells (Greenwood et al. 2005), might be compatible with a role for the endosperm in the absorption and transport of nutrients to the growing embryo. As shown in Fig. 7.5, transfer cells in the placentochalazal region of the maize endosperm are composed of extensively anastomosing wall proliferations made up of an enormous quantity of cell wall materials and associated plasma membrane. Similar anatomical modifications of cell walls concerned with moving nutrients to the endosperm are found in the cells of the aleurone layer adjacent to the placental bundle in *Setaria lutescens* (Poaceae; Rost and Lersten 1970), aleurone cells facing the placental sap in *Sorghum bicolor* (Poaceae; Maness and McBee 1986), crease aleurone cells and the bordering nucellus in barley (Cochrane and Duffus 1980), and aleurone cells sandwiched between the basal starchy endosperm and the nucellus in *Echinochloa utilis* (Poaceae; Zee and O'Brien 1971). An important caveat is that, while it is reasonable to conclude that nutrients that accumulate in cereal endosperms are used in the synthesis of storage products, it has not been determined whether they are also used for embryo nutrition.

The possible function of the endosperm as a conduit for metabolites from ovular tissues is supported by numerous reports of highly specialized modifications of the endosperm into outgrowths known as haustoria, which penetrate the tissues of the ovule. Whole mount preparations of the entire endosperm have shown that haustoria are notorious for their structural complexity. They arise at either the micropylar or the chalazal end, or at both ends of the developing endosperm, and appear as unbranched or branched tubular processes that

Fig. 7.5 Electron micrograph of transfer cells of the placento-chalazal region of the maize endosperm, showing accumulation of cell wall material. *MT* Crushed maternal cells, *N* irregularly shaped nucleus, *R* ribbon of wall material, *TC* transfer cell. Bar 7 µm. (Reprinted from Davis et al. 1990)

meander through adjacent parts of the ovule, such as integuments, nucellus, and chalazal tissues. The morphological nature of endosperm haustoria has been periodically updated with descriptive accounts of additions to the list (Maheshwari 1950; Raghavan 1976; Vijayaraghavan and Prabhakar 1984), and these references should be consulted to gain a sense of the complexity of these structures at the light microscopic level. Study of the haustoria at the electron microscopic level is in its infancy and is restricted to observations made on *Glycine max* (Dute and Peterson 1992) and *Rhinanthus minor* (Scrophulariaceae; Nagl 1992). The ultrastructural characteristic common to chalazal endosperm haustoria in both species is the presence of pronounced wall ingrowths into the cytoplasm, the signature feature of cells that play a role in transferring metabolites. The concentration of a large number of mitochondria in the vicinity of the wall projections in *R. minor* conjures up the suggestion of active transport through the plasma membrane (Nagl 1992). The chalazal endosperm chamber in the Brassicaceae is modified at least in two different ways to function as an incipient haustorium. As mentioned earlier, in *Arabidopsis*, the globular cyst at the chalazal chamber has branching root-like basal processes; *Lepidium virginicum*, another member of the Brassicaceae, has an elongate stalk-like basal part (Nguyen et al. 2000; Brown et al. 2004). An important motivation for deciphering the function of the chalazal endosperm projections and the chalazal endosperm cysts of the Brassicaceae is their similarity to the endosperm haustoria in *G. max* and *R. minor*. As in the chalazal haustorium of *R. minor*, the wall ingrowths found in the chalazal endosperm processes in *Arabidopsis* are closely associated with numerous mitochondria. The functioning of the chalazal endosperm chamber in members of the Brassicaceae probably involves an interaction with the metabolites of the cells of the nucellus, furthering the notion that the wall invaginations can best be viewed as specialized adaptations for the uptake and processing of these metabolites and their release into the central cell. Particularly informative in this context is an analysis of the one-way trafficking of soluble phytin salts between the developing endosperm and embryo of *Arabidopsis*, using high-pressure freeze substitution techniques (which preserve water-soluble salts) and biochemical methods. This work identified the ER and vacuolar subcompartments of the chalazal endosperm as sites for transient storage of Mn-enriched and Zn-enriched phytic acid salts, respectively, before their mobilization to the growing embryo. Structural and biochemical observations support the view that the Mn-salt is transferred from the endosperm to bent-cotyledon stage embryos, and that the Zn-salt is transferred to globular-stage embryos (Otegui et al. 2002). While it possible that embryos might use Mn-salts for the synthesis of metalloproteins, how Zn-salts are used in embryo metabolism is not clear. Although this work provides a detailed outline of how heavy metal salts might be targeted to the embryo, it appears that additional unknown processes are at work in facilitating this one-way traffic *in planta*.

Based on the premise that the endosperm represents a transient sink for compounds transported from other parts of the plant for embryo nutrition, Hirner et al. (1998) have shown that in ovules of *Arabidopsis*, the amino acid transporter gene *AAP1* (*AMINO ACID PERMEASE1*) is first localized in the endosperm and later in the embryo. Although the amino acids transported may be used later for the synthesis of storage proteins by the embryo, the endosperm expression of the *AAP1* gene, followed by its embryo expression during early stages of embryogenesis is considered as the first molecular evidence of nutritional interaction between the endosperm and embryo.

7.3.2
Physiological Considerations

Any approach to an understanding of the role of the endosperm in embryo nutrition requires some knowledge of the chemical composition of the tissue. The quantity of endosperm, especially the nuclear endosperm, that would be available for any meaningful biochemical analysis is extremely limited in ovules of most plants. An exception is the liquid endosperm of coconut, which has a long and successful tradition of use as an adjuvant in plant tissue culture media, in the form of coconut milk or coconut water. It was noted in Chap. 2 that the first successful culture of proembryos was accomplished by supplementing the mineral salt-sucrose medium with coconut water, and that this experiment has served as a paradigm for subsequent embryo culture investigations. Chemical analysis of coconut water has led to an understanding of some of the critical substances involved in stimulating growth of cultured embryos and plant organs, as well as tissues in general. As tabulated elsewhere (Raghavan 1976), a variety of inorganic ions (Cl, Cu, Fe, K, Na, Mg, P, and S), amino acids and related compounds (alanine, γ-aminobutyric acid, arginine, asparagine, glutamine, aspartic acid, cystine, glutamic acid, glycine, histidine, homoserine, hydroxyproline, isoleucine, leucine, lysine, methionine, ornithine, phenylalanine, pipecolic acid, proline, serine, threonine, tryptophan, tyrosine, and valine), organic acids (citric acid, malic acid, pyrrolidine carboxylic acid, quinic acid, and shikimic acid), vitamins (biotin, folic acid, nicotinic acid, pantothenic acid, pyridoxine, riboflavin, and thiamine), plant hormones (auxins, gibberellins, and cytokinins), sugars (sucrose, glucose, fructose, and mannitol), and sugar alcohols (*myo*-inositol, *scyllo*-inositol, and sorbitol) form part of the essential ingredients of coconut water. Although no single compound or group of compounds have been shown to duplicate the growth-promoting activity of coconut water in tissue culture media, it is the general belief that plant hormones, either alone or in combination with sugars, account for the growth-promoting effects of coconut water in plant tissue cultures.

While the use of coconut water in plant tissue culture media grew in popularity throughout the 1960s, information on the chemical composition of endosperms of other plants has not kept pace with the expanded use of coconut water. Among the limited number of endosperm tissues analyzed chemically, maize endosperm at the milky stage contains several amino acids and plant hormones, including the cytokinin zeatin. Endosperms of other cereals, such as barley, rice, rye, and wheat, are known to contain auxins. A potent source of endosperm as a growth promoter in cultured plant tissues is the liquid content of the vesicular embryo sac of *Aesculus woerlitzensis* (horse chestnut; Hippocastanaceae), from which chlorogenic acid, *myo*-inositol, and auxin have been identified. A fruitful interaction of chemical analyses of the endosperm and embryo culture has led to the notion that the in vivo function of the endosperms of *Datura stramonium*, *D. tatula*, *Juglans regia*, *Allanblackia parviflora* [Guttiferae (Clusiaceae)], and *Cucumis sativus* (Cucurbitaceae) is concerned with the supply of nutrients to the growing embryo. Consistent with this idea, it is conceivable that detection of substances with gibberellin-like properties by bioassays in endosperms of several plants might reflect the utilization of the hormone by the growing embryo (Raghavan 1976).

Sugar is a ubiquitous component of the endosperm, and addition of sucrose at an appropriate concentration to the tissue culture medium is now seen as a rational approach to provide a carbon energy source to sustain continued growth of the explanted organ or tissue. However, the close correlation found between changes in the sugar concentration and changes in the osmotic value of the ovular sap of certain plants makes it plausible that the high concentration of sugars found in the amorphous liquid endosperm in which young embryos are constantly bathed may serve as an osmoticum exercising extracellular control over embryo growth, rather than as a carbon energy source (Ryczkowski 1962).

7.3.3
Genetic Considerations

Some very useful information about the role of the endosperm in embryo nutrition has been gathered from standard plant breeding practices followed in agricultural and horticultural research institutions throughout the world to generate new varieties of crop plants. It is well-known that crosses between unrelated species and genera ("wide crosses") of plants can lead to failure of seed set. The prevailing view is that, in attempts to introduce beneficial foreign genes across interspecific and intergeneric barriers, some deleterious genes that interfere with the growth of the embryo and endosperm in the ovule are also introduced. Although double fertilization occurs normally in wide crosses, embryo lethality and endosperm failure soon follow, leading to the collapse of seeds. In developmental terms, Renner (1914) first showed that, in reciprocal crosses between *Oenothera biennis* and *O. muricata* (Onagraceae), and between *O. biennis* and *O. lamarckiana*, the embryo did not proceed beyond a few-celled stage. Failure of embryo growth is preceded by the disintegration of the endosperm beginning soon after fertilization, thus depriving the embryo of access to nutrients. Embryo and endosperm development progressed further in some hybrid ovules but survival was generally erratic, with most ovules enclosing underdeveloped embryo and endosperm maturing into shrunken aborted seeds. As reviewed previously (Raghavan 1977), these basic details of embryo and endosperm development have been confirmed in a number of nonviable crosses. The apparent simplicity of these observations, and the order in which they occur, i.e., endosperm disintegration immediately precedes embryo lethality, suggest that hybrid embryos may starve themselves to death due to the inability of the physiologically disturbed endosperm to supply the exacting nutrients for embryo growth or to absorb them from the surrounding maternal tissues and transmit them to the embryo. Despite the exquisite details given of the anatomical changes accompanying endosperm failure in wide crosses, it is impossible to identify the specific nutrient factors unavailable from the endosperm that cause the failure of embryo growth. In one productive approach, the idea that reduced cytokinin biosynthesis in the endosperm is responsible for abortion of the embryo was borne out by a comparative study of the levels of compounds with cytokinin activity in ovules of selfed *Phaseolus vulgaris* and a *P. vulgaris* × *P. acutifolius* hybrid. Whereas the concentration of cytokinin-like compounds in the endosperm of selfed ovules is high, and is closely correlated with periods of cell division activity in the embryo, the level of the hormone in the hybrid ovule is greatly reduced (Nesling and Morris 1979). This, in turn, might suggest a link between embryo abortion and the failure of the endosperm to supply cytokinins to the embryo.

In some nonviable hybrids, endosperm failure might take place in discrete ways that require the participation of ovular tissues such as the nucellus. Anatomical dissection of ovules from crosses between *Nicotiana rustica* × *N. tabacum* has provided a mechanistic model of how retarded growth of the endosperm causes embryo abortion. Here, embryo failure is associated with a failure of differentiation of vascular tissues that supply nutrients to the endosperm. Because of this, competition between the endosperm and the nucellus to accumulate nutrients is tipped in favor of the latter. For this model to be sensible, the consequent abnormal distribution of nutrients in the ovule, especially the accumulation of nutrients in the nucellus, is believed to cause its proliferative growth and embryo abortion (Brink and Cooper 1941). Monitoring the progress of the primary endosperm nucleus in reciprocal crosses involving diploid and tetraploid races of *Lycopersicon pimpinellifolium*, and in crosses of these races with *L. peruvianum*, has complemented experimental analysis of the eventual fate of the embryo and endosperm in hybrid ovules. A common theme that has emerged is that, after an initial period of almost identical increases in endosperm cell number in both successful and abortive crosses, endosperm growth in the unsuccessful crosses is retarded. The weak growth of the endosperm in nonviable hybrids is also accompanied by abnormal cytological changes leading to cellular disintegration. In the abortive ovules generated in one cross, as early as 144 h after pollination, cells of the endosperm at the chalazal end become vacuolate and less dense than those at the micropylar end. Although the endosperm survives for some time, it undergoes no further divisions. Instead, as a result of cell wall dissolution and fusion of the protoplasm and of nuclei with those of contiguous cells, the endosperm is reduced to a few giant cells surrounding the embryo. In some abortive ovules, the innermost layer of the integument, known as the endothelium, surrounding the endosperm thickens and becomes hypertrophied (Cooper and Brink 1945). Defects underpinning endosperm abnormalities related to nutrient transport in abortive crosses in maize have been traced to the characteristic wall ingrowths of the transfer cell layer, which are almost completely suppressed in the endosperm of grains generated in a cross between normal diploid and auotetraploid maize (Charlton et al. 1995). As in interspecific and intergeneric crosses, seed failure resulting from matings between races differing in chromosome number, or which create unbalanced genomic ratio in the endosperm, also appears to be due to disturbances in the nutritive milieu of the ovule, which presents itself as a hostile environment for embryo growth. Although several hypotheses have surfaced over the years to explain endosperm failure in these crosses, it now appears that parental imprinting of genes involved in endosperm development has a major contributory role in this misfortune (Gutierrez-Marcos et al. 2003).

The failure of endosperm development and its secondary effects on the maternal tissues of ovules in unsuccessful crosses are obviously in accord with the view that gene action deprives hybrid embryos of a continuing supply of nutrients to complete development. The postulated nutritional dependence of the embryo on the endosperm invites a simple experimental test: can the cryptic potentiality of the hybrid embryo to complete development be realized if it is allowed to grow in an artificial medium supplied with nutrient substances that are normally identified with the endosperm? A positive answer to this question came first from the experiments of Laibach (1925), who demonstrated that progeny can be obtained from nonviable seeds of *Linum perenne* × *L. austriacum* hybrid by excision and culture of embryos before they begin to disintegrate. From this it appears unlikely that there is an inherent disparity in the genetic constitution of the hybrid embryo that stymies its growth within the ovule. Since this pioneering work, embryo culture methods have been used to obtain transplantable seedlings from aborted seeds of interspecific and intergeneric crosses, which are traditionally condemned as being incapable of further growth. The tabulation of Collins and Grasser (1984) shows that hybrid embryo rescue operations have been successfully mounted in nearly 70 wide crosses involving approximately 35 genera and 120 species and, undoubtedly, several new cases deserve to be added to this list to make it current. Other developments in this area of research have shown that, in certain cases, continued growth of the hybrid embryo is secured by implanting it on a normal endosperm, which is then cultured on a synthetic medium, thereby initiating a nurse culture. In contrast to embryo culture, the implantation method overcomes the limitations of an artificial medium in promoting the growth of abortive embryos and shows in a direct way the role of the endosperm as a nutrient (Raghavan 1984). An important factor to be reckoned with in the successful culture of hybrid embryos is their age at excision, as embryos that abort at very early stages of development are difficult to isolate and pose greater risks of mutilation and damage than embryos at later stages of development. The nutrient requirements of younger embryos are also more exacting than those of older embryos. The problems associated with successful culture of small hybrid embryos can be overcome by culturing ovules and ovaries. Examples of ovule and ovary cultures for the successful recovery of hybrids from wide crosses are described by Raghavan (1984). From these multifaceted approaches, it appears increasingly clear that aborted embryos in wide hybrids are best thought of as merely those for which a nutritional supply for growth has been discontinued.

The appearance of haustorial structures on the endosperm, the presence of growth-promoting substances in this tissue, and rescue of hybrid embryos from ovules lacking a functional endosperm by tissue culture approaches, present evidence of the potential role of the endosperm in the nutrition of the embryo. These observations, considered in conjunction with the various ultrastructural modifications of the embryo and suspensor cells (Chap. 2) for the absorption and translocation of metabolites from the surrounding endosperm, make a compelling case for the nutrition of the embryo by the endosperm. Apart from nurturing the developing embryo, another primary function of the endosperm is to provide nutrition for the seedling during seed germination. This important role of the endosperm is outside the scope of this book and has not been covered in this chapter. A few additional functions of the endosperm, admittedly speculative, have been listed by Lertsen (2004).

7.4
Concluding Comments

Major advances have occurred over the past 10 years concerning our understanding of the structural events associated with the transition of the endosperm from a free nuclear to a cellular tissue. The cellularization process established by classical histological observations has fostered a series of innovative studies to reveal the role of cytoskeletal elements in the reorganization and polarization of the cytoplasm around the free nuclei to form initially a homogeneous mass of cells, and later the separation of the aleurone layer and starchy cells. Progress in understanding the basis for DNA amplification in endosperm nuclei has been slow to begin with, but the notion that, at the molecular level, endoreduplication is due to an increase in the activity of S-phase cyclin-dependent kinase in tandem with a loss of M-phase cyclin-dependent kinase seems to have taken hold. Endoreduplication of the endosperm nuclei is thought to increase the metabolic activity of cells for the rapid synthesis and accumulation of storage products, but hard evidence for this view is yet to be obtained. The endosperm has long been a vulnerable target for investigations into its role in providing nutrients for the developing embryo, but not much evidence based on modern studies has become available to support the prevailing generalizations. As will be described in the following chapter, molecular answers to questions about the genetic interactions between endosperm, embryo, and maternal tissues of the ovule are beginning to emerge, making this an exciting time for research on the endosperm.

REFERENCES

Alvarez-Buylla ER, Liljegren SJ, Pelaz S, Gold SE, Burgeff C, Ditta GS, Vergara-Silva F, Yanofsky MF (2000) MADS-box gene evolution beyond flowers: expression in pollen, endosperm, guard cells, roots and trichomes. Plant J 24:457–466

Bajer A (1965) Cine micrographic analysis of cell plate formation in endosperm. Exp Cell Res 37:376–398

Baroux C, Fransz P, Grossniklaus U (2004) Nuclear fusions contribute to polyploidization of the gigantic nuclei in the chalazal endosperm of *Arabidopsis*. Planta 220:38–46

Bethke PC, Lonsdale JE, Fath A, Jones RL (1999) Hormonally regulated programmed cell death in barley aleurone cells. Plant Cell 11:1033–1045

Bharathan G (1999) Endosperm development in angiosperms. In: Tandon RK, Prithipalsingh (eds) Biodiversity, taxonomy, and ecology. Scientific Publishers (India), New Delhi, pp 167–180

Bhatnagar SP, Sawhney V (1981) Endosperm – its morphology, ultrastructure and histochemistry. Int Rev Cytol 73:55–102

Boisnard-Lorig C, Colon-Carmona A, Bauch M, Hodge S, Doerner P, Bancharel E, Dumas C, Haseloff J, Berger F (2001) Dynamic analyses of the expression of the HISTONE::YFP fusion protein in *Arabidopsis* show that syncytial endosperm is divided into mitotic domains. Plant Cell 13:495–509

Bosnes M, Olsen O-A (1992) The rate of nuclear gene transcription in barley endosperm syncytia increases sixfold before cell-wall formation. Planta 186:376–383

Bosnes M, Weideman F, Olsen O-A (1992) Endosperm differentiation in barley wild-type and *sex* mutants. Plant J 2:661–674

Brachet J (1985) Molecular cytology, vol 1. Academic Press, Orlando, FL

Brink RA, Cooper DC (1941) Incomplete seed failure as a result of somatoplastic sterility. Genetics 26:487–505

Brink RA, Cooper DC (1947) The endosperm in seed development. Bot Rev 13:423–541

Brown RC, Lemmon BE, Olsen O-A (1994) Endosperm development in barley: microtubule involvement in the morphogenetic pathway. Plant Cell 6:1241–1252

Brown RC, Lemmon BE, Olsen O-A (1996a) Polarization predicts the pattern of cellularization in cereal endosperm. Protoplasma 192:168–177

Brown RC, Lemmon BE, Olsen O-A (1996b) Development of the endosperm in rice (*Oryza sativa* L): cellularization. J Plant Res 109:301–313

Brown RC, Lemmon BE, Stone BA, Olsen O-A (1997) Cell wall (1→3)- and (1→3, 1→4)-β-glucans during early grain development in rice (*Oryza sativa* L.). Planta 202:414–426

Brown RC, Lemmon BE, Nguyen H, Olsen O-A (1999) Development of endosperm in *Arabidopsis thaliana*. Sex Plant Reprod 12:32–42

Brown RC, Lemmon BE, Nguyen H (2002) Endosperm development. In: O'Neill SD, Roberts JA (eds) Plant reproduction. Annual plant reviews, vol 6. Sheffield Academic Press, Sheffield, pp 193–220

Brown RC, Lemmon BE, Nguyen H (2003) Events during the first four rounds of mitosis establish three developmental domains in the syncytial endosperm of *Arabidopsis thaliana*. Protoplasma 222:167–174

Brown RC, Lemmon BE, Nguyen H (2004) Comparative anatomy of the chalazal endosperm cyst in seeds of the Brassicaceae. Bot J Linn Soc 144:375–394

Brunori A, Forino LMC, Frediani M (1989) Polyploid DNA contents in the starchy endosperm nuclei and seed weight in *Triticum aestivum*. J Genet Breed Rome 43:131–134

Cavallini A, Natali L, Balconi C, Rizzi E, Motto M, Cionini G, D'Amato F (1995) Chromosome endoreduplication in endosperm cells of two maize genotypes and their progenies. Protoplasma 189:156–162

Charlton WL, Keen CL, Merriman C, Lynch P, Greenland AJ, Dickinson HG (1995) Endosperm development in *Zea mays*; implications of gametic imprinting and paternal excess in regulation of transfer layer development. Development 121:3089–3097

Choi S-B, Wang C, Muench DG, Ozawa K, Franceschi VR, Wu Y, Okita TW (2000) Messenger RNA targeting of rice seed storage proteins to specific ER subdomains. Nature 407:765–767

Chojecki AJS, Bayliss MW, Gale MD (1986) Cell production and DNA accumulation in the wheat endosperm, and their association with grain weight. Ann Bot 58:809–817

Cochrane MP, Duffus CM (1980) The nucellar projection and modified aleurone in the crease region of developing caryopses of barley (*Hordeum vulgare* L. var. *distichum*). Protoplasma 103:362–375

Collins GB, Grosser JW (1984) Culture of embryos. In: Vasil IK (ed) Cell culture and somatic cell genetics of plants. Academic Press, Orlando, FL pp 247–257

Cooper DC, Brink RA (1945) Seed collapse following matings between diploid and tetraploid races of *Lycopersicon pimpinellifolium*. Genetics 30:376–401

Creech RG (1965) Genetic control of carbohydrate synthesis in maize endosperm. Genetics 52:1175–1186

D'Amato F (1984) Role of polyploidy in reproductive organs and tissues. In: Johri BM (ed) Embryology of angiosperms. Springer, Berlin Heidelberg New York, pp 519–566

Datta SK (1999) Transgenic cereals: *Oryza sativa* (rice). In: Vasil IK (ed) Molecular improvement of cereal crops. Kluwer, Dordrecht, pp 149–187

Davis RW, Smith JD, Cobb BG (1990) A light and electron microscope investigation of the transfer cell region of maize caryopses. Can J Bot 68:471–479

DeMason DA (1997) Endosperm structure and development. In: Larkins BA, Vasil IK (eds) Cellular and molecular biology of plant seed development. Kluwer, Dordrecht, pp 73–115

Dilkes BP, Dante RA, Coelho C, Larkins BA (2002) Genetic analyses of endoreduplication in *Zea mays* endosperm: evidence of sporophytic and zygotic maternal control. Genetics 160:1163–1177

Doan DNP, Linnestad C, Olsen O-A (1996) Isolation of molecular markers from the barley endosperm coenocyte and the surrounding nucellus cell layers. Plant Mol Biol 31:877–886

Dute RR, Peterson CM (1992) Early endosperm development in ovules of soybean, *Glycine max* (L.) Merr. (Fabaceae). Ann Bot 69:263–271

Eames AJ (1961) Morphology of the angiosperms. McGraw-Hill, New York

Esau K (1965) Plant anatomy, 2nd edn. Wiley, New York

Fineran BA, Wild DJC, Ingerfeld M (1982) Initial wall formation in the endosperm of wheat, *Triticum aestivum*: a reevaluation. Can J Bot 60:1776–1795

Floyd SK, Friedman WE (2000) Evolution of endosperm developmental patterns among basal flowering plants. Int J Plant Sci 161:S57–S81

Floyd SK, Friedman WE (2001) Developmental evolution of endosperm in basal angiosperms: evidence from *Amborella* (Amborellaceae), *Nuphar* (Nymphaeaceae), and *Illicium* (Illiciaceae). Plant Syst Evol 228:153–169

Floyd SK, Lerner VT, Friedman WE (1999) A developmental and evolutionary analysis of embryology in *Platanus* (Platanaceae), a basal eudicot. Am J Bot 86:1523–1537

Friedman WE (1998) The evolution of double fertilization and endosperm: an "historical" perspective. Sex Plant Reprod 11:6–16

Friedman WE, Williams JH (2004) Developmental evolution of the sexual processes in ancient flowering plant lineages. Plant Cell 16:S119–S132

Geeta R (2003) The origin and maintenance of nuclear endosperms: viewing developing through a phylogenetic lens. Proc R Soc B270:29–35

Giese H (1992) Replication of DNA during barely endosperm development. Can J Bot 70:313–318

Gordon-Kamm WJ, Baszczynski CL, Bruce WB, Tomes DT (1999) Transgenic cereals – *Zea mays* (maize). In: Vasil IK (ed) Molecular improvement of cereal crops. Kluwer, Dordrecht, pp 189–253

Grafi G (1998) Cell cycle regulation of DNA replication: the endoreduplication perspective. Exp Cell Res 244:372–378

Grafi G, Larkins BA (1995) Endoreduplication in maize endosperm: involvement of M phase-promoting factor inhibition and induction of S phase-related kinases. Science 269:1262–1264

Grafi G, Burnett RJ, Helentjaris T, Larkins BA, DeCaprio JA, Sellers WR, Kaelin WG Jr (1996) A maize cDNA encoding a member of the retinoblastoma protein family: involvement in endoreduplication. Proc Natl Acad Sci USA 93:8962–8967

Greenwood JS, Helm M, Gietl C (2005) Ricinosomes and endosperm transfer cell structure in programmed cell death of the nucellus during *Ricinus* seed development. Proc Natl Acad Sci USA 102:2238–2243

Guitton A-E, Page DR, Chambrier P, Lionnet C, Faure J-E, Grossniklaus U, Berger F (2004) Identification of new members of fertilisation independent seed polycomb group pathway involved in the control of seed development in *Arabidopsis thaliana*. Development 131:2971–2981

Gutierrez-Marcos JF, Pennington PD, Costa LM, Dickinson HG (2003) Imprinting in the endosperm: a possible role in preventing wide hybridization. Philos Trans R Soc Ser B 358:1105–1111

Herr JM Jr (1999) Endosperm development in *Arabidopsis thaliana* (L.) Heynh. Acta Biol Cracov Ser Bot 41:103–109

Herz M, Brunori A (1985) Nuclear DNA contents in the endosperm of developing grain of hexaploid Triticales and parental species. Z Pflanzenphysiol 95:336–341

Hirner B, Fischer WN, Rentsch D, Kwart M, Frommer WB (1998) Developmental control of $H^{+/}$/amino acid permease gene expression during seed development of *Arabidopsis*. Plant J 14:535–544.

Hu S, Zhu C, Zee SY (1983) Transfer cells in suspensor and endosperm during early embryogeny of *Vigna sinensis*. Acta Bot Sin 25:1–7

Johansson M, Walles B (1994) Functional anatomy of the ovule in broad bean (*Vicia faba* L.): ultrastructural seed development and nutrient pathways. Ann Bot 74:233–244

Jolliffe NA, Brown JC, Neumann U, Vicré M, Bachi A, Hawes C, Ceriotti A, Roberts LM, Frigerio L (2004) Transport of ricin and 2S albumin precursors to the storage vacuoles of *Ricinus communis* endosperm involves the Golgi and VSR-like receptors. Plant J 39:821–833

Kerr JFR, Gobé GC, Winterford CM, Harmon BV (1995) Anatomical methods in cell death. Methods Cell Biol 46:1–27

Khoo U, Wolf MJ (1970) Origin and development of protein granules in maize endosperm. Am J Bot 57:1042–1050

Kim CS, Woo Y, Clore AM, Burnett RJ, Carneiro NP, Larkins BA (2002) Zein protein interactions, rather than the asymmetric distribution of zein mRNAs on endoplasmic reticulum membranes, influence protein body formation in maize endosperm. Plant Cell 14:655–672

Koltunow AM (1993) Apomixis: embryo sacs and embryos formed without meiosis or fertilization in ovules. Plant Cell 5:1425–1437

Koltunow AM, Grossniklaus U (2003) Apomixis: a developmental perspective. Annu Rev Plant Biol 54:547–574

Kowles RV, Phillips RL (1988) Endosperm development in maize. Int Rev Cytol 112:97–136

Kowles RV, Srienc F, Phillips RL (1990) Endoreduplication of nuclear DNA in the developing maize endosperm. Dev Genet 11:125–132

Kowles RV, Yerk GL, Haas KM, Phillips RL (1997) Maternal effects influencing DNA endoreduplication in developing endosperm of *Zea mays*. Genome 40:798–805

Kranz E, von Wiegen P, Quader H, Lörz,H (1998) Endosperm development after fusion of isolated, single maize sperm and central cells in vitro. Plant Cell 10:511–524

Laibach F (1925) Das Taubwerden von Bastardsamen und die künstliche Aufzucht früh absterbender Bastardembryonen. Z Bot 17:417–459

Larkins BA, Dilkes BP, Dante RA, Coelho CM, Woo Y, Liu Y (2001) Investigating the hows and whys of DNA endoreduplication. J Exp Bot 52:183–192

Larkins BA, Hurkman WJ (1978) Synthesis and deposition of zein in protein bodies of maize endosperm. Plant Physiol 62:256–263

Leblanc O, Pointe C, Hernandez M (2002) Cell cycle progression during endosperm development in Zea mays depends on parental dosage effects. Plant J 32:1057–1066

Leiva-Neto JT, Grafi G, Sabelli PA, Dante RA, Woo Y, Maddock S, Gordon-Kamm WG, Larkins BA (2004) A dominant negative mutant of cyclin-dependent kinase A reduces endoreduplication but not cell size or gene expression in maize endosperm. Plant Cell 16:1854–1869

Lersten NR (2004) Flowering plant embryology. Blackwell, Ames, IA

Levanony H, Rubin R, Altschuler Y, Galili G (1992) Evidence for a new route of wheat storage proteins to vacuoles. J Cell Biol 119:1117–1128

Li X, Franceschi VR, Okita TW (1993) Segregation of storage protein mRNAs on the rough endoplasmic reticulum membranes of rice endosperm cells. Cell 72:869–879

Lopes MA, Larkins BA (1993) Endosperm origin, development, and function. Plant Cell 5:1383–1399

Luo M, Bilodeau P, Dennis ES, Peacock WJ, Chaudhury A (2000) Expression and parent-of-origin effects for *FIS2*, *MEA*, and *FIE* in the endosperm and embryo of developing *Arabidopsis* seeds. Proc Natl Acad Sci USA 97:10637–10642

Lur H-S, Setter TL (1993) Role of auxin in maize endosperm development. Plant Physiol 103:273–280

Maheshwari P (1950) An introduction to the embryology of angiosperms. McGraw-Hill, New York

Maness NO, McBee GC (1986) Role of placental sap in endosperm carbohydrate import in *Sorghum* caryopses. Crop Sci 26:1201–1207

Mares DJ, Norstog K, Stone BA (1975) Early stages in the development of wheat endosperm. I. The change from free nuclear to cellular endosperm. Aust J Bot 23:311–326

Marinos NG (1970) Embryogenesis of the pea (*Pisum sativum*) I. The cytological environment of the developing embryo. Protoplasma 70:261–279

McClintock B (1978) Development of the maize endosperm as revealed by clones. In: Subtelny S, Sussex IM (eds) The clonal basis of development. Academic Press, New York, pp 217–237

Mól R, Betka A, Wojciechowicz M (1995) Induction of autonomous endosperm in *Lupinus luteus*, *Helleborus niger* and *Melandrium album* by in vitro culture of unpollinated ovaries. Sex Plant Reprod 8:273–277

Morgan DO (1997) Cyclin-dependent kinases: engines, clocks, and microprocessors. Annu Rev Cell Dev Biol 13:261–291

Musial K, Przywara L (1998) Influence of irradiated pollen on embryo and endosperm development in kiwifruit. Ann Bot 82:747–756

Nagl W (1992) The polytenic endosperm haustorium of *Rhinanthus minor* (Scrophulariaceae): functional ultrastructure. Can J Bot 70:1997–2004

Nesling FAV, Morris DA (1979) Cytokinin levels and embryo abortion in interspecific *Phaseolus* crosses. Z Pflanzenphysiol 91:345–358

Newcomb W (1978) The development of cells in the coenocytic endosperm of the African blood lily *Haemanthus katherinae*. Can J Bot 56:483–501

Newcomb W, Fowke LC (1973) The fine structure of the change from the free-nuclear to cellular condition in the endosperm of chickweed *Stellaria media*. Bot Gaz 134:236–241

Newcomb W, Steeves TA (1971) *Helianthus annuus* embryogenesis: embryo sac wall projections before and after fertilization. Bot Gaz 132:367–371

Nguyen H, Brown RC, Lemmon BE (2000) The specialized chalazal endosperm in *Arabidopsis thaliana* and *Lepidium virginicum* (Brassicaceae). Protoplasma 212:99–110

Nguyen H, Brown RC, Lemmon BE (2001) Patterns of cytoskeletal organization reflect distinct developmental domains in endosperm of *Coronopus didymus* (Brassicaceae). Int J Plant Sci 162:1–14

Nguyen H, Brown RC, Lemmon BE (2002) Cytoskeletal organization of the micropylar endosperm in *Coronopus didymus* L. (Brassicaceae). Protoplasma 219:210–220

Olsen O-A (2001) Endosperm development: cellularization and cell fate specification. Annu Rev Plant Physiol Plant Mol Biol 52:233–267

Olsen O-A (2004) Nuclear endosperm development in cereals and *Arabidopsis thaliana*. Plant Cell 16:S214–S227

Olsen O-A, Brown RC, Lemmon BE (1995) Pattern and process of wall formation in developing endosperm. Bioessays 17:803–812

Opsahl-Ferstad H-G, le Deunff E, Dumas C, Rogowsky PM (1997) *ZmEsr*, a novel endosperm-specific gene expressed in a restricted region around the maize embryo. Plant J 12:235–246

Otegui M, Staehelin LA (2000a) Cytokinesis in flowering plants: more than one way to divide a cell. Curr Opin Plant Biol 3:493–502

Otegui M, Staehelin LA (2000b) Syncytial-type cell plates: a novel kind of cell plate involved in endosperm cellularization of *Arabidopsis*. Plant Cell 12:933–947

Otegui MS, Mastronarde DN, Kang B-H, Bednarek SY, Staehelin LA (2001) Three-dimensional analysis of syncytial-type cell plates during endosperm cellularization visualized by high resolution electron tomography. Plant Cell 13:2033–2051

Otegui MS, Capp R, Staehelin LA (2002) Developing seeds of *Arabidopsis* store different minerals in two types of vacuoles and in the endosperm reticulum. Plant Cell 14:1311–1323

Raghavan V (1976) Experimental embryogenesis in vascular plants. Academic Press, London

Raghavan V (1977) Applied aspects of embryo culture. In: Reinert J, Bajaj YPS (eds) Applied and fundamental aspects of plant cell, tissue, and organ culture. Springer, Berlin Heidelberg New York, pp 375–397

Raghavan V (1984) Variability through wide crosses and embryo rescue. In: Vasil IK (ed) Cell culture and somatic cell genetics of plants. Academic Press, Orlando, FL pp 613–633

Raghavan V (1997) Molecular embryology of flowering plants. Cambridge University Press, New York

Ramachandran C, Raghavan V (1989) Changes in nuclear DNA content of endosperm cells during grain development in rice (*Oryza sativa*). Ann Bot 64:459–468

Randolph LF (1936) Developmental morphology of the caryopsis in maize. J Agric Res 53:881–916

Renner O (1914) Befruchtung und Embryobildung bei *Oenothera lamarckiana* und einigen verwandten Arten. Flora 107:115–150

Rost TL, Lersten NR (1970) Transfer aleurone cells in *Setaria lutescens* (Gramineae). Protoplasma 71:403–408

Ryczkowski M (1962) Changes in the concentration of sugars in developing ovules. Acta Soc Bot Pol 31:53–65

Sangduen N, Kreitner GL, Sorensen EL (1983) Light and electron microscopy of embryo development in perennial and annual *Medicago* species. Can J Bot 61:837–849

Sano Y (1984) Differential regulation of waxy gene expression in rice endosperm. Theor Appl Genet 68:467–473

Schel JHN, Kieft H, van Lammeren AAM (1984) Interactions between embryo and endosperm during early developmental stages of maize carypses (*Zea mays*). Can J Bot 62:2842–2853

Schmid M, Simpson D, Gietl C (1999) Programmed cell death in castor bean endosperm is associated with the accumulation and release of a cysteine endopeptidase from ricinosomes. Proc Natl Acad Sci USA 96:14159–14164

Schneitz K, Hülskamp M, Pruitt RE (1995) Wild-type ovule development in *Arabidopsis thaliana*: a light microscope study of cleared whole-mount tissue. Plant J 7:731–749

Schulz P, Jensen WA (1974) *Capsella* embryogenesis: the development of the free nuclear endosperm. Protoplasma 80:183–205

Schulz P, Jensen WA (1977) Cotton embryogenesis: the early development of the free nuclear endosperm. Am J Bot 64:384–394

Sørensen MB, Chaudhury AM, Robert H, Bancharel E, Berger F (2001) Polycomb group genes control pattern formation in plant seed. Curr Biol 11:277–281

Stangeland B, Salehian Z, Aalen R, Mandal A, Olsen O-A (2003) Isolation of GUS marker lines for genes expressed in *Arabidopsis* endosperm, embryo and maternal tissues. J Exp Bot 54:279–290

Sun Y, Dilkes BP, Zhang C, Dante RA, Carneiro NP, Lowe KS, Jung R, Gordon-Kamm WJ, Larkins BA (1999a) Characterization of maize (*Zea mays* L.) Wee1 and its activity in developing endosperm. Proc Natl Acad Sci USA 96:4180–4185

Sun Y, Flannigan BA, Setter TL (1999b) Regulation of endoreduplication in maize (*Zea mays* L.) endosperm. Isolation of a novel B1-type cyclin and its quantitative analysis. Plant Mol Biol 41:245–258

Swamy BGL, Parameswaran N (1963) The helobial endosperm. Biol Rev 38:1–50

Than ME, Helm M, Simpson DJ, Lottspeich F, Huber R, Gietl C (2004) The 2.0 Å crystal structure and substrate specificity of the KDEL-tailed cysteine endopeptidase functioning in programmed cell death of *Ricinus communis* endosperm. J Mol Biol 336:1103–1116

Thompson RD, Hueros G, Becker H-A, Maitz M (2001) Development and function of seed transfer cells. Plant Sci 160:775–783

Tian G-W, You R-L, Guo F-L, Wang X-C (1998) Microtubular cytoskeleton of free endosperm nuclei during division in wheat. Cytologia 63:427–433

van Lammeren AAM, Kieft H, Ma F, van Veenendaal WLH (1996) Light microscopical study of endosperm formation in *Brassica napus* L. Acta Soc Bot Pol 65:267–272

Vasil IK, Vasil V (1999) Transgenic cereals: *Trititcum aestivum* (wheat). In: Vasil IK (ed) Molecular improvement of cereal crops. Kluwer, Dordrecht, pp 133–147

Vijayaraghavan MR, Prabhakar K (1984) The endosperm. In: Johri BM (ed) Embryology of angiosperms. Springer, Berlin Heidelberg New York, pp 319–376

Wang M, Oppedijk BJ, Lu X, van Duijn B, Schilperoort RA (1996) Apoptosis in barley aleurone during germination and its inhibition by abscisic acid. Plant Mol Biol 32:1125–1134

Wijowska M, Kuta E, Przywara L (1999) Autonomous endosperm induction by in vitro culture of unfertilized ovules of *Viola odorata* L. Sex Plant Reprod 12:164–170

Wredle U, Walles B, Hakman I (2001) DNA fragmentation and nuclear degradation during programmed cell death in the suspensor and endosperm of *Vicia faba*. Int J Plant Sci 162:1053–1063

Yeung EC, Cavey MJ (1988) Cellular endosperm formation in *Phaseolus vulgaris*. I. Light and scanning electron microscopy. Can J Bot 66:1209–1216

Young TE, Gallie DR (1999) Analysis of programmed cell death in wheat endosperm reveals differences in endosperm development between cereals. Plant Mol Biol 39:915–926

Young TE, Gallie DR (2000) Programmed cell death during endosperm development. Plant Mol Biol 44:283–301

Young TE, Gallie DR, DeMason DA (1997) Ethylene-mediated programmed cell death during maize endosperm development of wild-type and *shrunken2* genotypes. Plant Physiol 115:737–751

Zee S-Y, O'Brien TP (1971) Aleurone layer transfer cells and other structural features of the spikelet of millet. Aust J Biol Sci 24:391–395

Zhao J, Grafi G (2000) The high mobility group I/Y protein is hypophosphorylated in endoreduplicating maize endosperm cells and is involved in alleviating histone H1-mediated transcriptional repression. J Biol Chem 275:27494–27499

8 Genetics and Molecular Biology of the Endosperm – A Tale of Two Model Systems

Thus not only are both endosperm and embryo of biparental origin in cross-pollinated species but the two structures differ in hereditary organization. The endosperm is 3x, having received a double complement of inheritance from the pistillate parent. The embryo is 2x. Genetic diversity within the seed is further increased by the fact that, since the maternal tissues and the embryo belong to different sporophytic generations, they may be unlike in genotype. ... The problems which the endosperm presents stem from the peculiarities of its origin and genetic endowment and its intercalary position between the old and the new sporophytes.

R.A. Brink, D.C. Cooper 1947

8.1   Specification of Form in the Endosperm of *Arabidopsis*   174
8.1.1   Endosperm Development without Double Fertilization   175
8.1.2   Parental Gene Dosage in Endosperm Development   176
8.2   Genetics and Molecular Biology of the Cereal Endosperm   178
8.2.1   Embryo-surrounding Region and Transfer Layer   178
8.2.2   Aleurone Cells and Starchy Endosperm   180
8.3   Concluding Comments   182
References   182

Endosperm is not produced instantly as a ready-made tissue following double fertilization; rather, its development is a progressive process that comprises various steps, each depending upon a few genes. To place these steps in their proper perspective, the introduction to this chapter begins with brief comments on the cytological, genetic, and molecular changes that occur in the endosperm to make it a mature tissue. First are the repeated rounds of mitosis, followed by the cellularization and histodifferentiation that carve out the principal regions of the endosperm. During the latter period of histodifferentiation, there is a surge in the nuclear DNA content of the cells that results in high DNA values due to endoreduplication. Temporally correlated with endoreduplication, starch and storage proteins are rapidly synthesized and accumulate in the endosperm. Identification of the genes, and their protein products, involved in endosperm histodifferentiation, beginning with the syncytial stage, has been made possible by analysis of mutants affected in endosperm development. In the formation of the starchy cells, aleurone cells, transfer cells, and cells of the embryo-surrounding region, which constitute the four major functional regions of the cereal endosperm, deposition and interpretation of positional information play important roles. Nevertheless, the link between the establishment of these cell types in the endosperm and their genetic specification is not fully understood. The purpose of this chapter is to present an overview of current information relating to those episodes in endosperm development not considered in the previous chapter, drawing principally on work done on *Arabidopsis*, with its transient, minimal endosperm, and cereal grains, with their persistent, significant quantities of endosperm. Investigations undertaken with these systems have met with notable success as they provide instructive comparisons that illuminate the hitherto unrecognized role of parental genomes and will, no doubt, continue to provide insightful information on the underlying mechanisms of endosperm development in flowering plants.

Reviews that highlight the genetic and molecular biology of endosperm development have been published by Chaudhury et al. (2001), Becraft et al. (2001), Berger (2003), Olsen (2004b), and Costa et al. (2004).

8.1
Specification of Form in the Endosperm of *Arabidopsis*

Based on analysis of mutations that disrupt normal seed development, considerable progress has been made in the last few years toward an understanding of the role of gene action in the generation of form in the endosperm of *Arabidopsis*, and in the interaction of the endosperm with the embryo and maternal tissues of the ovule. Reference was made in Chaps. 3 and 5 to mutations that disrupt embryogenesis in *Arabidopsis*; some of the same mutations also show remarkable parallels in impairing development of the endosperm, a tissue bearing little resemblance to the embryo either in its morphology or chromosome number. One such mutation is *tor*, which causes premature arrest of free nuclear divisions in the endosperm; only about 25% of the number of free nuclei produced in the wild-type endosperm is generated in the mutant endosperm (Menand et al. 2002). However, it appears unlikely that the *TOR* gene is a key regulator of the cell cycle during the syncytial phase of the endosperm. The *KN* gene and genes included in the *TTN* and *PILZ* groups, which have been identified through analysis of mutations affecting cytokinesis in the embryo, also impair endosperm cellularization, implying that these events in the two products of double fertilization share components of the same genetic machinery. Cytokinetic defects in embryo cells of *keu* and *kn* mutants were described in Chap. 3. Comparative studies have shown that, whereas endosperm cellularization is not affected in seeds of the *keu* mutant, most seeds of *kn* monogenic, as well as of *kn/keu* double mutants, produce non-cellularized endosperm that survives as a syncytium. During endosperm cellularization, the syntaxin family of proteins encoded by the *KN* gene are thought to play a vital role in cytokinesis, serving as key molecules in the docking and fusion of vesicles at membranes (Lauber et al. 1997; Sørensen et al. 2002). An understanding of the involvement of the protein products of the *TTN* (Liu and Meinke 1998; Springer et al. 2000; Tzafrir et al. 2002) and *PILZ* (Mayer et al. 1999) groups of genes in an essential role in endosperm development stems from the observation that embryo-lethal mutants of these genes have endosperms with giant nuclei. An endosperm phenotype with dramatically enlarged nuclei similar to the *TTN* and *PILZ* groups of mutants is also displayed by the *orc2* embryo-lethal mutant (Collinge et al. 2004). As described in Chap. 5, the varied protein products of the *TTN* and *PILZ* groups of genes play pivotal roles in cell division (McElver et al. 2000; Liu et al. 2002; Tzafrir et al. 2002; Steinborn et al. 2002). Characteristic of regulatory molecules, individual proteins probably participate in specific steps threading the cascade of mitosis and crafting of the cytokinetic apparatus associated with endosperm development. Whereas virtually all of the mutants referred to above with defective endosperm are also embryo-defective, an exclusive role in endosperm cellularization has been attributed to the *SPÄTZLE* gene. As shown in Fig. 8.1 (a–e), abnormalities in endosperm development observed in ovules of the *spätzle* mutant include failure of the nuclear-cytoplasmic domains around nuclei to separate, attachment of nuclei to incompletely separated nuclear envelopes, nuclear fusion, and formation of multinucleate nuclear-cytoplasmic domains surrounding normally developing heart-shaped and torpedo-shaped embryos (Sørensen et al. 2002). A commentary by Dickinson (2003) underscores the need to fill in the gaps of uncertainty regarding the role of the *SPÄTZLE* gene in the formation of the nuclear-cytoplasmic domains and phragmoplasts during cellularization of the endosperm. When this gene is

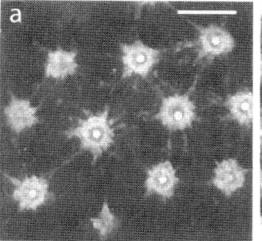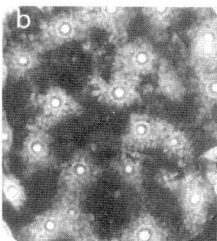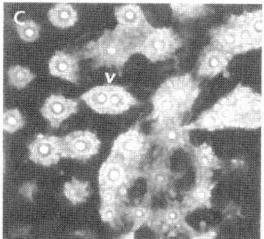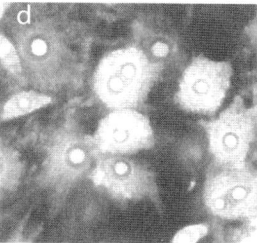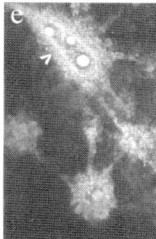

Fig. 8.1a–e Role of the SPÄTZLE gene in the cellularization of endosperm of *Arabidopsis*. **a** Peripheral endosperm nuclei at the heart-shaped stage of the embryo showing lack of cellularization. **b** Failure of the nuclear-cytoplasmic domains to separate following nuclear divisions. **c** Failure of nucleo-cytoplasmic domains to separate followed by attachment of nuclei by incompletely separated nuclear envelopes (*arrowhead*). **d** Formation of multinucleate nuclear-cytoplasmic domains. **e** Fusion of nuclei leading to the formation of large nuclei with multiple nucleoli (*arrowhead*). *Bar* 20 µm. (Reprinted from Sørensen et al. 2002)

cloned and its protein product identified, an answer can also be sought to the question as to why the embryo is not affected in the *spätzle* mutant.

8.1.1
Endosperm Development without Double Fertilization

The genetic basis of endosperm development in *Arabidopsis* has been illuminated greatly by the isolation of female gametophyte mutants in which unfertilized ovules display certain aspects of reproductive processes found only in fertilized ovules. Formation of an incomplete endosperm without fertilization is the signature feature of *fie*, the first such mutant isolated. The effect of the mutation is dramatically revealed by comparing the frequency of multinucleate central cells in unpollinated heterozygous *FIE/fie* mutant ovules with that in the corresponding wild-type ovules (see Plate 12, Fig. g,h). In contrast to the 3–5% of wild-type central cells with more than one nucleus during a 6-day interval after emasculation, 47% of mutant central cells had two or more nuclei over the same period. Besides its effect on endosperm development, the mutation also activates transformation of the integuments into seed coats (Ohad et al. 1996). Lack of a fully formed endosperm, however, seems to blur the effect of the mutation, and invites the possible involvement of other genes in endosperm development. Later additions to the list known as the *fis* class of mutants, in which endosperm development is uncoupled from double fertilization, are *fis1*, *fis2*, *fis3* (Chaudhury et al. 1997), *mea* (Grossniklaus et al. 1998), *f644* (Kiyosue et al. 1999), *dme* (Choi et al. 2002), *msi1* (Köhler et al. 2003), *msi1-2*, and *bga* (Guitton et al. 2004). Based on the observation that, in the *msi1* mutant, the polar fusion nucleus starts to divide without fusion with the sperm nucleus even after successful pollination, a case can be made that different alleles of the *fis* mutants that compromise or enhance the mutational effects are likely to emerge in the future (Köhler et al. 2003). Autonomous endosperm development is not, however, displayed by *cap* mutants, whose normal-looking embryo sacs are devoid of a homogenous population of endosperm nuclei and a chalazal endosperm cyst (Grini et al. 2002).

As described in Chap. 5, all of the original *FIS* genes except *FIS2* encode proteins that are homologous to the polycomb group proteins (Grossniklaus et al. 1998; Luo et al. 1999; Ohad et al. 1999). Moreover, the *FIS* genes are imprinted such that only the maternal copy of the gene is expressed in the endosperm, whereas the paternal copy is not; in other words, parental *FIS* genes are inherited in a silenced state on the paternally inherited chromosomes of the wild-type endosperm (Vielle-Calzada et al. 1999; Kinoshita et al. 1999; Luo et al. 2000; Yadegari et al. 2000). One test of imprinting is to monitor the activity of these genes in the endosperm of transgenic lines of *Arabidopsis* containing their promoters linked to a *GUS* reporter gene. As shown by Luo et al. (2000), in contrast to the endosperm-specific activity of the maternally derived *FIS2::GUS*, *MEA::GUS*, and *FIE::GUS* constructs, no GUS expression is seen when these constructs are introduced through pollen grains. These results also indirectly indicate that endosperm development before fertilization is suppressed by maternal expression of the wild-type function of *FIS* genes.

8.1.2
Parental Gene Dosage in Endosperm Development

Because *Arabidopsis*, like most other sexually reproducing flowering plants, requires both paternal and maternal genes for the development of the endosperm, imprinting is best understood in crosses between plants of different ploidies or belonging to different species. This rests on the assumption that a ratio of two maternal genomes (2m) to one paternal genome (1p), both endosperm-specific, is crucial for endosperm development, and that endosperm in seeds from intraploidy, interploidy, and interspecific crosses would develop abnormally if there is any deviation from this ratio (Vinkenoog and Scott 2001). Scott et al. (1998) found that, in *Arabdiopsis*, where imprinting can be manipulated, $2x \times 4x$ crosses (diploid mother and tetraploid father, 2m:2p) inflicting a paternal genomic excess, result in a massively overgrown endosperm and a correspondingly large seed, whereas $4x \times 2x$ crosses (4m:1p endosperm) inflicting a maternal genomic excess, produce a precociously cellularizing endosperm within a small seed. Viable seeds containing triploid embryos are produced from crosses in either direction. Crosses between a diploid and a hexaploid maintain this trend with similar but extreme reciprocal phenotypes and aborted seeds. In all cases, gene dosage imbalance affects primarily the timing of cellularization of the endosperm and its proliferative potential. When the embryo and endosperm ploidy is increased without disrupting the 2m:1p ratio such as in $2x \times 2x$, $4x \times 4x$, or $6x \times 6x$ crosses, normal development of the embryo and endosperm ensues. In a series of interspecific crosses between diploid and tetraploid *A. thaliana* and $4x$ *A. arenosa*, it was found that the endosperm from $2x$ *A. thaliana* × $4x$ *A. arenosa* crosses is phenotypically characteristic of a paternal excess cross, with features such as lack of cellularization and prolonged proliferation, whereas increasing the maternal genome, as in $4x$ *A. thaliana* × $4x$ *A. arenosa* crosses, reduces endosperm proliferation to the normal level (Bushell et al. 2003). These results reinforce the importance of the 2m:1p ratio for the development of the endosperm in *Arabidopsis*, and imply that endosperm development requires the activity of imprinted genes or parent-of-origin chromosome dosage.

An influential theory that has been invoked to explain genomic imprinting in the development of the endosperm is the parental conflict theory (Haig and Westoby 1989, 1991). This theory assumes that in intraploidy, interploidy, and interspecific crosses, where maternal-origin genes suppress and paternal-origin genes promote endosperm development, there is an inherent conflict between the maternal and paternal interests in resource allocation from the plant to the seed. This means that a large endosperm with many nutrient-rich cells resulting from a paternal genome excess cross is due to the allocation of more resources from the plant to the seed than the small endosperm resulting from maternal excess cross, thus making sense of why genes that promote unrestricted maternal resource allocation should be expressed paternally. On the other hand, genes that promote a restricted distribution of resources from the mother plant to the seed will be expressed maternally, but not paternally. Since the endosperm regulates maternal nutrient fluxes to the embryo, it also becomes the site of imprinting, or parent-of-origin gene expression (Gehring et al. 2004). It has, however, been argued that parental imprinting alone cannot explain the fate of the endosperm in some of the mutants discussed above (von Wangenheim and Peterson 2004).

To identify genes whose expression is disturbed in the maternal excess effect phenotypes, Garcia et al. (2003) have isolated two mutants, *haiku1* (*iku1*) and *iku2* that produce small seeds displaying precocious cellularization of the endosperm. The resemblance of mutant seeds to those endowed with increased maternal dosage in interploidy crosses might indicate a role for the *IKU* gene in the maternal excess phenotypes and in the hypomethylation of the paternal genome described below.

Following up on work implicating DNA methylation (methylation of cytosine in the gene sequence and associated transcriptional repression) in imprinting in mammals, mechanisms that modify imprinted loci are being uncovered by investigations into the role of DNA methylation in parent-of-origin effects in the development of *Arabidopsis* endosperm. Adams et al. (2000) studied the effects of hypomethylation in crosses involving plants in which DNA methylation is substantially reduced by introducing a maintenance DNA

METHYLTRANSFERASE1 antisense (*MET1* a/s) construct. This work showed that the effects of interploidy crosses resulting in increased maternal or paternal gene dosage are phenocopied by pollinating wild-type ovules with pollen carrying a demethylated genome or by pollination of a hypomethylated plant with wild-type pollen, respectively. If methylation is involved in gene-silencing, the prediction from these crosses is that in a wild-type × *MET1* a/s cross, the normally maternal-specific alleles would be derepressed on the paternal chromosomes, thus phenocopying the cross that provides an overdose of the maternal genome, whereas hypomethylation of the ovule parent, as in the *MET1* a/s × wild-type cross, would derepress the paternally expressed genes imprinted on the maternal chromosomes and phenocopy the paternal excess cross. These predictions are borne out in experimental results, thus suggesting a role for DNA methylation in parent-of-origin effects, in the sense that the inactive allele is the methylated one. A model of the results from interploidy crosses, and of the effect of global DNA hypomethylation on parental imprinting in *Arabidopsis*, is shown in Plate 13, Fig. a,b. In similar experiments with *fis* mutants, which exhibit endosperm phenotypes diagnostic of paternal excess crosses, it was found that pollination of mutant ovules with pollen from low methylation plants restores the normal pattern of endosperm development and rescues seed viability. In the case of *mea* and *fis2* mutants, the rescue occurs even in the absence of functional paternally-derived *MEA* and *FIS2* alleles, respectively (Luo et al. 2000; Vinkenoog et al. 2000). Thus, methylation of DNA as an essential component of the imprinting mechanism provides a simple conceptual framework to explain endosperm development in *Arabidopsis*.

A complete understanding of the mechanism by which the DNA methylation machinery regulates imprinting in *Arabidopsis* endosperm continues to remain elusive, as interactions of imprinted genes with those that maintain cytosine methylation expand the complexity of the process. Along with *MEA*, a late-flowering gene, *FWA*, which encodes a homeodomain transcription factor, has become the focus of much interest because it displays imprinted expression in the endosperm. Molecular data have established that *FWA* gene expression is confined to the developing endosperm and is coincident with a loss of DNA methylation of the direct repeat sequences of the 5' region of the gene; in contrast, the promoter repeats remain methylated in the embryo and seed coat, in which *FWA* gene is not expressed (Kinoshita et al. 2004). A functional DNA glycosylase encoded by the *DME* gene seems to play a key role in activating maternal expression of imprinted *MEA* and *FWA* genes in the endosperm by reducing methylated cytosine residues in the promoter. How this might happen is hinted at by observations such as inheritance of lesions in embryo and endosperm development in the *dme* mutant allele by the female gametophyte, initial expression of the *DME* gene in the unfused polar nuclei in the central cell and the absence of expression in the primary endosperm nucleus and the developing endosperm, lack of expression of *MEA* and *FWA* transgenes in the central cell and in the endosperm inheriting the *dme*-mutant allele, activation of the normally silenced paternal *MEA* allele by ectopic *DME* expression in the endosperm, and the speculation that DME protein might modify the genic chromatin structure by excising 5-methylcytosine residues (Choi et al. 2002; Kinoshita et al. 2004). The *MET1* gene, which maintains cytosine methylation, has also been implicated in imprinting of the *MEA* gene from the observation that mutations that suppress *dme*-mediated seed abortion phenotypes are due to lesions in the *MET1* gene. Since the *DME* gene activates and the *MET1* gene suppresses *MEA* gene expression, the data have been construed to favor the hypothesis that imprinting of the *MEA* gene is controlled in the female gametophyte by an antagonism between the two DNA modifying enzymes, MET1 methyltransferase and DNA glycosylase (Xiao et al. 2003).

Genetics and molecular biology have provided a good sense of the cellular processes that control endosperm development in *Arabidopsis*. Although all the pitfalls associated with the life of general conclusions based on investigations on a traditional model system well-known for its genetic and/or experimental tractability apply to the investigations reviewed above, future work is poised to address the many mysteries surrounding the mechanism of genomic imprinting in this unusual triploid tissue in *Arabidopsis* and, by extrapolation, in economically important cereal grains.

8.2 Genetics and Molecular Biology of the Cereal Endosperm

Among cereal grains, powerful genetic approaches were used to elucidate the dynamics of endosperm development in maize, long before comparable techniques came to be used in other cereals and in *Arabidopsis*. In a series of crosses, Kermicle (1970) showed that when the allele *R* (red) for pigmentation of the aleurone cells is inherited from the male parent, irrespective of the dose transmitted, the aleurone phenotype is patchy (mottled), in contrast to the solidly red-colored aleurone when *R* is transmitted through the female gametophyte. This mode of inheritance of aleurone pigmentation, initially considered as a maternal effect, was later established as a case of imprinting (Kermicle 1978). Following this work, the long arm of chromosome 10 of maize (Lin 1982), and genes encoding zeins (Lund et al. 1995a), α-tubulin (Lund et al. 1995b), a post-transcriptional regulator of zeins (*dzr1*; Chaudhuri and Messing 1994), maize homologs of *FIE* genes (Danilevskaya et al. 2003; Grimanelli et al. 2005), and two novel genes named *NO-APICAL MERISTEM (NAM) RELATED PROTEIN1* (*NRP1*) (Guo et al. 2003) and *MATERNALLY EXPRESSED GENE1* (*MEG1*) (Gutiérrez-Marcos et al. 2004) have been shown to be imprinted in the maize endosperm; a possible role for methylation in regulating the expression of the zein, α-tubulin, and *MEG1* genes was also indicated. Similar to the interploidy crosses in *Arabidopsis* described in the previous section, the development of the endosperm is greatly impaired in reciprocal crosses between diploid and tetraploid maize. Using the *indeterminate gametophyte* (*ig*) mutation, which produces abnormal numbers of polar nuclei that participate in fertilization as the female parent in crosses in maize, Lin (1984) generated grains with a range of endosperm karyotypes that varied both in total ploidy and in the balance of paternal and maternal genomes. Although there were some exceptions, the normal endosperm was invariably found to result from the union of two polar nuclei and one sperm with a 2m:1p ratio. From a geneticist's vantage point, this work definitively showed that maternally and paternally inherited alleles function differently, suggesting that parentally imprinted genes are involved in development of the endosperm. Some other aspects of gene dosage in the development of maize endosperm have been reviewed by Birchler (1993).

The endosperm of cereal grains serves well as a model system of a tissue constituted of four distinct cell types, namely, the embryo-surrounding region, basal endosperm transfer layer, aleurone cells, and starchy cells. Outstanding features of the cereal endosperm are that each differentiated cell type falls into a clearly recognizable discrete category, and that their fate is specified almost simultaneously during the waning period of the free nuclear phase or soon after cellularization is initiated (Becraft 2001; Olsen 2004b). Although the genetic programs that underlie specification of these cell types are different, taken as a whole, it is evident that a variety of interrelated activities are involved in integrating the information and shaping the mature endosperm.

8.2.1 Embryo-surrounding Region and Transfer Layer

The embryo-surrounding region has been particularly well-studied in maize, and is regulated by a characteristic set of genes. The first indications of this were obtained by Opsahl-Ferstad et al. (1997) who showed that transcripts of the gene *EMBRYO SURROUNDING REGION* (*Esr*), isolated by differential display between early developmental stages of the endosperm and embryo, are expressed in a restricted region of the endosperm close to the suspensor, later spreading to the endosperm cells surrounding the entire embryo and eventually settling in the cells close to the lower part of the suspensor (see Plate 13, Fig. c–f). Subsequent work established that *Esr* includes three highly homologous genes, *Esr1*, *Esr2*, and *Esr3*, which show similar expression patterns. These genes encode proteins that share partial sequence homology with the CLV3 protein of *Arabidopsis*, and that are released into the cell wall; the properties of the proteins are thus compatible with their function as the ligand of a receptor-like kinase (Bonello et al. 2000, 2002). Transcripts of two unrelated genes, *ZEA MAYS ANDROGENIC EMBRYOS1* (*ZmAE1*) and *ZmAE3*, which are expressed during embryogenic transformation of pollen grains of maize, have also been shown to be expressed in the endosperm surrounding the embryo (Magnard et al. 2000). Clearly, the early processes of delineation of the embryo-surrounding region of the endosperm and formation of pollen embryos need to be understood more completely to explain these findings. There is no counterpart to the embryo-surrounding region as a defined tissue in other parts of the plant in which to seek instructive comparisons about its origin, development, and function.

Progress in our understanding of the regulation of development of cells of the basal endosperm transfer layer can be attributed to the realization that these cells are the source and amplifiers of signals for solute transfer from the mother plant to the endosperm of cereal grains. The major solutes transferred are amino acids, sucrose, and monosaccharides, and their uptake by the growing endosperm is a critical factor in grain filling (Thompson et al. 2001). The basal endosperm transfer layer, or its equivalent, is ultrastructurally well characterized in maize, wheat, and barley. The unique part of the maize endosperm where transfer layers are found is the placentochalazal region; the two to three layers of cells of the endosperm close to the chalaza have extensive wall ingrowths in a decreasing gradient toward the inner cells (Schel et al. 1984; Gao et al. 1998). During wheat grain development, the nucellar projection cells that serve as the main route for solutes transported through the vascular tissues into the endosperm cavity become decorated with wall ingrowths and function as transfer cells (Wang et al. 1994), whereas in barley the transfer layer forms over the nucellar projection cells (Olsen et al. 1999). In barley endosperm, transcripts of the *ENDOSPERM1* (*END1*) gene begin to accumulate in the free nuclei confined to the area above the nucellar projection cells, and transcript accumulation continues into the fully cellularized endosperm (see Plate 14, Fig. a,b). Because the endosperm cells above the nucellar projection cells are destined to form the transfer layer, the accumulation of the *END1* gene transcripts in the free nuclei should be critical in indicating the existence of a positional signal for transfer cell differentiation at an early stage (Doan et al. 1996; Olsen et al. 1999; Becraft 2001).

In maize, there are several examples of genes that are highly expressed in the basal endosperm transfer layer. Probably the most well-characterized maize gene belongs to the *BASAL ENDOSPERM TRANSFER LAYER* (*BETL*) family, represented by *BETL1* to *BETL4*. In situ hybridization using *BETL* gene probes (see Plate 14, Fig. c–e) showed that these gene transcripts are expressed specifically in the transfer cells during early- to mid-stage endosperm development, with signal strength decreasing with grain maturity (Hueros et al. 1995, 1999b). Attesting to the functional potential of the *BETL1* gene, a promoter segment of the gene fused to a reporter gene is found to be expressed in the transfer layer cells of transgenic maize plants (Hueros et al. 1999a). The genetic characterization of a defective maize kernel mutant termed *reduced grain filling1* (*rgf1*), which was identified in a screen for mutations that reduced starch and fresh weight accumulation in the grain, suggested that a secondary effect of the mutation might be to impair the function of transfer cells. This was found to be the case, as expression of both *BETL1* and *BETL2* genes are significantly reduced in the transfer cells of the mutant grain (Maitz et al. 2000). Although the four *BETL* genes encode small polypepetides, they probably mediate diverse biological processes, such as *BETL1* in the structural specialization of the cell wall, *BETL1* and *BETL3* in plant defense-related antimicrobial functions, and *BETL4* as a trypsin inhibitor. Following the identification of additional cDNAs related to the *BETL2* gene, this gene family is now known as *BASAL LAYER ANTIFUNGAL PROTEINS* (*BAP*); data garnered from in vitro antifungal activity assays have indicated that BAP proteins possess fungicidal activity (Hueros et al. 1995, 1999b; Serna et al. 2001).

Consistent with the function of the transfer cells in the hydrolysis and resynthesis of sucrose entering the endosperm, transcripts of a cell wall invertase gene isolated from maize kernels are expressed specifically in the basal endosperm transfer cells (Taliercio et al. 1999). A gene that begins to be transcribed in the chalazal pole of the embryo sac of maize soon after fertilization, as well as in the free nuclei destined to form the endosperm transfer cells, has been identified as *ZmMRP1* (for *Zea mays MYB-RELATED PROTEIN1*). It encodes a protein whose regulatory function lies in the presence of nuclear localization signals and a MYB-related DNA-binding domain. Together, these findings suggest that the protein has the hallmarks of a transcription factor that might play a role in transfer cell differentiation, probably by triggering the diffusion of a signal from the maternal cells (Gómez et al. 2002). The group of genes known as *ZmEBE* (for *Zea mays embryo sac/*

basal endosperm transfer layer/embryo surrounding region), isolated by Magnard et al. (2003), appear to fulfill the expectation of all-purpose genes since they are expressed in both the transfer cells and embryo surrounding region of the endosperm as well as in the central cell before fertilization. These observations have raised the possibility that development of specialized endosperm domains is initiated already in the central cell even before fertilization. The newly characterized maize *MEG1* gene encodes a small, glycosylated, cysteine-rich polypeptide localized exclusively within the wall projections of transfer cells, and displaying other features typical of a transfer-cell specific gene, such as predominant expression in transfer cells during midstage of endosperm development, disappearance from the mature endosperm, and expression in transfer cells of endosperm of maize plants transformed with the promoter region of the gene fused with a reporter gene. The most important difference between the *MEG1* gene and other previously described transfer-cell-specific genes was unraveled by genetic characterization, which showed that the former displays maternal parent-of-origin expression at early stages of endosperm development, but reverts to biparental expression at later stages (Gutiérrez-Marcos et al. 2004). An important prerequisite for an understanding of the mechanism regulating the formation of the basal endosperm transfer layer domain is the identification of mutants with defects in the development of this tissue. Along with defects in pattern formation in the embryo and in the free-nuclear and cellular stages of endosperm development, the *globby1-1* (*glo1-1*) mutation in maize causes localized disruptions in the organization of the transfer cell layers (Costa et al. 2003). Study of this mutation, while not offering a definitive interpretation of the mechanism regulating the formation of transfer cell layers, is a useful guide for future dissection of the developmental signals that operate during the early stages of endosperm development in maize. Taken as a whole, studies on transfer cells have provided the best evidence for the existence of localized genetic factors for their differentiation.

8.2.2
Aleurone Cells and Starchy Endosperm

When, in the 1960s, the role of aleurone cells in response to a GA signal emanating from the embryo in the synthesis of α-amylase involved in the digestion of starch in the endosperm during germination of barley grains began to be unraveled, it was billed as a remarkable scientific landmark by being the first documented case of a hormone-induced change in gene expression in a plant. The work also focused renewed attention on the fact that, compared to the cells of the starchy endosperm, the aleurone layer is constituted of living cells. As described in Chap. 7, the role of microtubules in cellularization of the endosperm was established first in barley by the precise account of formation of nuclear-cytoplasmic domains leading to the periclinal division of the alveolar nuclei into an outer layer of aleurone layer initials and an inner layer of starchy cells. Based on the observation that the presence of cortical arrays and preprophase band of microtubules makes subsequent division of the aleurone initials different from the divisions of the starchy cell initials, it has been suggested that aleurone cell specification occurs after the first periclinal division of the alveolar nuclei (Olsen 2004b). At the molecular level, of the genes known to mark aleurone cell differentiation in barley, transcripts of the *LIPID TRANSFER PROTEIN2* (*LTP2*) gene are detected exclusively in aleurone cells shortly after they are delimited in the endosperm; the promoter region of this gene drives expression of a reporter gene exclusively in the aleurone layers of the developing barley endosperm in transient assays, and of the endosperm of transgenic rice (Kalla et al. 1994). A steadily growing, and clearly not yet definitive, list of marker genes for aleurone cell differentiation in barley endosperm includes *B22E* (Klemsdal et al. 1991), *pZE40* (Smith et al. 1992), *CHITINASE26* (*CHI26*) (Leah et al. 1994), *OLE1*, *OLE2* (Aalen 1995), and *PEROXYREDOXIN1* (*PER1*) (Stacy et al. 1999).

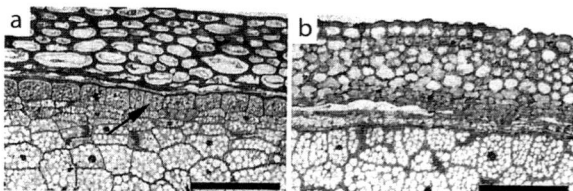

Fig. 8.2a,b Aleurone layer development cell in wild type and mutant maize. **a** Section of the wild-type grain showing a single layer of aeurone cells (*arrow*). **b** Section of the grain of the intermediate *dek1-792* mutant in which the aleurone layer in the endosperm is replaced by starchy cells. *Bars* 100 µm. (Reprinted from Becraft et al. 2002)

An important finding that has emerged from genetic analysis of maize endosperm cell lineage is that aleurone and starchy endosperm cells share a common lineage throughout development. This view, which has a bearing on the specification of aleurone cell fate, is contrary to the previously accepted position that aleurone and starchy endosperm cells are derived from separate lineages (Becraft and Asuncion-Crabb 2000). Mature aleurone cells of maize contain anthocyanins, which account for the wide variety of grain colors. Insights into the molecular basis of aleurone cell specification in maize endosperm have been provided by analyses of a small number of mutants impaired in aleurone cell development, and by characterization of the protein products of the mutated genes. One of the first mutants isolated is *crinkly4* (*cr4*), which displays aleurone mosaicism leading to the formation of patches of pigmented and nonpigmented regions in the aleurone layer. Characteristically, the peripheral cells in the nonpigmented regions of the mutant assume the identity of starchy endosperm, and not of aleurone cells, suggesting that the *CR4* gene is necessary for the acquisition of aleurone cell fate (Becraft et al. 1996). A second maize gene required for aleurone cell specification is *DEK1*; a loss of function in this gene blocks the formation of aleurone cells and causes a switch in the fate of peripheral cells into starchy cells from the time aleurone cells are specified in the wild-type endosperm (Fig. 8.2a,b). Conversely, aleurone cell identity is restored in cells already differentiated as starchy endosperm in grains in which the *DEK1* function is also restored by using a *Mu*-induced allele (Becraft and Asuncion-Crabb 2000; Becraft et al. 2002). These results show that, in maize endosperm, aleurone cell fate is not fixed until late in development and that positional cues specify and maintain aleurone cell fate. Cloning of the *CR4* and *DEK1* genes and identification of their protein products have made it possible to piece together a speculative molecular mechanism underlying aleurone cell specification in maize. The *CR4* gene encodes a receptor protein kinase similar to a mammalian tumor necrosis factor receptor, making it a likely candidate for the reception of signals for aleurone cell specification (Becraft et al. 1996; Olsen et al. 1998). As noted in Chap. 5, the *DEK1* gene encodes a membrane protein similar to animal calpains. The evidence is not yet all in, but is consistent with a possible role for DEK1 protein in maintaining the aleurone cell fate specified by the CR4 receptor kinase in a signal transduction pathway (Lid et al. 2002). The isolation of another maize mutant, named *supernumerary aleurone1* (*sal1*), which carries up to seven layers of aleurone cells in the endosperm of defective kernels compared with the single layer in wild-type grains, shows that the *SAL1* gene is involved in specifying the aleurone cell layers in the endosperm (see Plate 14, Fig. f–h). The identification of the gene product as a protein peripherally implicated in membrane vesicle trafficking has raised additional interesting questions about the exact roles of the proteins encoded by the *CR4*, *DEK1* and *SAL1* genes in aleurone layer signaling (Shen et al. 2003). A similar question about an aleurone-specific gene, *Vpp1*, isolated from maize kernels is germane. This gene encodes a type of vacuolar H^+-translocating inorganic pyrophosphatase, a typical house-keeping protein located in the vacuolar membrane and probably involved in maintaining cell turgor (Wisniewski and Rogowsky 2004). The phenotypic consequences of other mutations in the aleurone layer of maize, such as defects in pigmentation (Gavazzi et al. 1996), defects in cell division pattern (Kessler et al. 2002), cellular disorganization (Lid et al. 2004), and penetration into the starchy layer (Olsen 2004a), are varied but striking.

As noted earlier, a dramatic change in the fate of the inner layer of cells of the endosperm evoked by a periclinal division of the alveolar nuclei results in the formation of the starchy endosperm. As the inner layer of cells divide in various planes, they begin to accumulate storage proteins to become the dominant part of the cereal endosperm. The *dek* mutants of maize were identified on the basis of a severe reduction of starchy endosperm in the grain along with profound defects in embryo development. The first group of *dek* mutants were induced by chemical treatment and included mostly kernel mutants with varying degrees and types of lesions in the endosperm (Neuffer and Sheridan 1980). Later, Scanlon et al. (1994) isolated by transposon insertion additional *dek* mutants that included a wide range of phenotypes, most commonly those with reduced endosperm size and empty pericarp or papery kernel. Phenotypic analysis of one of the mutants in the latter group designated as *discolored1* (*dsc1*) has shown that the endosperm fails to develop completely and later undergoes degradation and necrosis. Transcripts of the cloned gene are detected specifically in the kernels coincident with the first appearance of endosperm defects; this indicates a requirement for *DSC1* gene function for continued endosperm development (Scanlon and Myers 1998). In the *emp2* mutant of this group, the endosperm tissue that survives in the grain is found to be reabsorbed, or has become necrotic, before grain maturity (Fu et al. 2002).

In barley, defects in the starchy endosperm resulting in shrunken kernels are caused by *seg* (for *shrunken endosperm caused by the maternal genotype*), *dex* (for *defective endosperm expressing xenia*), and *sex* (for *shrunken endosperm expressing xenia*) mutations (Bosnes et al. 1987). An anatomical survey of eight *seg* mutants revealed that some of the mutants exhibited premature termination of grain filling due to the necrosis and degradation of maternal tissues such as the chalaza and the nucellar projections of the pericarp, and failure of antipodal degeneration in the embryo sac, whereas others exhibited abnormalities in endosperm growth pattern such as lack of central endosperm cells and distorted, disorganized, or uneven growth of the endosperm (Felker et al. 1985, 1987). A similar survey of collections of *sex* and *dex* mutants has shown varying degrees of defects in the formation of starchy endosperm, ranging from minor deviations from the wild-type to complete loss of starchy endosperm traits beginning with the arrest of free nuclear divisions (Bosnes et al. 1987, 1992).

8.3
Concluding Comments

Current research on the genetics and molecular biology of the endosperm, mainly in *Arabidopsis* and to a lesser extent in cereal grains, has generated important information on genes that prevent autonomous endosperm development in unfertilized ovules and on the control check points leading to nonequivalence in the function of the maternal and paternal contributions to the development of the endosperm. From this analysis, the endosperm has turned out to be an important site of imprinting following double fertilization. Although many questions remain concerning the genetics and molecular biology of endosperm development in the model systems considered in this chapter, one is especially compelling. To understand clearly the positional cues and signal transduction pathways in the specification of cell fate in the endosperm, it will be necessary to fill in missing links such as the distribution of important cell fate determinants for specifying aleurone and starchy cells. On the applied side, further analysis of the role of parental imprinting might provide unrivalled opportunities to overcome the effect of the endosperm acting as a barrier during wide hybridization of plants.

REFERENCES

Aalen RB (1995) The transcripts encoding two oleosin isoforms are both present in the aleurone and in the embryo of barley (*Hordeum vulgare* L.) seeds. Plant Mol Biol 28:583–588

Adams S, Vinkenoog R, Spielman M, Dickinson HG, Scott RJ (2000) Parent-of-origin effects on seed development in *Arabidopsis thaliana* require DNA methylation. Development 127:2493–2502

Becraft PW (2001) Cell fate specification in the cereal endosperm. Semin Cell Dev Biol 12:387–394

Becraft PW, Asuncion-Crabb Y (2000) Positional cues specify and maintain aleurone cell fate in maize endosperm development. Development 127:4039–4048

Becraft PW, Stinard PS, McCarty DR (1996) CRINKLY4: a TNFR-like receptor kinase involved in maize endosperm differentiation. Science 273:1406–1409

Becraft PW, Brown RC, Lemmon BE, Olsen O-A, Opsahl Ferstad HG (2001) Endosperm development. In: Bhojwani SS, Soh WY (eds) Current trends in the embryology of angiosperms. Kluwer, Dordrecht, pp 353–374

Becraft PW, Li K, Dey N, Asuncion-Crabb Y (2002) The maize *dek1* gene functions in embryonic pattern formation and cell fate specification. Development 129:5217–5225

Berger F (2003) Endosperm: the crossroad of seed development. Curr Opin Plant Biol 6:42–50

Birchler JA (1993) Dosage analysis of maize endosperm development. Annu Rev Genet 27:181–204

Bonello J-F, Opsahl-Ferstad H-G, Perez P, Dumas C, Rogowsky PM (2000) *Esr* genes show different levels of expression in the same region of the maize endosperm. Gene 246:219–227

Bonello J-F, Sevilla-Lecoq S, Berne A, Risueño M-C, Dumas C, Rogowsky PM (2002) Esr proteins are secreted by the cells of the embryo surrounding region. J Exp Bot 53:1559–1568

Bosnes M, Harris E, Aigeltinger L, Olsen O-A (1987) Morphology and ultrastructure of 11 barley shrunken endosperm mutants. Theor Appl Genet 74:177–187

Bosnes M, Weideman F, Olsen O-A (1992) Endosperm differentiation in barley wild-type and *sex* mutants. Plant J 2:661–674

Brink RA, Cooper DC (1947) The endosperm in seed development. Bot Rev 13:423–541

Bushell C, Spielman M, Scott RJ (2003) The basis for natural and artificial postzygotic hybridization barriers in *Arabidopsis* species. Plant Cell 15:1430–1442

Chaudhuri S, Messing J (1994) Allele-specific parental imprinting of *dzr1*, a posttranscriptional regulator of zein accumulation. Proc Natl Acad Sci USA 91:4867–4871

Chaudhury AM, Ming L, Miller C, Craig S, Dennis ES, Peacock WJ (1997) Fertilization-independent seed development in *Arabidopsis thaliana*. Proc Natl Acad Sci USA 94:4223–4228

Chaudhury AM, Koltunow A, Payne T, Luo M, Tucker MR, Dennis ER, Peacock WJ (2001) Control of early seed development. Annu Rev Cell Dev Biol 17:677–699

Choi Y, Gehring M, Johnson L, Hannon M, Harada JJ, Goldberg RB, Jacobsen SE, Fischer RL (2002) DEMETER, a DNA glycosylase domain protein, is required for endosperm gene imprinting and seed viability in *Arabidopsis*. Cell 110:33–42

Collinge MA, Spillane C, Köhler C, Gheyselinck J, Grossniklaus U (2004) Genetic interaction of an origin recognition complex subunit of the polycomb group gene *MEDEA* during seed development. Plant Cell 16:1035–1046

Costa LM, Gutierrez-Marcos JF, Brutnell TP, Greenland AJ, Dickinson HG (2003) The *globby1-1* (*glo1-1*) mutation disrupts nuclear and cell division in the developing maize seed causing alterations in endosperm cell fate and tissue differentiation. Development 130:5009–5017

Costa LM, Gutièrrez-Marcos JF, Dickinson HG (2004) More than a yolk: the short life and complex times of the plant endosperm. Trends Plant Sci 9:507–514

Danilevskaya ON, Hermon P, Hantke S, Muszynski MG, Kollipara K, Ananiev EV (2003) Duplicated *fie* genes in maize: expression pattern and imprinting suggest distinct functions. Plant Cell 15:425–438

Dickinson H (2003) Plant cell cycle: cellularization of the endosperm needs spätzle. Curr Biol 13:R146–R148

Doan DNP, Linnestad C, Olsen O-A (1996) Isolation of molecular markers from the barley endosperm coenocyte and the surrounding nucellus cell layers. Plant Mol Biol 31:877–886

Felker FC, Peterson DM, Nelson OE (1985) Anatomy of immature grains of eight maternal effect shrunken endosperm barley mutants. Am J Bot 72:248–256

Felker FC, Peterson DM, Nelson OE (1987) Early grain development of the *seg2* maternal-effect shrunken-endosperm mutant of barley. Can J Bot 65:943–948

Fu S, Meeley R, Scanlon MJ (2002) *empty pericarp2* encodes a negative regulator of the heat shock response and is required for maize embryogenesis. Plant Cell 14:3119–3132

Gao R, Dong S, Fan J, Hu C (1998) Relationship between development of endosperm transfer cells and grain mass in maize. Biol Plant 41:539–546

Garcia D, Saingery V, Chambrier P, Mayer U, Jürgens G, Berger F (2003) *Arabidopsis haiku* mutants reveal new controls of seed size by endosperm. Plant Physiol 131:661–1670

Gavazzi G, Dolfini S, Allegra D, Castiglioni P, Todesco G, Hoxha M (1996) *Dap* (*D*efective *a*leurone *p*igmentation) mutations affect maize aleurone development. Mol Gen Genet 256:223–230

Gehring M, Choi Y, Fischer RL (2004) Imprinting and seed development. Plant Cell 16:S203–S213

Gómez E, Royo J, Guo Y, Thompson R, Hueros G (2002) Establishment of central endosperm expression domains: identification and properties of a maize transfer-cell-specific transcription factor, *ZmMRP-1*. Plant Cell 14:599–610

Grimanelli D, Perotti E, Ramirez J, Leblanc O (2005) Timing of the maternal-to-zygotic transition during early seed development in maize. Plant Cell 17:1061–1072

Grini PE, Jürgens G, Hülskamp M (2002) Embryo and endosperm development is disrupted in the female gametophytic *capulet* mutants of *Arabidopsis*. Genetics 162:1911–1925

Grossniklaus U, Vielle-Calzada J-P, Hoeppner MA, Gagliano WB (1998) Maternal control of embryogenesis by *MEDEA*, a polycomb group gene in *Arabidopsis*. Science 280:446–450

Guitton A-E, Page DR, Chambrier P, Lionnet C, Faure J-E, Grossniklaus U, Berger F (2004) Identification of new members of fertilization independent seed polycomb group pathway involved in the control of seed development in *Arabidopsis thaliana*. Development 131:2971–2981

Guo M, Rupe MA, Danilevskaya ON, Yang X, Hu Z (2003) Genome-wide mRNA profiling reveals heterochronic allelic variation and a new imprinted gene in hybrid maize endosperm. Plant J 36:30–44

Gutiérrez-Marcos JF, Costa LM, Biderre-Petit C, Khbaya B, O'Sullivan DM, Wormald M, Perez P, Dickinson HG (2004) *maternally expressed gene1* is a novel maize endosperm transfer cell-specific gene with a maternal parent-of-origin pattern of expression. Plant Cell 16:1288–1301

Haig D, Westoby M (1989) Parent-specific gene expression and the triploid endosperm. Am Nat 134:147–155

Haig D, Westoby M (1991) Genomic imprinting in endosperm: its effect on seed development in crosses between species, and between different ploidies of the same species, and its implications for the evolution of apomixis. Philos Trans R Soc London Ser B 333:1–13

Hueros G, Varotto S, Salamini F, Thompson RD (1995) Molecular characterization of *BET1*, a gene expressed in the endosperm transfer cells of maize. Plant Cell 7:747–757

Hueros G, Gomez E, Cheikh N, Edwards J, Weldon M, Salamini F, Thompson RD (1999a) Identification of a promoter sequence from the *BETL1* gene cluster able to confer transfer-cell-specific expression in transgenic maize. Plant Physiol 121:1143–1152

Hueros G, Royo J, Maitz M, Salamini F, Thompson RD (1999b) Evidence for factors regulating transfer cell-specific expression in maize endosperm. Plant Mol Biol 41:403–414

Kalla R, Shimamoto K, Potter R, Nielsen PS, Linnestad C, Olsen O-A (1994) The promoter of the barley aleurone-specific gene encoding a putative 7 kDa lipid transfer protein confers aleurone cell-specific expression in transgenic rice. Plant J 6:849–860

Kermicle JL (1970) Dependence of the *R*-mottled aleurone phenotype in maize on mode of sexual transmission. Genetics 66:69–85

Kermicle JL (1978) Imprinting of gene action in maize endosperm. In: Walden (ed) Maize breeding and genetics. Wiley, New York, pp 357–371

Kessler S, Seiki S, Sinha N (2002) *Xcl1* causes delayed oblique periclinal cell divisions in developing maize leaves, leading to cellular differentiation by lineage instead of position. Development 129:1859–1869

Kinoshita T, Yadegari R, Harada JJ, Goldberg RB, Fischer RL (1999) Imprinting of the *MEDEA* polycomb gene in the *Arabidopsis* endosperm. Plant Cell 11:1945–1952

Kinoshita T, Miura A, Choi Y, Kinoshita Y, Cao X, Jacobsen SE, Fischer RL, Kakutani T (2004) One-way control of *FWA* imprinting in *Arabidopsis* endosperm by DNA methylation. Science 303:521–523

Kiyosue T, Ohad N, Yadegari R, Hannon M, Dinneny J, Wells D, Katz A, Margossian L, Harada JJ, Goldberg RB, Fischer RL (1999) Cotntrol of fertilization-independent endosperm development by the *MEDEA* polycomb gene in *Arabidopsis*. Proc Natl Acad Sci USA 96:4186–4191

Klemsdal SS, Hughes W, Lönneborg A, Aalen RB, Olsen O-A (1991) Primary structure of a novel barley gene differentially expressed in immature aleurone layers. Mol Gen Genet 228:9–16

Köhler C, Hennig L, Bouveret R, Gheyselinck J, Grossniklaus U, Gruissem W (2003) *Arabidopsis* MSI1 is a component of the MEA/FIE *polycomb* group complex and required for seed development. EMBO J 22:4804–4814

Lauber MH, Waizenegger I, Steinmann,T, Schwarz H, Mayer U, Hwang I, Lukowitz W, Jürgens G (1997) The *Arabidopsis* KNOLLE protein is a cytokinesis-specific syntaxin. J Cell Biol 139:1485–1493

Leah R, Skriver K, Knudsen S, Ruud-Hansen J, Raikhel NV, Mundy J (1994) Identification of an enhancer/silencer sequence directing the aleurone-specific expression of barley chitinase gene. Plant J 6:579–589

Lid SE, Gruis D, Jung R, Lorentzen JA, Ananiev E, Chamberlin M, Niu X, Meeley R, Nichols S, Olsen O-A (2002) The *defective kernel 1* (*dek1*) gene required for aleurone cell development in the endosperm of maize grains encodes a membrane protein of the calpain gene superfamily. Proc Natl Acad Sci USA 99:5460–5465

Lid SE, Al RH, Krekling T, Meeley RB, Ranch J, Opsahl-Ferstad H-G, Olsen O-A (2004) The maize *disorganized aleurone layer 1* and *2* (*dil1, dil2*) mutants lack control of the mitotic division plane in the aleurone layer of developing endosperm. Planta 218:370–378

Lin B-Y (1982) Association of endosperm reduction with parental imprinting in maize. Genetics 100:475–486

Lin B-Y (1984) Ploidy barrier to endosperm development in maize. Genetics 107:103–115

Liu C, McElver JA, Tzafrir I, Joosen R, Wittich P, Patton D, van Lammeren AAM, Meinke D (2002) Condensin and cohesin knockouts in *Arabidopsis* exhibit a *titan* seed phenotype. Plant J 29:405–415

Liu C, Meinke DW (1998) The *titan* mutants of *Arabidopsis* are disrupted in mitosis and cell cycle control during development. Plant J 16:12–31

Lund G, Ciceri P, Viotti A (1995a) Maternal-specific demethylation and expression of specific alleles of zein genes in the endosperm of *Zea mays* L. Plant J 8:571–581

Lund G, Messing J, Viotti A (1995b) Endosperm-specific demethylation and activation of specific alleles of α-tubulin genes of *Zea mays* L. Mol Gen Genet 246:716–722

Luo M, Bilodeau P, Koltunow A, Dennis ES, Peacock WJ, Chaudhury AM (1999) Genes controlling fertilization-independent seed development in *Arabidopsis thaliana*. Proc Natl Acad Sci USA 96:296–301

Luo M, Bilodeau P, Dennis ES, Peacock WJ, Chaudhury A (2000) Expression and parent-of-origin effects for *FIS2*, *MEA*, and *FIE* in the endosperm and embryo of developing *Arabidopsis* seeds. Proc Natl Acad Sci USA 97:10637–10642

Magnard J-L, le Deunff E, Domenech J, Rogowsky PM, Testillano PS, Rougier M, Risueño MC, Vergne P, Dumas C (2000) Genes normally expressed in the endosperm are expressed at early stages of microspore embryogenesis in maize. Plant Mol Biol 44:559–574

Magnard J-L, Lehouque G, Massonneau A, Frangne N, Heckel T, Gutierrez-Marcos JF, Perez P, Dumas C, Rogowsky PM (2003) ZmEBE genes show a novel, continuous expression pattern in the central cell before fertilization and in specific domains of the resulting endosperm after fertilization. Plant Mol Biol 53:821–836

Maitz M, Santandrea G, Zhang Z, Lal S, Hannah LC, Salamini F, Thompson RD (2000) *rgf1*, a mutation reducing grain filling in maize through effects on basal endosperm and pedicel development. Plant J 23:29–42

Mayer U, Herzog U, Berger F, Inzé D, Jürgens G (1999) Mutations in the *PILZ* group genes disrupt the microtubule cytoskeleton and uncouple progression from cell division in *Arabidopsis* embryo and endosperm. Eur J Cell Biol 78:100–108

McElver J, Patton D, Rumbaugh M, Liu C, Yang LJ, Meinke D (2000) The *TITAN5* gene of *Arabidopsis* encodes a protein related to ADP ribosylation factor family of GTP binding proteins. Plant Cell 12:1379–1392

Menand B, Desnos T, Nussaume L, Berger F, Bouchez D, Meyer C, Robaglia C (2002) Expression and disruption of the *Arabidopsis TOR* (target of rapamycin) gene. Proc Natl Acad Sci USA 99:6422–6427

Neuffer MG, Sheridan WF (1980) Defective kernel mutants of maize. Genetic and lethality studies. Genetics 95:929–944

Ohad N, Margossian L, Hsu Y, Williams C, Repetti P, Fischer RL (1996) A mutation that allows endosperm development without fertilization. Proc Natl Acad Sci USA 93:5319–5324.

Ohad N, Yadegari R, Margossian L, Hannon M, Michaeli D, Harada JJ, Goldberg RB, Fischer RL (1999) Mutations in *FIE*, a WD polycomb group gene, allow endosperm development without fertilization. Plant Cell 11:407–415

Olsen O-A (2004a) Dynamics of maize aleurone cell formation: the "surface" rule. Maydica 49:37–40

Olsen O-A (2004b) Nuclear endosperm development in cereals and *Arabidopsis thaliana*. Plant Cell 16:S214–S227

Olsen O-A, Lemmon B, Brown R (1998) A model for aleurone development. Trends Plant Sci 3:168–169

Olsen O-A, Linnestad C, Nichols SE (1999) Developmental biology of the cereal endosperm. Trends Plant Sci 4:253–257

Opsahl-Ferstad H-G, le Deunff E, Dumas C, Rogowsky PM (1997) *ZmEsr*, a novel endosperm-specific gene expressed in a restricted region around the maize embryo. Plant J 12:235–246

Scanlon MJ, Myers AM (1998) Phenotypic analysis and molecular cloning of *discolored-1* (*dsc1*), a maize gene required for early kernel development. Plant Mol Biol 37:483–493

Scanlon MJ, Stinard PS, James MG, Myers AM, Robertson DS (1994) Genetic analysis of 63 mutations affecting maize kernel development isolated from *Mutator* stocks. Genetics 136:281–294

Schel JHN, Kieft H, van Lammeren AAM (1984) Interaction between embryo and endosperm during early developmental stages of maize carypses (*Zea mays*). Can J Bot 62:2842–2853

Scott RJ, Spielman M, Bailey J, Dickinson HG (1998) Parent-of-origin effects on seed development in *Arabidopsis thaliana*. Development 125:3329–3341

Serna A, Maitz M, O'Connell T, Santandrea G, Thevissen K, Tienens K, Hueros G, Faleri C, Cai G, Lottspeich F, Thompson RD (2001) Maize endosperm secretes a novel antifungal protein into adjacent maternal tissue. Plant J 25:687–698

Shen B, Li C, Min Z, Meeley RB, Tarczynski MC, Olsen O-A (2003). *sal1* determines the number of aleurone cell layers in maize endosperm and encodes a class E vacuolar sorting protein. Proc Natl Acad Sci USA 100:6552–6557

Smith LM, Handley J, Li Y, Martin H, Donovan L, Bowles DJ (1992) Temporal and spatial regulation of a novel gene in barley embryos. Plant Mol Biol 20:255–266

Sørensen MB, Mayer U, Lukowitz W, Robert H, Chambrier P, Jürgens G, Somerville C, Lepiniec L, Berger F (2002) Cellularisation in the endosperm of *Arabidopsis thaliana* is coupled to mitosis and shares multiple components with cytokinesis. Development 129:5567–5576

Springer PS, Holding DR, Groover A, Yordan C, Martienssen RA (2000) The essential Mcm7 protein PROLIFERA is localized to the nucleus of dividing cells during the G_1 phase and is required maternally for early *Arabidopsis* development. Development 127:1815–1822

Stacy RAP, Nordeng TW, Culiáñez-Macià FA, Aalen RB (1999) The dormancy-related peroxiredoxin anti-oxidant, PER1, is localized to the nucleus of barley embryo and aleurone cells. Plant J 19:1–8

Steinborn K, Maulbetsch C, Priester B, Trautmann S, Pacher T, Geiges B, Küttner F, Lepiniec L, Stierhof Y-D, Schwarz H, Jürgens G, Mayer U (2002) The *Arabidopsis PILZ* group genes encode tubulin-folding cofactor orthologs required for cell division but not cell growth. Genes Dev 16:959–971

Taliercio EW, Kim J-Y, Mahe A, Shanker S, Choi J, Cheng W-H, Prioul J-L, Chourey PS (1999) Isolation, characterization and expression analyses of two cell wall invertase genes in maize. J Plant Physiol 155:197–204

Thompson RD, Hueros G, Becker H-A, Maitz M (2001) Development and function of seed transfer cells. Plant Sci 160:775–783

Tzafrir I, McElver JA, Liu C, Yang LJ, Wu JQ, Martinez A, Patton DA, Meinke DW (2002) Diversity of TITAN functions in *Arabidpsis* seed development. Plant Physiol 128:38–51

Vielle-Calzada J-P, Thomas J, Spillane C, Coluccio A, Hoeppner MA, Grossniklaus U (1999) Maintenance of genomic imprinting at the *Arabidopsis medea* locus requires zygotic *DDM1* activity. Genes Dev 13:2971–2982

Vinkenoog R, Scott RJ (2001) Autonomous endosperm development in flowering plants: how to overcome the imprinting problem? Sex Plant Reprod 14:189–194

Vinkenoog R, Spielman M, Adams S, Fischer RL, Dickinson HG, Scott RJ (2000) Hypomethylation promotes autonomous endosperm development and restores postfertilization lethality in *fie* mutants. Plant Cell 12:2271–2282

von Wangenheim K-H, Peterson H-P (2004) Aberrant endosperm development in interploidy crosses reveals a timer of differentiation. Dev Biol 270:277–289

Wang HL, Offler CE, Patrick JW (1994) Nucellar projection transfer cells in the developing wheat grain. Protoplasma 182:39–52

Wisniewski J-P, Rogowsky PM (2004) Vacuolar H^+-translocating inorganic pyrophosphatase (Vpp1) marks partial aleurone cell fate in cereal endosperm development. Plant Mol Biol 56:325–337

Xiao W, Gehring M, Choi Y, Margossian L, Pu H, Harada JJ, Goldberg RB, Pennell RI, Fischer RL (2003) Imprinting of the *MEA* polycomb gene is controlled by antagonism between MET1 methyltransferase and DME glycosylase. Dev Cell 5:891–901

Yadegari R, Kinoshita T, Lotan O, Cohen G, Katz A, Choi Y, Katz A, Nakashima K, Harada JJ, Goldberg RB, Fischer RL, Ohad N (2000) Mutations in the *FIE* and *MEA* genes that encode interacting polycomb proteins cause parent-of-origin effects on seed development by distinct mechanisms. Plant Cell 12:2367–2381

9 Non-zygotic Embryo Development – Embryogenesis without Sex

Thus the totipotency of carrot cells is now established beyond question, and it is also very clear that the environment of the ovule and the embryo sac is dispensable if the proper nutrition and external conditions are furnished. Instead of the zygote being, developmentally speaking, a unique cell it is really to be regarded as a very general one; the zygote of an angiosperm now seems to be adequately described as a "diploid cell which can grow, in a medium which will make it grow, and in a space which will protect it and allow it to grow". From this point of view, the early sequence of cell patterns in embryos seems not to be so peculiar to embryogeny as might otherwise be supposed. When the stages in growth of carrot embryoids from free cells are examined they are found to recapitulate normal embryogeny in a surprisingly faithful manner.

F.C. Steward 1968

9.1   Apomixis   188
9.1.1   Case Studies of Diplosporous and Aposporous Apomicts   189
9.1.2   Adventive Embryogenesis   190
9.1.3   Molecular Genetics of Apomixis   192
9.2   Somatic Embryogenesis   192
9.2.1   A History of the Recent Past   192
9.2.2   Somatic Embryogenesis in Carrot and other Model Systems   194
9.2.3   Embryonic Proteins and Regulation of Gene Expression   199
9.3   Pollen Embryogenesis   200
9.3.1   Responsive Stage of Pollen Development and Pollen Embryogenic Potential   201
9.3.2   Cytology of Pollen Embryogenesis   203
9.3.3   Molecular Biology of Pollen Embryogenesis   204
9.4   Concluding Comments   205
    References   206

Flowering plants display many and varied forms of vegetative reproduction, such as by roots and leaves and by aerial stems known as runners or stolons, and underground stems known as rhizomes, corms, bulbs, and tubers. Two features define progeny that are produced vegetatively: they are identical to the single parent from which they arise, and they do not go through an embryogenic phase on their way to becoming a new plant. Indeed, these methods form the basis of clonal propagation, which is widely employed in agricultural and horticultural practices. Asexual (non-zygotic) development of embryos enclosed within seeds as a way of propagation of progeny, and induction of embryos by culture of plant organs for clonal multiplication, are also now considered to be relatively prevalent in flowering plants. The term apomixis, which appears to have an old pedigree, has long been used to describe all forms of asexual reproduction in plants, including multiplication by vegetative propagules. However, this broad definition of apomixis is rarely used nowadays and the current usage of apomixis employed in this book restricts it to the formation of a seed enclosing an embryo produced from the maternal gametophytic or sporophytic cells of the ovule, circumventing the processes of meiosis and double fertilization (Nogler 1984; Bicknell and Koltunow 2004). Two later episodes in the non-zygotic embryo development saga in flowering plants occurred with the discoveries of somatic embryogenesis and pollen (or microspore) embryogenesis. The concept of somatic embryogenesis was launched with the demonstration by tissue culture techniques that single somatic cells nurtured in a suspension culture can give rise to fertile plants simulating stages starkly reminiscent of zygotic embryogenesis. Although pollen grains of flowering plants are programmed for terminal differentiation to produce pollen tubes and male gametes, culture of anthers or isolated pollen grains at an appropriate stage of development in a mineral salt medium with or without hormonal supplements was shown to induce repeated divisions in the pollen grains. The multicellular pollen grains thus formed go through an embryogenic phase of development to form haploid plants; this constitutes the phenomenon of pollen embryogenesis.

In the context of double fertilization, combined with the ability to form embryos without fertilization, the gradually increasing knowledge about the developmental and molecular aspects of apomixis, somatic embryogenesis, and pollen embryogenesis makes these topics suitable for consideration in this final chapter of the book. Over the years, each of these topics has amassed its own vast literature and so some subjective judgment has gone into the selection of model systems and issues considered here.

9.1 Apomixis

Because some fundamental cytological features are shared with sexual reproduction, a brief background of the apomictic pathway is necessary to assess the precise cytological changes that initiate embryo development without meiosis and fertilization, and to put in proper perspective recent developments in the field. Reported cases of apomixis, conservatively estimated at 300–400 species of flowering plants distributed within 35–40 families, have begun to fulfill the promise that this is a natural process for seed production without fertilization; among eudicots, apomixis is well represented in the Asteraceae and Rosaceae, and among monocots, Poaceae has the largest number of reported cases (Carman

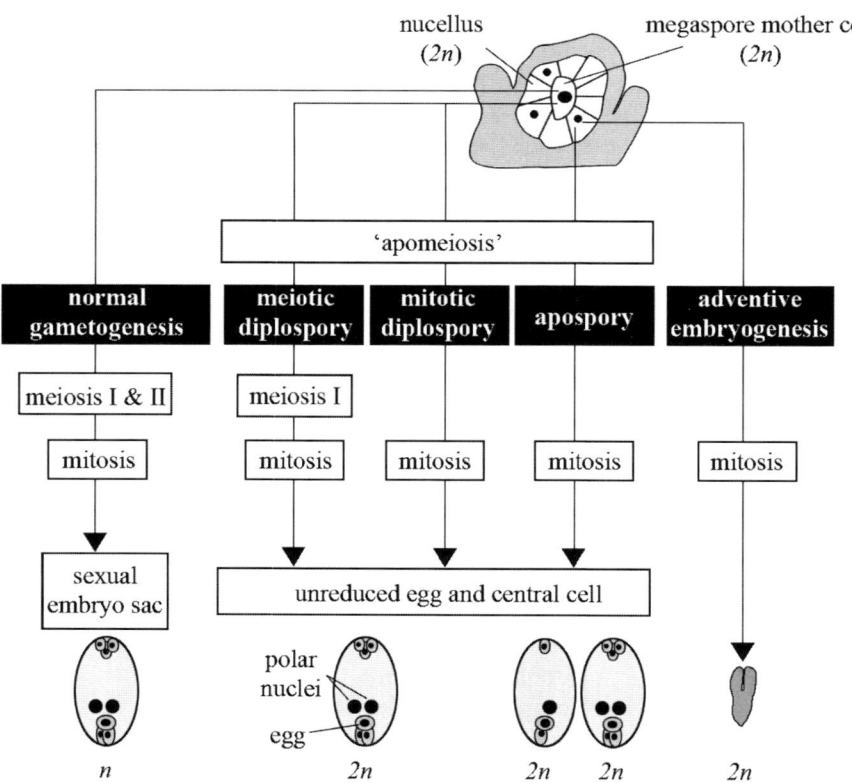

Fig. 9.1
Diagrams showing normal sexual and major apomictic pathways. In the sexual pathway, double fertilization results in the formation of a diploid embryo and triploid endosperm. In the diplosporous and aposporous types of apomixis, the embryo develops by parthenogenesis of the diploid egg cell. In most apomicts, development of the endosperm follows pollination and/or fertilization, but in some the endosperm develops autonomously without fertilization. In adventive embryogenesis, an asexual embryo is produced from a sporophytic cell outside the normal sexual embryo sac. (Modified from Spielman et al. 2003)

1997; Richards 1997). Acceptance of the view that, in apomictic reproduction, the embryo has its origin directly in an unreduced cell lineage, leads to the consideration of three main pathways of apomixis, namely diplospory, apospory, and adventive embryogenesis. In the first two pathways, which are designated as gametophytic apomixis, well-developed embryo sacs are formed from the unreduced megaspore mother cell (diplospory), or from a somatic cell of the ovule such as the nucellus (apospory). In both diplospory and apospory, meiosis is circumvented by a process known as apomeiosis, resulting in the formation of unreduced embryo sacs. Although many variations in the development of embryo sacs arising from apomeiosis have been described, a further division of diplospory into the meiotic type, where the megaspore mother cell fails to complete meiosis, and the mitotic type, where this cell completely bypasses meiosis, is widely recognized. In both diplosporous and aposporous types of apomicts, fertilization of the egg is avoided by parthenogenic initiation of embryogenesis; the embryo might also arise autonomously from one of the other cells of the embryo sac. During adventive embryogenesis, a somatic cell of the ovule directly gives rise to an embryo. The three pathways of apomixis described here, along with the pathway of sexual female gametogenesis, are summarized in Fig. 9.1. Cytological aberrations in megasporogenesis and megagametogenesis in these and other different forms of apomixis are described in reviews by Nogler (1984), Koltunow (1993), and Savidan (2000).

9.1.1
Case Studies of Diplosporous and Aposporous Apomicts

Past research on the diplosporic and aposporic types of apomixis has traditionally maintained a strong focus on embryo sac development. Named after the genera in which apomictic gametogenesis was first described, there are a number of well-documented comparisons of megagametogenesis in sexual and apomictic plants of the same species. The basic feature of the Taraxacum type (described in *Taraxacum megalorrhizon*; Asteraceae) illustrating meiotic diplospory is, as expected, the consistent derangement of meiosis in the megaspore mother cell. Although this cell initially enters the meiotic prophase, normal pairing of homologous chromosomes does not take place due to asynapsis. Consequently, the orphaned univalents are scattered over the mitotic spindle at the metaphase stage of meiosis I. The result is that, during meiosis I, a restitution nucleus is formed and this nucleus divides mitotically to form a dyad with somatic chromosome numbers. One of the cells of the dyad, generally the one at the micropylar end, degenerates and the chalazal cell functions as the embryo sac mother cell. This cell undergoes three more mitotic divisions to form an eight-nucleate embryo sac of the Polygonum type (see Chap. 1), which is also found in the sexual relative (Battaglia 1948; Nogler 1984). Recent additions to the Taraxacum type of diplosporous apomict are *Arabis holboellii* and *A. gunnisoniana*, which, as members of the Brassicaceae, like *Arabidopsis*, are considered to have the potential to develop into model plants for molecular studies of apomixis (Naumova et al. 2001; Taşkin et al. 2004). In the Antennaria type, illustrating mitotic diplospory, the megaspore mother cell does not go into meiosis but, after a long interphase, it divides mitotically and becomes an unreduced binucleate functional megaspore. Two further mitotic divisions lead to the formation of a typical Polygonum-type of embryo sac (Stebbins 1932; Nogler 1984). Contrary to the condition observed in sexual plants, in most aposporic apomicts, asynchronous development of several embryo sacs, each having its origin in an aposporus initial cell of an ovule, is frequent. The best-studied example of an aposporous apomict is *Hieracium* (Asteraceae). In apomictic accessions of *H. aurantiacum* and *H. piloselloides*, cells that give rise to unreduced embryo sacs are seen close to the functional megaspore of the meiotic tetrad, whose disintegration is initiated by the appearance of apomictic initials. Following directional growth of the aposporous initial toward the micropylar end of the ovule, and its differentiation into a megaspore mother cell, a single embryo sac is formed in the same position occupied by the developing sexual embryo sac. It is believed that the megaspore mother cell does not undergo meiosis, or does so infrequently. In *H. aurantiacum*, embryo sacs containing a few nuclei sometimes fuse together and form a disorganized, functional embryo sac with a highly variable nuclear complement distinct from the Polygonum type embryo sac observed in the sexual plant (Koltunow et al. 1998, 2000).

Although much of the mystery surrounding apomixis has been dispelled by precise cytological studies, it is still unclear whether there are any reliable clues to establish apomictic cell-type identity in an ovule. The characterization of temporary deposits of callose on the walls of megaspore mother cells of the monosporic and bisporic types of embryo sacs, and its absence in the corresponding cells of several diplosporous apomicts exhibiting meiotic diplospory (Carman et al. 1991; Peel et al. 1997) and mitotic diplospory (Leblanc et al. 1995), have revealed a feature of apomixis different from the usual sexual reproduction; however, the fact that both the absence of callose (Araujo et al. 2000; Tucker et al. 2001) and the presence of normal levels of callose (Naumova et al. 1993; Naumova and Willemse 1995) have been reported in aposporous species seems to indicate that the absence of callose may not be a secure developmental marker for all types of apospory.

To complete the apomictic life cycle without altering the chromosome number of the sporophytic generation, the diploid egg develops into an embryo without fusion with a sperm nucleus. As described in *Hieracium*, apomictic embryo development might appear not to differ much from that of the sexual embryo, except for the frequent formation of more than one embryo in a seed and irregularities in cotyledon development (Koltunow et al. 1998). In apomictic plants, before or coincident with the parthenogenic division of the egg, the polar nuclei, unencumbered by fusion with a second sperm cell, also begin to divide to form the endosperm. During its autonomous development, the endosperm shows varying degrees of ploidy, such as the $2n$ and $4n$ cells that arise from free and fused polar nuclei, respectively (Stebbins and Jenkins 1939). Development of the endosperm in most apomicts follows pollination and/or fertilization by the process known as pseudogamy. Although endosperm development going through free nuclear and cellular phases has been described in *Hieracium*, a requirement for fusion of the polar nuclei with a sperm for endosperm formation has not been established (Koltunow et al. 1998). However, data from flow cytometric analysis of seeds of *Arabis holboellii* and *A. gunnisoniana* have been interpreted as indicating the occurrence of pseudogamous endosperm development following fusion of the unreduced polar nuclei with an unreduced or reduced sperm (Matzk et al. 2000; Naumova et al. 2001; Taşkin et al. 2004).

As described in Chap. 8, in-depth studies of *fis*-class mutants of *Arabidopsis* have generated a considerable body of data on the differential effects of genes derived from paternal and maternal genomes on endosperm development. The relevance of these mutants to endosperm development in apomicts is that mutations in the three *FIS*-class genes (*MEA*, *FIS2*, and *FIE*) cause some endosperm proliferation in the ovule, even in the absence of fertilization, in a way reminiscent of apomixis. However, unlike in apomixis, in *fis* mutants, the embryo sac is formed by meiotic division and, since a full-term fertilization-independent embryo is not formed, seed abortion is the invariable final outcome in these mutants. The observation that events associated with endosperm formation are sufficient to trigger fruit development in *fis* mutants, together with evidence for genomic imprinting in endosperm formation in the mutants, have provoked renewed interest in signal activation at fertilization, control of embryo growth and endosperm formation, the role of the maternal tissues of the ovule, and in genomic imprinting in seed development in apomicts (Grossniklaus et al. 2001).

9.1.2
Adventive Embryogenesis

In adventive embryogenesis, which is treated differently from diplospory and apospory due to the absence of embryo sac formation, the embryo develops from a somatic cell of the ovule such as the integuments or nucellus, and is initiated as a bud-like outgrowth by mitotic divisions of the progenitor cell. In their early stages of development, these outgrowths resemble adventitious buds or globular embryos. Later, they differentiate and display facsimiles of subsequent stages of zygotic embryogenesis (Koltunow 1993). The nucellar type of adventive embryogenesis is more common than the integumentary type, and *Citrus* (Rutaceae) has emerged as an excellent system to study morphogenesis of nucellar embryogenesis. One feature that makes *Citrus* ideal for this type of study is that sexual and apomictic processes coexist in the same ovule, allowing on-the-spot comparisons between the two. Although several anatomical investigations have provided insights into the development of asexual embryos, especially in *Citrus* cultivars, the observations made by Wilms et al. (1983), Wakana and Uemoto (1987, 1988), Naumova (1993), and Koltunow et al. (1995) will be highlighted here to provide a general account of this phenomenon. Despite some previous uncertainties, it now appears that the stimulus of pollination or fertilization is not necessary for nucellar embryogenesis, and that nucellar embryos are initiated autonomously in unpollinated and unfertilized ovules (Wakana and Uemoto 1987; Koltunow et al. 1995; Sharma and Thorpe 1995). The potential embryo initials are distinguished from the surrounding nucellar cells by their large nuclei and dense cytoplasm, but the molecular pathways that link cell determination with embryo formation in the absence of fertilization have not been identified (Wilms et al. 1983; Naumova 1993; Koltunow et al. 1995). Based on the formation of nucellar embryos in cultured ovules at different stages of development of the *Citrus* cultivar 'Valencia', it has been concluded that embryo initials are specified even before they are identified histologically in ovules. In unfertilized ovules of this cultivar, nucellar embryo initials are found either in the position normally occupied by the embryo sac or surrounding the crushed remnants of the embryo sac (Fig. 9.2a,b). In both fertilized and unfertilized ovules, embryogenic division of the nucellar cells is coincident with degeneration of the nucellus in the vicinity of the potential embryogenic nucellar cells. In addition, nucellar embryogenic cells in both fertilized and unfertilized ovules share initial isolation from maternal tissues by a thick wall, and reestablishment of contact with maternal nucellar cells. These observations suggest a role for the healthy or degenerated nucellar cells as a source of nutrients for the nascent embryogenic nucellar cells. Indicative of a need for a sufficient amount of nutrients of the type provided by the endosperm, it was noted that, unlike in fertilized ovules where nucellar embryos complete their development, nucellar embryogenesis is arrested at the globular stage in unfertilized ovules. In contrast to nonviable seeds, which enclose atrophied globular embryos generated by nucellar embryony in unfertilized ovules, the result of nucellar embryony in fertilized ovules is the production of polyembryonic seeds containing embryos at different stages of development potentiated to germinate and form seedlings (Koltunow et al. 1995). Despite the wealth of anatomical information, the extrinsic and intrinsic factors that direct certain cells of the nucellus to an embryo fate, as well as the molecular relationship between nucellar and sexual embryo developement, remain to be defined.

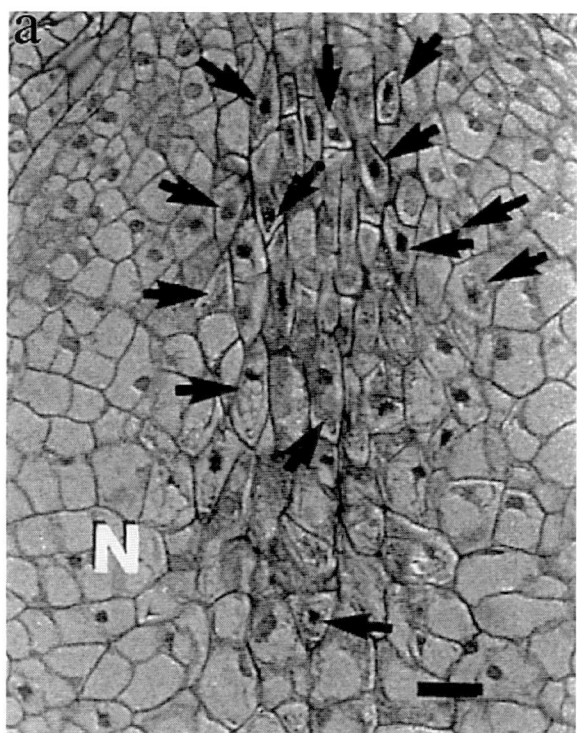

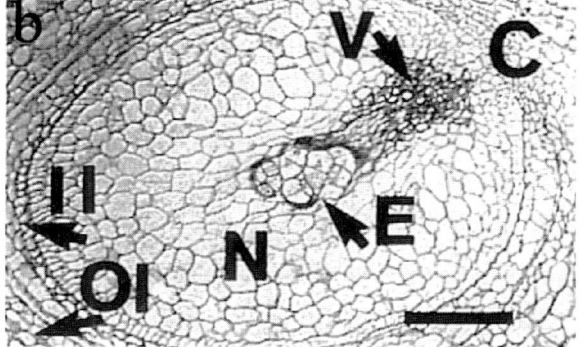

Fig. 9.2a,b Adventive embryogenesis in *Citrus*. **a** Section of part of the nucellus showing embryo initials (*arrows*) with thick walls. **b** Section of an ovule showing an embryogenic mass in the chalazal part of the ovule connected by a vascular network. *C* Chalazal end, *E* embryo, *II* inner integument, *N* nucellus, *OI* outer integument, *V* vascular network. *Bars* **a** 10 µm, **b** 100 µm. (Reprinted from Koltunow et al. 1995)

9.1.3
Molecular Genetics of Apomixis

A number of genetic crosses followed by analysis of hybrid progeny among aposporous apomicts have indicated that apospory is probably controlled by at least a few, and possibly many, genes. Similarly, analysis of hybrid progeny from crosses between sexual and apomictic species of *Citrus* has established that a single dominant gene controls the expression of the nucellar embryo phenotype. The available data do not, however, allow firm conclusions to be drawn about the inheritance of diplospory (Koltunow 1994; Grimanelli et al. 2001). Although genes triggering apomixis have not been isolated, differential and subtractive hybridization techniques have been successfully applied to identify cDNA clones or gene transcripts involved in apomictic and sexual pathways in *Pennisetum ciliare* (Vielle-Calzada et al. 1996), *Brachiaria brizantha* (Poaceae; Leblanc et al. 1997), *Hieracium piloselloides* (Guerin et al. 2000), and *Paspalum notatum* (Poaceae; Pessino et al. 2001). Based on an examination of the expression patterns of certain reproductive marker genes from *Arabidopsis*, such as *SPOROCYTELESS* (*SPL*), *SOMATIC EMBRYOGENESIS RECEPTOR KINASE1* (*SERK1*), and various *FIS* genes in transgenic sexual and apomictic lines of *Hieracium*, Tucker et al. (2003) have shown that initiation of apospory and autonomous embryo and endosperm development share gene expression and regulatory pathways with the corresponding stages of sexual reproduction. The work of Grimanelli et al. (2005) referred to in Chap. 5, using microarray technology to compare more than 5,000 unique sequences, revealed no significant modifications in transcript profiles during proembryo development in sexual and apomictic lines of maize. These collective results are instructive in emphasizing that apomixis and sexual reproduction are comparable but for the elimination of meiosis and fertilization in the former. As more and more research is conducted aimed at dissecting the genetic and molecular basis of apomixis, the hope is that researchers will eventually be able to ferret out the genes controlling apomixis and, by introducing these genes into sexual crops, change the latter into commercially viable apomicts for the propagation of desirable genotypes.

9.2
Somatic Embryogenesis

The discovery of somatic embryogenesis is usually traced to the investigations of Steward (1963) and Wetherell and Halperin (1963) who reported that, under certain experimental conditions, cultured cells and cell clusters of carrot are restructured in an embryogenic pathway, and regenerate facsimiles of zygotic embryos. Steward (1963; see also Steward et al. 1964) showed that when free cells sloughed off from immature embryos of carrot grown in a mineral salt medium supplemented with coconut water are plated on a solidified nutrient medium of the same composition, virtually every cell of the suspension yields an embryo-like structure, faithfully recapitulating the globular, heart-shaped, torpedo-shaped, and cotyledonary stages of zygotic embryogenesis. Following an initial demonstration that cells originating from a callus obtained from the root tissue of wild carrot nurtured in a medium containing coconut water form somatic embryos (Wetherell and Halperin 1963), later studies showed that it was possible to trigger somatic embryogenesis in cells originating from various organs of wild carrot including the root, peduncle, and petiole by inducing callus growth on explants in a medium containing a moderately high level of 2,4-D and transferring the callus to a medium containing a reduced level of the auxin (Halperin and Wetherell 1964; Halperin 1966). These observations provided the framework for a widely used protocol for inducing somatic embryogenesis in many plants by the simple expedient of inducing callus growth in explants cultured in a medium containing a high concentration of auxin and transferring the callus to a medium containing a reduced amount of the hormone or none at all.

9.2.1
A History of the Recent Past

The discovery of somatic embryogenesis has served plant biologists extremely well, as the synchronous embryogenic cell systems developed in carrot and other plants are increasingly being utilized as models of zygotic embryogenesis in attempts to characterize the molecular changes involved in the transformation of a single cell into an embryo. Starting with the first accounts of somatic embryogenesis in carrot, subsequent reports of somatic embryogenesis were initially confined to members of the carrot family (Umbelliferae) and later spread to members of a number of angiosperm and gymnosperm families. No consolidated listing of these plants is currently available, but separate listings of herbaceous eudicots (Brown et al. 1995), herbaceous monocots (Krishnaraj and Vasil 1995), woody angiosperms and gymnosperms (Dunstan et al. 1995), and angiosperms in general (Thorpe and Stasolla 2001), have been published. Among eudicots, Fabaceae stands out with the largest number of reports of somatic embryogenesis in about 82 species and hybrids, followed by Rutaceae with about 45 listings and Umbelliferae with 28 species. With about 96 listings, Poaceae has by far the largest number of species, hybrids, and transgenic plants showing somatic embryogenesis among monocots, followed distantly by Arecaceae with 16 listings (Thorpe and Stasolla 2001). Following the success achieved by Vasil and Vasil (1981) in inducing somatic embryogenesis in suspension cultures of *Pennisetum americanum* by the judicial choice of explants, particularly with regard to their physiological age at culture, use of selected hormonal additives and unorthodox manipulations of the medium at critical stages of culture, the same or modified protocols have been used to obtain prolific embryogenic cell suspension cultures of many cereals. Embryogenic cell cultures of various cereals have served as the source of totipotent protoplasts for the generation of somatic hybrids and transgenic plants by embryogenic episodes; transgenic cereals have also been recovered by somatic embryogenesis from cell suspension, callus or scutellar tissue of immature embryos following microprojectile-mediated delivery of DNA (Krishnaraj and Vasil 1995). Identical expression patterns of transcripts of critical genes such as *EXTRACELLULAR PROTEIN2* (*EP2*), *LEC1*, and *KN1* during zygotic and somatic embryogenesis in representatives of eudicots and monocots support the view that these embryogenic events proceed through similar developmental pathways (Sterk et al. 1991; Zhang et al. 2002; Yazawa et al. 2004).

Somatic embryogenesis is routinely induced in *Arabidopsis* by culturing zygotic embryos in a medium containing 2,4-D. A protocol for large-scale production of somatic embryos involves the culture of zygotic embryos at the bent cotyledon-stage in a medium containing 2,4-D to induce the formation of embryogenic callus on cotyledons and transfer of 10-day-old cultures to an auxin-free medium to promote the formation of somatic embryos (see Plate 15, Fig. a–d). By successive subculture of the callus in a medium containing a high concentration of auxin, it was possible to maintain the callus in a state of embryogenic competence for a long period of time (Pillon et al. 1996; Ikeda-Iwai et al. 2002; Raghavan, 2004). Osmotic or heavy metal stress is also conducive to inducing somatic embryogenesis from the seedling shoot apex of *Arabidopsis* in the presence of auxin (Ikeda-Iwai et al. 2003). Mordhorst et al. (1998) showed that seeds of *Arabidopsis* monogenic mutants *primordial timing* (*pt*) and *clv*, and *pt/clv* double mutant directly germinated in a liquid medium containing 2,4-D regenerate stable embryogenic cultures and somatic embryos from seedlings, thus circumventing the tedious dissection of immature embryos. What is striking about the mutant seed embryos is the presence of an unusually large shoot apical meristem, leading to the suggestion that somatic embryogenesis relies upon the presence of noncommitted cells of the enlarged meristem. However, the role of an active shoot apical meristem in the production of embryogenic cells and somatic embryos is not entirely clear, as embryos isolated from mutants defective in the formation of the shoot apical meristem, such as *stm*, *wus*, and *zll/pnh*, also readily form somatic embryos in the same medium that favors somatic embryogenesis in wild-type *Arabidopsis* (Mordhorst et al. 2002).

Molecular approaches in *Arabidopsis* have led to the identification of genes that play a role in the transition of vegetative cells to embryogenic cells in the absence of 2,4-D. For example, transgenic plants engineered by ectopic expression of the *LEC1* and *LEC2* genes appear abnormal, with wide phenotypic variations, displaying embryogenic programs by regenerating embryo-like structures or somatic embryos. Somatic embryos induced ectopically also express embryo-specific genes, such as those encoding the 2S storage protein and oleosin (Lotan et al. 1998; Stone et al. 2001). Using an efficient and powerful methodology involving overexpression of a *WUS*-type gene designated as *PLANT GROWTH ACTIVATOR6* (*PGA6*), Zuo et al. (2002) obtained high-frequency somatic embryogenesis from vegetative tissues and zygotic embryos of *Arabidopsis*, even when cultured the absence of 2,4-D. Applying a similar methodology, expression of the full-length MADS-box gene *AGL15* has been shown to enhance production of somatic embryos on zygotic embryos of transgenic *Arabidopsis* cultured in a hormone-free medium, leading to long-term maintenance of the system in embryogenic mode (Harding et al. 2003). Other genes, such as *Arabidopsis thaliana SERK1* (*AtSERK1*; Hecht et al. 2001) and *BABY BOOM* (*BBM*) from *Brassica napus* (Boutilier et al. 2002), whose ectopic expression confers enhanced embryogenic competence on wild-type *Arabidopsis* seedlings, have also been identified. The expression of the *AtSERK1* gene in germline cells and early-stage zygotic embryos of *Arabidopsis* might indicate that the same signal transduction pathway operates during embryogenesis from somatic cells and germ cells (Hecht et al. 2001). The identification of the protein product of this gene as a receptor-like protein kinase consisting of a leucine zipper motif, leucine-rich repeats, a proline-rich region, a transmembrane region, and an intracellular kinase domain, has opened up new possibilities for its biochemical and molecular characterization in the signaling cascade (Shah et al. 2001a, 2001b, 2002). In another work, the primary root of an *Arabidopsis* mutant designated as *pickle* (*pkl*) has been shown to possess embryogenic tissues, and to produce embryo-like structures (Ogas et al. 1997). Based on the identity of the protein products of genes such as *LEC* that, as transcription factors, bestow embryogenic competence on wild-type *Arabidopsis*, it is reasonable to conclude that these genes modulate somatic embryogenesis by promoting embryogenic transition of somatic cells or by maintaining their embryogenic identity. These observations confirm the crucial role that further investigations on *Arabidopsis* will play in revealing the molecular basis of somatic embryogenesis.

9.2.2
Somatic Embryogenesis in Carrot and other Model Systems

Following the discovery of somatic embryogenesis, carrot has provided a very useful model system for the analysis of the role of auxin in maintaining embryogenic competence in cells, and has been instrumental in the identification of embryonic proteins and secreted molecules that affect cell fate. Initial attempts were directed at developing protocols for obtaining somatic embryos reproducibly and in large numbers from carrot. A widely used method begins with the culture of root or hypocotyl segments excised from aseptically germinated seedlings of carrot on the surface of solidified high-nitrogen-containing Murashige-Skoog medium supplemented with sucrose, *myo*-inositol, a cytokinin and 2,4-D. A piece of the callus regenerated on the explant is transferred to an agitated liquid medium of the same composition, but with a reduced level of 2,4-D to produce a suspension culture consisting of single cells and proliferating cell clusters known as proembryogenic masses. The cell population can be stably maintained in this medium for several months by repeated subculture. Embryogenesis is induced by transferring an aliquot of the suspension to a medium totally deprived of 2,4-D (induction medium). After growth in the induction medium for 4–5 days, the proembryogenic masses are transformed into globular embryos, followed in rapid succession by the initiation of cotyledons and establishment of bipolarity, formation of shoot and root apices, and embryo maturation. By appropriate culture manipulation, conditions for obtaining high yields of synchronously developing somatic embryos, starting with a suspension of cell clusters or a population of potentially embryogenic single cells, have been established (Fujimura and Komamine 1979; Giuliano et al. 1983; Nomura and Komamine 1985). A somatic embryogenesis system developed from selected genotypes of *Medicago sativa* has been found to be particularly useful in studying the molecular biology of reactivation of cell division and cell cycle regulation during embryogenic transformation of somatic cells. The protocol involves development of a callus on the explant nurtured in a medium containing a weak auxin such as NAA and the cytokinin kinetin, maintenance of the tissue as a suspension of microcalli in the same medium, initiation of embryogenesis by a short pulse treatment with 2,4-D, and subsequent growth of somatic embryos in a hormone-free medium (Dudits et al. 1991). An innovative somatic embryogenesis system developed for a hybrid clone of *Cichorium* (Asteraceae) has taken advantage of the inclusion of glycerol in the induction medium to delay embryogenic divisions in cells of leaf explants followed by their transfer to a medium without glycerol to trigger synchronized embryogenic divisions of induced cells (Robatche-Claive et al. 1992). Although somatic embryogenesis has been induced on calluses derived from a variety of explants of several members of the Poaceae (Krishnaraj and Vasil 1995), the slow growth of the callus and the rapid loss of its embryogenic potency, have limited the use of this system in biochemical and molecular investigations. However, orchardgrass, *Dactylis glomerata* (Poaceae), in which highly embryogenic genotypes have been developed, has a high capacity for somatic embryogenesis, which occurs directly from the mesophyll cells of leaves. Young leaves of *D. glomerata* also display a gradient of embryogenic response, with the most basal portion giving rise to both embryogenic callus and somatic embryos, while the more distal segments of the leaf form exclusively somatic embryos (Conger et al. 1983; Conger and Hanning 1991).

Identification and characterization of embryogenically competent cells of carrot in a suspension culture consisting of a heterogenous mixture of several different cell types and cell clusters with ill-defined morphology is a vexing problem. A gene – *Daucus carota SERK* (*DcSERK1*) – isolated from a population of embryogenically competent single cells of carrot has served as a molecular marker to signify the acquisition of embryogenic competence by cells. The protein product of this gene, like that of the *AtSERK1* gene, is a leucine-rich repeat transmembrane receptor-like kinase. The expression of *SERK1*, as determined by in situ hybridization, was found to correlate qualitatively and quantitatively with the presence of embryogenically competent single cells in the suspension (see Plate 16, Fig. a–g). By tracking single competent cells in a culture transformed with a *SERK1* promoter-luciferase gene construct, it was confirmed that the transgene-expressing cells indeed develop into somatic embryos (Schmidt et al. 1997). The *DcSERK1* gene was also found to be a good molecular marker of cells competent to form somatic embryos directly on cultured leaf explants of *D. glomerata* (Somleva et al. 2000). A *SERK* gene isolated from initial embryogenic cells produced on cultured staminodes of cacao (*Theobroma cacao*; Sterculiaceae) was not only detected in these cells, but was also expressed in secondary somatic embryos regenerated from them (Santos et al. 2005). A high level of expression of the *Medicago truncatula SERK1* (*MtSERK1*) gene is associated with somatic embryogenesis in *M. truncatula*, but is not specific for the process (Nolan et al. 2003). In contrast, *Zea mays SERK* genes, *ZmSERK1* and *ZmSERK2*, isolated from maize, are expressed nonspecifically in both embryogenic and nonembryogenic callus cultures (Baudino et al. 2001), whereas the expression levels of a corresponding gene isolated from cultured sunflower embryos do not correlate with somatic embryogenesis in sunflower (Thomas et al. 2004). So, it is premature to anoint the *SERK* gene as a reliable molecular marker of somatic embryogenesis.

ROLE OF AUXIN

As pointed out earlier, a sequence of transfer of cells from a medium containing 2,4-D to a medium lacking the auxin essentially defines the protocol for inducing somatic embryogenesis in carrot. A key question relating to the role of auxin in somatic embryogenetic processes in carrot is whether cells are programmed for embryogenesis before they encounter auxin in the medium. Although it has been difficult to dissociate the role of auxin in promoting callus growth from its role in conferring embryogenic competence on cells, the demonstration that epidermal cells of carrot hypocotyl acquire the capacity to form embryos only after exposure to auxin for at least 12–24 h, engenders the notion that auxin treatment is necessary to bestow embryogenic competence on cells (Masuda et al. 1995). The main focus of many physiological investigations on the function of auxin in somatic embryogenesis in carrot has been on the dilemma posed by the requirement for auxin to make cells embryogenically competent, and yet, the subsequent inhibition of embryogenesis by auxin. This paradox has been exacerbated by reports of embryogenic development of carrot cells by culture of mericarp (one-sided half of the ovary) and mechanically wounded zygotic embryos in a totally auxin-free medium under conditions of low pH and with a low concentration of NH_4^+ as the sole source of nitrogen (Smith and Krikorian 1989, 1990), and of seedlings in media containing heavy metal ions (Kiyosue et al. 1990), and by subjecting whole seedlings or parts of seedlings to salt stress (Kiyosue et al. 1989), osmotic stress (Kamada et al. 1993), or heat stress (Kamada et al. 1994). These results have led to the suggestion that cells express their innate embryogenic potential when they encounter physical or chemical stress. This view is also supported by a report that carrot seedlings cultured in a medium containing ABA as the sole growth hormone regenerate somatic embryos directly from the epidermal cells of the hypocotyl (Nishiwaki et al. 2000). Here, ABA, well-known for its role in stress signal transduction, might be considered as a signal transducer in stress-induced somatic embryogenesis. These successful attempts to develop carrot somatic embryogenesis systems in auxin-free media offer new opportunities for analyzing the induction of somatic embryos from a new perspective.

CA^{2+}-MEDIATED SIGNALING DURING SOMATIC EMBRYOGENESIS

The finding that the embryogenic potential of carrot cells increases substantially in the presence of moderately high Ca^{2+} concentrations in the medium, and that a rise in extracellular Ca^{2+} counteracts the inhibitory effect of 2,4-D on somatic embryogenesis, provided suggestive evidence for its possible role as a second messenger in the hormone-regulated switch of somatic cells into embryos (Jansen et al. 1990). Two subsequent independent investigations using fluorescent dyes revealed changes in the distribution of free cytosolic calcium during embryogenesis without alterations in membrane-associated calcium concentration. These observations, supported by the inhibitory effects inflicted by the Ca^{2+} channel blocker verapamil or the Ca^{2+} ionophore A23187 on the embryogenic potential of cell suspensions, have highlighted the importance for embryogenic development of free cells and cell clusters of maintaining a gradient of exogenous calcium (Overvoorde and Grimes 1994; Timmers et al. 1996). Whether the ubiquitous calcium-binding protein calmodulin has a functional role in somatic embryogenesis in carrot is not yet known. A first indication could be derived from reports of the modulation of calmodulin levels during carrot cell growth and somatic embryogenesis (Oh et al. 1992); localization of Ca^{2+}-calmodulin complexes in meristematic regions of developing carrot somatic embryos (Overvoorde and Grimes 1994); and the isolation of carrot somatic embryo cDNA clones encoding protein kinases possessing similar or divergent forms of calmodulin-like regulatory domains (Suen and Choi 1991; Lindzen and Choi 1995).

The search for the involvement of Ca^{2+}-regulated protein kinases in somatic embryogenesis has resulted in the isolation of a cDNA clone encoding a calmodulin-like protein kinase from cultured *Medicago sativa* cells; indicative of its role in somatic embryognesis, it was relatively easy to show an increase in transcript levels during embryogenic transformation of cells grown in a hormone-free medium following a 1 h pulse of 2,4-D (Davletova et al. 2001). Pharmacological experiments have provided evidence for a role for Ca^{2+} as a second messenger during somatic embryogenesis in cell clusters originating from the endosperm of sandalwood (*Santalum album*; Santalaceae); identification in the soluble protein extracts of embryogenic cultures of two members of a novel family of calcium-dependent protein kinases that are independent of calmodulin as intermediates in this signaling process is likely to reveal important insights into both the operation and evolution of the Ca^{2+}-mediated signaling pathway during somatic embryogenesis in sandalwood (Anil and Rao 2000; Anil et al. 2000).

ROLE OF EXTRACELLULAR PROTEINS

Perhaps one of the most intriguing observations made in carrot and a few other embryogenic cell suspension systems is that, as cells metabolize nutrient substances of the medium to form somatic embryos, they release a diverse array of molecules, including polysaccharides, proteoglycans, and polypeptides, that condition the medium and favor cell proliferation and somatic embryogenesis. In the light of this observation, biochemical analysis of the excreted molecules, combined with simple culture techniques, became a powerful way of investigating somatic embryogenesis in a fresh context with new players. Implicating the extracellular proteins identified as glycoproteins in somatic embryogenetic process in carrot was the demonstration that cell lines impaired in somatic embryogenesis fail to excrete one or more of these proteins and that their embryogenic potency is partially restored by the addition of a protein preparation from embryogenic cell lines (de Vries et al. 1988). The role of excreted proteins in promoting somatic embryogenesis in carrot is supported by the finding that a mutational inhibition of protein synthesis and somatic embryogenesis in a temperature-sensitive embryogenic cell line is alleviated by the addition of the protein mixture to the medium (Lo Schiavo et al. 1990); this theme has been strengthened by the use of a 38 kDa glycoprotein purified from the conditioned medium to reverse the inhibition of somatic embryogenesis caused by the glycosylation inhibitor tunicamycin (Cordewener et al. 1991). Of the other extracellular proteins characterized from carrot cell suspensions, proteins designated as EP1 and EP4 are not released by embryogenic cells, but are secreted by nonembryogenic cells. EP1 protein is localized in the pectic material of the cell wall of nonembryo-

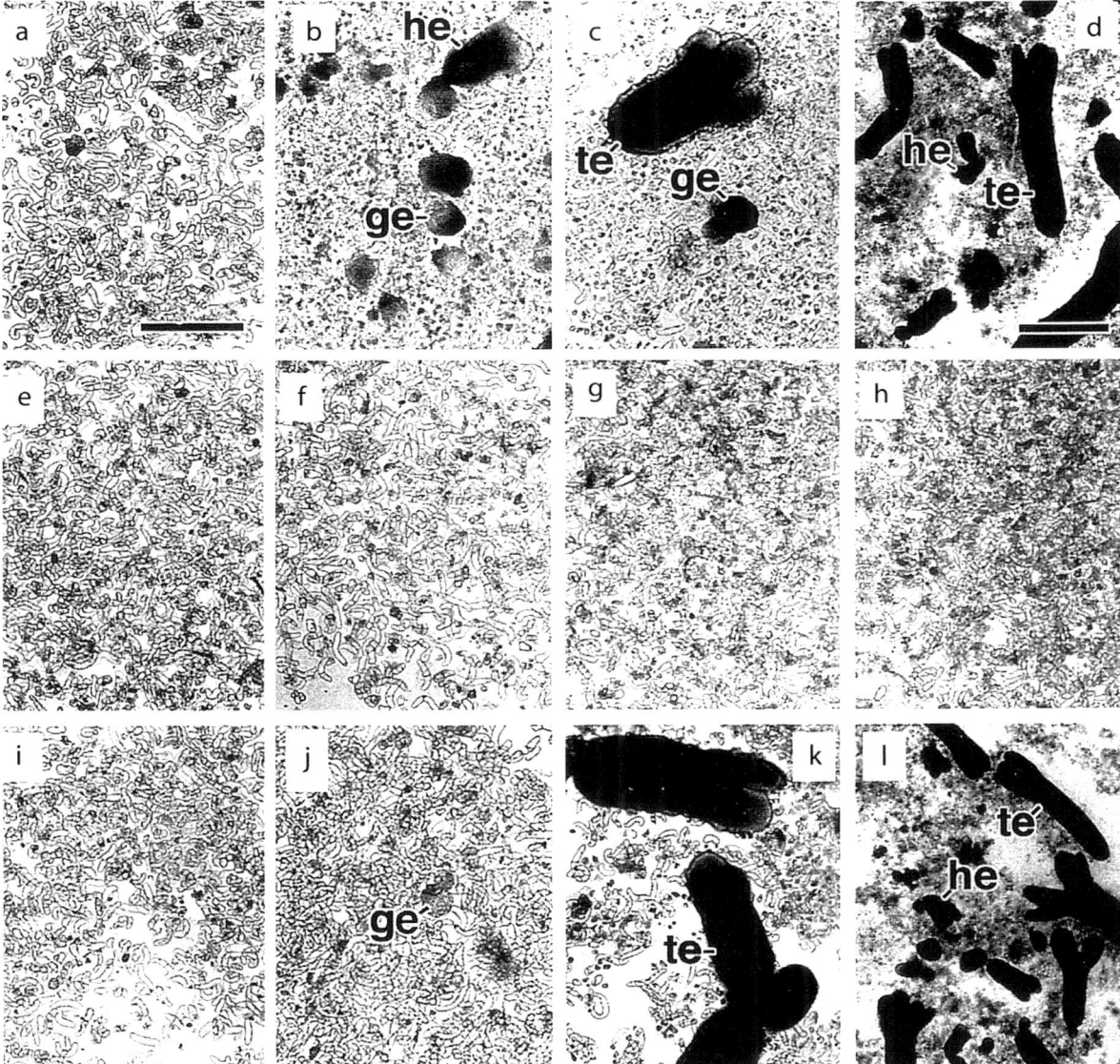

Fig. 9.3a–l Embryogenic development of JIM8(+) and JIM8(-) cells of carrot. **a–d** Culture of JIM8(+) in hormone-free medium for 7, 14, 21, and 35 days, respectively. **e–h** Culture of JIM8(-) cells in hormone-free medium for 7, 14, 21, and 35 days, respectively. **i–l** culture of JIM8(-) cells in hormone-free medium conditioned by cultures of unsorted cells containing JIM8(+) cells for 7 days; growth periods are **i** 7, **j** 14, **k** 21, and **l** 35 days. *ge* Globular stage somatic embryo, *he* heart-shaped somatic embryo, *te* torpedo-shaped somatic embryo. Bars **a** 50 µm (for **a–c**, **e–g**, and **i–k**), **d** 100 µm (for **d**, **h**, and **l**). (Reprinted from McCabe et al. 1997)

genic single cells. Nonembryogenic cells containing the EP4 protein are mostly clustered rather than single cells, with the walls separating adjacent cells showing high levels of EP4 localization (van Engelen et al. 1991, 1995). According to Sterk et al. (1991), EP2 is a protein secreted by embryogenic cells and somatic embryos that shows homology to a lipid transfer protein. The abundance of transcripts of the *EP2* gene expressed in the epidermal cells of developing somatic and zygotic embryos of carrot has led to the view that EP2 protein might be involved in the transport of cutin monomers to the epidermal cells, which are the traditional sites of cutin biosynthesis. Expression of *EP2* gene transcripts in the proembryogenic mass of cells has suggested a role for the protein product of the gene as a marker of the embryogenic potential of cells. In support of this view, it was found that the expression pattern of an *AtLTP1* promoter-luciferase gene construct in the proembryogenic cells of transgenic carrot is identical to that of the endogenous carrot *EP2* gene revealed by whole-mount in situ hybridization. Moreover, cell tracking also established that somatic embryos are invariably formed from cell clusters expressing the *AtLTP1*-luciferase gene construct (Toonen et al. 1997b). Purification of the secreted protein (EP3) responsible for reversing the inhibition of somatic embryogenesis in the temperature-sensitive carrot cell line referred to earlier has been identified as a glycosylated acidic endochitinase Class IV, which is apparently secreted by mutant cells at a reduced level during a transient period of growth at the nonpermissive temperature (de Jong et al. 1992; Kragh et al. 1996). Since carrot EP3 is expressed in the nonembryogenic cells of the suspension culture, and in the integument and selected cells of the endosperm of carrot seeds, it has been surmised that EP3 endochitinase performs a nursing role during somatic embryogenesis (van Hengel et al. 1998).

Arabinogalactan proteins identified as secreted molecules in carrot cell suspension cultures have been shown to confer embryogenic potential on established nonembryogenic cell lines (Kreuger and van Holst 1993; van Hengel et al. 2001). However, failure to standardize the cell-type composition of suspension cultures has led to conflicting reports on the effectiveness of the addition of monoclonal-antibody fractionated arabinogalactan proteins in somatic embryogenesis in carrot (Kreuger and van Holst 1995; Toonen et al. 1997a). Arabinogalactan proteins can also confer embryogenic potential on protoplasts of carrot with reduced capacity for somatic embryogenesis (van Hengel et al. 2001). By grouping discrete populations of single cells differing in morphology and embryogenic competence in a carrot cell suspension by their ability to recognize the monoclonal antibody JIM8 (which reacts with a carbohydrate epitope of arabinogalactan protein) in their cell wall and, in conjunction with the use of secondary antibodies to label and sort out pure populations of JIM8 (-) and JIM8 (+) cells, it was shown that JIM8 (-) cells, which do not develop into somatic embryos in any medium, do so when they are cultured in a medium conditioned by the culture of JIM8 (+) cells or a cell mixture containing JIM8(+) cells (Fig. 9.3a–l). This observation has suggested a role for cell interaction involving release of soluble signals of the arabinogalactan type by cells in the JIM8 (+) population in control of a transient stage in the early developmental pathway of somatic embryogenesis in carrot (McCabe et al. 1997).

It appears from these investigations that the type of proteins secreted plays a role in determining the developmental state of cells of the carrot cell suspension culture, specifically, whether they are diverted in the embryogenic pathway or bide their time by repeated divisions. Other somatic embryogenesis systems in which extracellular proteins have been identified are those of barley (Nielsen and Hansen 1992), grapevine (*Vitis*) hybrid (Coutos-Thevenot et al. 1992), *Citrus aurantium* (Gavish et al. 1991, 1992), *Chicorium* (Helleboid et al. 1998), and *Asparagus officinalis* (Liliaceae; Takeda et al. 2003). The grapevine cell suspension may be described as that bearing closest resemblance to the carrot system; as in the latter, two extracellular proteins secreted by embryogenic grapevine cells nurtured in an auxin-free medium are a 32-kDa cationic peroxidase and a 10-kDa lipid transfer protein (Coutos-Thevenot et al. 1992, 1993). By secreting endochitinases into the medium, embryogenic cells of barley also bear some similarities to carrot cells (Kragh et al. 1991). Embryogenic competence in cultured leaf segments of *Chicorium* is characterized by the appearance of β-1,3-glucanses in the medium. Expression of a cloned glucanse gene, probably functioning in degradation of callose localized around embryogenic cells, as early as 1 day after transfer of induced leaf explants to a glycerol-free medium is correlated with somatic embryogenesis (Helleboid et al. 2000). The inhibition of somatic embryogenesis in *Cichorium* by β-d-glucosyl Yariv reagent, which binds specifically to arabinogalactans, has also implicated the latter in somatic embryogenesis in this system (Chapman et al. 2000). The normal progress of somatic embryogenesis in *C. aurantium* does not, however, depend on secreted proteins, which appear to be inhibitory for the process (Gavish et al. 1992).

9.2.3
Embryonic Proteins and Regulation of Gene Expression

The carrot system has been widely used as a model to study the biochemical and molecular changes associated with somatic embyogenesis. Much of the evidence favors the view that transfer of cells from a medium containing 2,4-D to an auxin-free medium modulates embryogenesis by the synthesis of new mRNA and proteins. Yet, it has not appeared unequivocally clear that gene activity for embryogenic induction is initiated in cells upon their transfer to a medium lacking auxin. The issue is whether proteins synthesized in cells grown in the auxin-free medium are encoded on newly formed mRNA or on mRNA transcribed when cells are bathed in the auxin-containing medium. Wilde et al. (1988) found striking similarities between the in vitro translation products of mRNA populations of proembryogenic masses and torpedo-shaped somatic embryos of carrot, leading to the suggestion that a gene expression program for somatic embryogenesis is initiated when cells grow in auxin-containing medium. Based on a comparison of the spectrum of proteins synthesized by carrot cells growing for 12 days in the presence or absence of 2,4-D in the medium, in an earlier work Sung and Okimoto (1981) found no pronounced differences in the nearly 200 or so polypeptides spotted on gels, except for two additional proteins, E1 and E2 (designated as embryonic proteins) in embryogenic cells grown in the absence of auxin. The surprising finding is that, regardless of the presence or absence of 2,4-D in the medium, these two proteins are synthesized by cells as early as 4 h of growth in the fresh medium but, in the presence of auxin, they gradually diminish and finally disappear. Synthesis of embryonic proteins appears to be an early event of embryogenic induction triggered by 2,4-D, although by its very presence in the medium, auxin also inhibits the continued synthesis of these proteins and the execution of the embryogenic program by cells. In *Cichorium* leaves in which somatic embryos are formed directly by the transformation of mesophyll cells, synthesis of the first embryonic proteins, indicative of cell reprogramming, occurs within 2 days of culture of the explant, before any overt morphological or cytological signs of change are detected. During a 5-day induction period, during which the mesophyll cells acquire embryogenic competence, at least 15 new proteins that appear transiently or accumulate steadily have been identified (Boyer et al. 1993). Although a role for embryonic proteins in somatic embryogenesis in *Cichorium* is more explicit than in carrot, it does not provide a perspective on how synthesis of these proteins provokes a program of embryogenic differentiation of mesophyll cells. This issue also remains unclear in other somatic embryogenesis systems such as rice, *Dactylis glomerata*, *Trifolium rubens*, *T. pratense* (Fabaceae), soybean, *Nicotiana plumbaginifolia*, *Digitalis lanata* (Scrophulariaceae) and *Coffea arabica* (Rubiaceae), in which embryonic proteins have been characterized. Recently, Imin et al. (2005) employed high-resolution proteomic analysis to compare the changing protein profiles during somatic embryogenesis in leaf explants of wild-type and a highly embryogenic mutant line of *Medicago truncatula*. Of the 16 proteins differentially expressed during an 8-week culture period, two proteins – thioredoxin H, present in high concentrations during the induction phase of somatic embryos, and 1-cysteine peroxiredoxin, peaking during late embryogenesis – appear to have a modest role in some aspects of embryogenic transformation. Identification of new proteins by the use of proteomics technology will likely render the molecular circuitry of somatic embryogenesis in this and other systems more complex than previously assumed.

The carrot system has been used by a number of investigators to identify genes associated with the commitment of somatic cells to an embryogenic fate, and with progressive differentiation of the committed cells into somatic embryos (Zimmerman 1993; Rao, 1996; Chugh and Khurana 2002, for reviews).

Given the overlap between the in vivo synthesized proteins of nonembryogenic and embryogenically – induced cells, it is likely that only minor changes in gene expression programs accompany embryogenic induction, and that understanding the molecular regulation of somatic embryogenesis in carrot may hinge on some rare class of genes. In an attempt to identify genes that have a role in the initiation of embryogenic development in somatic cells, Aleith and Richter (1990), using the traditional approach of differential screening, found that transcripts of several clones isolated from carrot cells cultured in an auxin-free medium accumulate transiently in cells from 3 days up to 16 days after transfer, coinciding with the development of globular or heart-shaped somatic embryos. Involvement of one of the isolated clones in somatic embryogenesis was confirmed by the activity expressed by its promoter sequences in somatic embryos produced in transgenic carrot (Holk et al. 1996). A gene isolated by Sato et al. (1995) by subtractive differential screening was expressed as early as 1 day after transfer of embryogenic cell clusters to an auxin-free medium. A gene designated as *CARROT EARLY SOMATIC EMBRYOGENESIS1* (*C-ESE1*) can be considered to respond to an early signal of embryogenic development as it is expressed in embryogenic cells within 8 h of their transfer to an auxin-free medium (Takahata et al. 2004). Despite the fact that the deduced protein products of this and other genes showed some resemblance to certain cell wall proteins, their relevance to the potentiation of embryogenic developement remains doubtful in view of the lack of a clear function.

Genes expressed at high levels during late stages of somatic embryogenesis in carrot include those encoding LEA proteins (Borkird et al. 1988; Wurtele et al. 1993) and the eukaryotic translation elongation factor 1α (Kawahara et al. 1992), homeobox genes (Kawahara et al. 1995; Hiwatashi and Fukuda 2000), and the carrot homologue of the *Arabidopsis LEC1* gene (Yazawa et al. 2004). Comparative studies of gene expression patterns in nonembryogenic and embryogenic cells of other systems have led to the demonstration of up-regulation of genes associated with cell cycle activity in *Medicago sativa* (Hirt et al. 1991), MADS-box genes in *Cucumis sativus* (Filipecki et al. 1997) and maize (Heuer et al. 2001), genes involved in nuclear regulatory functions in *Dactylis glomerata* (Alexandrova and Conger 2002), and genes encoding 'germin-like' oxalate oxidase in wheat (Caliskan et al. 2004). Although these studies provide important validation of the occurrence of a substantial reprogramming of the gene expression pattern during the developmental switch from somatic cells to embryos, identification of those critical rare genes whose up-regulation can be considered to be causal for embryogenesis continues to elude us.

It has been proposed that the transition of somatic cells to proembryogenic masses or somatic embryos might not involve changes in the most abundant proteins or mRNAs, but is rather programmed by the down-regulation of some genes expressed in somatic cells. In support of this view, characterization of the expression and regulation of a collection of 38 genes isolated by a subtraction-probe strategy using mRNA from carrot seedlings to screen embryo-enhanced genes from somatic embryos has shown that most of the genes are not only expressed in the callus, but some are even expressed at higher levels in the callus than in somatic embryos (Lin et al. 1996). However, this work has not fished out any rare genes whose down-regulation modulates the transition of somatic cells to embryos. Speculation on the molecular mechanism of somatic embryogenesis does not end here, as the role of other tractable changes such as DNA methylation, important in the regulation of gene expression in eukaryotic systems, is now being investigated (Chakrabarty et al. 2003).

9.3
Pollen Embryogenesis

The pollen grain, or microspore, in flowering plants is the product of a reduction division of the pollen mother cell in the anther, and is the first cell of the male gametophyte. The essential features of male gametogenesis in flowering plants are an asymmetric division of the pollen grain into a large vegetative cell and a small generative cell (termed first pollen mitosis), germination of the pollen grain to produce a pollen tube, growth of the pollen tube, and division of the generative cell into the two sperm cells involved in double fertilization. After the pollen grain has fulfilled its function in fertilization-related events, this would appear to be the end of the story, but not so. Guha and Maheshwari (1964) showed that, when anthers of *Datura innoxia* at the pollen grain stage were cultured in a mineral salt medium supplemented with casein hydrolyzate, IAA and kinetin, or with coconut water, grape juice or plum juice, embryo-like structures appeared from the sides of the anther in 6–7 weeks. A subsequent ontogenetic study established that, in cultured anthers, a variable but substantial number of pollen grains enlarge and divide repeatedly, forming multicellular units within the exine. Later, the exine gives way, freeing the contents, which organize into typical bipolar embryos with the haploid or gametic number of chromosomes (Guha and Maheshwari 1966). This process, which bypasses sexual reproduction to produce haploid embryos, was described in subsequent investigations on other plants as pollen embryogenesis, androgenesis, or haploid embryogenesis.

The successful induction of pollen embryogenesis in cultured anthers tended to overlook the influence of the anther wall and tapetum in triggering embryogenic division of pollen grains during a critical period after culture. Study of pollen embryogenesis unhindered by the presence of somatic tissues of the anther was first achieved by culturing pollen grains isolated from cold-stressed anthers of *D. innoxia* in a liquid medium conditioned by an extract of cultured anthers of the same species (Nitsch and Norreel 1973). In later years, procedures have been streamlined to induce high-frequency embryogenesis in isolated pollen grains of *Nicotiana tabacum*, *B. napus*, and *Triticum aestivum* cultured in media of known chemical composition to replace the anther extract. The method developed for *N. tabacum* begins with density gradient centrifugation to obtain a homogeneous population of embryogenic pollen grains. If these pollen grains are first subjected to a starvation diet by culture in a medium lacking sucrose and glutamine, and then transferred to an enriched medium, they divide in the embryogenic pathway in high numbers (Kyo and Harada 1986). Touraev et al. (1996a) subsequently showed that combining growth in a starvation medium with heat stress at 33°C was superior to starvation alone in obtaining high yield of embryogenic pollen grains and pollen embryos from cultured tobacco pollen (see Plate 16, Fig. h–k). Still later, starvation conditions for embryogenic divisions have been simulated by culturing tobacco pollen in a sucrose-containing medium at a pH of 8.0–8.5 (Barinova et al. 2004). The critical step in inducing embryogenic divisions in a homogeneous suspension of pollen grains of *B. napus* is a high-temperature shock by culture in the dark at 33°C for 3 days followed either by a more comfortable temperature of 25°C or by incubation in colchicine for 18-42 h and subsequent culture in the regular medium, both at 25°C (Huang 1992; Zhao et al. 1996). For efficient production of embryos from wheat pollen, a combination of treatments involving starvation of cultured anthers in a minimal medium at 33°C and separation of enlarged embryogenic microspores by density gradient centrifugation, followed by their culture in an enriched medium conditioned by immature wheat ovaries was found most suitable (Touraev et al. 1996b). The realization that the fate of terminally differentiated pollen grain can be diverted to one of immortality by imposition of starvation and temperature stress as described in *B. napus*, tobacco, and wheat, or by stress-eliciting stimuli such as mannitol, calcium, and ABA as described in barley (Hoekstra et al. 1997), provides a new perspective on pollen embryogenesis, making traditional concepts of hormone action in embryogenic transformation of pollen grains less satisfactory as an explanation.

The discovery of pollen embryogenesis has spawned investigations on economically important crops to produce isogenic doubled haploid plants in quantity for genetic and breeding experiments, and the increasing demand for doubled haploids for these purposes in other plants continues to propel improvements in the currently used anther and pollen culture methods. A review by Touraev et al. (2001) presents a consolidated account of the genetic, cell biological, and molecular aspects of pollen embryogenesis, whereas Reynolds (1997) and Pechan and Smykal (2001) have reviewed some of the cytological and molecular investigations.

9.3.1 Responsive Stage of Pollen Development and Pollen Embryogenic Potential

At present, the list of species in which pollen embryogenesis by anther or pollen culture techniques has been reported includes those from a large number of families, including major crop plants (Jain et al. 1996, 1997). Based on these studies, a wide range of extracellular and intracellular factors, such as physiological stage and conditions of growth of donor plants, plant age, genotype of donor plants, stage of pollen development, pretreatment of flower buds, imposition of stress, temperature and photoperiodic regimes of culture, composition of the nutrient medium, and composition of the culture vessel atmosphere, appear to affect the induction of embryogenic development and yield of pollen embryos. Of these, by far the most important is the discrete developmental window at which pollen grains become embryogenic, and the increasing number of reports in which pollen embryogenesis is induced by anther and pollen culture methods make it necessary to consider the importance of this factor in the induction process. For most species, the stage of pollen development at which embryogenic induction can occur appears to lie between the early unicellular and the bicellular stage of the pollen grain. In the widely investigated tobacco, anthers are generally responsive when they are cultured at stages beginning with the liberation of microspores from the tetrad and ending with bicellular pollen grains, but embryos are readily formed in large numbers from anthers cultured at the unicellular pollen grain stage, or as pollen grains begin to divide (Sunderland and Wicks 1971). Isolated pollen grains of tobacco cultured at stages ranging from the mid-unicellular to the bicellular (Heberle-Bors and Reinert 1979; Kyo and Harada 1986; Touraev et al. 1996a), of wheat cultured at the uninucleate to premitotic stages (Touraev et al. 1996b), and of *B. napus* cultured at the unicellular stage close to the first pollen mitosis stages (Huang 1992; Zhao et al. 1996), yield embryos with high frequency. Much of the quantitative data on embryo formation as a function of pollen developmental stage are notable for the attention they focus on the unicellular stage, or a stage on the verge of the first pollen mitosis, as most responsive to culture.

For optimization of culture conditions of isolated anthers or of pollen grains to obtain pollen embryos reproducibly and in large numbers, identification of embryogenic pollen grains in a population is important. We are far from understanding the reasons why only certain pollen grains in cultured anthers or in isolated pollen cultures become embryogenic and produce embryos, while others complete male gametogenesis or disintegrate in culture. One view is that pollen grains become competent to form embryos by changes in their size and by covert cytoplasmic changes that occur naturally during anther development, and that their subsequent culture provides an appropriate environment for expression of this predetermined potential. The production of two types of pollen grains in the anther, known as dimorphism, is a way of life for the pollen population produced by many plants. Pollen grains that differ from the main population with respect to their small size and reduced affinity for cytoplasmic stains are believed to be those with embryogenic potential. The cytological changes that occur in a small population of pollen grains, which suppress their gametophytic program and govern subsequent divisions in the embryogenic pathway, have not yet been identified. A stringent line of evidence in support of the origin of embryos from variant pollen grains in barley and tobacco anthers is the observed correlation between the number of such pollen grains in the anther and the number of embryos or multicellular pollen grains formed in representative samples of cultured anthers (Dale 1975; Horner and Street 1978; Horner and Mott 1979). In isolated pollen cultures of tobacco, embryogenic divisions were confined to the variant pollen grains collected from donor plants grown in short days under a low temperature regime and purified by density gradient centrifugation (Heberle-Bors and Reinert 1980).

The physiological and biochemical changes that occur in isolated pollen grains during stress treatment preparatory to transfer to a favorable culture medium have linked embryogenic competence of pollen grains to changes in their cytoplasmic organization. Structural reorganization of the cytoplasm in starved and temperature-stressed pollen grains of tobacco involved vacuolation and formation of criss-crossing cytoplasmic strands (Touraev et al. 1996a). The available data suggest that heat-stressed pollen grains of *B. napus* become embryogenically determined by a variety of cellular cues, such as the synthesis of a thick fibrillar wall adjacent to the intine, movement of the nucleus to a central position, appearance of large globules in the cytoplasm (Zaki and Dickinson 1990), formation of a planar rather than a lens-shaped cross wall during the first pollen mitosis (Telmer et al. 1993), the presence of a perinuclear array of microfilaments and a preprophase band of microtubules preparatory to the first pollen mitosis (Gervais et al. 2000), changes in nuclear pore complex density (Straatman et al. 2000), and a decrease in the number and size of coiled bodies containing small nuclear ribonucleoprotein particles involved in transcription and their appearance in the nucleoplasm (Straatman and Schel 2001). Beginning with the culture of isolated wheat pollen grains immediately before the first pollen mitosis, embryogenic pollen grains become enlarged and internally star-like, due to the presence of a centrally located nucleus suspended by cytoplasmic threads (Touraev et al. 1996b). Tracking these pollen grains individually has unambiguously demonstrated their division in the embryogenic pathway (Indrianto et al. 2001).

9.3.2
Cytology of Pollen Embryogenesis

In the context of pollen embryogenesis, the unicellular pollen grain, once considered as the first cell of the gametophytic phase, is now recognized as playing the role of the progenitor cell of the pollen embryo. In most cases, the first division of this cell in the embryogenic pathway is identical to the division that occurs during male gametogenesis and results in the formation of the large vegetative cell and small generative cell. Much of our knowledge of the pathways by which embryos are formed by the division of one or both these cells comes from cytological analyses of model plants (Fig. 9.4). One of the most common pathways, and one that has been well-investigated, is the repeated division of the vegetative cell to form the embryo. As first demonstrated in tobacco, by the 6th day of anther culture, the switch in the developmental program of the vegetative cell becomes evident when it loses its morphogenetic individuality and is partitioned by a series of internal walls until a mass of cells typical of somatic cell size is produced. Upon its release from the exine, the multicellular mass becomes highly dynamic as it faithfully recapitulates stages of zygotic embryos before appearing outside the anther wall as a plantlet. A restrictive feature of this embryogenic division sequence of the vegetative cell, known as the A pathway, is that the generative cell either disintegrates or undergoes only a few divisions without contributing to the formation of the embryo (Sunderland and Wicks 1971). The division of the vegetative cell in the embryogenic pathway in cultured pollen grains of tobacco has been supported by microspectrophotometric and autoradiographic studies showing renewed DNA replication in the vegetative cell following starvation treatment (Žárský et al. 1992).

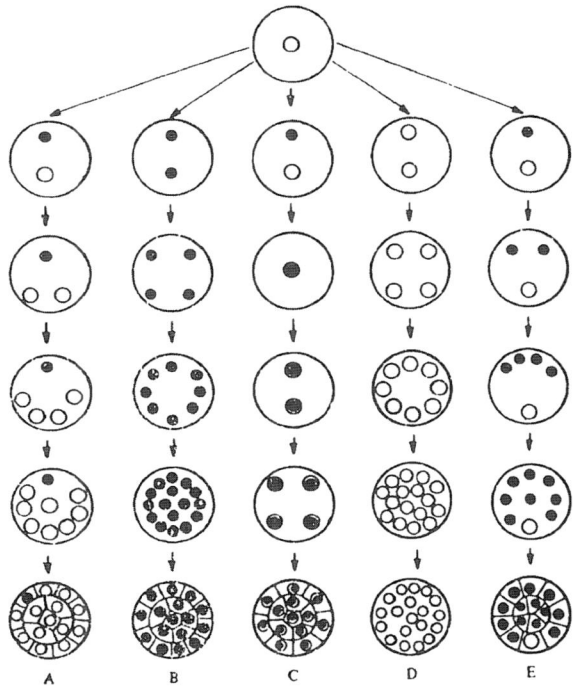

Fig. 9.4 Diagrammatic representation of the different pathways of pollen embryogenesis. In *A*, *C*, and *E*, *solid circle* represents nucleus of the generative cell or its division products; *open circle* represents nucleus of the vegetative cell or its division products. In *C*, *solid circle* enclosed in *open circle* indicates fusions between the two nuclei. In *B* and *D*, *solid* or *open circle* represents symmetrical nuclei born out of the first pollen mitosis and their division products. (Reprinted from Raghavan 1986)

Periodic cytological examination of cultured anthers and anther segments of *Hyoscyamus niger* (Solanaceae) led to its emergence as a model plant in which embryos are formed by repeated divisions of the generative cell. This appeared to be a surprising rarity because of the undisputed formation of embryos via division of the vegetative cell in other plants investigated. In *H. niger*, following formation of the vegetative and generative cells by an asymmetric division of the pollen grain, further divisions are confined to the generative cell, which initially generates a group of cells within the exine. Here, the vegetative cell does not divide, or undergoes only a few divisions; in other cases, the vegetative cell or its division products constitute a suspensor-like structure on the organogenic part of the embryo formed by derivatives of the generative cell. In anther cultures of *H. niger*, embryos are also formed by repeated divisions of both generative and vegetative cells (Raghavan 1976, 1978). Given that the generative and vegetative cells are in the G1 phase, it is not surprising to see that their division in the embryogenic pathway is associated with autoradiographically detectable DNA synthesis, either in the generative cell alone or in both generative and vegetative cells (Raghavan 1977). The route to embryo formation involving the generative cell or both generative and vegetative cells is known as the E pathway.

Much of the foundation work delimiting the C pathway of pollen embryogenesis was carried out in cultured anthers of *Datura innoxia*. Here, the vegetative and generative cells, rather than dividing independently, fuse with one another, and both nuclei divide simultaneously on a common spindle. The fusion might occur between one or two haploid vegetative cell nuclei and a haploid or endoreduplicated generative cell nucleus, so that nuclei with nonhaploid modal chromosome numbers are produced. The frequent occurrence of embryos with different ploidy levels from the same cultured anther has been traced to the repeated divisions of these fusion products (Sunderland et al. 1974).

In some anther culture systems, a symmetrical division of the pollen grain to produce two identical cells or nuclei after the first pollen mitosis sets the stage for the production of embryos. As described in *Atropa belladonna* (Solanaceae), division products of both cells or nuclei contribute to the formation of the embryo (Rashid and Street 1973); this pathway is known as the B pathway. Another scenario that appears to be a variation of the B pathway has been described in cultured anthers of wheat, and is designated the D pathway. This pathway involves repeated divisions of the two identical nuclei formed from the first pollen mitosis to form a cluster of free nuclei. The fate of these nuclei has not been followed further to determine whether they form cells and whether the cellular mass forms an embryo (Zhu et al. 1978).

Pollen embryos formed in cultured anthers and in isolated pollen cultures originate by multiple division pathways, although what prompts pollen grains to choose a particular division sequence is not clear. One view is that the phase of the cell cycle of the first mitosis in which pollen grains are held at the time of culture is important in determining the specific division sequence, whereas the administration of a temperature stress enhances the frequency of occurrence of certain division pathways (Sunderland et al. 1979).

9.3.3
Molecular Biology of Pollen Embryogenesis

Given the small number of pollen grains in cultured anthers that become embryogenic, early insights into the molecular changes that occur during pollen embryogenesis came from cytochemical and autoradiographic studies. With the use of isolated pollen cultures, study of the gene expression pattern during embryogenic transformation of pollen grains entered a new era, leading to the isolation of genes activated in the process. A relevant question in some of the early studies was whether the initial pattern of gene expression observed in pollen grains of cultured anthers reflects elimination of the existing gametophytic program or is linked to the initiation of a novel embryogenic program. Capitalizing on the stainable RNA content of pollen grains of cultured tobacco anthers as a marker for distinguishing between potentially embryogenic and normal gametophytic pollen grains, it was found that, compared to the low level of RNA in the former, the gametophytic pollen grains display a four- to sixfold increase in RNA content. These observations have been interpreted as indicating that suppression of the gametophytic program in embryogenic pollen grains is necessary to ensure that genes for embryogenesis are fully expressed without being masked by the simultaneous expression of genes for pollen maturation and germination (Bhojwani et al. 1973). In contrast, compared to nonembryogenic pollen grains, embryogenic pollen grains identified in cultured anthers of *Datura innoxia* show an increased stainability for cytoplasmic RNA prior to the first pollen mitosis (Sangwan-Norreel 1978). These contradictions cannot easily be resolved because of the diverse pathways of pollen embryogenesis observed in cultured anthers.

Autoradiographic investigations of RNA synthesis during gametophytic development of pollen grains of *H. niger* and their embryogenic transformation in cultured anthers have found a general correlation between transcriptional activity of the nucleus of the generative cell and its division in the embryogenic pathway (Raghavan 1979; Reynolds and Raghavan 1982). That transcriptional regulation controls embryogenic divisions of the generative cell is also reflected in the continued accumulation of mRNA in this cell and its derivatives (Raghavan 1981). Cell-free systems have been developed in other investigations to identify, from in vitro translation profiles, abundant mRNAs that are synthesized during embryogenic induction of cultured pollen grains. For example, Pechan et al. (1991) found that a temperature stress at 32°C, which induces embryogenic divisions in isolated pollen cultures of *B. napus*, also triggers the synthesis of several mRNAs not present in freshly isolated pollen grains. Similarly, in vitro translation of mRNAs isolated from embryogenic tobacco pollen grains grown in starvation medium revealed the appearance of two abundant mRNAs that were not present in mid-binucleate pollen grains. Since no new proteins were detected in the embryogenic pollen grains, it appears that the starvation-induced mRNAs accumulate in a transcriptionally inactive form until the pollen grains are transferred to an enriched medium (Garrido et al. 1993). Thus, pollen embryogenic induction could be envisioned as involving a significant change in the gene expression program of the immature pollen grains.

Although changes in protein synthetic activity accompanying embryogenic induction have been described in *H. niger* (Raghavan 1984), tobacco (Kyo and Harada 1990; Kyo and Ohkawa 1991), and *B. napus* (Cordewener et al. 1994, 1995, 2000), it has not been possible to identify specific proteins associated with a functional role in pollen embryogenesis. Observations such as the high phosphorylation state of a heat-shock protein (HSP-70) in embryogenic pollen grains, and the appearance of secreted proteins in the embryogenic medium, have potential implications in elucidating the signaling pathways during the temperature-stress that triggers pollen embryogenesis in *B. napus* (Codewener et al. 2000).

Characterization of genes expressed specifically during early stages of pollen embryogenesis promises to provide insights into the molecular mechanisms underlying the developmental switch of pollen grains from gametophytic- to embryogenic-type development. A gene that is activated during early embryogenic development of pollen grains in cultured anthers of wheat has been shown to encode a cysteine-rich metallothionein-like protein that functions in plant cell metabolism by binding toxic and nontoxic metal ions (Reynolds and Kitto 1992). Suggestive of its role as an early molecular marker of pollen embryogenesis, transcripts of the gene are detected in embryogenic pollen grains within hours after culture of anthers (Reynolds and Crawford 1996). The systematic testing of candidate genes by mRNA differential display has resulted in the isolation of two genes (*ZmAE1* and *ZmAE3*) from 5-day-old embryogenic maize pollen cultures. Functional analysis of the *ZmAE3* gene showed that fusion of its potential promoter fragment with a *GUS* reporter gene results in transient promoter activity in early-stage pollen embryos. Besides their pollen embryo-specific expression, both genes showed endosperm-specific expression. This has led to the view that pollen embryogenesis in maize might involve not only the development of the embryo but also the establishment of an endosperm-like tissue (Magnard et al. 2000; Sevilla-Lecoq et al. 2003). From this review of pollen embryogenesis, it seems we have now come full circle, beginning with the embryogenic division of the pollen grain and ending with the provision for a tissue for nurture of the embryo.

9.4 Concluding Comments

The aspects of apomixis, somatic embryogenesis, and pollen embryogenesis considered in this final chapter provide an overview of the three main pathways by which the embryo and, infrequently, the endosperm are produced in flowering plants, bypassing meiosis and double fertilization. The pervasiveness of these spontaneously occurring or induced alternative methods of embryogenesis in a wide selection of plants belonging to both eudicots and monocots illustrates that these strategies are typical rather than exceptional in the life of flowering plants. The numerous studies alluding to the economic importance of apomixis in modifying seed production in crop plants by introducing apomictic traits, of somatic embryogenesis in clonal propagation and production of synthetic seeds, and of pollen embryogenesis in the production of double haploids and in gene transfer through haploid plants, predict that future experimental agenda will include making headway towards introducing these methods into practical agricultural practices. However, the genetic basis of apomixis in established natural apomicts remains largely unknown and only initial insights into the nature of the genes relevant to apomictic reproduction have emerged. The key morphological and physiological events of somatic embryogenesis in several widely investigated systems are now well-known, but we have barely begun to understand the connections between somatic embryogenesis and genes isolated from embryogenic cells. Although genetic loci that determine the embryogenic potential of pollen grains of some important crop plants have been mapped, no genes that play a central role in the embryogenic transformation of pollen grains have been isolated. The fact that we now have an appreciation of the complexity of the three routes of non-zygotic embryogenesis in plants is a testimony to the advances already made, but many questions that are equally important for our genetic and molecular understanding of apomixis, somatic embrogenesis, and pollen embryogenesis, remain unanswered.

REFERENCES

Aleith F, Richter G (1990) Gene expression during induction of somatic embryogenesis in carrot cell suspensions. Planta 183:17–24

Alexandrova KS, Conger BV (2002) Isolation of two somatic embryogenesis-related genes from orchardgrass (*Dactylis glomerata*). Plant Sci 162:301–307

Anil VS, Rao KS (2000) Calcium-mediated signaling during sandalwood somatic embryogenesis. Role of exogenous calcium as second messenger. Plant Physiol 123:1301–1311

Anil VS, Harmon AC, Rao KS (2000) Spatio-temporal accumulation and activity of calcium-dependent protein kinases during somatic embryogenesis, seed development, and germination in sandalwood. Plant Physiol 122:1035–1043

Araujo ACG, Mukhambetzhanov S, Pozzobon MT, Santana EF, Carneiro VTC (2000) Female gametophyte development in apomictic and sexual *Brachiaria brizantha* (Poaceae). Rev Cytol Biol Vég 23:13–26

Barinova I, Clément C, Martiny L, Baillieul F, Soukupova H, Heberle-Bors E, Touraev A (2004) Regulation of developmental pathways in cultured microspores of tobacco and snapdragon by medium pH. Planta 219:141–146

Battaglia E (1948) Ricerche sulla parameiosi restituzionale nel genere *Taraxacum*. Caryologia 1:1–47

Baudino S, Hansen S, Brettschneider R, Hecht VFG, Dresselhaus T, Lörz H, Dumas C, Rogowsky PM (2001) Molecular characterization of two maize LRR receptor-like kinases, which belong to the *SERK* family. Planta 213:1–10

Bhojwani SS, Dunwell JM, Sunderland N (1973) Nucleic-acid and protein contents of embryogenic tobacco pollen. J Exp Bot 24:863–871

Bicknell RA, Koltunow AM (2004) Understanding apomixis: recent advances and remaining conundrums. Plant Cell 16:S228–S245

Borkird C, Choi JH, Jin Z-H, Franz G, Hatzopoulos P, Chorneau R, Bonas U, Pelegri F, Sung ZR (1988) Developmental regulation of embryonic genes in plants. Proc Natl Acad Sci USA 85:6399–6403

Boutilier K, Offringa R, Sharma VK, Kieft H, Ouellet T, Zhang L, Hattori J, Liu C-M, van Lammeren AAM, Miki BLA, Custers JBM, van Lookeren Campagne MM (2002) Ectopic expression of BABY BOOM triggers a conversion from vegetative to embryonic growth. Plant Cell 14:1737–1749

Boyer C, Hilbert J-L, Vasseur J (1993) Embryogenesis-related protein synthesis and accumulation during early acquisition of somatic embryogenesis competence in *Cichorium*. Plant Sci 93:41–53

Brown DCW, Finstad KI, Watson EM (1995) Somatic embryogenesis in herbaceous dicots. In: Thorpe TA (ed) In vitro embryogenesis in plants. Kluwer, Dordrecht, pp 345–415

Caliskan M, Turet M, Cuming AC (2004) Formation of wheat (*Triticum aestivum* L.) embryogenic callus involves peroxide-generating germin-like oxalate oxidase. Planta 219:132–140

Carman JG (1997) Asynchronous expression of duplicate genes in angiosperms may cause apomixis, bispory, tetraspory, and polyembryony. Biol J Linn Soc 61:51–94

Carman JG, Crane CF, Riera-Lizarazu O (1991) Comparative histology of cell walls during meiotic and apomeiotic megasporogenesis in two hexaploid Australasian *Elymus* species. Crop Sci 31:1527–1532

Chakrabarty D, Yu KW, Paek KY (2003) Detection of DNA methylation changes during somatic embryogenesis of Siberian ginseng (*Eleuterococcus senticosus*). Plant Sci 165:61–68

Chapman A, Blervacq A-S, Vasseur J, Hilbert J-L (2000) Arabinogalactan-proteins in *Cichorium* somatic embryogenesis: effect of β-glucosyl Yariv reagent and epitope localization during embryo development. Planta 211:305–314

Chugh A, Khurana P (2002) Gene expression during somatic embryogenesis – recent advances. Curr Sci 83:715–730

Conger BV, Hanning GE (1991) Registration of embryogen-P orchardgrass germplasm with a high capacity for somatic embryogenesis from in vitro cultures. Crop Sci 31:855

Conger BV, Hanning GE, Gray DJ, McDaniel JK (1983) Direct embryogenesis from mesophyll cells of orchardgrass. Science 221:850–851

Cordewener J, Booij H, van der Zandt H, van Engelen F, van Kammen A, de Vries SC (1991) Tunicamycin-inhibited carrot somatic embryogenesis can be restored by secreted cationic peroxidase isoenzymes. Planta 184:478–486

Cordewener JHG, Busink R, Traas JA, Custers JBM, Dons HJM, van Lookeren Campagne MM (1994) Induction of microspore embryogenesis in *Brassica napus* L. is accompanied by specific changes in protein synthesis. Planta 195:50-56

Cordewener JHG, Hause G, Görgen E, Busink R, Hause B, Dons HJM, van Lammeren AAM, van Lookeren Campagne MM, Pechan P (1995) Changes in synthesis and localization of members of the 70-kDa class of heat-shock proteins accompany the induction of embryogenesis in *Brassica napus* L. microspores. Planta 196:747-755

Cordewener J, Bergervoet J, Liu C-M (2000) Changes in protein synthesis and phosphorylation during microspore embryogenesis in *Brassica napus*. J Plant Physiol 156:156-163

Coutos-Thevenot P, Maes O, Jouenne T, Mauro MC, Boulay M, Deloire A, Guern J (1992) Extracellular protein patterns of grapevine cell suspensions in embryogenic and non-embryogenic situations. Plant Sci 86:137-145

Coutos-Thevenot P, Jouenne T, Maes O, Guerbette F, Grosbois M, le Caer JP, Boulay M, Deloire A, Kader JC, Guern J (1993) Four 9-kDa proteins excreted by somatic embryos of grapevine are isoforms of lipid transfer proteins. Eur J Biochem 217:885-889

Dale PJ (1975) Pollen dimorphism and anther culture in barley. Planta 127:213-220

Davletova S, Mészáros T, Miskolczi P, Oberschall A, Török K, Magyar Z, Dudits D, Deák M (2001) Auxin and heat shock activation of a novel member of the calmodulin like domain protein kinase gene family in cultured alfalfa cells. J Exp Bot 52:215-221

de Jong AJ, Cordewener J, Lo Schiavo F, Terzi M, Vandekerckhove J, van Kammen A, de Vries SC (1992) A carrot somatic embryo mutant is rescued by chitinase. Plant Cell 4:425-433

de Vries SC, Booij H, Janssens R, Vogels R, Saris L, LoSchiavo F, Terzi M, van Kammen A (1988) Carrot somatic embryogenesis depends on the phytohormone-controlled presence of correctly glycosylated extracellular proteins. Genes Dev 2:462-476

Dudits D, Bögre L, Györgyey J (1991) Molecular and cellular approaches to the analysis of plant embryo development from somatic cells in vitro. J Cell Sci 99:473-482

Dunstan DI, Tautorus TE, Thorpe TA (1995) Somatic embryogenesis in woody plants. In: Thorpe TA (ed) In vitro embryogenesis in plants. Kluwer, Dordrecht, pp 471-538

Filipecki MK, Sommer H, Malepszy S (1997) The MADS-box gene *CUSI* is expressed during cucumber somatic embryogenesis. Plant Sci 125:63-74

Fujimura T, Komamine A (1979) Synchronization of somatic embryogenesis in a carrot cell suspension culture. Plant Physiol 64:162-164

Garrido D, Eller N, Heberle-Bors E, Vicente O (1993) De novo transcription of specific mRNAs during the induction of tobacco pollen embryogenesis. Sex Plant Reprod 6:40-45

Gavish H, Vardi A, Fluhr R (1991) Extracellular proteins and early embryo development in *Citrus* nucellar cell cultures. Physiol Plant 82:606-616

Gavish H, Vardi A, Fluhr R (1992) Suppression of somatic embryogenesis in *Citrus* cell cultures by extracellular proteins. Planta 186:511-517

Gervais C, Newcomb W, Simmonds DH (2000) Rearrangement of the actin filament and microtubule cytoskeleton during induction of microspore embryogenesis in *Brassica napus* L. cv. Topas. Protoplasma 213:194-202

Giuliano G, Rosellini D, Terzi M (1983) A new method for the purification of the different stages of carrot embryoids. Plant Cell Rep 2:216-218

Grimanelli D, Leblanc O, Perotti E, Grossniklaus U (2001) Developmental genetics of gametophytic apomixis. Trends Genet 17:597-604

Grimanelli D, Perotti E, Ramirez J, Leblanc O (2005) Timing of the maternal-to-zygotic transition during early seed development in maize. Plant Cell 17:1061-1072

Grossniklaus U, Spillane C, Page DR, Köhler C (2001) Genomic imprinting and seed development: endosperm formation with and without sex. Curr Opin Plant Biol 4:21-27

Guerin J, Rossel JB, Robert S, Tsuchiya T, Koltunow A (2000) A *DEFICIENS* homologue is down-regulated during apomictic initiation in ovules of *Hieracium*. Planta 210:914-920

Guha S, Maheshwari SC (1964) In vitro production of embryos from anthers of *Datura*. Nature 204:497

Guha S, Maheshwari SC (1966) Cell division and differentiation of embryos in the pollen grains of *Datura* in vitro. Nature 212:97-98

Halperin W (1966) Alternative morphogenetic events in cell suspensions. Am J Bot 53:443-453

Halperin W, Wetherell DF (1964) Adventive embryony in tissue cultures of the wild carrot, *Daucus carota*. Am J Bot 51:274-283

Harding EW, Tang W, Nichols KW, Fernandez DE, Perry SE (2003) Expression and maintenance of embryogenic potential is enhanced through constitutive expression of *AGAMOUS-Like 15*. Plant Physiol 133:653-663

Heberle-Bors E, Reinert J (1979) Androgenesis in isolated pollen cultures of *Nicotiana tabacum*: dependence upon pollen development. Protoplasma 99:237-245

Heberle-Bors E, Reinert J (1980) Isolated pollen cultures and pollen dimorphism. Naturwissenschaft 67:311

Hecht V, Vielle-Calzada J-P, Hartog MV, Schmidt EDL, Boutilier K, Grossniklaus U, de Vries SC (2001) The *Arabidopsis SOMATIC EMBRYOGENESIS RECEPTOR KINASE1* gene is expressed in developing ovules and embryos and enhances embryogenic competence in culture. Plant Physiol 127:803-816

Helleboid S, Bauw G, Belingheri L, Vasseur J, Hilbert J-L (1998) Extracellular β-1,3-glucanases are induced during early somatic embryogenesis in *Cichorium*. Planta 205:56-63

Helleboid S, Chapman A, Hendriks T, Inzé D, Vasseur J, Hilbert J-L (2000) Cloning of β-1,3-glucanases expressed during *Cichorium* somatic embryogenesis. Plant Mol Biol 42:377-386

Heuer S, Hansen S, Bantin J, Brettschneider R, Kranz E, Lörz H, Dresselhaus T (2001) The maize MADS box gene *ZmMADS3* affects node number and spikelet development and is co-expressed with *ZmMADS1* during flower development, in egg cells, and early embryogenesis. Plant Physiol 127:33-45

Hirt H, Páy A, Györgyey J, Bakó L, Németh K, Bögre L, Schweyen R, Heberle-Bors E, Dudits D (1991) Complementation of a yeast cell cycle mutant by an alfalfa cDNA encoding a protein kinase homologous to p34^{cdc2}. Proc Natl Acad Sci USA 88:1636-1640

Hiwatashi Y, Fukuda H (2000) Tissue-specific localization of mRNA for carrot homeobox genes, *CHBs*, in carrot somatic embryos. Plant Cell Physiol 41:639–643

Hoekstra S, van Bergen S, van Brouwershaven IR, Schilperoort RA, Wang M (1997) Androgenesis in *Hordeum vulgare* L: effects of mannitol, calcium and abscisic acid on anther pretreatment. Plant Sci 126:211–218

Holk A, Kaldenhoff R, Richter G (1996) Regulation of an embryogenic carrot gene (*DC 2.15*) and identification of its active promoter sites. Plant Mol Biol 31:1153–1161

Horner M, Mott RL (1979) The frequency of embryogenic pollen grains is not increased by in vitro anther culture in *Nicotiana tabacum* L. Planta 147:156–158

Horner M, Street HE (1978) Pollen dimorphism – origin and significance in pollen plant formation by anther culture. Ann Bot 42:763–771

Huang B (1992) Genetic manipulation of microspores and microspore-derived embryos. In Vitro Cell Dev Biol 28P:53–58

Ikeda-Iwai M, Satoh S, Kamada H (2002) Establishment of a reproducible tissue culture system for the induction of *Arabidopsis* somatic embryos. J Exp Bot 53:1575–1580

Ikeda-Iwai M, Umehara M, Satoh S, Kamada H (2003) Stress-induced somatic embryogenesis in vegetative tissues of *Arabidopsis thaliana*. Plant J 34:107–114

Imin N, Nizamidin M, Daniher D, Nolan KE, Rose RJ, Rolfe BG (2005) Proteomic analysis of somatic embryogenesis in *Medicago truncatula*. Explant cultures grown under 6-benzylaminopurine and 1-naphthaleneacetic acid treatments. Plant Physiol 137:1250–1260

Indrianto A, Barinova I, Touraev A, Heberle-Bors E (2001) Tracking individual wheat microspores in vitro: identification of embryogenic microspores and body axis formation in the embryo. Planta 212:163–174

Jain SM, Sopory SK, Veilleux RE (1996) In vitro haploid production in higher plants, vol 1–3. Kluwer, Dordrecht

Jain SM, Sopory SK, Veilleux RE (1997) In vitro haploid production in higher plants, vol 4, 5. Kluwer, Dordrecht

Jansen MAK, Booij H, Schel JHN, de Vries SC (1990) Calcium increases the yield of somatic embryos in carrot embryogenic suspension cultures. Plant Cell Rep 9:221–223

Kamada H, Ishikawa K, Saga H, Harada H (1993) Induction of somatic embryogenesis in carrot by osmotic stress. Plant Tissue Cult Lett 10:38–44

Kamada H, Tachikawa Y, Saitou T, Harada H (1994) Heat stress induction of carrot somatic embryogenesis. Plant Tissue Cult Lett 11:229–232

Kawahara R, Sunabori S, Fukuda H, Komamine A (1992) A gene expressed preferentially in the globular stage of somatic embryogenesis encodes elongation- factor 1α in carrot. Eur J Biochem 209:157–162

Kawahara R, Komamine A, Fukuda H (1995) Isolation and characterization of homeobox-containing genes of carrot. Plant Mol Biol 27:155–164

Kiyosue T, Kamada H, Harada H (1989) Induction of somatic embryogenesis by salt stress in carrot. Plant Tissue Cult Lett 6:162–164

Kiyosue T, Takano K, Kamada H, Harada H (1990) Induction of somatic embryogenesis in carrot by heavy metal ions. Can J Bot 68:2301–2303

Koltunow AM (1993) Apomixis: embryo sacs and embryos formed without meiosis or fertilization in ovules. Plant Cell 5:1425–1437

Koltunow AM (1994) Apomixis – other pathways for reproductive development in angiosperms. In: Williams EG et al. (eds) Genetic control of self-incompatibility and reproductive development in flowering plants. Kluwer, Dordrecht, pp 486–512

Koltunow AM, Soltys K, Nito N, McClure S (1995) Anther, ovule, seed, and nucellar embryo development in *Citrus sinensis* cv. Valencia. Can J Bot 73:1567–1582

Koltunow AM, Johnson SD, Bicknell RA (1998) Sexual and apomictic development in *Hieracium*. Sex Plant Reprod 11:213–230

Koltunow AM, Johnson SD, Bicknell RA (2000) Apomixis is not developmentally conserved in related, genetically characterized *Hieracium* plants of varying ploidy. Sex Plant Reprod 12:253–266

Kragh KM, Jacobsen S, Mikkelsen JD, Nielsen KA (1991) Purification and characterization of three chitinases and one β-1,3-glucanase accumulating in the medium of cell suspension cultures of barley (*Hordeum vulgare* L.). Plant Sci 76:65–77

Kragh KM, Hendriks T, de Jong AK, Lo Schiavo F, Bucherna N, Højrup P, Mikkelsen JD, de Vries SC (1996) Characterization of chitinases able to rescue somatic embryos of the temperature-sensitive carrot variant *ts11*. Plant Mol Biol 31:631–645

Kreuger M, van Holst G-J (1993) Arabinogalactan proteins are essential in somatic embryogenesis of *Daucus carota* L. Planta 189:243–248

Kreuger M, van Holst G-J (1995) Arabinogalactan-protein epitopes in somatic embryogenesis of *Daucus carota* L. Planta 197:135–141

Krishnaraj S, Vasil IK (1995) Somatic embryogenesis in herbaceous monocots. In: Thorpe TA (ed) In vitro embryogenesis in plants. Kluwer, Dordrecht, pp 417–470

Kyo M, Harada H (1986) Control of the developmental pathway of tobacco pollen in vitro. Planta 168:427–432

Kyo M, Harada H (1990) Specific phosphoproteins in the initial period of tobacco pollen embryogenesis. Planta 182:58–63

Kyo M, Ohkawa T (1991) Investigation of subcellular localization of several phosphoproteins in embryogenic pollen grains of tobacco. J Plant Physiol 137:525–529

Leblanc O, Peel MD, Carman JG, Savidan Y (1995) Megasporogenesis and megagametogenesis in several *Tripsacum* speies (Poaceae). Am J Bot 82:57–63

Leblanc O, Armstead I, Pessino S, Ortiz JPA, Evans C, do Valle C, Hayward MD (1997) Non-radioactive mRNA fingerprinting to visualise gene expression in mature ovaries of *Brachiaria* hybrids derived from *B. brizantha*, an apomictic tropical forage. Plant Sci 126:49–58

Lin X, Hwang G-JH, Zimmerman JL (1996) Isolation and characterization of a diverse set of genes from carrot somatic embryos. Plant Physiol 112:1365–1374

Lindzen E, Choi JH (1995) A carrot cDNA encoding an atypical protein kinase homologous to plant calcium-dependent protein kinases. Plant Mol Biol 28:785–797

Lo Schiavo F, Giuliano G, de Vries SC, Genga A, Bollini R, Pitto L, Cozzani F, Nuti-Ronchi V, Terzi M (1990) A carrot cell variant temperature sensitive for somatic embryogenesis reveals a defect in the glycosylation of extracellular proteins. Mol Gen Genet 223:385–393

Lotan T, Ohto M, Yee KM, West MAL, Lo R, Kwong RW, Yamagishi K, Fischer RL, Goldberg RB, Harada JJ (1998) *Arabidopsis* LEAFY COTYLEDON1 is sufficient to induce embryo development in vegetative cells. Cell 93:1195–1205

Magnard J-L, le Deunff E, Domenech J, Rogowsky PM, Testillano PS, Rougier M, Risueño MC, Vergne P, Dumas C (2000) Genes normally expressed in the endosperm are expressed at early stages of microspore embryogenesis. Plant Mol Biol 44:559–574

Masuda H, Oohashi S-I, Tokuji Y, Mizue Y (1995) Direct embryo formation from epidermal cells of carrot hypocotyls. J Plant Physiol 145:531–534

Matzk F, Meister A, Schubert I (2000) An efficient screen for reproductive pathways using mature seeds of monocots and dicots. Plant J 21:97–108

McCabe PF, Valentine TA, Forsberg LS, Pennell RI (1997) Soluble signals from cells identified at the cell wall establish a developmental pathway in carrot. Plant Cell 9:2225–2241

Mordhorst AP, Voerman KJ, Hartog MV, Meijer EA, van Went J, Koornneef M, de Vries SC (1998) Somatic embryogenesis in *Arabidopsis thaliana* is facilitated by mutations in genes repressing meristematic cell divisions. Genetics 149:549–563

Mordhorst AP, Hartog MV, El Tamer MK, Laux T, de Vries SC (2002) Somatic embryogenesis from *Arabidopsis* shoot apical meristem mutants. Planta 214:829–836

Naumova TN (1993) Apomixis in angiosperms. Nucellar and integumentary embryony. CRC Press, Boca Raton, FL

Naumova TN, Willemse MTM (1995) Ultrastructural characterization of apospory in *Panicum maximum*. Sex Plant Reprod 8:197–204

Naumova T, den Nijs APM, Willemse MTM (1993) Quantitative analysis of aposporous parthenogenesis in *Poa pratensis* genotypes. Acta Bot Neerl 42:299–312

Naumova TN, van der Laak J, Osadtchiy J, Matzk F, Kravtechenko A, Bergervoet J, Ramulu KS, Boutilier K (2001) Reproductive development in apomictic populations of *Arabis holboellii* (Brassicaceae). Sex Plant Reprod 14:195–200

Nielsen KA, Hansen IB (1992) Appearance of extracellular proteins associated with somatic embryogenesis in suspension cultures of barley (*Hordeum vulgare* L.). J Plant Physiol 139:489–497

Nishiwaki M, Fujino K, Koda Y, Masuda K, Kikuta Y (2000) Somatic embryogenesis induced by the simple application of abscisic acid to carrot (*Daucus carota* L.) seedlings in culture. Planta 211:756–759

Nitsch C, Norreel B (1973) Effet d'un choc thermique sur le pouvoir embryogène du pollen de *Datura innoxia* cultivé dans l'anthère ou isolé de l'anthère. C R Acad Sci Paris 276:303–306

Nogler GA (1984) Gametophytic apomixis. In: Johri BM (ed) Embryology of angiosperms. Springer, Berlin Heidelberg New York, pp 475–518

Nolan KE, Irwanto RR, Rose RJ (2003) Auxin up-regulates *MtSERK1* expression in both *Medicago truncatula* root-forming and embryogenic cultures. Plant Physiol 133:218–230

Nomura K, Komamine A (1985) Identification and isolation of single cells that produce somatic embryos at a high frequency in a carrot suspension culture. Plant Physiol 79:988–991

Ogas J, Cheng J-C, Sung ZR, Somerville C (1997) Cellular differentiation regulated by gibberellin in the *Arabidopsis thaliana pickle* mutants. Science 277:91–94

Oh S-K, Steiner H-Y, Dougall DK, Roberts DM (1992) Modulation of calmodulin levels, calmodulin methylation, and calmodulin binding proteins during carrot cell growth and embryogenesis. Arch Biochem Biophys 297:28–34

Overvoorde PJ, Grimes HD (1994) The role of calcium and calmodulin in carrot somatic embryogenesis. Plant Cell Physiol 35:135–144

Pechan PM, Smykal P (2001) Androgenesis: affecting the fate of the male gametophyte. Physiol Plant 111:1–8

Pechan PM, Bartels D, Brown DCW, Schell J (1991) Messenger-RNA and protein changes associated with induction of *Brassica* microspore embryogenesis. Planta 184:161–165

Peel MD, Carman JG, Leblanc O (1997) Megasporocyte callose in apomictic buffelgrass, Kentucky bluegrass, *Pennisetum squamulatum* Fresen, *Tripsacum* L., and weeping lovegrass. Crop Sci 37:724–732

Pessino SC, Espinoza F, Martínez EJ, Ortiz JPA, Valle EM, Quarín CA (2001) Isolation of cDNA clones differentially expressed in flowers of apomictic and sexual *Paspalum notatum*. Hereditas 134:35–42

Pillon E, Terzi M, Baldan B, Mariani P, Lo Schiavo F (1996) A protocol for obtaining embryogenic cell lines from *Arabidopsis*. Plant J 9:573–577

Raghavan V (1976) Role of the generative cell in androgenesis in henbane. Science 191:388–389

Raghavan V (1977) Pattern of DNA synthesis during pollen embryogenesis in henbane. J Cell Biol 73:521–526

Raghavan V (1978) Origin and development of pollen embryoids and pollen calluses in cultured anther segments of *Hyoscyamus niger* (henbane). Am J Bot 65:984–1002

Raghavan V (1979) An autoradiographic study of RNA synthesis during pollen embryogenesis in *Hyoscyamus niger* (henbane). Am J Bot 66:784–795

Raghavan V (1981) Distribution of poly(A)-containing RNA during normal pollen development and during induced pollen embryogenesis in *Hyoscyamus niger*. J Cell Biol 89:593–606

Raghavan V (1984) Protein synthetic activity during normal pollen development and during induced pollen embryogenesis in *Hyoscyamus niger*. Can J Bot 62:2493–2513

Raghavan V (1986) Embryogenesis in angiosperms. A developmental and experimental study. Cambridge University Press, New York

Raghavan V (2004) Role of 2,4-dichlorophenoxyacetic acid (2,4-D) in somatic embryogenesis in cultured zygotic embryos of *Arabidopsis*: cell expansion, cell cyling, and morphogenesis during continuous exposure of embryos to 2,4-D. Am J Bot 91:1743–1756

Rao KS (1996) Embryogenesis in flowering plants: recent approaches and prospects. J Biosci 21:827–841

Rashid A, Street HE (1973) The development of haploid embryoids from anther cultures of *Atropa belladonna* L. Planta 113:262–270

Reynolds TL (1997) Pollen embryogenesis. Plant Mol Biol 33:1–10

Reynolds TL, Crawford RL (1996) Changes in abundance of an abscisic acid-responsive, early cysteine-labeled metallothionein transcript during pollen embryogenesis in bread wheat (*Triticum aestivum*). Plant Mol Biol 32:823–829

Reynolds TL, Kitto SL (1992) Identification of embryoid-abundant genes that are temporally expressed during pollen embryogenesis in wheat anther cultures. Plant Physiol 100:1744–1750

Reynolds TL, Raghavan V (1982) An autoradiographic study of RNA synthesis during maturation and germination of pollen grains of *Hyoscyamus niger*. Protoplasma 111:177–188

Richards AJ (1997) Plant breeding systems, 2nd edn. Chapman and Hall, London

Robatche-Claive A-S, Couillerot J-P, Dubois J, Dubois T, Vasseur J (1992) Embryogenèse somatique directe dans les feuilles du *Cichorium* hybride "474" : synchronization de l'induction. C R Acad Sci Paris 314:371–377

Sangwan-Norreel BS (1978) Cytochemical and ultrastructural peculiarities of embryogenic pollen grains and of young androgenic embryos in *Datura innoxia*. Can J Bot 56:805–817

Santos MO, Romano E, Yotoko KSC, Tinoco MLP, Dias BBA, Aragão FJL (2005) Characterisation of the cacao *somatic embryogenesis receptor-like kinase* (*SERK*) expressed during somatic embryogenesis. Plant Sci 168:723–729

Sato S, Toya T, Kawahara R, Whittier RF, Fukuda H, Komamine A (1995) Isolation of a carrot gene expressed specifically during early-stage somatic embryogenesis. Plant Mol Biol 28:39–46

Savidan Y (2000) Apomixis: genetics and breeding. Hortic Rev 18:13–86

Schmidt EDL, Guzzo F, Toonen MAJ, de Vries SC (1997) A leucine-rich repeat containing receptor-like kinase marks somatic plant cells competent to form embryos. Development 124:2049–2062

Sevilla-Lecoq S, Deguerry F, Matthys-Rochon E, Perez P, Dumas C Rogowsky PM (2003) Analysis of *ZmAE3* upstream sequences in maize endosperm and androgenic embryos. Sex Plant Reprod 16:1–8

Shah K, Gadella TWJ Jr, van Erp H, Hecht V, de Vries SC (2001a) Subcellular localization and oligomerization of the *Arabidopsis thaliana* somatic embryogenesis receptor kinase 1 protein. J Mol Biol 309:641–655

Shah K, Vervoort J, de Vries SC (2001b) Role of threonines in the *Arabidopsis thaliana* somatic embryogenesis receptor kinase I activation loop in phosphorylation. J Biol Chem 276:41263–41269

Shah K, Russinova E, Gadella TWJ Jr, Willemse J, de Vries SC (2002) The *Arabidopsis* kinase-associated protein phosphatases controls internalization of the somatic embryogenesis receptor kinase 1. Genes Dev 16:1707–1720

Sharma KK, Thorpe TA (1995) Asexual embryogenesis in vascular plants in nature. In: Thorpe TA (ed) In vitro embryogenesis in plants. Kluwer, Dordrecht, pp 17–72

Smith DL, Krikorian AD (1989) Release of somatic embryogenic potential from excised zygotic embryos of carrot and maintenance of proembryonic cultures in hormone-free medium. Am J Bot 76:1832–1843

Smith DL, Krikorian AD (1990) Somatic embryogenesis of carrot in hormone-free medium: external pH control over morphogenesis. Am J Bot 77:1634–1647

Somleva MN, Schmidt EDL, de Vries SC (2000) Embryogenic cells in *Dactylis glomerata* L. (Poaceae) explants identified by cell tracking and SERK expression. Plant Cell Rep 19:718–726

Spielman M, Vinkenoog R, Scott RJ (2003) Genetic mechanisms of apomixis. Philos Trans R Soc Ser B 358:1095–1103

Stebbins GL Jr (1932) Cytology of *Antennaria* II. Parthenogenetic species. Bot Gaz 94:322–345

Stebbins GL Jr, Jenkins JA (1939) Aposporic development in the north American species of *Crepis*. Genetica 21:191–224

Sterk P, Booij H, Schellekens GA, van Kammen A, de Vries SC (1991) Cell-specific expression of the carrot EP2 lipid transfer protein gene. Plant Cell 3:907–921

Steward FC (1963) The control of growth in plant cells. Sci Am 209(4):104–113

Steward FC (1968) Growth and organization in plants. Addison-Wesley, Reading, MA

Steward FC, Mapes MO, Kent AE, Holsten RD (1964) Growth and development of cultured plant cells. Science 143:20–27

Stone SL, Kwong LW, Yee KM, Pelletier J, Lepiniec L, Fischer RL, Goldberg RB, Harada JJ (2001) *LEAFY COTYLEDON2* encodes a B3 domain transcription factor that induces embryo development. Proc Natl Acad Sci USA 98:11806–11811

Straatman KR, Schel JHN (2001) Distribution of splicing proteins and putative coiled bodies during pollen development and androgenesis in *Brassica napus* L. Protoplasma 216:191–200

Straatman KR, Nijsse J, Kieft H, van Aelst AC, Schel JHN (2000) Nuclear pore dynamics during pollen development and androgenesis in *Brassica napus*. Sex Plant Reprod 13:43–51

Suen K-L, Choi JH (1991) Isolation and sequence analysis of a cDNA clone for a carrot calcium-dependent protein kinase: homology to calcium/calmodulin-dependent protein kinases and to calmodulin. Plant Mol Biol 17:581–590

Sunderland N, Wicks FM (1971) Embryoid formation in pollen grains of *Nicotiana tabacum*. J Exp Bot 22:213–226

Sunderland N, Collins GB, Dunwell JM (1974) The role of nuclear fusion in pollen embryogenesis of *Datura innoxia* Mill. Planta 117:227–241

Sunderland N, Roberts M, Evans LJ, Wildon DC (1979) Multicellular pollen formation in cultured barley anthers. J Exp Bot 30:1133–1144

Sung ZR, Okimoto R (1981) Embryonic proteins in somatic embryos of carrot. Proc Natl Acad Sci USA 78:3683–3687

Takahata K, Takeuchi M, Fujita M, Azuma J, Kamada H, Sato F (2004) Isolation of putative glycoprotein gene from early somatic embryos of carrot and its possible involvement in somatic embryo development. Plant Cell Physiol 45:1658–1668

Takeda H, Kotake T, Nakagawa N, Sakurai N, Nevins DJ (2003) Expression and function of cell wall-bound cationic peroxidase in *Asparagus* somatic embryogenesis. Plant Physiol 131:1765–1774

Taşkin KM, Turgut K, Scott RJ (2004) Apomictic development in *Arabis gunnisoniana*. Israel J Plant Sci 52:155–160

Telmer CA, Newcomb W, Simmonds DH (1993) Microspore development in *Brassica napus* and the effect of high temperature on division in vitvo and in vitro. Protoplasma 172:154–165

Thomas C, Meyer D, Himber C, Steinmetz A (2004) Spatial expression of a sunflower *SERK* gene during induction of somatic embryogenesis and shoot organogenesis. Plant Physiol Biochem 42:35–42

Thorpe TA, Stasolla C (2001) Somatic embryogenesis. In: Bhojwani SS, Soh WY (eds) Current trends in the embryology of angiosperms. Kluwer, Dordrecht, pp 279–336

Timmers ACJ, Reiss H-D, Bohsung J, Traxel K, Schel JHN (1996) Localization of calcium during somatic embryogenesis of carrot (*Daucus carota* L.). Protoplasma 190:107–118

Toonen MAJ, Schmidt EDL, van Kammen A, de Vries SC (1997a) Promotive and inhibitory effects of diverse arabinogalactan proteins on *Daucus carota* L. somatic embryogenesis. Planta 203:188–195

Toonen MAJ, Verhees JA, Schmidt EDL, van Kammen A, de Vries SC (1997b) AtLTP1 luciferase expression during carrot somatic embryogenesis. Plant J 12:1213–1221

Touraev A, Ilham A, Vicente O, Heberle-Bors E (1996a) Stress-induced microspore embryogenesis in tobacco: an optimized system for molecular studies. Plant Cell Rep 15:561–565

Touraev A, Indrianto A, Wratschko I, Vicente O, Heberle-Bors E (1996b) Efficient microspore embryogenesis in wheat (*Triticum aestivum* L.) induced by starvation at high temperature. Sex Plant Reprod 9:209–215

Touraev A, Pfosser M, Heberle-Bors E (2001) The microspore: a haploid multipurpose cell. Adv Bot Res 35:53–109

Tucker MR, Paech NA, Willemse MTM, Koltunow AMG (2001) Dynamics of callose deposition and β-1,3-glucanase expression during reproductive events in sexual and apomictic *Hieracium*. Planta 212:487–498

Tucker MR, Araujo A-CG, Paech NA, Hecht V, Schmidt EDL, Rossell JB, de Vries SC, Koltunow AMG (2003) Sexual and apomictic reproduction in *Hieracium* subgenus *Pilosella* are closely interrelated developmental pathways. Plant Cell 15:1524–1537

van Engelen FA, Sterk P, Booij H, Cordewener JHG, Rook W, van Kammen A, de Vries SC (1991) Heterogeneity and cell type-specific localization of a cell wall glycoprotein from carrot suspension cells. Plant Physiol 96:705–712

van Engelen FA, de Jong AJ, Meijer EA, Kuil CW, Meyboom JK, Dirkse WG, Booij H, Hartog MV, Vandekerckhove J, de Vries SC, van Kammen A (1995) Purification, immunological characterization and cDNA cloning of a 47 kDa glycoprotein secreted by carrot suspension cells. Plant Mol Biol 27:901–910

van Hengel AJ, Tadesse Z, Immerzeel P, Schols H, van Kammen A, de Vries SC (2001) N-acetylglucosamine and glucosamine-containing arabinogalactan proteins control somatic embryogenesis. Plant Physiol 125:1880–1890

van Hengel AJ, Guzzo F, van Kammen A, de Vries SC (1998) Expression pattern of the carrot *EP3* endochitinase genes in suspension cultures and in developing seeds. Plant Physiol 117:43–53

Vasil V, Vasil IK (1981) Somatic embryogenesis and plant regeneration from suspension cultures of pearl millet (*Pennisetum americanum*). Ann Bot 47:669–678

Vielle-Calzada J-P, Nuccio ML, Budiman ML, Thomas TL, Burson BL, Hussey MA, Wing RA (1996) Comparison gene expression in sexual and apomictic ovaries of *Pennisetum ciliare* (L.) Link. Plant Mol Biol 32:1085–1092

Wakana A, Uemoto S (1987) Adventive embryogenesis in *Citrus* I. The occurrence of adventive embryos without pollination or fertilization. Am J Bot 74:517–530

Wakana A, Uemoto S (1988) Adventive embryogenesis in *Citrus* (Rutaceae) II. Postfertilization development. Am J Bot 75:1033–1047

Wetherell DF, Halperin W (1963) Embryos derived from callus tissue cultures of the wild carrot. Nature 200:1336–1337

Wilde HD, Nelson WS, Booij H, de Vries SC, Thomas TL (1988) Gene-expression programs in embryogenic and non-embryogenic carrot cultures. Planta 176:205–211

Wilms HJ, van Went JL, Cresti M, Ciampolini F (1983) Adventive embryogenesis in *Citrus*. Caryologia 36:65–78

Wurtele ES, Wang H, Durgerian S, Nikolau BJ, Ulrich TH (1993) Characterization of a gene that is expressed early in somatic embryogenesis of *Daucus carota*. Plant Physiol 102:303–312

Yazawa K, Takahata K, Kamada H (2004) Isolation of the gene encoding carrot leafy cotyledon1 and expression analysis during somatic and zygotic embryogenesis. Plant Physiol Biochem 42:215–223

Zaki MAM, Dickinson HG (1990) Structural changes during the first divisions of embryos resulting from anther and free microspore culture in *Brassica napus*. Protoplasma 156:149–162

Žárský V, Garrido D, Říhová L, Tupý J, Vicente O, Heberle-Bors E (1992) Derepression of the cell cycle by starvation is involved in the induction of tobacco pollen embryogenesis. Sex Plant Reprod 5:189–194

Zhang S, Wong L, Meng L, Lemaux PG (2002) Similarity of expression patterns of *knotted1* and *ZmLEC1* during somatic and zygotic embryogenesis in maize (*Zea mays* L.). Planta 215:191–194

Zhao J-P, Simmonds DH, Newcomb W (1996) Induction of embryogenesis with colchicine instead of heat in microspores of *Brassica napus* L. cv. Topas. Planta 198:433–439

Zhu Z, Sun J, Wang J (1978) Cytological investigation on androgenesis of *Triticum aestivum*. Acta Bot Sin 20:6–12

Zimmerman JL (1993) Somatic embryogenesis: a model for early development in higher plants. Plant Cell 5:1411–1423

Zuo J, Niu Q-W, Frugis G, Chua N-H (2002) The *WUSCHEL* gene promotes vegetative-to-embryonic transition in *Arabidopsis*. Plant J 30:349–359

COLOR PLATES

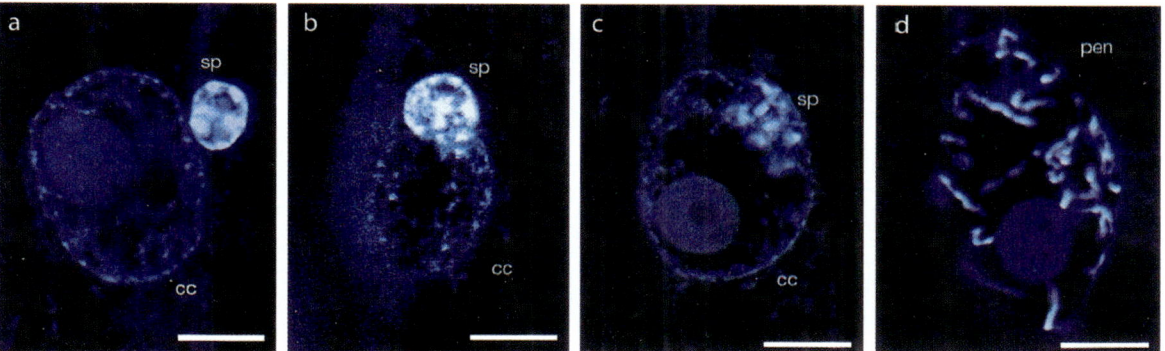

Plate 1, Fig. a–d The second fertilization event in *Nuphar polysepalum* as seen in fluorescent micrographs of sections of four ovules at 13 and 22 h after pollination. **a** Sperm nucleus approaching the haploid nucleus of the central cell. **b** Beginning of nuclear fusion. **c** Sperm nucleus being engulfed by the nucleus of the central cell. **d** Primary endosperm nucleus, 22 h after pollination, in early prophase. *cc* Central cell, *pen* primary endosperm nucleus, *sp* sperm nucleus. *Bars* 10 µm. (Reprinted from Williams and Friedman (2002) Nature 415:522–526)

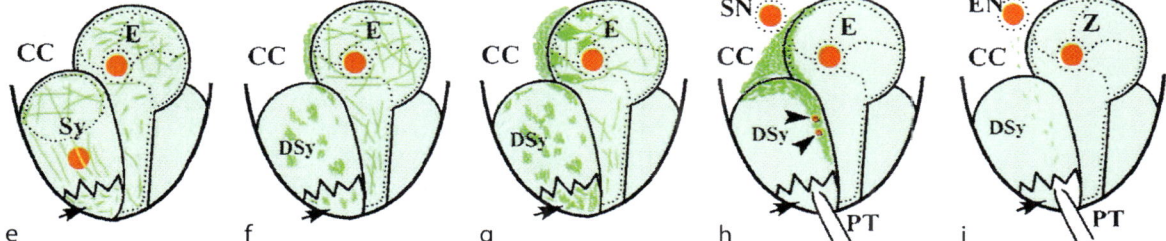

Plate 1, Fig. e–i Diagrams showing changes in the organization of actin filaments in the egg apparatus of *Torenia fournieri* before and after fertilization. **e** Actin filaments are longitudinally aligned in the micropylar cytoplasm and some organize into a network in the chalazal cortex of the synergid. *Arrow* Dense actin patch found near the filiform apparatus of the synergid. Short actin filaments are randomly distributed throughout the egg cytoplasm. **f** After anthesis, synergid degeneration is accompanied by degradation of actin filaments. Some actin filaments appear in the intercellular gap between the egg and central cell, forming an actin band. *Arrow* Actin filaments present as a cap in the filiform apparatus. Actin filaments become elongate and are organized into a distinct network in the egg cytoplasm. **g** After pollination, actin filaments degrade into patches in the peripheral cytoplasm of the egg cell at the same time as the actin corona in the interface between the egg and central cell becomes distinct. *Arrow* Prominent actin cap in the filiform apparatus. **h** A conspicuous actin corona forms after the pollen tube discharges its contents into a degenerating synergid. *Arrowheads* Male gametes about to fuse with the target female cells, *arrow* the filiform apparatus from which the actin cap has disappeared. **i** Disintegration of the actin corona after fertilization; no actin is detected in the filiform apparatus (*arrow*). *CC* Central cell, *DSy* degenerating synergid, *E* egg cell, *EN* primary endosperm nucleus, *PT* pollen tube, *SN* secondary (polar fusion) nucleus, *Sy* synergid, *Z* zygote. (Reprinted from Fu et al. (2000) Sex Plant Reprod 12:315–322)

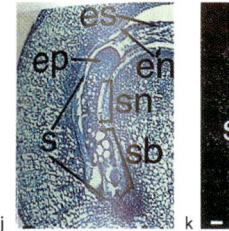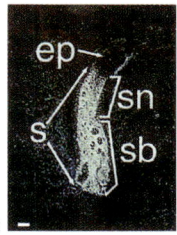

Plate 1, Fig. j–l Localization of two mRNAs in the embryos of *Phaseolus coccineus*; images are light micrographs taken after in situ hybridization. **j** Ovule 7 days after pollination, showing the embryo-suspensor complex. **k** Ovule of the same age hybridized with ^{32}P-labeled G564 mRNA probe. **l** Ovule of the same age hybridized with ^{32}P-labeled G541 mRNA probe. *en* Endothelium, *ep* embryo proper, *es* endosperm, *s* suspensor, *sb* basal region of the suspensor, *sn* neck of the suspensor. *Bars* 50 µm. (Reprinted from Weterings et al. (2001) Plant Cell 13:2409–2425)

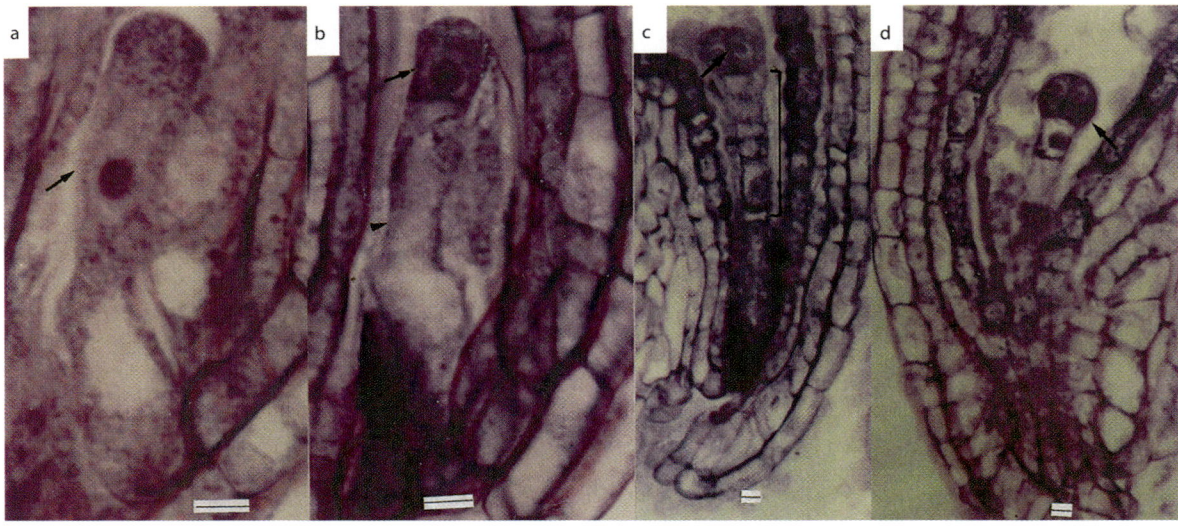
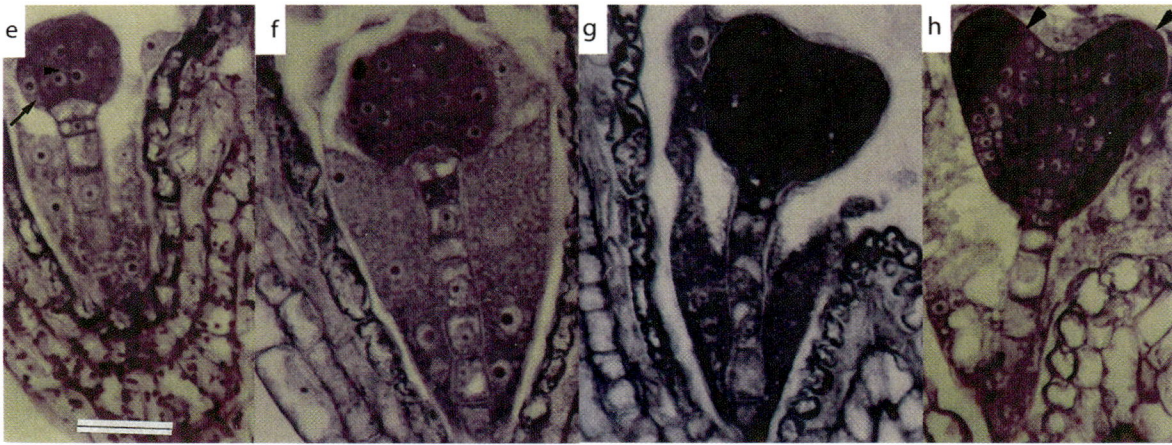

Plate 2, Fig. a–h Sections of *Arabidopsis* ovule showing stages in the development of the embryo. **a** Zygote (*arrow*) primed to divide asymmetrically. **b** The first asymmetric division of the zygote forming a small terminal cell (*arrow*) and a large basal cell (*arrowhead*). **c** The first longitudinal division of the terminal cell (*arrow*). The basal cell has formed a suspensor (*square bracket*) of six cells. **d** Octant-stage embryo (*arrow*) consisting of two tiers of four cells each. **e** Sixteen-celled embryo with eight external cells of the protoderm (*arrow*) and eight inner cells of the procambium and ground meristem (*arrowhead*). **f** Globular embryo. **g** Triangular or early heart-shaped stage embryo. **h** Mid-heart-shaped embryo showing the emerging cotyledons (*arrowheads*)

Plate 3, Fig. i–k (continues Plate 2, Fig. a–h) i Early torpedo-shaped embryo; *arrow* incipient shoot apical meristem. **j** Bent-cotyledon stage embryo; *arrow* root pole. **k** Mature embryo. *c* Cotyledons, *h* hypocotyl, *hp* derivatives of the hypophysis, *r* root apical meristem, *s* shoot apical meristem. *Bars* **a–d** 10 µm; **e**, **k** 50 µm (*bar* in **e** applies also to **f–i**); **j** 40 µm

Plate 3, Fig. l Diagrams showing cell fate determination in the octant-stage embryo of *Arabidopsis* leading to the formation of tissues and organs in the torpedo-shaped embryo. *Left* Delimitation of the upper (*green*) and lower (*red*) tiers of cells in the octant-stage embryo. The hypophysis is colored *blue*. *Right* A torpedo-shaped embryo showing the formation of the shoot apical meristem (*light green*) and part of the cotyledons (*green*) from the upper tier of cells, and the rest of the cotyledons (*red*), hypocotyl, and radicle, including most of the root apical meristem and meristem initials from the lower tier. The hypophysis contributes to the formation of the quiescent center (*yellow*) and root cap columella (*blue*). Clonal boundaries of tissues and organs of the torpedo-shaped embryo derived from the octant-stage embryo are indicated by *lines*

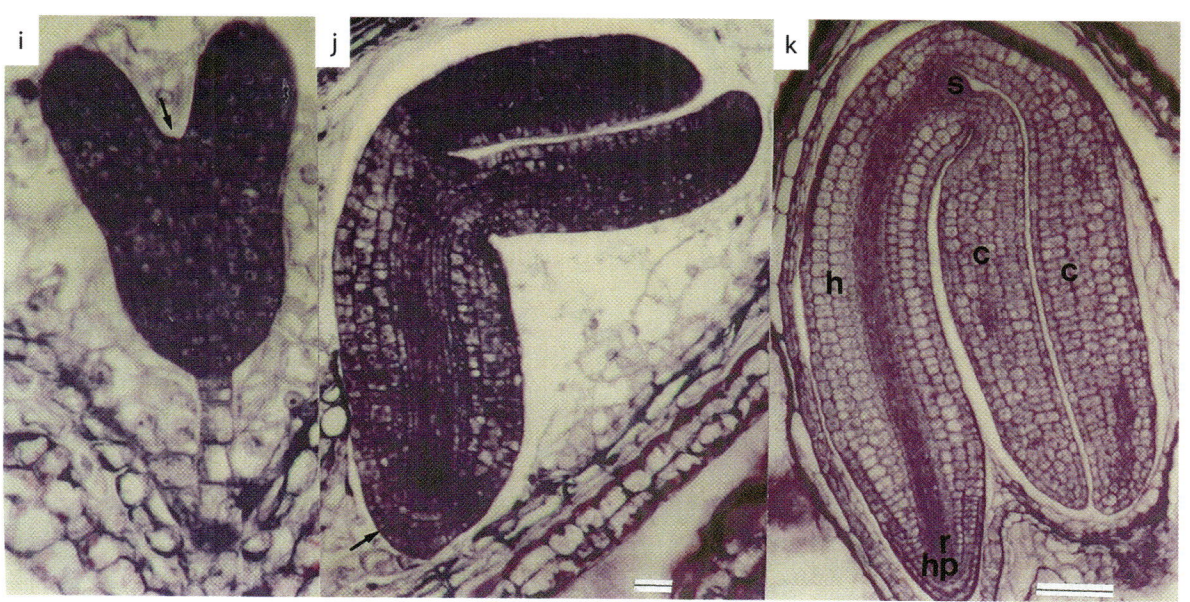

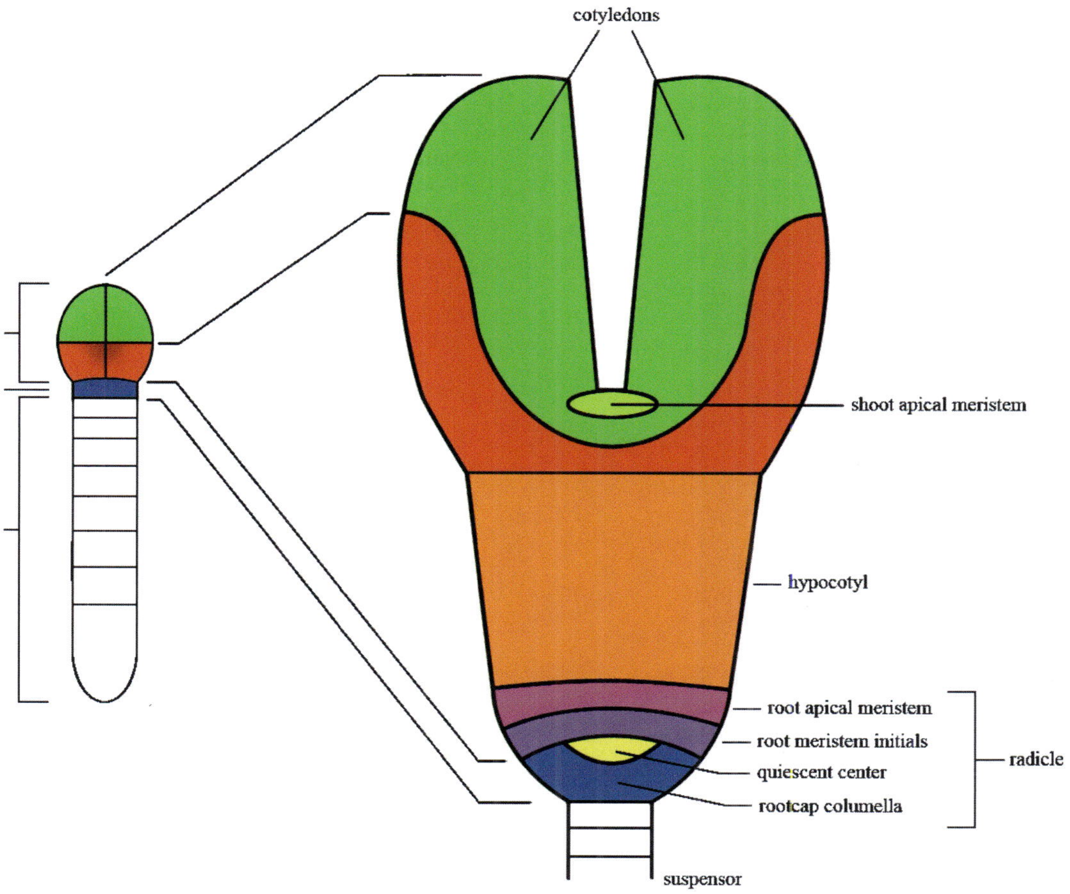

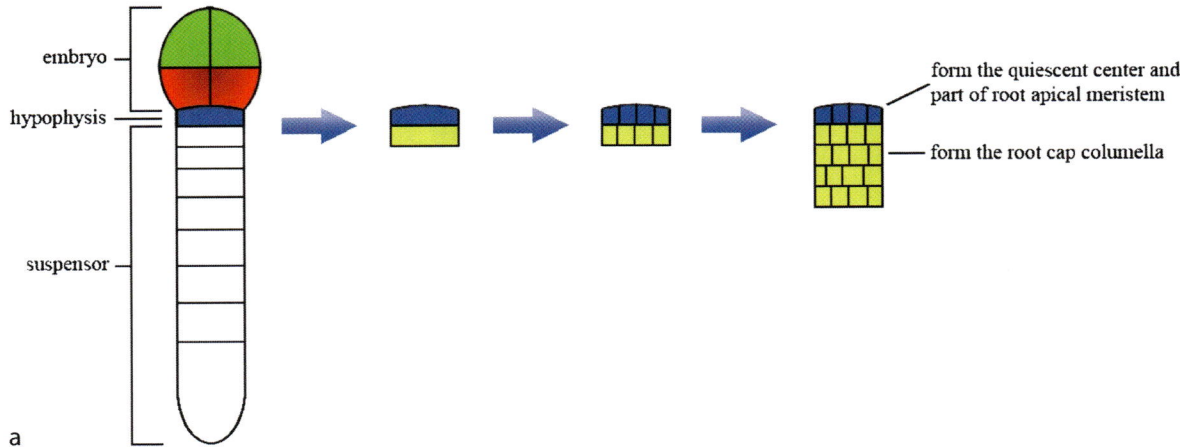

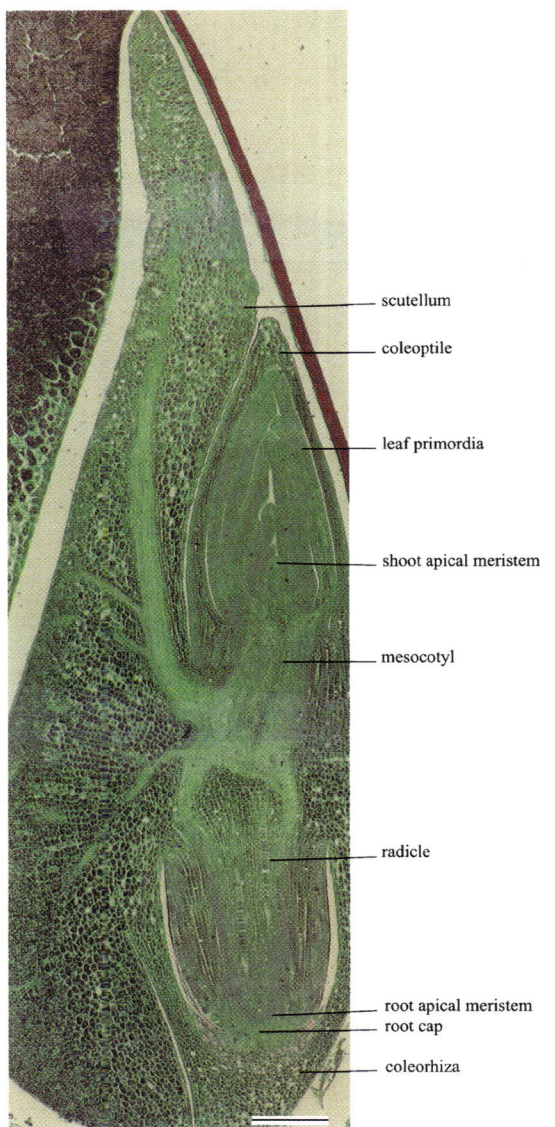

Plate 4, Fig. a Diagrams showing the division of the hypophysis in *Arabidopsis*. From *left* to *right* Embryo with the suspensor (the hypophysis is colored *blue*), division of the hypophysis to form a lens-shaped upper cell (*blue*) and a lower cell (*yellow*), vertical divisions of both cells to form two layers of four cells each, the upper four cells form the quiescent center. Horizontal divisions of lower cells to form four superimposed layers of four cells each; the root cap columella is generated by these cells

Plate 4. Fig. b Section of a mature maize embryo. *Bar* 500 μm

Plate 5, Fig. a–i Localization of *KNOTTED1* (*KN1*) mRNA (**a–e**) and proteins (**f–i**) during embryogenesis in maize. Embryos are oriented with the anterior face to the *right*. mRNA was localized by in situ hybridization and is indicated by blue staining in the sections. **a** Embryo 8 days after pollination. **b** Embryo 10 days after pollination. **c** Embryo 13 days after pollination. **d, e** Embryos 20 days after pollination showing close-ups of the shoot and root apical meristems, respectively. Protein was localized with a polyclonal antibody and is visualized as black staining over nuclei. Sections were counterstained with basic fuchsin to show unlabeled nuclei and other cellular structures in pink. **f** Embryo 8 days after pollination. **g** Embryo 10 days after pollination. **h** Embryo 13 days after pollination. **i** Embryo 14 days after pollination. *c* Coleoptile, *l* leaf primordium, *rm* root apical meristem, *sm* shoot apical meristem. *Bars* 50 μm (*bars* in **a–d** apply to **f–i**, respectively). (Reprinted from Smith et al. (1995) Dev Genet 16:344–348

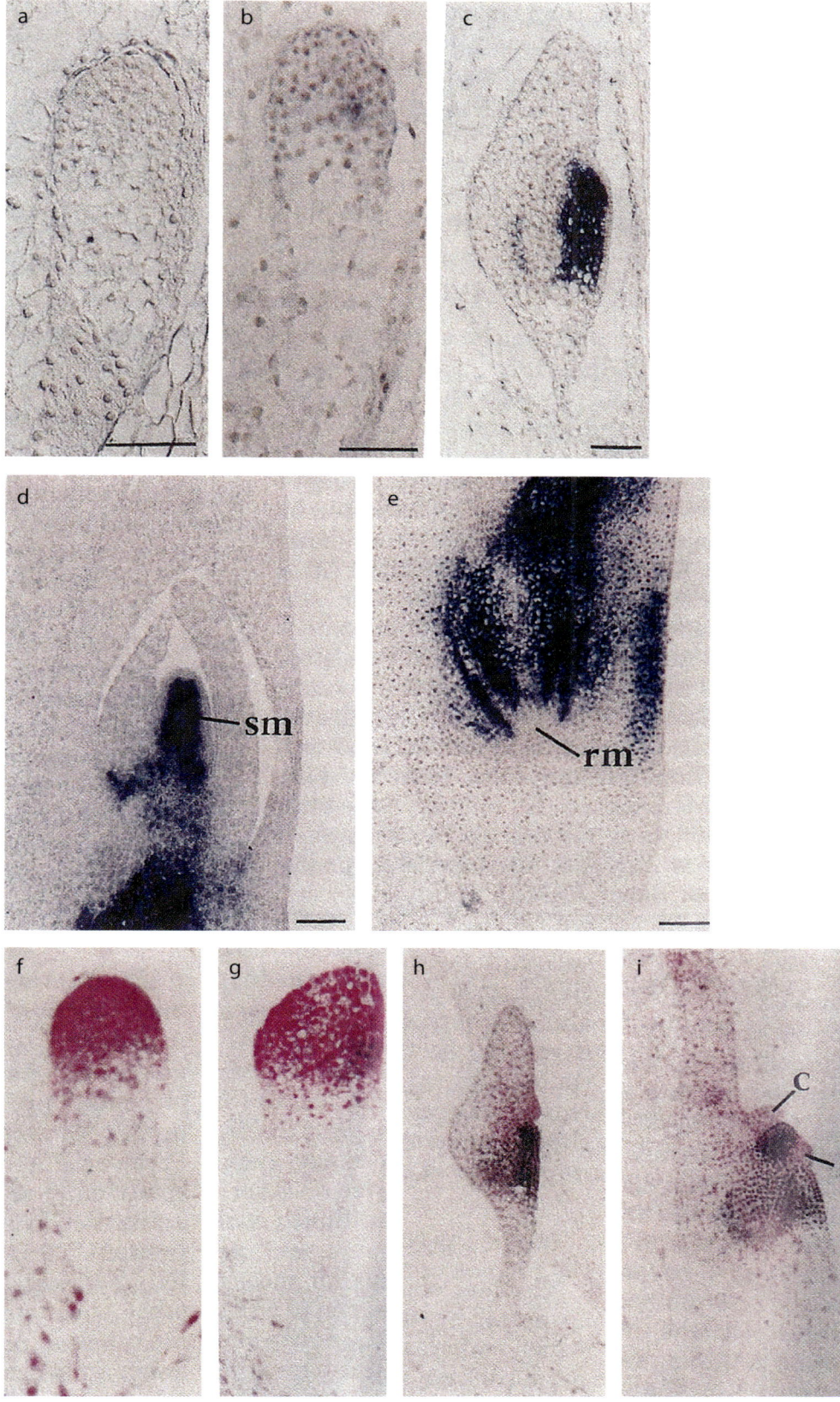

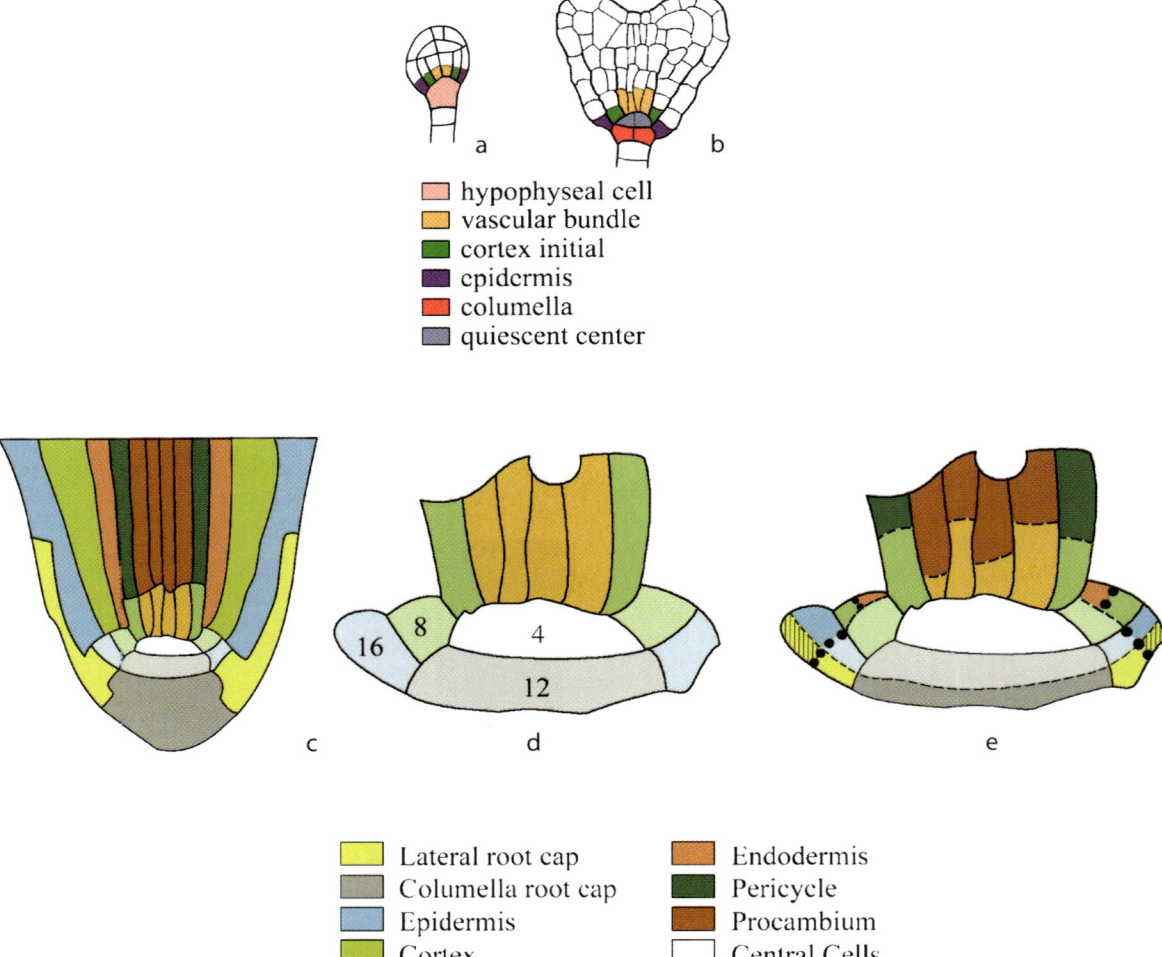

Plate 6, Fig. a–e Fate map of the embryonic root of *Arabidopsis*. **a** Globular embryo. **b** Heart-shaped embryo in which all cells of the incipient root meristem are present. (Reprinted from van den Berg et al. (1995) Nature 378:62–65. **c** Central region of the root meristem of a mature embryo showing the position of initials for all tissue layers surrounding the central cells. **d** Magnified view of the core meristem shown in **c** Fixed numbers of different initials found in the mature embryo are indicated. **e** Divisions of the initials to reestablish identically grouped initials and derivatives. *Dashed line* first division, *dots* second division, *striped* third (radial) division. (Reprinted from Dolan et al. (1993) Development 119:71–84

Plate 7, Fig. a–r Expression of the *SCARECROW* (*SCR*) gene during embryogenesis in *Arabidopsis*. **a–c** Early globular-stage embryos. **d–f** Late globular-stage embryos. **g–i** Triangular-stage embryos. **j–l** Midheart-shaped stage embryos. **m–o** Torpedo-shaped stage embryos. **p–r** Hypocotyl and root region of nearly mature embryos. **a, d, g, j, m, p** Longitudinal sections of wild-type embryos. **b, e, h, k, n, q** Sections of embryos following in situ hybridization with an *SCR* antisense gene probe. **c, f, i, l, o, r** Confocal images of green fluorescent protein (GFP) expression driven by the *SCR* promoter. Schematic drawings on the right show tracings made from the leftmost panels of the progenitors of the ground tissue, their derivatives and hypophysis/central cells. *SCR* gene expression is indicated by *blue shading*. The shift of *SCR* gene expression to the innermost cell layer after periclinal divisions is indicated by *arrows*. The boundary of the upper and lower tiers of cells delineated in the globular-stage embryo is indicated by *dark lines*. Bars **d–l** 25 μm, **m–r** 50 μm. (Reprinted from Wysocka-Diller et al. (2000) Development 127:595–603

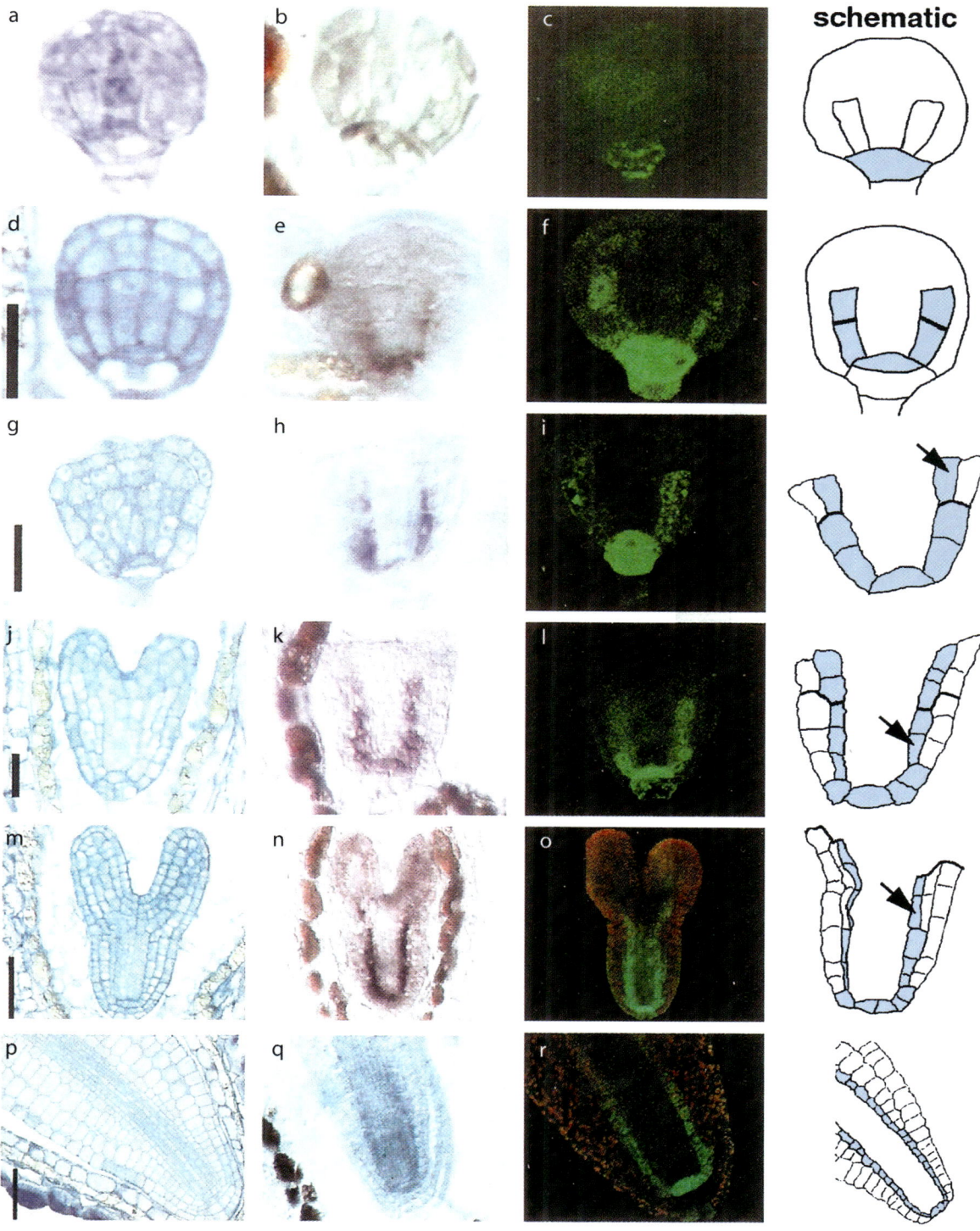

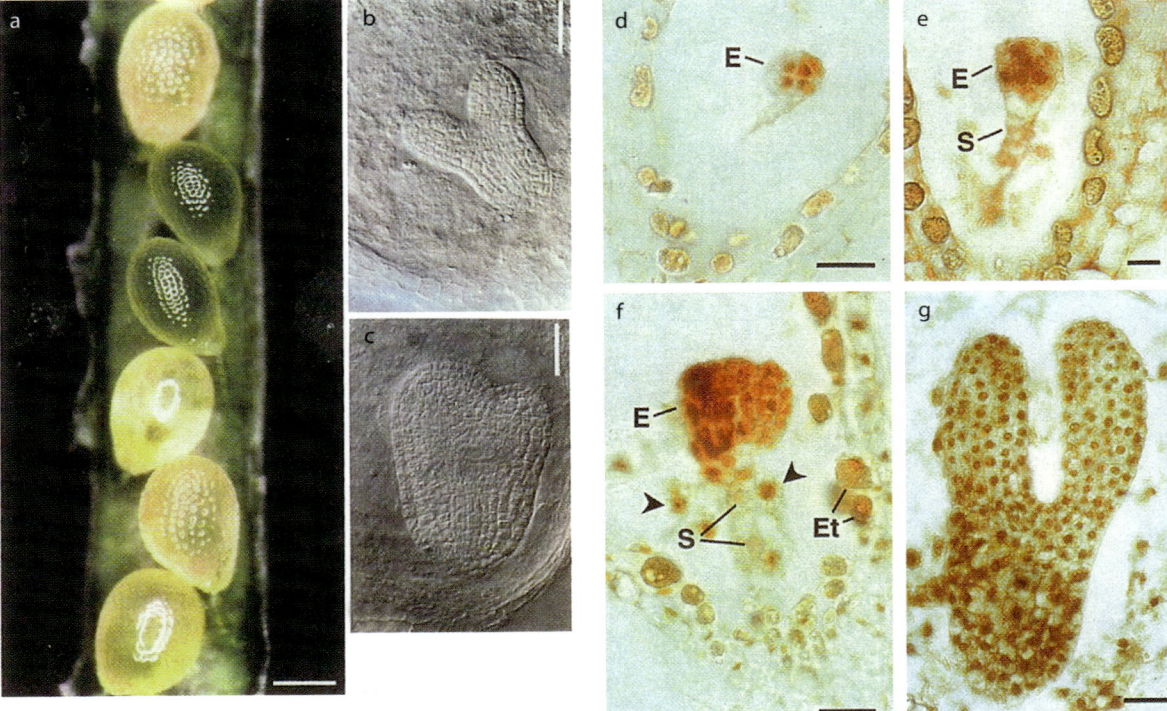

Plate 8, Fig. a–c Phenotypes of the *medea* (*mea*) mutant. **a** Silique of a selfed heterozygous *mea-1/MEA* plant showing aborted seeds derived from female gametophytes carrying the *mea-1* allele. **b** Late heart-shaped wild-type embryo. **c** Late heart-shaped embryo that inherited a maternal *mea-1* allele. Morphogenesis is delayed and results in a larger than normal heart-shaped embryo that eventually aborts. *Bars* **a** 200 µm; **b**, **c** 50 µm. (Reprinted from Grossniklaus et al. (2001) Curr Opin Plant Biol 4:21–27

Plate 8, Fig. d–g Sections of *Arabidopsis* ovules treated with antiserum against AGAMOUS-like (AGL15), showing the localization of AGL15 protein. **d**, **e** Wild-type proembryo and globular embryo, respectively. **f** Wild-type embryo at the transition stage; *arrowheads* labeled endosperm nuclei. **g** Wild-type torpedo-shaped embryo. *E* Embryo, *Et* endothelium, *S* suspensor. *Bars* **d**, **f**, **g** 20 µm; **e** 10 µm. (Reprinted from Perry et al. (1996) Plant Cell 8:1977–1989

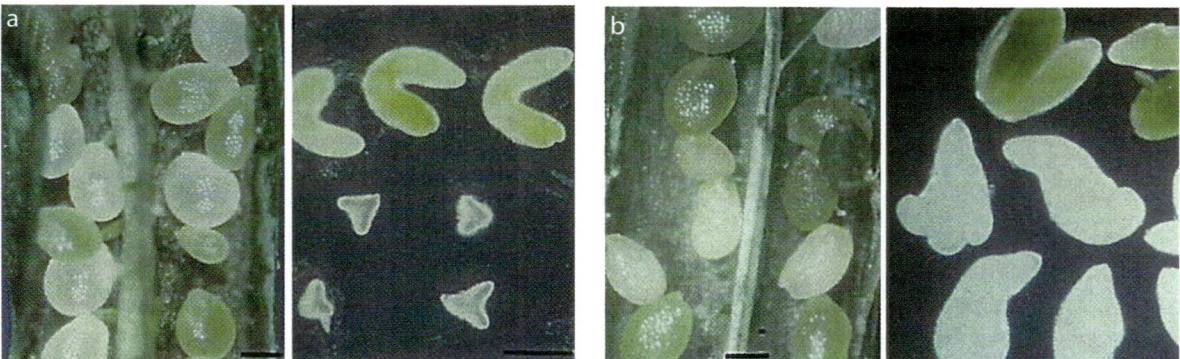

Plate 9, Fig. a,b Photographs of abnormal embryo phenotypes in a *PEI1* antisense line of *Arabidopsis*. *Left* Part of the opened siliques with green ovules harboring normal embryos and white ovules enclosing abnormal embryos; *right* embryos excised from green ovules (top row) and white ovules (bottom two rows). **a** Young antisense ovules and embryos. **b** Late-stage antisense ovules and embryos from the same plant as in **a**. *Bars* 200 µm. (Reprinted from Li and Thomas (1998) Plant Cell 10:383–398

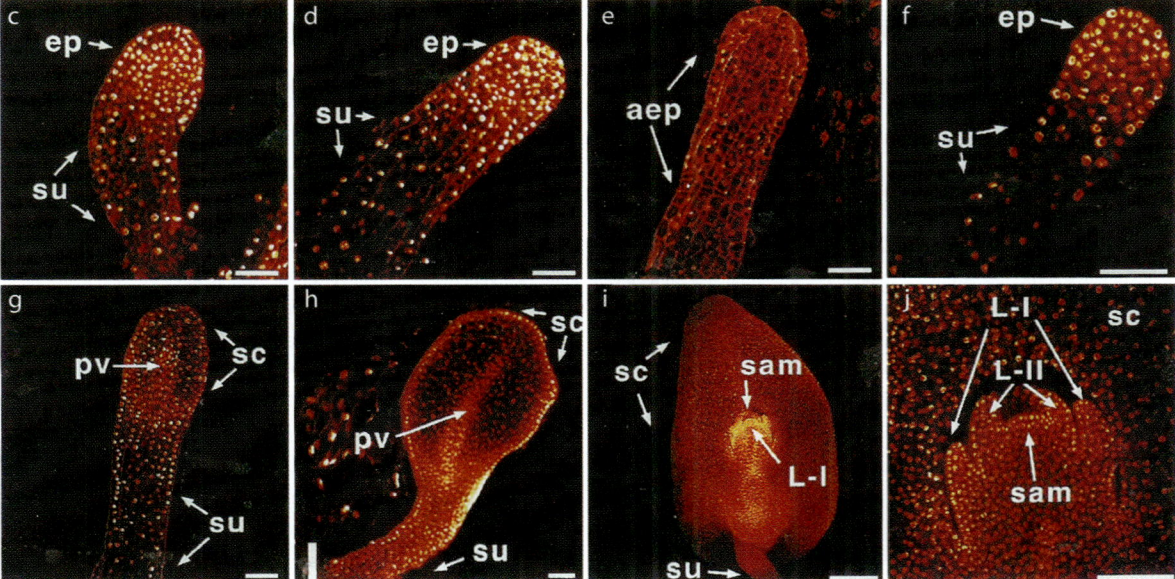

Plate 9, Fig. c–j Confocal laser scanning microscopic images of propidium-iodide-stained embryos of maize mutants *emb**-*8518* (**c–e**) and *emb**-*8537* (**f–j**). **c** Embryo 10 days after pollination. **d** Embryo 15 days after pollination. **e** Embryo 20 days after pollination; mutant embryos are arrested at an early transition stage, 10 days after pollination. **f** Embryo 10 days after pollination. **g** Embryo 15 days after pollination. **h** Embryo 20 days after pollination. Scutellum begins to form at 15 days and is well-developed at 20 days. **i, j** Embryos at 25 days after pollination. The coleoptile is missing (**i**) and leaves fail to cover the meristem (**j**). *aep* Aberrant embryo proper, *ep* embryo proper, *pv* provascular tissue, *sam* shoot apical meristem, *sc* scutellum, *su* suspensor, *L-1* and *L-II* leaves. *Bars* **c–h** 50 µm; **i** 250 µm; **j** 100 µm. (Reprinted from Elster et al. (2000) Dev Genes Evol 210:300–310

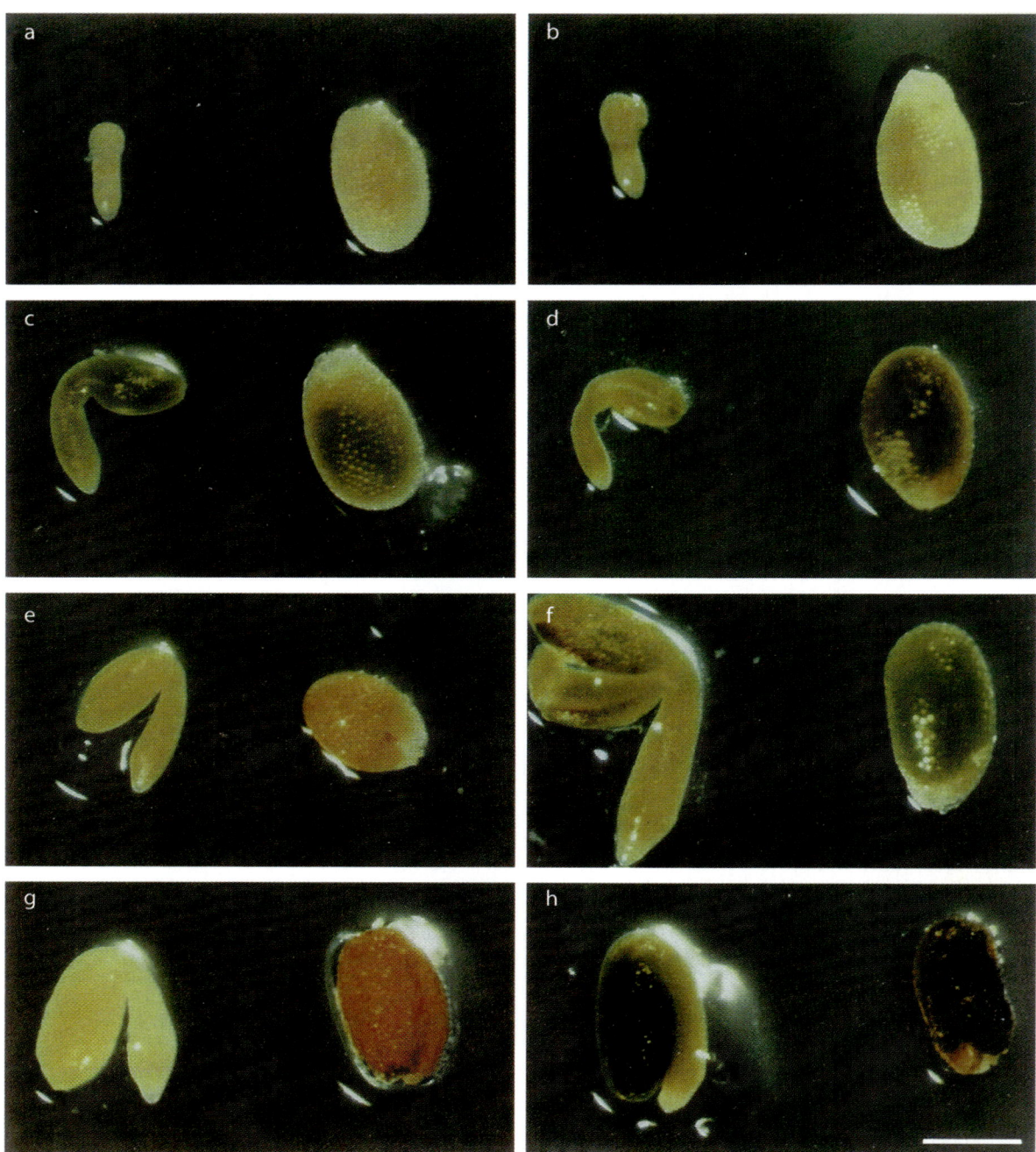

Plate 10, Fig. a–h Seed development in the wild-type (**a, c, e, g**) and *aba-insensitive3* (*abi3*) mutant (**b, d, f, h**) of *Arabidopsis*. Embryos dissected from seeds of wild-type and mutant harvested at 4 (**a, b**), 8 (**c, d**), 12 (**e, f**), and 16 (**g, h**) days after flowering are shown. Embryos were dissected after imbibing seeds on agar plates and photographed immediately. *Bar* 300 μm. (Reprinted from Nambara et al. (1995) Development 121:629–636

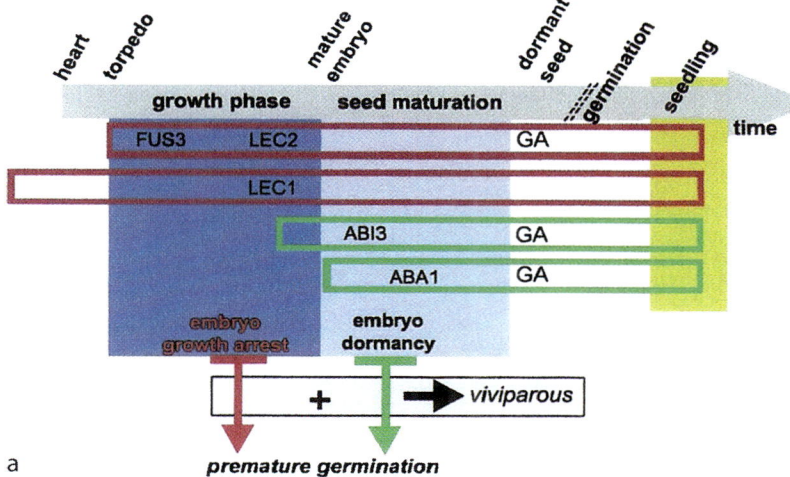

Plate 11, Fig. a A model for developmental arrest and premature germination of *Arabidopsis* seeds. During the growth phase, expression of *FUSCA3* (*FUS3*), *LEAFY COTYLEDON1* (*LEC1*), and *LEC2* genes in the embryo results in growth arrest. Mutations in these genes lead to reduced embryo growth arrest and premature germination. Early during the seed maturation phase, the activity of the *ABI3* and *ABA-DEFICIENT1* (*ABA1*) genes causes embryo dormancy, and mutations in these genes lead to premature germination. Premature germination is regulated by gibberellic acid (GA)-dependent and GA-independent pathways. Vivipary and premature germination inside the silique occur in *embryo growth arrest* and *embryo dormancy* double mutants. *Red* and *green boxes* indicate gene expression and function. (Reprinted from Raz et al. (2001) Development 128:243–252

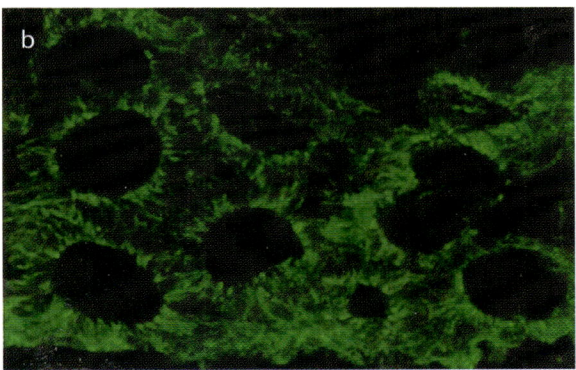

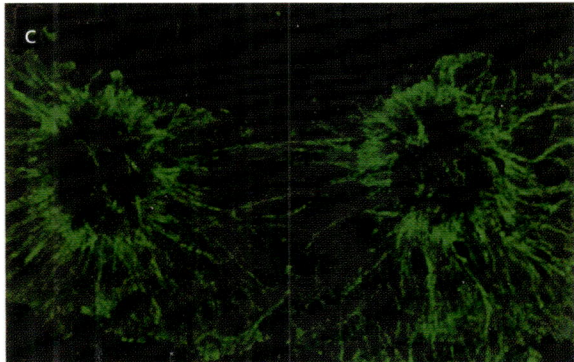

Plate 11, Fig. b–d Role of microtubules in the cellularization of free nuclei of bareley endosperm. Microtubules are stained by immunofluorescence; chromosomes and/or nuclei are unstained and appear black. **b** Formation of the radial system of microtubules that organize the cytoplasm and maintain the nuclei in an evenly spaced pattern. **c** Interaction of the microtubules of two adjacent nuclei. **d** Location of walls indicated by the unstained zones at the perimeters of the radial microtubules originating from the nuclei. *Bars* **b** 7 µm; **c, d** 5 µm. (Reprinted from Brown et al. (1994) Plant Cell 6:1241–1252

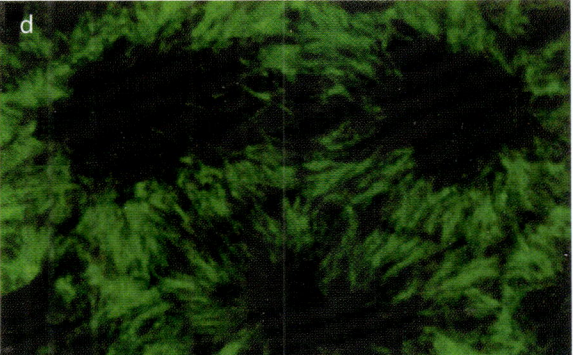

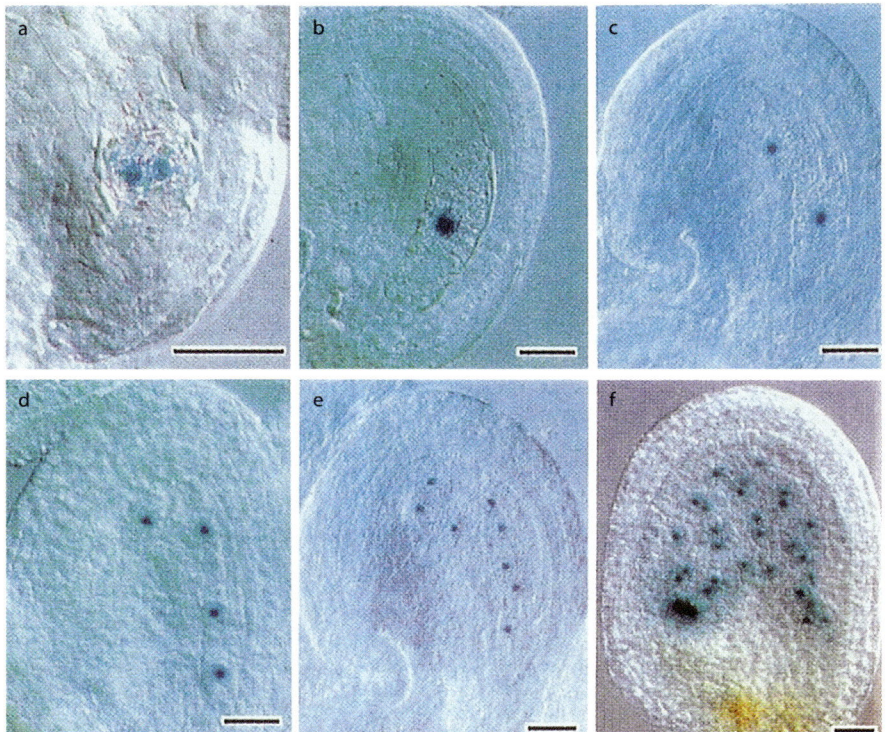

Plate 12, Fig. a–f β-Glucuronidase (GUS) expression in the endosperm nuclei of transgenic ovules of *Arabidopsis* harboring *FERTILIZATION-INDEPENDENT SEEDS2 (FIS2):: GUS* constructs. **a, b** Unpollinated ovules showing GUS activity in the central cell nucleus. **c–f** GUS activity in endosperm nuclei formed in ovules following double fertilization; **c** 2 nuclei, **d** 4 nuclei, **e** 8 nuclei, **f** 32 nuclei. *Bars* 50 mm. (Reprinted from Luo et al. (2000) Proc Natl Acad Sci USA 97:10637–10642

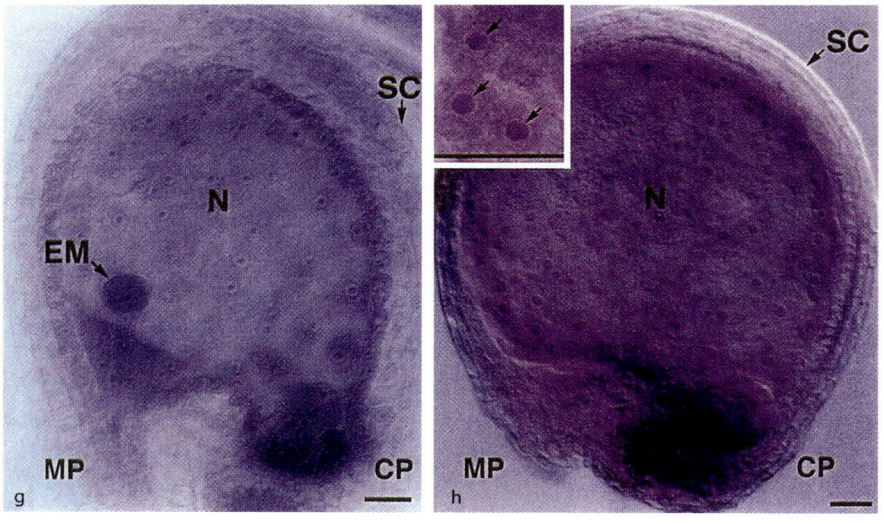

Plate 12, Fig. g,h Embryo, endosperm, and seed coat development in wild-type and mutant *Arabidopsis* plants. **g** Whole mount of an ovule of a flower from a wild-type plant, 2 days after self-pollination, showing a globular embryo surrounded by nuclear endosperm. **h** Whole mount of an ovule of an emasculated flower from a heterozygous *FIE/fie* plant, 7 days after emasculation. *Inset* Endosperm nuclei (*arrows*). *CP* Chalazal pole, *EM* embryo, *MP* micropylar pole, *N* endosperm, *SC* seed coat. *Bars* 25 μm. (Reprinted from Ohad et al. (1996) Proc Natl Acad Sci USA 93:5319–5324

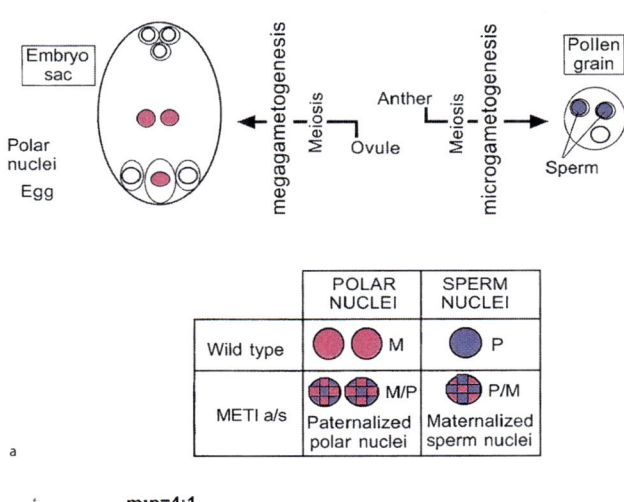
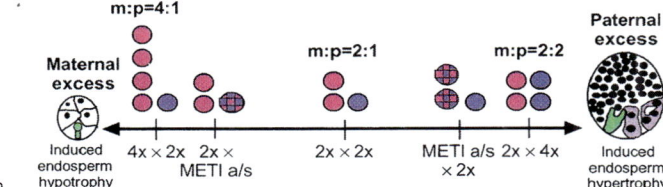

Plate 13, Fig. a,b Diagrams showing the effect of global DNA hypomethylation on parental imprinting in *Arabidopsis*. **a** Formation of egg and sperm. The endosperm formed in the wild-type after double fertilization contains a ratio of two maternal genomes to one paternal genome. In the maternal genome, maternal-specific imprinted genes are active, whereas paternal-specific genes are repressed. Imprinted genes contributed by the paternal genome have a complementary expression pattern. When maternal genomes are contributed by a *METHYL TRANSFERASE1* (*MET1*) a/s parent, the paternal-specific genes are largely derepressed, producing a paternalized genome. Similarly, a *MET1* a/s pollen parent is expected to contribute a maternalized genome. **b** Interploidy crosses result in seeds with extra maternal or paternal genomes, and therefore extra doses of active maternal or paternal alleles of imprinted loci. Maternal or paternal excess results in small seeds with small endosperms, or large seeds with overgrown endosperms, respectively. A diploid *MET1* a/s parent does not contribute extra genomes but appears to contribute extra doses of active maternal- or paternal-specific genes, resulting in phenotypes similar to those produced by parental imbalance. (Reprinted from Adams et al. (2000) Development 127:2493–2502

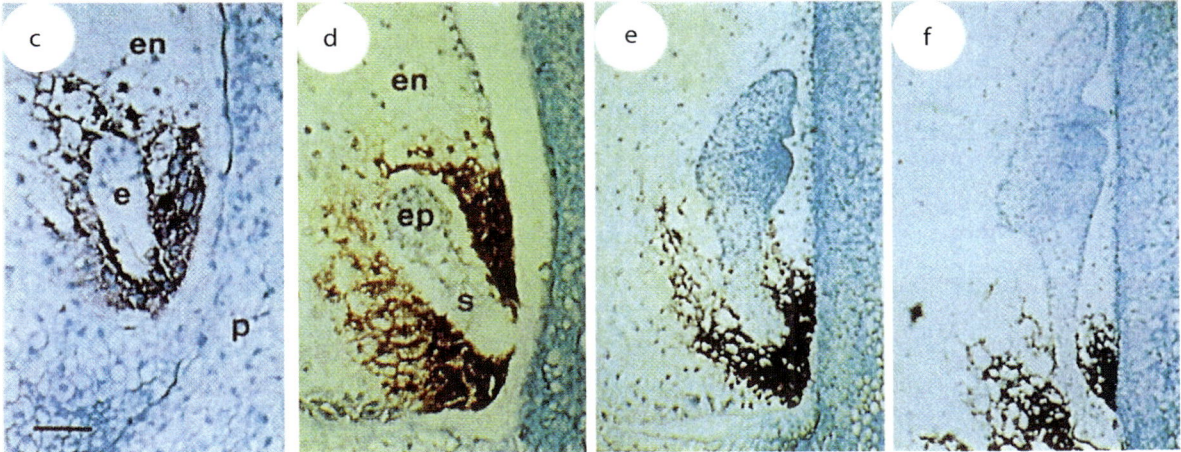

Plate 13, Fig. c–f In situ hybridization expression of *EMBRYO SURROUNDING REGION* (*ESR*) transcripts in maize endosperm. **c** Longitudinal section of the embryo-endosperm region of the grain 5 days after pollination hybridized with the RNA probe. **d** Section of the grain 7 days after pollination. **e** Section of the grain 9 days after pollination. **f** Section of the grain 12 days after pollination. Here the signal is confined to the region around the lower part of the suspensor. *e* Early-stage embryo; *en* endosperm; *ep* embryo proper, with the suspensor (*s*) separate; *p* pericarp. Bar 15 μm. (Reprinted from Opsahl-Ferstad et al. (1997) Plant J 12:235–246

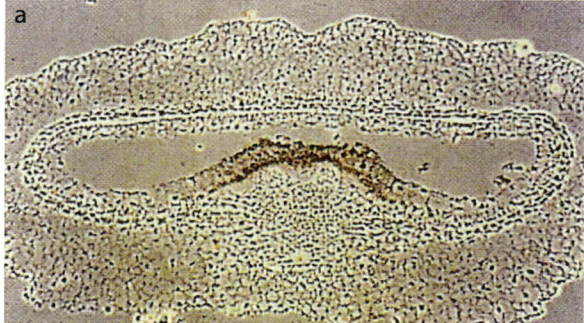
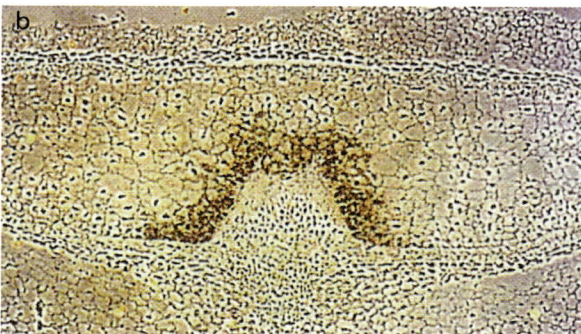

Plate 14, Fig. a,b In situ hybridization expression of transcripts of the *ENDOSPERM1* (*END1*) gene over the nucellar projection cells of barley endosperm. **a** Transcript expression is initiated in the free nuclei above the nucellar projection cells. **b** Later, expression spreads to the cells of the basal transfer cell layers of the endosperm located above the nucellar projection cells. (Reprinted from Olsen et al. (1999) Trends Plant Sci 4:253–257

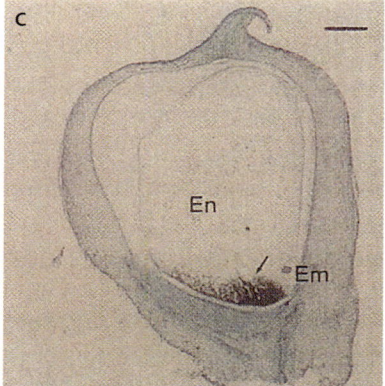
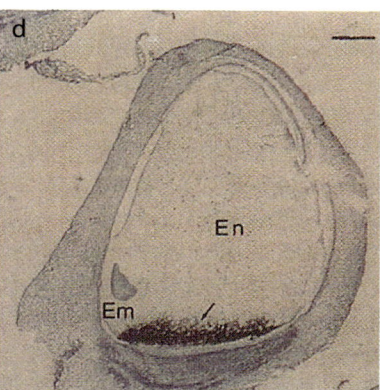
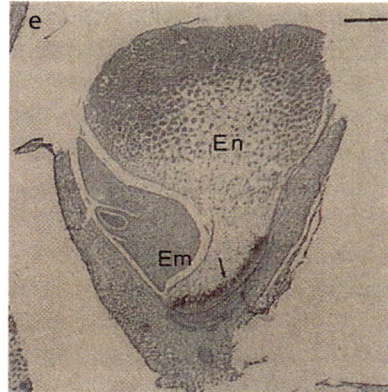

Plate 14, Fig. c–e In situ hybridization sections of maize kernel with a probe specific for the *BASAL ENDOSPERM TRANSFER LAYER* (*BETL*) gene. **c** Section of a kernel at 12 days after pollination. The signal extends over at least three cell layers in the basal endosperm. **d** Section of a kernel at 16 days after pollination. **e** Section of a kernel at 22 days after pollination showing a decrease in signal intensity; *arrows* hybridization signals in the endosperm transfer cells. *em* Embryo, *en* endosperm. *Bars* 200 μm. (Reprinted from Hueros et al. (1995) Plant Cell 7:747–757

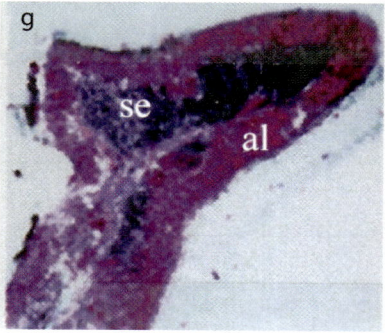
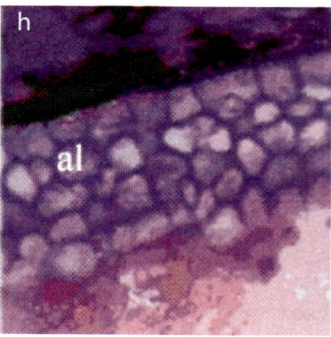

Plate 14, Fig. f–h Phenotype of *supernumerary aleurone1* (*sal1*) mutant endosperm of maize. **f** Homozygous *sal-1* defective kernels. Two wild-type grains are shown on the *right*. **g** Section of the endosperm of a defective kernel. The cells of the supernumerary aleurone layers (*al*) are stained red, whereas the starchy endosperm (*se*) is stained dark blue. **h** Part of the multilayered aleurone of the mutant kernel. (Reprinted from Shen et al. (2003) Proc Natl Acad Sci USA 100:6552–6557

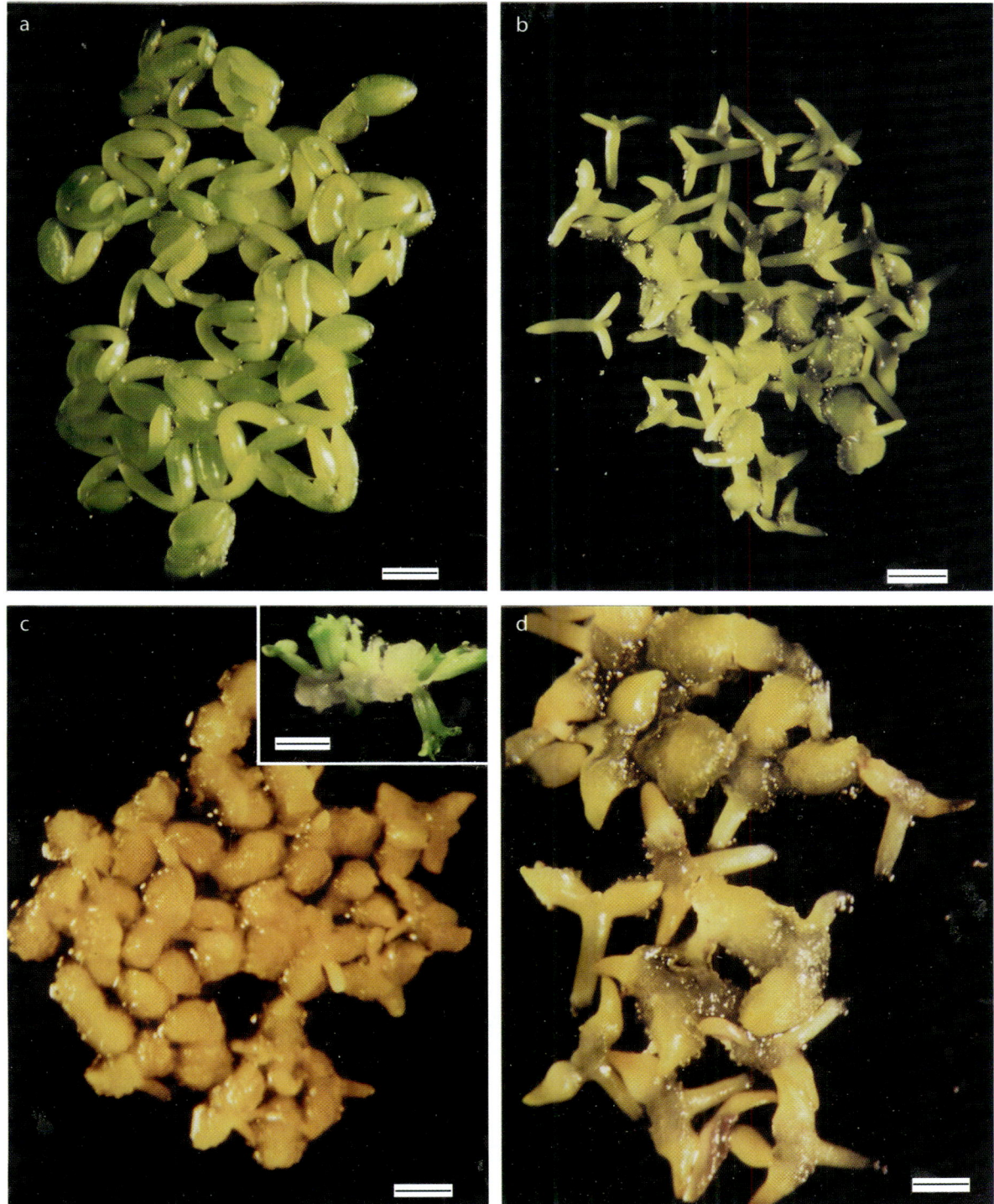

Plate 15, Fig. a–d Somatic embryogenesis in cultured zygotic embryos of *Arabidopsis*. **a** Bent cotyledon-stage embryos at the time of culture in a medium containing 2,4-dichlorophenoxyacetic acid (2,4-D). **b** Beginning of callus growth on cotyledons 4 days after culture. **c** Advanced callus growth on cotyledons 7 days after culture. **d** Massive callus growth with early-stage somatic embryos (seen in sections) on cotyledons 10 days after culture. *Inset* Somatic embryos arising from the callus transferred at 10 days after culture in an auxin-containing medium to a medium lacking auxin for 14 days. *Bars* **a** 400 μm, **b** 750 μm, **c** 1 mm, **d** 1.25 mm, *inset* 2 mm

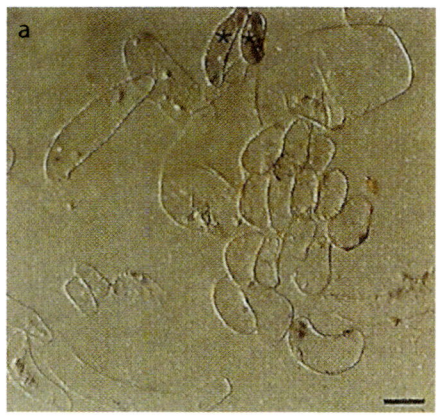
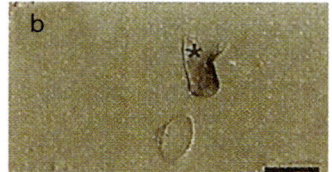

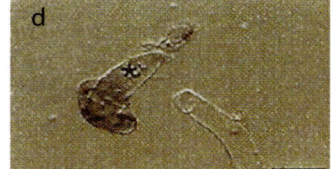
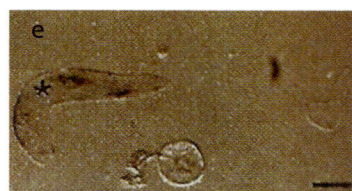
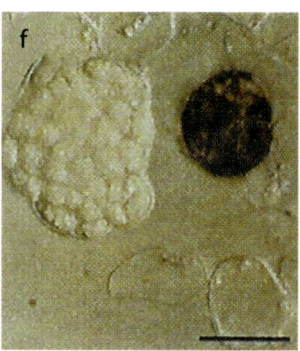
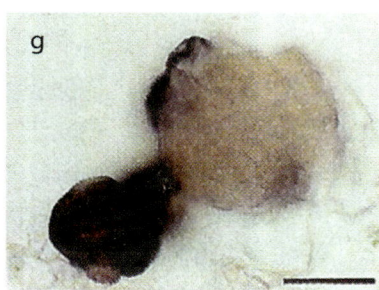

Plate 16, Fig. a–g Expression of the *Daucus carota* SOMATIC EMBRYOGENESIS RECEPTOR KINASE (*DcSERK*) gene in embryogenically competent carrot cells, and early-stage embryos formed from such cells. **a–e** Gene expression visible as a purple precipitate in individual cells (*asterisks*) obtained by mechanical fragmentation of hypocotyl segments treated with 2,4-D. **f** Gene expression in a small cluster of embryogenic cells. **g** Gene expression in a small globular embryo. *Bars* 50 μm. (Reprinted from Schmidt et al. (1997) Development 124:2049–2062

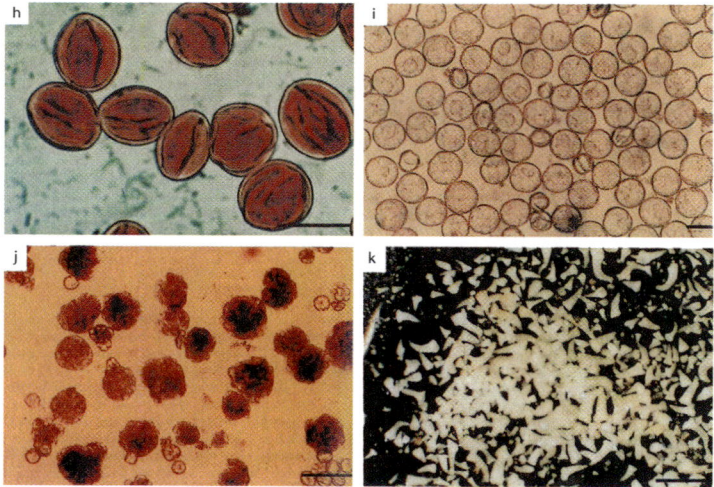

Plate 16, Fig. h–k Embryogenesis in isolated pollen cultures of tobacco. **h** Freshly isolated, acetocarmine-stained pollen grains. **i** Embryogenic pollen grains formed after culture for 6 days at 33°C in starvation medium. **j** Globular embryos formed from embryogenic pollen grains 4 weeks after transfer to an induction medium. **k** Torpedo-shaped embryos formed from embryogenic pollen grains 6 weeks after transfer to an induction medium. *Bars* **h**, **i** 25 μm; **j** 100 μm, **k** 2 mm. (Reprinted from Touraev et al. (1996) Plant Cell Rep 15:561–565

Index

Italic page numbers indicate figures.

A

ABA-deficient (*aba*) mutants/genes   136–138, 140, 145
ABA-INSENSITIVE (*ABI*) genes/mutants   121, 136–138, 140–141, 145, *222*
ABNORMAL LEAF SHAPE (*ALE1*) gene   37
Abscisic acid (ABA)   49, 91, 114, 132, 146, 201
— role in embryo dormancy   138–145
— role in embryo maturation   134–138
— role in precocious germination   135–136
— role in somatic embryogenesis   196
Acacia retinodes   84
Actin, role in double fertilization   15–16
— role in egg polarity   31
Activator (*Ac*)   43
Aesculus woerlitzensis   165
AGAMOUS-Like (*AGL*), *MADS*-box genes   108–109, 157, 194, *220*
Agrobacterium   115
Agrobacterium tumefaciens   95
AINTEGUMENTA (*ANT*) gene   39
Alectra vogelii   45
Alisma lanceolatum   89
Alisma plantago-aquatica   89
Allanblackia parviflora   165
Amborella trichopoda   7–8
AMINO ACID PERMEASE1 (*AAP1*) gene   164
Androgenesis   201
Anemone nemorosa   5
Anther culture   21, 188, 201–205
Antipodals   2, 7–8
Antirrhinum majus   117
APETALA2 (*AP2*) gene   110
Apicobasal pattern   37, 39–40, 57, 58, 69–72, 75
Apomixis   8, 20, 102, 109, 187, *188*–190, 205–206
— adventive embryogenesis   188–*191*
— apospory   188–190
— diplospory   188–190
— molecular genetics of   192
Apoptosis   93

Arabidopsis   6–8, 14, 17, 21–22, 30, *32*, 34, 42–43, 46–47, 51, 57–69, 71–76, 82, 87, 92, 94–*95*, 96–97, 103–106, 108–110, 115–121, 123, 132–134, 136–145, 152–154, 164, 173, 178, 189, 190, 192, 200
— embryogenesis in   35–40, *214–215*
— endosperm development in   156–158, 174–177
— *LEAFY COTYLEDON* (*LEC*) genes of   111–112
— somatic embryogenesis in   193, 194
— storage reserves in embryos of   40–41, 193
Arabidopsis arenosa   176
Arabidopsis Dynamin-like Proteins1 (*adl1*) mutant   119
Arabidopsis Minute-like (*aml1; ATRPS5*) gene   116–117
Arabidopsis shaggy-related protein kinase *etha* (*ASKη*) gene   36, 110
Arabidopsis thaliana   2, 120, 176
Arabidopsis thaliana cullin1 (*Atcul1*) mutant/gene   120
Arabidopsis thaliane Em (*AtEm1, AtEm6*) genes   144–145
Arabidopsis thaliana HOMEOBOX8 (*Athb-8*) gene   108
Arabidopsis thaliana LIPID TRANSFER PROTEIN1 (*AtLTP1*) gene   37–38, 198
Arabidopsis thaliana MERISTEM L1 LAYER (*ATML1*) gene   36
Arabidopsis thaliana PIN4 (*AtPIN4*) gene   68
Arabidopsis thaliana 2S ALBUMIN (*At2S3*) gene   133–134, 137
ARABIDOPSIS THALIANA SEED (*ATS1, ATS3*) genes   133–134
Arabidopsis thaliana SERK1 (*AtSERK1*) gene   194–195
Arabidopsis thaliana Skp-like1 (*ask1, ask2*) mutants   120
Arabis gunnisoniana   189–190
Arabis holboellii   189–190
ARGONAUTE (*AGO1*) gene   108
ASKdzeta (*ASKζ*) gene   110
Asparagus officinalis   198
ASYMMETRIC LEAVES1 (*AS1*) gene   39, 62
Atropa belladonna   204
Auxin (see also 2,4-dichlorophenoxy acetic acid, indoleacetic acid, naphthaleneacetic acid)   31, 49
— role in somatic embryogenesis   194–196
— role in zygotic embryogenesis   39, 46–47, 68–71, 94
auxin-binding protein1 (*abp1*) mutant   37
auxin-resistant6 (*axr6*) mutant/gene   70–71
Auxin transport inhibitors   31, 46–47
Avena sativa   49
AX92 gene   113

B

B22E gene   180
BABY BOOM (BBM) gene   194
Barley   5, 12, 16, 49, 50–51, 134, 143–144, 153–155, 163, 165, 179, 180, 182, 198, 201–202
BASAL ENDOSPERM TRANSFER LAYER (BETL1) genes   179, *226*
BASAL LAYER ANTIFUNGAL PROTEINS (BAP) gene   179
Beta vulgaris   12
Betula pubescens   59
Biotin mutants (*bio1*, *bio2*)   115–116
Birch   59
BLADE-ON-PETIOLE1 (BOP1) gene   108
bodenlos (*bdl*) mutant/gene   46, 69–71
borgia (*bga*) mutant   104, 175
Brachiaria brizantha   192
Brasenia schreberei   8
Brassica campestris   14, 141
Brassica juncea   46, 50, 70
Brassica napus   109, 113–114, 132–133, 156, 194, 201–202, 205
BREVIPEDICELLUS (BP) gene   108
brittle (*bt*) mutant   160
Bulbophyllum   82
Bulbophyllum mysorense   83
Burmannia pusilla   45

C

C1 gene of anthocyanin pathway   141
Ca^{2+}
— as chemotropic attractant   15
— changes during fertilization   19
— in calreticulin   109
— in egg polarity   31
— signaling   111, 137, 196
cab (chlorophyll *a/b* binding protein) gene   137
Cacao   195
Calanthe   82
Calanthe veitchii   6
Calmodulin   19, 107, 111, 196
Caltha palustris   5
CAPRICE (CPC) gene   73
Capsella bursa-pastoris   20, 30, 34–35, 49–51, 66, 81–*83*, 86, *87*–88, *93*–94, 153
capulet (*cap1*, *cap2*) mutants   106, 175
Carboxypeptidase   114
Carrot   46, 144, 192, 193
— somatic embryogenesis in   194–200
CARROT EARLY SOMATIC EMBRYOGENESIS1 (C-ESE1) gene   200
Castor bean   144, 162
Cauliflower mosaic virus (CaMV) 35S promoter   43, 73, 141
Central cell   2, 6–9, 11, 15–19, 32, 47, 105, 156–157, 164, 175, 180, 188
cephalopod (*cph*) mutant   70
CHAMPIGNON (CHO) gene/mutant   118–119
CHAPERONIN-60α gene   117
CHITINASE26 (CHI26) gene   180
Cicer soongaricum   84–85
Cichorium   194, 198, 199
Cistanche tubulosa   45
Citrus   190–192
Citrus aurantium   198–199
clavata (*clv*) mutants/genes   61–65, 67, 193
Clematis viticella   5
CLV-like (CLE19) gene   67–68
Coconut   49, 164
— endosperm of (coconut milk, coconut water)   49–50, 164–165, 192, 201
Cocos nucifera   49
Coelogyne   82
Colocasia esculenta   49
Columella   37–38, 66–68
comatose (*cts*) mutant/gene   140
Comparative embryology   8, 20, 21
CORONA (CNA) gene   65
Coronopus didymus   156
Cortex, origin of   66–67, 74
Cotton   5, 12, 14, 16, 30, 32–33, 47, 51, 106–*107*, 114, 133–134, *135*–136, 143–144, 153, 162
crinkly4 (*cr4*) mutant/gene   181
Crotalaria verrucosa   84, 85
CRUCIFERIN C (CRC) gene   133–134, 137
Cucumis sativus   165, 200
CUP-SHAPED COTYLEDON (CUC) genes   62–63
CYCLIN (CYC) genes   108, 118
CYCLIN-DEPENDENT KINASE (CDK) gene   118, 159
CycZme1, mitotic cyclin of maize   159
Cymbidium   85
Cymbidium sinense   85
Cymbidium bicolor   86
Cypripedium insigne   6
Cytisus laburnum   84–85, 91
cytochrome *c*-subunit (*cox1*) gene   11
cytokinesis-defective (*cyd*) mutant   72

Cytokinesis-defective mutants of *Arabidopsis* 73, 117–120
Cytokinesis defects during embryogenesis 72–73, 117–120
Cytokinin 49, 51, 91–92, 165, 166, 194

D

D-type of cyclin *(CYCD)* gene 110
Dactylis glomerata 195, 199, 200
Datura 58
Datura innoxia 48, 201, 204
Datura stramonium 48–50, 165
Datura tatula 50, 165
Daucus carota 46
Daucus carota SERK (DcSERK1) gene 195, *228*
DECREASE IN DNA METHLA-
  TION1 *(DDM1)* gene 106
DEFECTIVE EMBRYO
  AND MERISTEMS *(DEM)* gene 66
defective endosperm expressing xenia (dex) mutant 182
defective kernel (dek) mutants 122, 180–181
defective seedling (des) mutants 122
Delphinium elatum 5
demeter (dme) mutant/gene 104, 175, 177
Dendrobium 82
Dendrobium barbatulum 83
Dendrobium nobile 6
Desiccation tolerance of embryos 131, 136–138
— role of carbohydrates in 141–143
— role of proteins in 143–145
2,4-Dichlorophenoxyacetic acid (2,4-D) 18, 46, 160, 192–194, 199
Digitalis lanata 199
Dimorphic sperm 12–13
Diplotaxis erucoides 82–83
discolored1 (dsc1) mutant/gene 181–182
distorted growth (dgr) mutant 65
DNA methylation 105–106, 176–177, 200
DOMINO1 (DOM1) gene 117
Double fertilization 2, *3–19*, 22–23, 106, 151, 162, 173–175, 188, 200, 205
Drosophila 159
— *EXTRA SEX COMBS (ESC)* gene of 104
— *SHAGGY* gene of 36
Drosophila melanogaster 104

E

EARLY METHIONINE-LABELED *(Em)* gene 141, 144
Echinochloa utilis 163

Egg 1–3, 7–11, 13–19, 29–33, 35, 51, 102–103, 105–106, 108–109, 156, 188, 190
— polarity of 30–31, 71
— ultrastructure of 30–31
EMB173 gene 104
Embryo 3, 9–11, 13, 18–21, 29–31, 34, 44–46, 57, 59–75, 81–82, 86–91, 93–97, 101, 103–107, 123, 142–144, 151, 154, 157, 164–165, 178, 180, 187, 190–195, 201–203, 205
— culture of 21, 48–51, 91–92, 121–122, 167
— — defective mutants of *Arabidopsis* 115, 120–121, 174
— — defective mutants of maize 121–123
— — defective mutants of rice 123
— — lethal mutants of *Arabidopsis* 115–117
— development in eudicots 35–41
— development in monocots 41–43
— division patterns of 34–35
— dormancy of 112, 131, 136–141, *223*
— maturation of 112, 131–138
— nutrition of 47–48
— storage proteins of 40–41, 93, 96, 132–137
EMBRYO-DEFECTIVE *(EMB30)* gene 71, 120–121
embryo-defective development1 (edd1)
  mutant/gene 96, 117
Embryo sac 1–*2*, 3–4, 6–8, 11, 13–15, 17, 47–49, 51, 75, 82, 103, 106, 108–109, 155, 162, 165, 175, 179, 188–190
— Polygonum type 2, 189
embryo-specific (emb) mutants 122, *221*
embryoless1 (eml1) mutant 123
Embryogenesis 19–22, 29, 34–35, 43, 45, 47, 57–63, 70–72, 75, 101–103, 105–107, 131, 133–136, 144, 174, 187, 191
— cell lineage during 20, 43–44, *215*
— gene expression during 108–115
— in pollen grains 200–205
— in somatic cells 192–200
— in wild crosses 165–167
— parthenogenic initiation of 189–190
EMBRYO SURROUNDING REGION *(Esr)* gene 178, *225*
EMBRYONIC ECTODERM
  DEVELOPMENT *(EED)* gene 104
Embryonic proteins 199
empty pericarp2 (emp2) mutant/gene 122, 182
Endoreduplication 41, 82, *158*–160, 167
— in suspensor cells 89–90
Endodermis, origin of 66–67, 74
Endosperm 3, 5, 8–11, 13, 18–22, 34, 45, 47–51, 82, 86–87, 90, 93, 103–105, 106, 122–123, 132, 144, *163*, 190–192, 205
— aleurone cells (layers) of 153–155, 157, 161–163, 178, *180*, 181
— basal endosperm transfer layer 154, 178–180
— chemical composition of 164–165
— cyst formation in 156–157, 164
— cytokinetic defects in mutants 174
— development in *Arabidopsis* 156–158, 174–177
— development in wide crosses 165–167

— development without double fertilization   175–177
— developmental types   153
— DNA amplification in   158–160
— embryo-surrounding region of   154, 178
— evolution of   10, 22, 152, 154
— free-nuclear to cellular transformation of   154, *155–156*
— genomic imprinting during development of   176–177
— haustoria of   158, 163–164
— of cereal grains   178–182
— programmed cell death of   161–162
— role in embryo nutrition   157, 162–167
— starchy cells of   153–155, 161, 163, 178, 180–181
— storage products of   156, 160–161
— transfer cells (cell layers) of   154, 166, 179, 180
ENDOSPERM1 (END1) gene   179, *226*
ENHANCER OF ZESTE gene   104
Ephedra   9–11
Ephedra nevadensis   9, 11
Ephedra trifurca   9–11, 17
Epidendrum scutella   14
ERA-RELATED GTPases (ERG) gene   117
Eranthis hiemalis   94
Erigeron strigosa   5
Eriocaulon robusto-brownianum   45
Eriocaulon xeranthemum   45
Eruca sativa   50, 89–91
Erythrina crista-galli   13
Erigeron philadelphicus   5
Euryale ferox   8
Experimental embryology   21
EXTRACELLULAR PROTEIN2 (EP2) gene   192
Extracellular proteins of embryogenesis   196–199
EXTRA COTYLEDON (XTC1, XTC2) genes   63
EXTRA SEX COMBS (ESC) gene   104

F

F644, gene/mutant   8, 104, 175
FACKEL (FK) gene/mutant   70–71
fass (fs) mutant/gene   40, 43, 46, 74–75, 120
Female germ unit   2
feronia (fer) mutant   14
FERTILIZATION-INDEPENDENT ENDOSPERM (FIE) genes/mutants   8–9, 104–105, 157, 175, 178, 190, *224*
FERTILIZATION-INDEPENDENT SEEDS2 (FIS2) genes/mutants   8–9, 104–105, 157, 177, 190, *224*
FILAMENTOUS FLOWER (FIL) gene   110
fis class mutants   8, 103, 106, 154, 175, 177, 190, 192
fist mutant   37
FLORAL BINDING PROTEIN (FBP7, FBP11) genes   103

Fritillaria   4
Fritillaria meleagris   5
Fritillaria tenella   3
Fucus   19, 31–32
— polarity in eggs of   31–32
FUSCA3 (FUS3) gene/mutant   111–112, 138, 140
FWA gene   177

G

Galanthus nivalis   6
Genomic imprinting   9, 22, 105–106, 154, 175, 190
— role of DNA methylation in   176–178, *225*
Geodorum   85
Geranium phaeum   89
Germination   112–115, 131–138, 140–144, 146, 158, 162
Gibberellin, gibberellic acid (GA)   49, 51, 74, 91, 165
— effect on embryo growth   49, 51, 91–92, 109, 140
— in suspensors   91
GIBBERELLIN-INSENSITIVE (GAI) gene   74
Ginkgo biloba   1
GLABRA2 (GL2) gene   73
globby1-1 (glo1-1) mutation   180
GLOBULAR ARREST1 (GLA1) gene   116
Gloxinia hybrida   3
glucosidase1 (gcs1) mutant   120
β-*GLUCURONIDASE (GUS)* gene   43, 157, 175, 205
Glycine max   37, 132, 164
Gnetum   9–10
Gnetum gnemon   9, 11, 16
gnom (gn) mutant/gene   46, 71, 75, 120, 121
Golden rice   22
GOLLUM (GLM) gene/mutant   74, 75
Gossypium hirsutum   5
Grapevine   198
Green fluorescent protein (GFP)   7, 74, 157
Guizotia oleiflora   5
gurke (gk) mutant/gene   69, 71

H

Haemanthus katherinae   153, 162
haiku (iku1, iku2) mutants   176
HALLIMASCH (HAL) gene/mutant   118, 119
Helianthus annuus   5, 162
Heliopsis patula   5
Helleborus foetidus   5
Hibiscus costatus   32

Hibiscus costatus-aculeatus 32
Hibiscus costatus-furcellatus 32
Hibiscus trionum 5
Hieracium 189–190
Hieracium aurantiacum 189
Hieracium piloselloides 189
Himantoglossum hircinum 3, 5
hinkel (hik) mutant 119
hobbit (hbt) mutant/gene 68, 71
Homeobox genes 42, 59–60
HOMEOBOX GENE OF *Oryza sativa* (HOS) 60
Hordeum bulbosum 102
Hordeum vulgare 5, 102
Horse chestnut 165
HYDRA (HYD) gene/mutant 40, 70
Hyoscyamus niger 203–205
Hypophysis 36–38, 57, 66–71, 74, 94, *216*

I

Illicium anisatum 8
Illicium floridanum 8
In vitro fertilization 13, 17, *18*–19, 49, 51, 108–109, 156
indeterminate gametophyte (ig) mutant 178
Indoleacetic acid (IAA) 31, 160
— effect on embryo growth 49–51
Invertase gene 179
Ipomoea purpurea 88
Isocitrate lyase (ICL) *113*–114, 135

J

JAGGED (JAG) gene 108
Jasione montana 6
Juglans 5
Juglans regia 165

K

Kadsura japonica 8
KANADI (KAN) gene 108, 110
Karyogamy 16, *17*–18
keule (keu) mutant/gene 72–73, 120–121, 174
KIESEL (KIS) gene 119
Kinase-associated protein phosphatase (KAPP) gene 64
Kinetin, effect on embryo growth 49–51
knolle (kn) mutant/gene 72–73, 118, 120, 174
knopf (knf) mutant 40

KNOTTED1 (KN1) gene 59–60, 193, *217*
Kunitz trypsin inhibitor (KTi) gene/protein 37, 133

L

lachrima mutant 122
LATE EMBRYOGENESIS ABUNDANT (LEA) proteins/genes 107, 112, 141, 143–*145*, 200
Lathyrus 89
Lathyrus angustifolia 84–85
LEAFY COTYLEDON *(LEC)* genes/mutants 39, 96, 109, 111–112, 136, 138, 140–141, 193–194, 200
LEC1-LIKE (L1L) gene/mutant 96, 111–112
Lepidium virginicum 164
Lilium 4
Lilium candidum 5
Lilium longiflorum 13
Lilium martagon 3–4
Lilium pyrenaicum 4
Linum austriacum 167
Linum catharticum 5
Linum perenne 167
Linum usitatissimum 14, 50
LIPID TRANSFER PROTEIN2 (LTP2) gene 180
Lupinus pilosus 84–85
Lycopersicon esculentum 66
Lycopersicon peruvianum 166
Lycopersicon pimpinellifolim 166

M

Macrosolen cochinsinensis 82–83
MADS-box genes 10, 103–104, 108–109, 157, 194, 200
Maize 5, 13–19, *41*, 44, 49–51, 59, 65, 102, 106, 108–111, 121–123, 134, *135*–136, 140, *142*–143, 153–154, 156, 158–163, 165–166, 178–181, 192, 200
— embryogenesis in 41–43, *216*
Malate synthase (MS) 113
Male germ unit 11, 13
Maternal effect genes 75, 103–105
MATERNALLY EXPRESSED GENE1 (MEG1) 178, 180
MATURATION (MAT) genes 133, *134*, 144
MEDEA (MEA) gene/mutant 8, 103–106, 157, 175, 177, 190, *220*
Medicago sativa 88, 162, 194, 196, 200
Medicao scutellata 88
Medicago truncatula 195, 199
Medicago truncatula SERK1 (MtSERK1) gene 195

Melandrium album 89
Melandrium rubrum 89
METHYLTRANSFERASE1 antisense (MET1 a/s) construct 177
mgoun (mgo1, mgo2) mutants 63
mickey (mic) mutant 40
Microarray analysis 102, 192
Microtubules
— role in cell division in cytokinesis-defective mutants 118–119
— role in endosperm cellularization 155–157, 180, *223*
— role in suspensor cell elongation 82, 85
— role in zygote elongation 32, 36
Mitochondrial genes 10
Molecular embryology 21
monopteros (mp) mutant/gene 46, 63, 70–71
Monotropa hypopitys 3, 5–6
Monotropa uniflora 6, 44, 63
multicopy suppressor of IRA *(msi1)* mutant 104, 175
Mutator (Mu) 122
myo-inositol-1-phosphate synthase gene *(RINO1)* 111

N

NAM RELATED PROTEIN1 (NRP1) gene 178
Naphthaleneacetic acid (NAA) 46, 49, 194
Narcissus poeticus 5
Neurospora crassa 116
Nicotiana plumbaginifolia 199
Nicotiana rustica 94, 166
Nicotiana tabacum 6, *15*, 166, 201
Nigella sativa 5
Nitella 6
no apical meristem (nam) mutant 65
Nuphar polysepalum 8, 10, *213*
Nymphaea stellata 8

O

Oat 49
Oenothera biennis 165
Oenothera lamarckiana 165
Oenothera muricata 165
OLEOSIN (OLEO1, OLEO2) genes 133–134, 137, 180, 193
Ononis fruticosa 84–85
Orchis latifolia 3, 5
Orchis maculata 5
Orchis mascula 5

ORIGIN RECOGNITION COMPLEX (ORC) gene 116
Orobanche aegyptiaca 45
Oryza sativa 21
ORYZA SATIVA HOMEOBOX1 (OSH1) genes 60, 65, 110
Oryza sativa KNOTTED1-like *(OsKn1)* genes 60, 110, 111
Ovary culture 51, 102, 153
Ovule culture 51, 102, 121, 153

P

Paeonia 44–45
Paeonia anomala *44–45*
Paeonia moutan 45
Paeonia wittmanniana 44–45
PAP85 gene 133–134, 137
Papaver nudicaule 33
Papaver somniferum 17, 51
Paspalum notatum 192
PASTICCINO (PAS) gene 40
Pea 22, 48, 72, 134, 162
Pearl millet 15
PEI1 gene 109, *221*
Pelargonium zonale 33
Pelvetia 31
Pennisetum americanum 193
Pennisetum ciliare 192
Pennisetum glaucum 15
Peraxilla tetrapetala 82–83
Pericycle, origin of 66–67, 74–75
Peristeria 82
Peristeria elata 83
PEROXYREDOXIN1 (PER1) gene 180
Petunia 16, 65
Petunia hybrida 103
PFIFFERLING (PFI) gene/mutant 118–119
PHABULOSA (PHB) gene 108, 110
Phaius tankervilliae 14–15, 82
Phaseolus 89, 91
Phaseolus acutifolius 89, 166
Phaseolus coccineus 37, 47, 50–51, *87–89*, *90–93*, *213*
— endoreduplication of suspensor nuclei 89–90
— polyteny in suspensor nuclei 89
Phaseolus hysterinus 89
Phaseolus lunatus 89
Phaseolus multiflorus 84–85, 89
Phaseolus mungo 89
Phaseolus tuberosus 89

Phaseolus vulgaris 47, 85, 88–89, 91–93, 132, 134, 153, 166
PHERES1 (PHE1) gene 104
pickle (pkl) mutant 194
pilz group of mutants/genes 118–*119*, 174
PIN-FORMED (PIN) gene/mutants 39, 68
PINHEAD (PNH) allele/mutant 61, 193
PINOCCHIO (PIC) gene/mutant 74
PINOID (PID) gene 39, 46
Pisum sativum 22, 88
PLANT GROWTH ACTIVATOR6 (PGA6) gene 193
PLETHORA (PLT1, PLT2) genes 68
Plumbago zeylanica 5, 12–14, 16, 30
Polar fusion nucleus 2, 4, 7–9, 15–16, 104, 156, 175
POLARIS (PLS) gene 37, 108
Pollen embryogenesis 21, 102, 109, 178, 188, 200–206
— by pollen culture 188, 201–202, 204–205, *228*
— cytology of *203*–204
— molecular biology of 204–205
— responsive stage of pollen development for 201–202
poltergeist (pol) mutant/gene 63
Polycomb proteins 104
Polygonum divaricatum 2
Polyteny 82, 89–90, 158
Populus deltoides 6
PORCINO (POR) gene/mutant 118–119
Precocious germination 49, 114, 134–136
Primary endosperm nucleus 18–19, 22, 47, 105, 151–154, 157, 162, 177
primordial timing (pt) mutant 193
Proembryo 11, 38, 45, 88, 109–110
— culture of 49–51
programmed cell death
— in endosperm *161*–162
— in scutellum and coleoptile 42
— in suspensor 92–93
PROLIFERA (PRL) gene 106
Protoderm-specific genes
— in *Arabidopsis* 37–38
— in rice 42
Pseudogamy 190
pZE40 gene 180

Q

Quercus gambelii 14
Quiescence of embryos 131–132
Quiescent center 37–38, 58, 66–68, 74

R

R (red) pigmentation gene 178
radially swollen1 (rsw1) mutant 40
Radish 144
Ranunculus flammula 5
Rapeseed 109, 134–135, 144
Raphanus sativus 144
raspberry (rsy1, rsy2, rsy3) mutants 96, 117
rbcL gene 10, 11
rDNA gene 11
reduced dormancy mutants 140
reduced grain filling1 (rgf1) mutant 179
REPRESSOR OF GA (RGA) gene 74
Reseda lutea 5
RESPONSIVE TO ABA21 (RAB21) gene 144
REVOLUTA (REV) gene 108, 110
Rhinanthus minor 164
Rice 21–22, 60, 65–66, 75, 108, 110–111, 121, 123, 144, 153, 155–156, 159–161, 165
Ricinus communis 144
Roc1 (rice outermost cell-specific1) gene 42
Root apical meristem, root apex 37–39, 42–43, 57–59, 61, 71, 110, 111
— organization of 66–69, *218*
— radial pattern of 37–40, 57–58, 69, 72–75
ROOT-SHOOT-HYPOCOTYL-DEFECTIVE (RSH) gene 120
rRNA gene 10
Rudbeckia grandiflora 5
Rudbeckia laciniata 5
Rudbeckia speciosa 5
Rye 159, 165

S

Santalum album 196
SCARECROW (SCR) gene/mutant 73–75, *219*
SCHIZORIZA (SCZ) gene/mutant 74–75
Schizosaccharomyces pombe 159
schlepperless (slp) mutant 117
Scilla bifolia 5
Secale cereale 159
Sedum ternatum 85–86
Seed embryo culture 48–49
Setaria lutescens 162
SHAGGY gene 36
SHEPHERD (SHD) gene 64

Shoot apex, shoot apical meristem of the embryo   37–39, 41–43, 46–47, 57–58, 66, 110–111, 123, 137, 193
— maintenance of   59–66
— relationship to cotyledons   39–40, 62–64
— specification of   65, 69–72
SHOOTLESS (SHL1-4) genes   65
SHOOT MERISTEMLESS (STM) gene/mutant   40, 61–65, 193
shootmeristemless (sml) mutant   65
short integument (sin1) mutant/gene   75, 103–104
SHORT ROOT (SHR) gene/mutant   73–75
shrunken (sh) mutant   160
shrunken endosperm caused by the maternal genotype (seg) mutant   182
shrunken endosperm expressing xenia (sex) mutant   182
Silique culture   139–140
Silphium integrifolium   5
Silphium laciniatum   5
Silphium terebinthinaceum   5
Silvetia   31
Sinapis alba   141
sirène (srn) mutant   14
SMALL SUBUNIT RIBOSOMAL PROTEIN S16 (SSR16) gene   116
SNAP33 gene   73
SNARE gene   73
Somatic embryogenesis   21, 102, 109, 112, 187–188
— genetic regulation of   199–200
— history of   192–194
— in *Arabidopsis*   193–194, *227*
— in carrot   194–196, *197–200*
— role of arabinogalactan proteins in   197–198
— role of embryonic proteins in   199–200
— role of secreted proteins in   196–199
SOMATIC EMBRYOGENESIS RECEPTOR KINASE1 (SERK1) gene   192
Sophora flavescens   84–85
Sorghum bicolor   162
Soybean   37, 47–48, 106–*107*, 133–135, 141, 143–144, 153, 199
Spathoglottis   82
Spathoglottis plicata   45, *83*
SPÄTZLE gene/mutant   174–*175*
Sperm   1, 9–18, 22, 46, 102, 156, 190, 200
— expression of polyubiquitin genes in   12
Spilanthes oleracea   5
Spinach   5, 14, 16
Spinacia oleracea   5
SPOROCYTELESS (SPL) gene   192
Stellaria media   88, 162
Stem cells   59, 61–68, 74

sterol methyl transferase1 (smt1, smt2) mutants/genes   70, 110
Striga gesnerioides   45
Sugar beet   12
sugary (su) mutant   160
suicide seeds   22
Sunflower   5, 14, 195
supernumerary aleurone1 (sal1) mutant   181, *226*
SUPPRESSOR OF LLP1 (SOL1, SOL2) genes   67–68
SUPPRESSOR OF VARIEGATION gene   104
Suspensor   20, 34–35
— endogenous growth hormones of   91–92
— genetic control of   94–96
— morphology of   82–88
— nuclear cytology of   88–90
— physiology of   90–94
— storage proteins of   93
suspensor (sus) mutant/gene   96–97
Suspensor mutants   95–97
Symplastic traffic
— between egg and central cell   32
— in the embryo   38
— in the shoot apical meristem   59
Synergid   2, 7–8, 13–15, 17, 30, 86, 105

T

Taraxacum megalorrhizon   189
TARGET OF RAPAMYCIN (TOR) gene/mutant   117, 174
TCP gene   108
Test-tube fertilization   17
Theobroma cacao   195
titan (ttn) group of mutans/genes   118–*119*, 174
Tobacco   6, 14–16, 19, 32, 37, 93, 144, 201–205
Tomato   66
tonneau (ton1, ton2) mutants/genes   36, 120
topless1 (tpl1) mutant   69
Torenia fournieri   6, 14–15, 32, *213*
Trapa natans   89
Transferred-DNA (T-DNA)   95, 97, 115, 177, 121, 157
TRANSPARENT TESTA GLABRA (TTG) gene/mutant   73, 112
trehalose phosphate synthase1 (tps1) gene   116
Trifolium pratense   199
Trifolium rubens   199
Tripsacum   102
TRITHORAX gene   104
Triticale   6, 16, 159
Triticum aestivum   6, 50, 159, 201

Triticum durum 159
Tropaeolum majus 89, 91–93
— suspensor structure of 85–86, 88
Trophoblast 82
Tulipa celsiana 5
Tulipa gesneriana 5
Tulipa sylvestris 5
Tunica saxifraga 89
Turnip 14, 30
twin (twn) mutants/genes 96–97

U

Ubiquitin 116

V

vacuoleless1 (vcl1) mutant 94, 120
Ventral canal nucleus 9
Vicia faba 88, 93–94, 133–133, 162
Vigna sinensis 162
Vitis vinifera 198
Vivipary 139–141
VIVIPAROUS1 (VP1) gene/protein 141–142
Vpp1 gene 181

W

waxy (wx) mutant 160
Welwitschia 10
WEREWOLF (WER) gene 73
Wheat 6, 14–16, 18, 46–47, 51, 109–110, 135, 141, 144, 153, 156, 159–162, 165, 179, 201–205
— wheat germ agglutinin 111

Wide crosses 165–167
WOODEN LEG (WOL) gene 74
WUSCHEL (WUS) gene/mutant 60–66, 193
WUSCHEL-related homeobox *(WOX)* genes 36

Y

YABBY (YAB) gene 110

Z

Zeama A1, B1, B2 genes 198
Zea mays 5, 33, 42
ZEA MAYS ANDROGENIC EMBRYOS1 (ZmAE1, ZmAE3) genes 178, 205
Zea mays embryo sac/basal endosperm transfer layer/embryo surrounding region *(ZmEBE)* gene 179–180
Zea mays MYB-RELATED PROTEIN *(ZmMRP1)* gene 179
Zea mays PLASTID RIBOSOMAL PROTEIN L35 *(ZmPRPL35-1)* gene 122
Zea mays SERK *(ZmSERK1, ZmSERK2)* genes 195
Z. mays Homeobox *(ZmHox1, ZmHox2)* genes 42, 110
Z. mays OUTER CELL LAYER3 *(ZmOCL3, ZmOCL4, ZmOCL5)* genes 42
Zein 160–161, 178
Zephyranthes 51
ZWILLE (ZLL) gene/mutant 61–62, 193
Zygote 1, 10–11, 16, 18–19, 21, 29–34, 36–37, 39, 42, 45, 47, 67, 71, 81, 87, 102, 105–106, 108–109, 117, 120, 162
— culture of 51
— polarity of 32–33, 42
— terminal and basal cells of 42, 69
— ultrastructure of 32–33